2nd International Conference on Advanced Engineering and Technology (ICATECH 2020)

Held online due to COVID-19

IOP Conference Series: Materials Science and Engineering
Volume 1010

Surabaya, Indonesia
26 September 2020

Editors:

Ming Jyh Chern
Kim Dong-Seong
Nima Vaziri
Michele Bici
Mat Syai'in
Dedy Zulnoor Hidayat
Agus Budianto
Syamsuri

ISBN: 978-1-7138-3319-2
ISSN: 1757-8981

Printed from e-media with permission by:

Curran Associates, Inc.
57 Morehouse Lane
Red Hook, NY 12571

Some format issues inherent in the e-media version may also appear in this print version.

Institute of Physics
Dirac House, Temple Back
Bristol BS1 6BE UK

Phone: 44 1 17 929 7481
Fax: 44 1 17 920 0979

techtracking@iop.org

Additional copies of this publication are available from:

Curran Associates, Inc.
57 Morehouse Lane
Red Hook, NY 12571 USA
Phone: 845-758-0400
Fax: 845-758-2633
Email: curran@proceedings.com
Web: www.proceedings.com

TABLE OF CONTENTS

IOP Conf. Series: Materials Science and Engineering 1010 (2021) 011001 doi:10.1088/1757-899X/1010/1/011001

PREFACE

It is with great pleasure that I introduce the proceedings of the **2nd International Conference on Advance Engineering and Technology (ICATECH 2020)** is an international seminar organized by Institut Teknologi Adhi Tama Surabaya (ITATS). ITATS is private university located in Surabaya, East Java, Indonesia. The conference is held on September 26th 2020.

This seminar has the main purpose to bring researcher and scholar to share their knowledge and experience in Material, Metallurgy, Energy, Design, Engineering, Applied Science, Information System and Technology area. The conference serves as an excellent opportunity to meet each other and to exchange ideas with theme of **"Empowering Research and Innovation for Sustainable Technology"**.

The conference is initially planned to be an offline meeting in the venue, but within the outbreak of COVID-19, it is changed into a virtual conference instead of being postponed. The conference is organized through online virtual mode using Internet Communication Technologies (ICT) via Zoom. Authors shared the screen and attendees could discuss their work via online platform. Overall, the conference was held successfully through the platform. Our conference committee and authors were dedicated to support the success of the seminar.

The conference program is consisted of keynote speaker session and oral presentation session. The Distinguished Professor from Taiwan and Korea were invited to be our keynote speaker regarding their latest research in their respective areas of expertise in 45 minutes of presentation. After keynote speaker session was done, it continued to oral presentation session for selected paper. The participants were divided into 3 room of Zoom with respect to topic areas and were given 15 minutes time to present their work, following by questions and answers. One excellent presenter will be chosen among them, depending on the technical merit, presentation file, Language fluency, and application etc for the best paper award. It is worth the option of virtual participation that allows expanding the geography significantly and increasing the number of participants. More than 200 participants participated in the conference and 66 papers from different countries are selected for presentation during the conference sessions.

We thank all of reviewers for their time and effort in reviewing the papers even in the COVID-19 pandemic situation, as well as the authors, who took into account recommendations of reviewers and improved their papers to ensure the publication high quality in IOP Conference Series: Materials Science and Engineering Series.

Syamsuri, ST., MT., PhD.
On behalf of ICATECH 2020 Organizing Committee
Institut Teknologi Adhi Tama Surabaya
E-mail: syamsuri@itats.ac.id

IOP Conf. Series: Materials Science and Engineering 1010 (2021) 011001 doi:10.1088/1757-899X/1010/1/011001

EDITOR

Prof. Ming Jyh CHERN
Mechanical Engineering Department of National Taiwan University Science and Technology
mjchern@mail.ntust.edu.tw

Prof. Kim Dong-Seong
School of Electronic Engineering Kumoh National Institute of Technology, South Korea
dskim@kumoh.ac.kr

Assoc. Prof. Nima Vaziri
Department of Physics, Islamic Azad University, Karaj Branch, Karaj, Iran
nima_vaziri@kiau.ac.id

Dr. Michele Bici
Mechanical Engineering Department Sapienza University of Rome, Italy
michele.bici@uniroma1.it

Dr. Mat Syai'in, ST., MT., PhD
Surabaya Shipbuilding State Polytechnic, Indonesia
matt.syaiin@gmail.com

Dedy Zulnoor Hidayat, ST., MT., PhD
Institut Teknologi Sepuluh Nopember, Surabaya, Indonesia
zulnoor@me.its.ac.id

Dr. Agus Budianto
Institut Teknologi Adhi Tama Surabaya, Indonesia
budichemical@itats.ac.id

Syamsuri, ST., MT., PhD
Institut Teknologi Adhi Tama Surabaya, Indonesia
syamsuri@itats.ac.id

IOP Conf. Series: Materials Science and Engineering 1010 (2021) 011001 doi:10.1088/1757-899X/1010/1/011001

SPECIAL THANKS TO OUR KEYNOTE SPEAKER

Prof. Ir. Nizam, M.Sc., DIC., Ph.D.
Director General of Higher Education
Ministry of Education and Culture, Indonesia

Prof. Ming-Jyh Chern
Professor and Associate Chair Department of Mechanical Engineering
National Taiwan University of Science and Technology, Taiwan

Prof. Dong-Seong Kim
Professor of EE and Director of CTRI and ICT-CRC
Kumoh National Institute of Technology

Ir. Eko Julianto, M.Sc., MRINA
Director of Politeknik Perkapalan Negeri Surabaya, East Java,
Indonesia

IOP Conf. Series: Materials Science and Engineering 1010 (2021) 011001 doi:10.1088/1757-899X/1010/1/011001

ORGANIZING COMMITTEE

Chairman	:	Syamsuri, PhD
Convenor	:	Dr. Agus Budianto
Secretary	:	Nareswarananindya, M.Ars
		Arintha Indah Dwi Syafiarti, M.Sc
		Hastawati Chrisna Suroso, M.Sc
		Laras L. Lestari, M.Sc
Technical Committee	:	Ahmad Anas Arifin, M.Sc
		Fajar Rizki Widiatmoko, M.Sc
		Mutiara Firdausi, MT.
		Zain Lillahulhaq, MT.
		Nanang Fakhrur Rozi, M.Kom
		Anwar Sodik, S.Kom, MT.
		Gusti Eka Yuliastuti, M.Kom
		Indra Komara, M.T.
		Hasan Syfik Maulana, M.Sc
		Anggi Yhurinda Perdana Putri, M.Kom
		Adib Pakarbudi, S.Kom, M.Kom.
		Isa Albana, M.Si.
		Ilmiatul Masfufiah, M.Sc
		Fairus Atikah, M.T.
		Farida, M.Kom.
		Lakon Utamakno, M.T.

IOP Conf. Series: Materials Science and Engineering 1010 (2021) 011001 doi:10.1088/1757-899X/1010/1/011001

SCIENTIFIC COMMIITTEE

Prof. Ming-Jyh Chern (National Taiwan University of Science and Technology, Taiwan)

Prof. Kim, Dong-Seong (Kumoh National Institute of Technology, South Korea)

Michele Bici, PhD. (Sapienza University of Rome, Italy)

Dr. Dimas Fajar (Institut Teknologi Sepuluh Nopember, Indonesia)

Dr. Zeplin J.H. Tarigan, ST., MMT. (Petra Christian University, Indonesia)

Dr. Drs. I Nyoman Sutapa, M.Sc.nat. (Petra Christian University, Indonesia)

Syamsuri, S.T., M.T., PhD. (Institut Teknologi Adhi Tama Surabaya, Indonesia)

Dr. Mat Syai'in, ST., MT., Ph.D (Institut Teknologi Adhi Tama Surabaya, Indonesia)

Dr.Eng. Rizal Mahmud, S.Pd., M.T. (Institut Teknologi Adhi Tama Surabaya, Indonesia)

Achmad Chusnun Ni'am, S.Si., MT, PhD (Institut Teknologi Adhi Tama Surabaya, Indonesia)

Dr. Yustia Wulandari Mirzayanti, ST., MT (Institut Teknologi Adhi Tama Surabaya, Indonesia)

Rinci Kembang Hapsari, S.Si., M.Kom (Institut Teknologi Adhi Tama Surabaya, Indonesia)

Dr. Moch. Junaidi Hidayat, ST., M.Ds (Institut Teknologi Adhi Tama Surabaya, Indonesia)

Dr. Lukmandono, ST., MT (Institut Teknologi Adhi Tama Surabaya, Indonesia)

Enggar Alfianto, S.Si., M.Si (Institut Teknologi Adhi Tama Surabaya, Indonesia)

Indra Komara, ST., MT (Institut Teknologi Adhi Tama Surabaya, Indonesia)

IOP Conf. Series: Materials Science and Engineering 1010 (2021) 011002 doi:10.1088/1757-899X/1010/1/011002

Peer review declaration

All papers published in this volume of *IOP Conference Series: Materials Science and Engineering* have been peer reviewed through processes administered by the Editors. Reviews were conducted by expert referees to the professional and scientific standards expected of a proceedings journal published by IOP Publishing.

- **Type of peer review:** Single-blind with two level review system
  - **i.** The conference editor reviewed the submitted papers relevance to the conference topics, checked whether the papers meet the IOP format requirements, and doing similarity check to ensure the percentage of similarity is not greater than 25%.
  - **ii.** The independent reviewers evaluated the submitted papers according to the following criteria as the relevance to meet the substance of study, novelty of work, quality and scientific knowledge, and adequate references etc.
- **Conference submission management system:**

  All the papers were submitted via EasyChair link and conference Email.
- **Number of submissions received:** 73 submissions received
- **Number of submissions sent for review:** 73 submissions were sent for review
- **Number of submissions accepted:** 66 papers accepted
- **Acceptance Rate (Number of Submissions Accepted / Number of Submissions**
- **Received X 100):** 90
- **Average number of reviews per paper:** 2
- **Total number of reviewers involved:** 16
- **Any additional info on review process:** -Nil-
- **Contact person for queries:**

Syamsuri, ST., MT., PhD.
Associate Professor
Phone: +62 819-3828-8275
Institut Teknologi Adhi Tama Surabaya
Gedung A Lantai 1, Kampus ITATS
Jl. Arief Rachman Hakim 100, Klampis Ngasem,
Surabaya, Jawa Timur, Indonesia 60117
E-mail: syamsuri@itats.ac.id, icatech.support@itats.ac.id
Website: http://icatech.itats.ac.id

Effect of welding electrode variation on dissimilar metal weld of 316l stainless steel and steel ST41

I A Pahlawan[1], A A Arifin[2], E Marliana[3], H Irawan[2]

[1]Department of Mechanical Engineering, Sapienza University of Rome, Italy
[2]Department of Mechanical Engineering, Institut Teknologi Adhi Tama Surabaya, Indonesia
[3]Department of Mechanical Engineering, Universitas 17 Agustus 1945 Surabaya, Indonesia

pahlawan.1757957@studenti.uniroma1.it

Abstract. The propeller shaft is part of the ship which transmitting rotation from engine to propeller. The shaft is also used in small-medium size of fisherman boat. In order to reduce the cost of production, dissimilar metal bases are used by joining Steel ST41 and 316L Stainless Steel in process. This research was conducted to analyze effect of welding electrode towards tensile strength, macrostructure, and microstructure of Steel ST41 and 316L Stainless Steel welded metal. Two variations of electrode that used in this paper were E309-L and E6013 with Shielded Metal Arc Welding process (SMAW). The other parameter such as current was set at 80 A and voltage was set at 20 V with 20 mm/min welding travel speed. The results of tensile test show average tensile strength max (σ_{max}) of E309-L and E6013 electrode were 581,42 MPa and 599,24 MPa respectively. These results show that welding process is increasing the maximum tensile strength from base material strength level. The microstructure test interpretation shows E309-L electrode has lower possibility to create delta ferrite in HAZ (Heat Affected Zone) rather than E6013 electrode. This means welding process with E6013 tends to have lower corrosion resistance.

1. Introduction

Sea water environment is contained NaCl as catalyst which could causes corrosion for some material, especially steel. It is important to use material with high corrosion resistance for ship components that have any direct intact to sea water. In the recent year stainless steel capable to protect the material from any corrosion because of its oxide layer. On the other hand, Stainless Steel is more expensive than carbon common steel in use. It leads to developing new dissimilar joints which has the quality of stainless steel at corrosion resistance and has low material cost [1]. In this case, small-medium size boat usually has stainless steel part at tip of shaft, due to the application of propeller inside the sea water. The quality of dissimilar joint depends on certain parameter, for example on joining configuration [2]. The other important parameter is welding position [3].

The welding parameter could leads heterogeneity in microstructure and degradation of material strength in period of time [4]. The consideration of fatigue in period of time is also important to achieve good quality of dissimilar joint [5]. The other parameter that needs to take into account is groove. Narrower groove means higher tensile strength that could be achieved [6]. In microstructure point of view, δ-Ferrite structure is important due to its contribution for preventing hot cracking after welding operation [7]. The formation of δ-Ferrite is affected by cooling rate after welding process [8].

Phase transformation also could alter the mechanical properties of material [9]. The other parameter is heat input of welding operation because it could determine the size of weld pool [10]. The type of welding technique also become a consideration for improving the quality of mechanical properties [11].

In the previous study by Hesham, the metals are CK45 and Stainless Steel 316L. On microstructure result, it showed that dissimilar joining process tend to create more δ-Ferrite structure. This condition could lead to reducing internal corrosion resistance. In comparison, this research was conducted by variating type of electrode E309-L and E6013 on Dissimilar Weld Metal (DWS) as an alternative to cut highly cost of fully austenite or clad joint in previous study comparison. The expected results of experiment are tensile strength, macrostructure, and microstructure of Steel ST41 and 316L Stainless Steel welded metal. The objective of this research is to analyze the correlation of electrode variation towards mechanical properties, macrostructure, and microstructure of material.

2. Materials and Methods

2.1 Materials

The materials used in this research were round bar of ST41 and 316L Stainless Steel with 22,225 mm (7/8 in) diameter. Welding groove was designed with 60° full circle. Fabrication of specimens was done by lathe machine. Chamfer in each material was designed to be 30°. The tip of chamfer was 1 mm in diameter. The weld was totally performed by certificated welder on lathe machine to perform good weld quality and to reduce deformation of bars. For detailed chemical composition and mechanical properties of base metals and weld metals (shown at Table 1).

Electrodes which used in this research were commercially by E6013 ENKA brand and NSN 309L Nikko Steel brand with same 2.6 mm diameter. Both of them were performed by same SMAW machine and same weld parameter with DCEN polarity, 80A current and 120 mm/min average travel speed to maintain heat input in range 800 J/mm. Welding process is performed until the groove was fully loaded. Then the process was continued by manufacturing it into ready test specimen. For detailed welding record are note in Table 2 and Table 3.

Table 1. The Composition of SGD 400-D (ST 41) and AISI 316L base material with E6013 and E309L weld material

	Chemical Composition %								**Mechanical Properties**	
Base metals	C	Si	Mn	S	P	Cr	Ni	Mo	Tensile Strength, MPa	
									Ult.	Yield
ST 41	0,146	0,187	0,589	0,034	0,015	-	-	-	316-550	250
AISI 316L	0,016	0,53	1,76	0,02	0,038	16,73	10,04	2,02	515	205
Weld metals	C	Si	Mn	S	P	Cr	Ni	Mo	Ult.	Yield
E6013	<0,2	0,25	<1,2	0,016	0,012	<0,2	-	<0,3	527	458
NSN-309L	0,015	0,40	2,0	0,010	0,012	23,50	13,5	0,25	590	400

IOP Conf. Series: Materials Science and Engineering 1010 (2021) 012001 doi:10.1088/1757-899X/1010/1/012001

Table 2. Record parameter for E 6013 electrode specimen

Layer Level	*Welding Process*	*Filler metal*		*Current*		*Volt age*	*Travel Speed*	*Heat Input*
		Class	*Ø size*	*Type*	A	V	mm/min	J/mm
1	SMAW	E 6013	2,6	DCEN	80	20	120	800
2	SMAW	E 6013	2,6	DCEN	80	20	120	800
3	SMAW	E 6013	2,6	DCEN	80	20	120	800
4	SMAW	E 6013	2,6	DCEN	80	20	100	960
5	SMAW	E 6013	2,6	DCEN	80	20	100	960

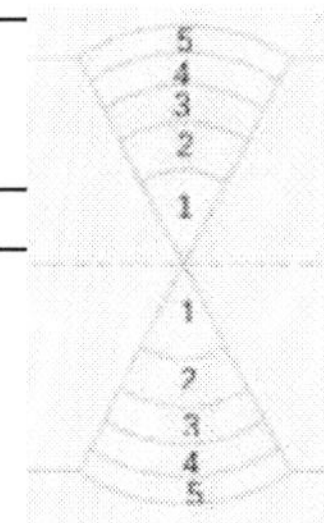

Table 3. Record parameter for of E309L electrode specimen

ayer Level	*Welding Process*	*Filler metal*		*Current*		*Volt age*	*Travel Speed*	*Heat Input*
		Class	*Ø size*	*Type*	A	V	mm/min	J/mm
1	SMAW	E309L	2,6	DCEN	80	20	120	800
2	SMAW	E309L	2,6	DCEN	80	20	120	800
3	SMAW	E309L	2,6	DCEN	80	20	120	800
4	SMAW	E309L	2,6	DCEN	80	20	100	960

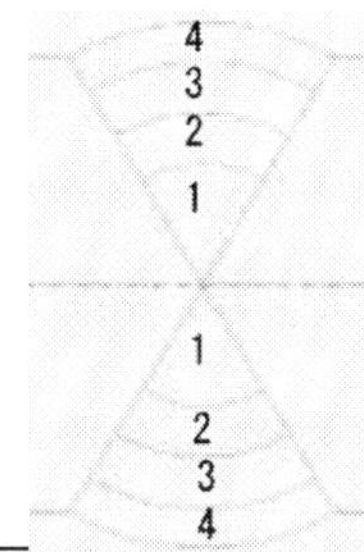

2.2 Specimen Testing Methods

Tensile Test was applied for evaluating the strength of material with respect to shaft condition during applied longitudinal force. Round bar specimen test was lathed for specific dimension required (ASTM E8). The standard of Inner diameter (D) is 12.5 +- 0.2 mm, Gage length (G) is 62.5 +- 0.1 mm, radius is min 10 mm, and Length of reduced section is 75 mm are represented in **Figure 1**. Tensile test was done by using common universal hydraulic machine, with quasi static speed. Yield point can be achieved by calculating with offset method 0,2 % of L_{max}. Next thing to do was measuring overall dimension before and after specimen was tested. By using equation below, stress and strain can be calculated by assigning P with load (N), A_0 with cross section area (mm^2), L_0 with initial length (mm^2), and L_1 with final length (mm^2).

$$\sigma_{eng} = \frac{P}{A_0} \; [MPa] \qquad (1)$$

$$\varepsilon_{eng} = \frac{L_1 - L_0}{L_0} \times 100\% \; [\%] \qquad (2)$$

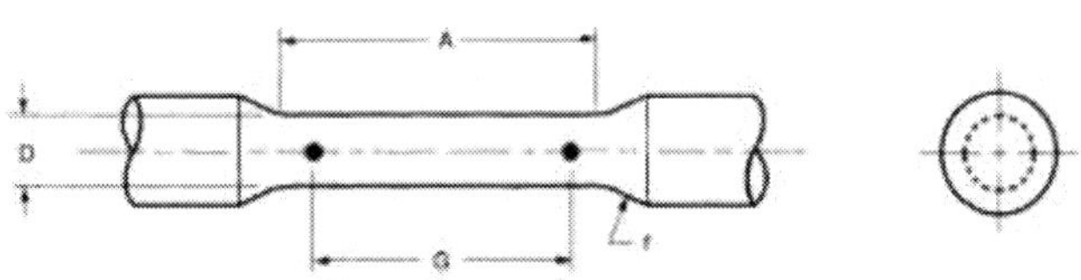

Figure 1. Round bar test specimen illustration

Metallography test were performed to analyze dilution or penetration depth and weld area of specimen. The specimen represented by 1 specimen on macro and 1 specimen on. Preparation is initiated by cutting specimen parallel to longitudinal line into two pieces. One side for macrograph and another one for micrograph. Specimen preparation were supported by mounting mold of resin to make better handgrip. The parallel surface was grinded and polished with several step grid to make surface smoothness up to 1-0,05 micron. Combination etching liquid for dissimilar material between stainless steel and carbon steel, Ethanol 95%:NHO3:HCl with ratio 1:1:½ were applied for short duration. The etched surface was cleaned by water and ethanol then dried. The picture was taken by common proportionate camera. CAD software was used to determine the area of weld and measuring specific area, and to calculate weld dilution with Equation 3.

$$\%Dilution\ (area) = \frac{Fused\ Base\ and\ Weld\ Metal\ Area}{Weld\ Area + Fused\ Base\ and\ Weld\ Metal\ Area} \times 100\ \% \quad (3)$$

The welding process is performed to achieve minimum standard dilution percentage between 10-20 % with welding parameter current 80A and electrode rod diameter 2.6 mm, neglecting some minor perturbations from environment.

Weld dissimilar material will lead to some specific problem occur, such as altering austenite structure into ferrite structure and could degrading quality of mechanical properties and corrosion resistance of stainless steel. This phenomenon is occurred in range of temperature between 600-900°C, where Chromium element easily attracted to grain boundary and lead to brittle carbide precipitation [12]. The emergence of δ-Ferrite certainly will reduce the ductility and the potential toughness. Meanwhile, austenite/ferrite bonding will become a place that has higher possibility to form a carbide participation $M_{23}C_6$ and sigma phase where embrittlement agent arises in stainless steel.

3. Discussions

3.1 Tensile test

The requirement for tensile test is the result of welding specimen tensile strength should not in lower level than minimum tensile strength of the base metal. On other hand, the crack has to be located in the weakest base metal, which for this case was on ST41 with minimum tensile strength (σ_u) 316 MPa. The lowest tensile strength test from 4 specimens result is 524 MPa. These results were already surpassing the minimum tensile strength (σ_u) acceptance of requirement as shown in **Figure 2**. In the Stress-Strain diagram **Figure 2**, the yield strength (σ_y) minimum from 4 specimens is 423,97 MPa (*by using offset method*), which means already surpassing the minimum base metal yield strength of 205 MPa. The initiation of necking and fracture was located in base metal of ST41 closed to Heat Affected Zone as shown in **Figure 3**.

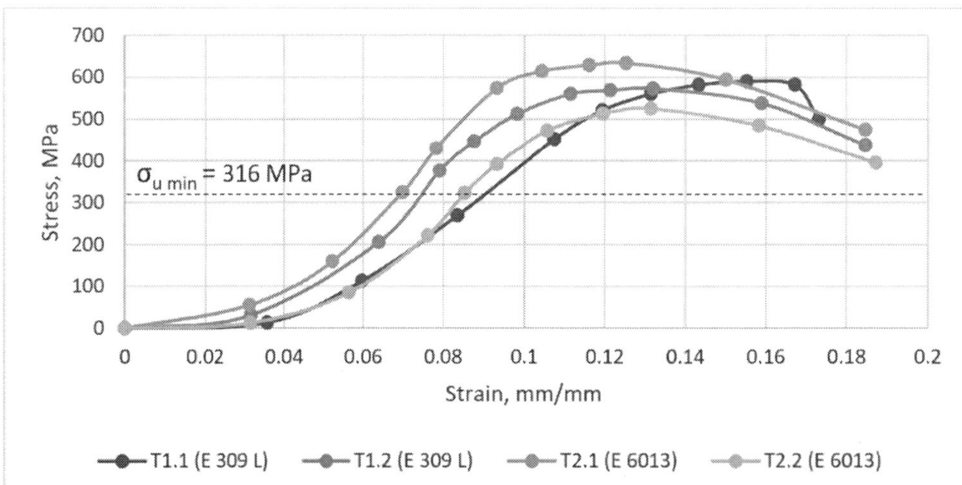

Figure 2. Stress – Strain engineering graph for each specimen tensile test result.

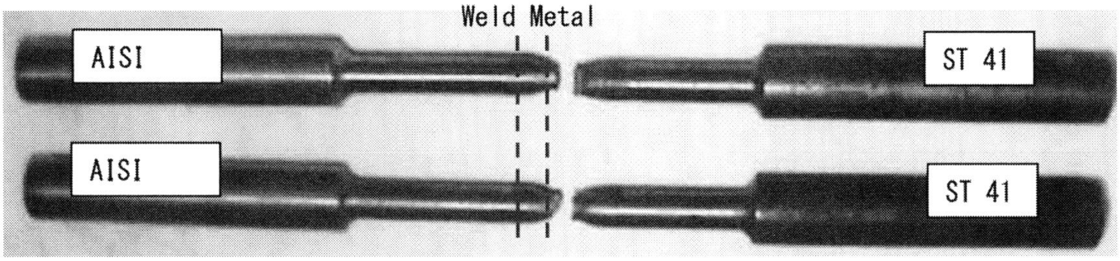

Figure 3. Tensile test specimen after tensile force applied.

3.2 Macro Test

Welded joint using E6013 electrode rod deposits more layer due to unstable puddle and it tends to create spatter spark during weld processes. It needs more weld layer and duration to fulfill complete grove. This phenomenon is affecting higher heat input and as a result is wider Heat Affected Zone (HAZ). Meanwhile, using E309L electrode rod creates more stable puddle and reduce duration of weld processes. It was filling groove faster even with the same parameter as before. It has lower heat input and followed by narrower Heat affected Zone (HAZ). Macrostructure were evaluated by calculate dilution area resulted from using E309L as shown in **Figure 4** and E6013 as shown in **Figure 5**, were up to 15,7% and 16,6% respectively. This result is fulfilling the criterion standard of dilution area 10 - 20 %. Raw results **Figure 4 (a)** and **Figure 5 (a)** were plotted to CAD software to measure the dilution distance as presented in **Figure 4 (b)** and **Figure 5 (b)**.

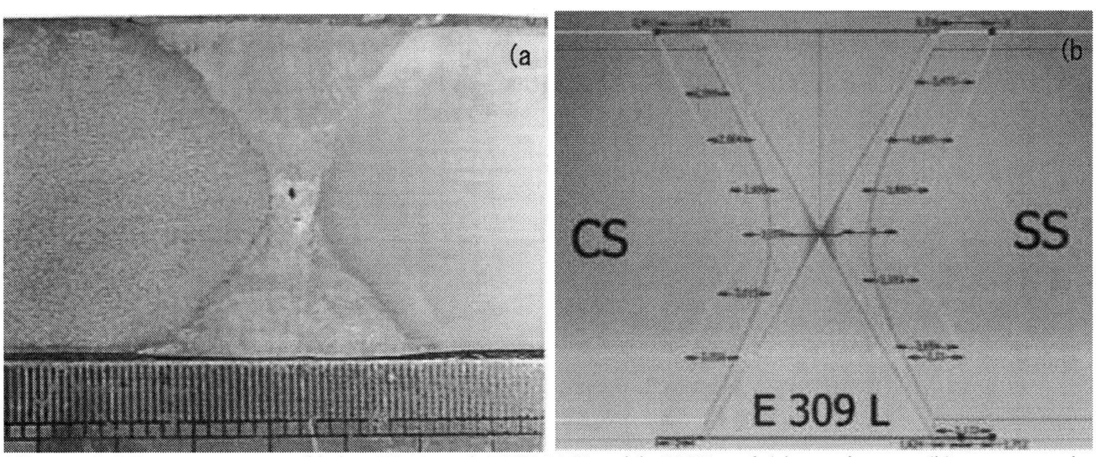

Figure 4. Macrostructure of joint ST41 and AISI 316L with 309L rod (a) raw image; (b) re-measuring groove and dilution.

IOP Conf. Series: Materials Science and Engineering 1010 (2021) 012001 doi:10.1088/1757-899X/1010/1/012001

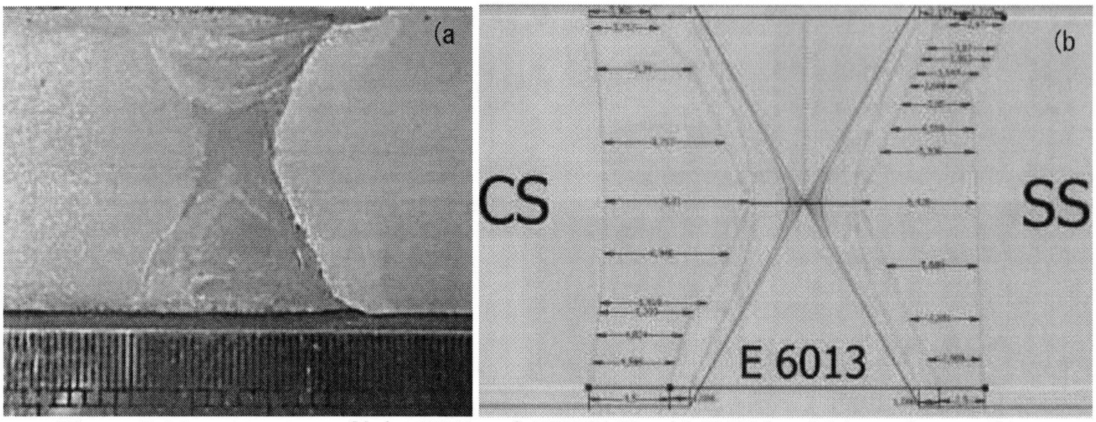

Figure 5. Macrostructure of joint ST41 and AISI 316L with E6013 rod (a) raw image; (b) re-measuring groove and dilution.

3.4 Micro Test

By using E309L electrode, could reduce the carbide participation risk and ferrite transformation in microstructure austenite base metal 316L as shown in **Figure 6(a)** and **(b)**, due to its low amount of carbon element, lower heat input and rapid deposit of weld. Relatively low amount of δ-Ferrite cluster (black sphere) were located only near the fusion line. Meanwhile by using E6013 electrode, could increase susceptible of carbide participation and ferrite transformation in HAZ of stainless steel AISI 316L. Furthermore, it could lead to high risk of carbide participation followed by intergranular corrosion, where shown in microstructure **Figure 7(a)** and **(b)**. It has higher amount of δ-Ferrite cluster. Meanwhile, microstructure in HAZ of Carbon Steel ST41 have no significantly affected by electrode used. Yet, grease or another protection are good enough to be alternative solution to give lower risk of corrosion and reduce the possibility to intergranular corrosion rise.

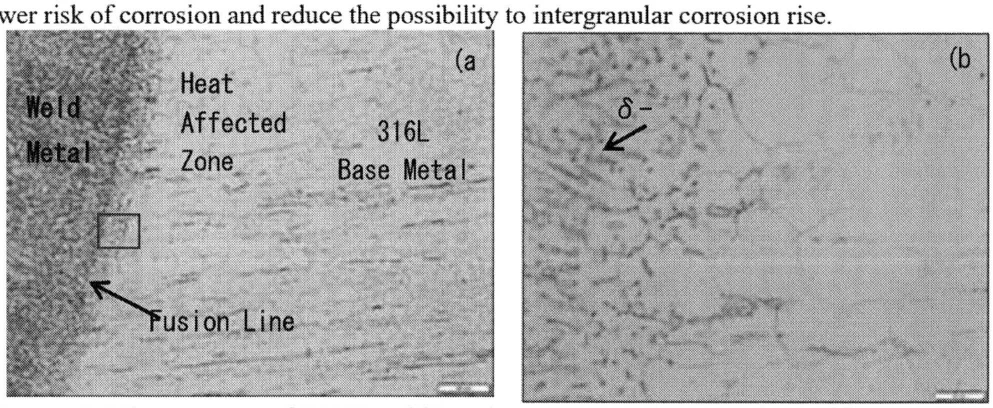

Figure 6. Microstructure of E309L weld metal to AISI 316L base metal magnification (a) 100 x (b) 500 x

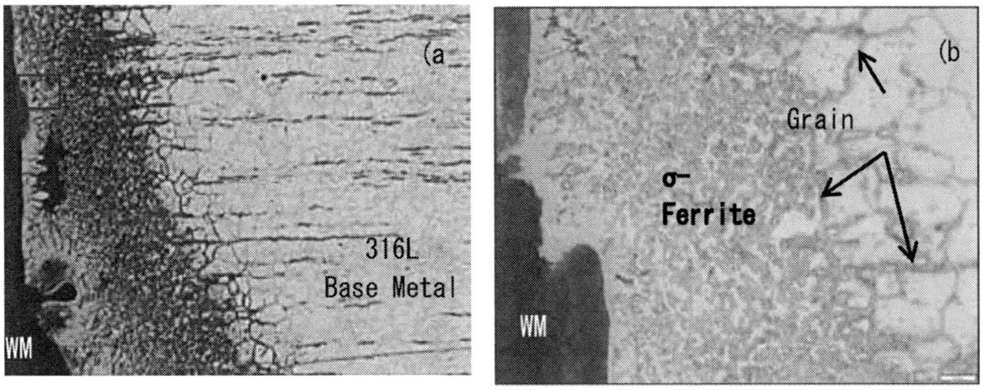

Figure 7. Microstructure of E6013 weld metal to AISI 316L base metal magnification (a) 100 x (b) 500 x

4. Conclusions

The results of tensile test for both variations of electrode are on higher level of base material ultimate tensile strength. This means by doing dissimilar joint of ST41 and 316 Stainless Steel with E6013 and E309L is increasing the quality of ultimate strength of material. Although, the disadvantage of using E6013 electrode is wide Heat Affected Zone (HAZ). It can decrease the mechanical properties of materials at some points. In term of dilution percentage, E6013 has higher result than E309L. This means welding process with E6013 has better penetration to base metal which can lead to decreasing probability of occurring defect of porosity. On microstructure result, it shows that joining process with E6013 tend to create more δ-Ferrite structure. This condition could lead to reducing internal corrosion resistance.

5. References

[1] Santoso W, Widodo EWR, Kuswanto A, et al. Analisis Kuat Arus, Jenis Logam Pengisi, dan Jenis Kampuh Las Terhadap Kekuatan Tarik dan Kekerasan Dissimilar Welding dengan Gas Tungsten Arc Welding (GTAW). *J Tek Mesin Inst Teknol padang* 2018; 8: 2–6.

[2] Rajput SK, Kumar A, Tripathi SS, et al. Investigation of microstructural behavior and mechanical properties of dissimilar weld joints of austenitic-ferritic stainless steel. *Mater Today Proc* 2020; 25: 778–784.

[3] Advaith V, Vaisikan A, Thirumalini S, et al. Comparative studies on pulsed GTAW and AGTAW on dissimilar alloy C-22 and AISI 316L weldments. *Mater Today Proc* 2020; 1–10.

[4] Awale DD, Ballal AR, Thawre MM. Dissimilar weld joints of P91 and 316LN for power plants Applications-A review. *Mater Today Proc*. Epub ahead of print 2020. DOI: 10.1016/j.matpr.2020.05.003.

[5] Wang WK, Zhang JX, Zhong J. Comparative evaluation of LCF behavior of dissimilar steels and welds in an ultra-supercritical turbine rotor at 280 °C. *Int J Fatigue* 2020; 139: 105771.

[6] Wahyu E, Widodo R, Setyowati VA, et al. PENGELASAN LOGAM TIDAK SEJENIS MATERIAL STAINLESS STEEL 304L DAN BAJA AISI 1040 DENGAN GAS TUNGSTEN ARC WELDING. 327–332.

[7] Setyowati VA. Pengaruh Pengelasan TIG pada Stainless Steel 304 dan 304L terhadap Sifat Mekanik, Karakterisasi XRD, dan EDX sebagai Material Pressure Vessel. *J Tek Mesin* 2011; 7: 2011.

[8] Wei P, Li H, Liu J, et al. The effect of water environment on microstructural characteristics, compositional heterogeneity and microhardness distribution of 16Mn/304L dissimilar welded joints. *J Manuf Process* 2020; 56: 417–427.

[9] Marliana E. Analisa Pengaruh Perubahan Fasa Terhadap Konstanta Pada Coil Spring Shape

Memory Alloy (SMA). 2020; 7: 9–16.
[10] Setyowati VA, Suheni S. Variasi Arus Dan Sudut Pengelasan Pada Material Austenitic Stainless Steel 304 Terhadap Kekuatan Tarik Dan Strukturmakro. *J IPTEK* 2016; 20: 29.
[11] Yelamasetti B, Rajyalakshmi G. Effect of TIG, pulsed TIG and Interpulse TIG welding techniques on weld strength of dissimilar joints between Monel 400 and AISI 316. *Mater Today Proc* 2019; 19: 755–760.
[12] Hesam Pouraliakbar, M. H. Designing of CK45 Carbon Steel and AISI 304 Stainless Steel Dissimilar Welds. *Materials Research* 2014; 106-114.

Numerical study of aerodynamics across three models car generation

N Vaziri[1], Syamsuri[2], Z Lillahulhaq[2], A A Arifin[2], F Z Achmad[2]

[1] Department of Physics, Islamic Azad University, Karaj, Tehran, Iran
[2] Department of Mechanical Engineering, Institut Teknologi Adhi Tama Surabaya Indonesia

syamsuri@itats.ac.id

Abstract. Many automotive companies are adding new devices and features to the latest generation of cars. Some changes such as the addition of an adds-on device and car design lead to changes in the aerodynamic characteristics of the vehicle. this study will be carried out on the changes in the aerodynamic properties of the 3 generations of Model cars: Model 1, Model 2 and Model 3 in 2 D Steady state simulation. The car simulated in various numbers of velocities in 40 km/h, 60 km/h ,80 km/h and 100 km/h. The characteristics of aerodynamic force around the car are discussed in this paper from the qualitative and quantitative data. The quantitative which used as measurable data are Coefficient of drag (C_D). Quantitative data has better visual explanation of the stream characteristic. The qualitative data which used in this paper are path line and velocity contours. The validation process is conducted with the previous research and the average error was 5%. The results showed that Model 2 has the most aerodynamic design than the other type due to the drag coefficient was lower than the other type.

1. Introduction

The development of automotive technology is getting faster, to create cars that are efficient and environmentally friendly. There are several steps to improve vehicle efficiency including reducing vehicle weight, using biodiesel fuel, using a hybrid engine and designing an aerodynamic body. The vehicle weight has increased because of the improvements in some factors such as safety, emissions control, and comfortability. The improvement of materials aspect is necessary to achieve weight savings without sacrificing performance, the property, manufacturability, and cost requirements for automotive structures [1]. Fuel consumption is much influenced by the aerodynamic force of the vehicle, especially the drag force. Approximately 30% to 50% of fuel energy is lost due to aerodynamic force drag. Several methods are used to reduce the force that hinders the vehicle by forming the underbody and directing exhaust gas to the rear side of the vehicle. A reduction in drag of up to 22% is achieved by decreasing the lower body flue gas mobility by up to 9%[2].

The aerodynamic flows around the car is causing aerodynamic force especially drag and lift forces. The lift force causes acceleration on the vehicle which reduce the friction between the wheels and road. The velocity reduction on vehicle is affected by drag. The lift and drag force are determined by several aspect, such as: velocity, cross-sectional area, car aerodynamics shape and weight. The dynamic aerodynamic force has influence on important factor in a vehicle such as safety, ergonomic and fuel consumption. The vehicle should be built as aerodynamic as possible to reduce drag force in order to

IOP Conf. Series: Materials Science and Engineering 1010 (2021) 012002 doi:10.1088/1757-899X/1010/1/012002

reduce fuel consumption. In order to get stability condition on high speed vehicle high speed, down force should be generate to keep the tires attach to the road surface [3].

Fluid flow that flows through the car body is divided into flow across the roof (upper flow) and flow through the bottom of the car (underneath). At the back of the car body a secondary flow is formed in the form of 2 coiled vortices, as shown in Figure 1 [4]. Vortex the top (A) is a vortex formed by the upper stream flow. The fluid fills the wake area behind the car and forms the upper vortex. The lower vortex (B) is a vortex that is formed as a result of the under-flow rolling to fill the wake area behind the car to become a twin of vortex. In some references the vortices A and B are also called circular torus [5]. Pillar vortices are vortices that are formed by a flow that flows from the side of the car.

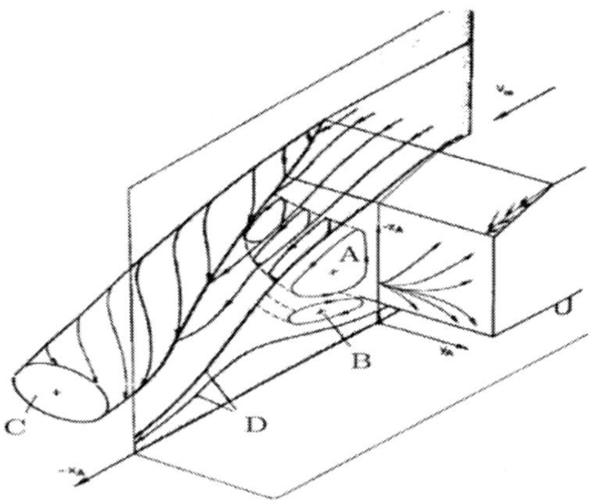

Figure 1. Twin of vortex behind the car body

Several devices such as vortex generator (VG), spoiler, or some modification is added to the car to get the ideal aerodynamic conditions of the vehicle. Spoilers are used to reduce air movement that impedes the vehicle. The spoiler mounted on the front of the car is used to flow air from the tires to the lower body. The rear spoiler attached to the trunk cover of passenger vehicles aims to reduce turbulence at the rear of the vehicle. This spoiler installation aims to increase the pressure under the car to the rear and reduce the lifting force on the rear trunk [6]. Vortex generators (VGs) which are installed in the end upper body of the car is able to control and delay the separation of the flow above the windshield of the car. The VG has great amount of energy to re-energizing the boundary which reduces the drag force of the car [7]. A diffuser is an arc-shaped section of the car underbody which improves the aerodynamic force by enhancing the transition between the high-velocity airflow underneath and the slower airflow of the ambient atmosphere. The diffuser provides a space for the underbody airflow to slow down and spread out. This condition prevents excessive flow separation and drag[8]. The addition of a device to the car causes changes to the aerodynamic properties of the vehicle.

Other accessories such as mirrors to the vehicle also allows changes to the aerodynamic properties of the vehicle. Many automotive companies are adding new devices and features to the latest generation of cars. In this research, this study will be carried out on the changes in the aerodynamic properties of the 3 generations of Model cars.

Some of those studies used the CFD method and were successful. Research using the CFD method includes: Syamsuri *et al.*, 2020 [9] conducted research on the effect of porous media on hydraulic jump characteristics by using smooth particle hydrodynamics. Another study that uses the CFD method is the SPH model for interacting of sloshing wave obstacle in shallow water tank was conducted by Syamsuri *et al.* in 2020 [10]. Research on simulations on cars has also been carried out by Obaidi and Otten 2018 [11]. The difference in the current research is that this study wants to see how the design of the model which is more aerodynamic is due to some of the geometric shapes are more curved.

IOP Conf. Series: Materials Science and Engineering 1010 (2021) 012002 doi:10.1088/1757-899X/1010/1/012002

Different models of car have unique aerodynamic characteristic due to the design, dimension and added on the device. The characteristics of aerodynamic force around the car are discussed in this paper from the qualitative and quantitative data. The quantitative which used as measurable data are Coefficient of drag (C_D). Quantitative data has better visual explanation of the stream characteristic. Another qualitative data which used also in this paper are path line and velocity contours.

2. Methods

This research simulated 2D of three generation Model cars in steady condition. The type of the cars which used in this simulation are Model 1, Model 2 and Model 3 which shown in Figure 2, 3 and 4. The meshing domain is also shown in Figure 2. The car simulated in various numbers of velocities in 40 km/h, 60 km/h ,80 km/h and 100 km/h. This simulation is conducted on a rectangular domain with virtual size which shown in Table 1. The boundary condition and parameters which used in this simulation is shown in Table 2. The mesh used in this research is Quad-Pave type which shown in figure 5. The validation of this simulation is based on numerical study of Model 3 car in 100 km/h.

Table 1. Virtual Domain Dimension.

	Length
Length in front of the car [m] (L1)	7.2 m
Length from the back of the car (L2)	14.4 m
Height [m] (H1)	7.2 m
Total length [m]	28.8

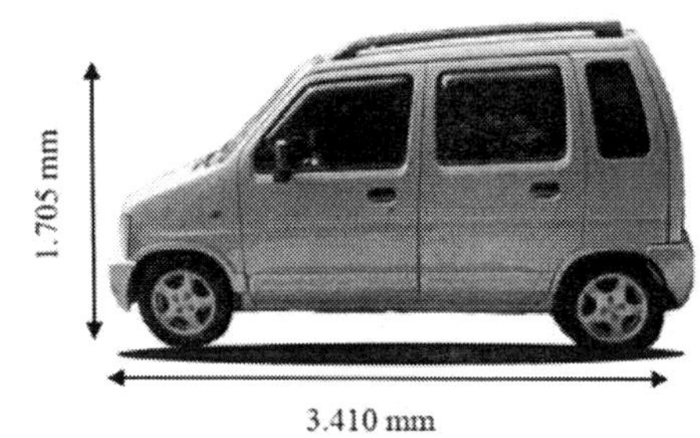

Figure 2 Model 1

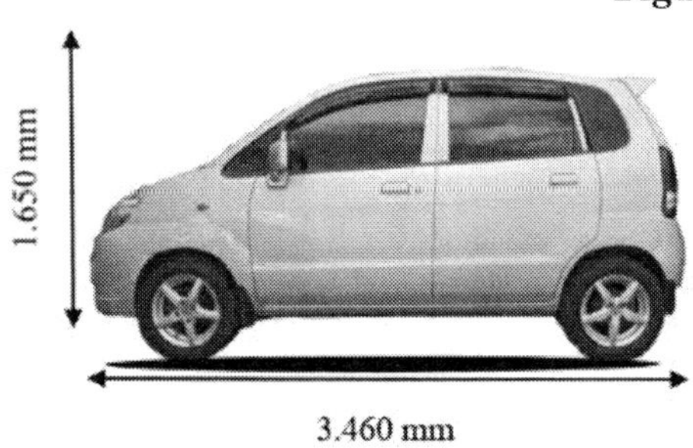

Figure 3. Model 2

Figure 4. Model 3

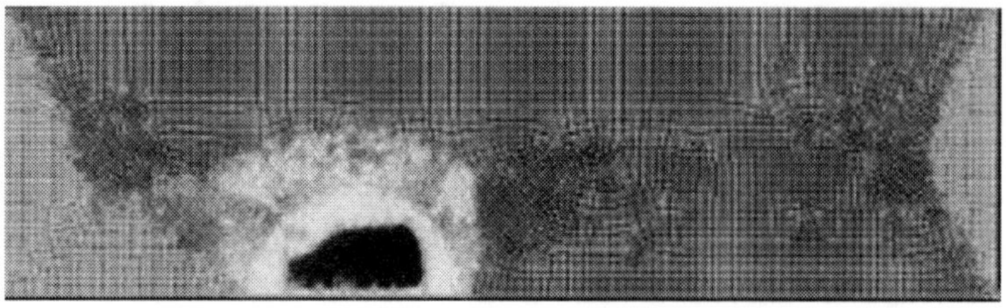

Figure 5. Meshing Quad-Pave

Table 2. Boundary condition and parameters.

	Area	Boundary
Boundary	Inlet	Velocity Inlet
	Outlet	Pressure Outlet
	Road	Wall - moving
	Top	Symmetric
	Input	Value
Parameters	Turbulence model	k-ε (2 eqn) Standard
	Velocity	40 km/h, 60 km/h ,80 km/h and 100 km/h
	Turbulence Intensity	10%
	Turbulence Length Scale	1 m

3. Result

3.1. Validation
Validation was carried out with previous researchers by Obaidi and Otten (2018) [12] with the same geometric shape and fluid flow velocity. Results from research by Obaidi and otten (2018) [12], shows that the drag coefficient of the car at a speed of 100 km / hr is equal to 0.32. While the simulation result is 0.34. In general, the error rate is 1.5%. This shows that research with numerical simulation has good agreement with previous researchers.

3.2. The effect of the XXXX Model car design coefficient of drag
The aerodynamic forces that affect the performance of the car consist of drag and lift forces. Table 3 shows the coefficient of drag which simulated on the three-generation model of the Model car.

The simulation is conducted on 4 various velocity: 40 km/h, 60 km/h ,80 km/h and 100 km/h. In table 3, the drag coefficient value on the three Model car models has increased along with the increase in flow speed. The simulation results show that the Model 1 car has the highest C_D value in all speed variations. Model 2 types of cars have increased the drag coefficient when the speed increases from 40 km / h to 60 km / h. Moreover, there is no increment of the drag coefficient in Model 2 types when the speed increases above 60 km / h. This condition is caused because Model 2 has the shape of the car body like a bow so that the coefficient of drag is smaller than the Model 1 and Model 3 which has the shape of a box. Although Model 3 is the latest car but has a drag coefficient that is larger than the previous generation of Model 2. Design Model 3 car carrying the classic model with a modern twist, which is a benchmark design Model 1. So that both the generation have close difference value in coefficient drag.

Table 3. The coefficient of drag in three generation car model.

	The coefficient of drag (C_D)			
	40 km/h	60 km/h	80 km/h	100 km/h
Model 1	0.407	0.408	0.408	0.410
Model 2	0.342	0.343	0.343	0.343
Model 3	0.383	0.407	0.408	0.409

3.3. The effect of the Model car design on the twin of vortex behind the car

The fluid that flows in the upper and lower parts of the body will be separated away from the car body. At the rear of the car body, a wake area is formed where the fluid does not pass. The wake area has a fairly low pressure compared to other regions. Fluid flow that is separated away from the body will experience backflow and flow to the wake area behind the body. This flow will roll up and form a vortex or a twin vortex.

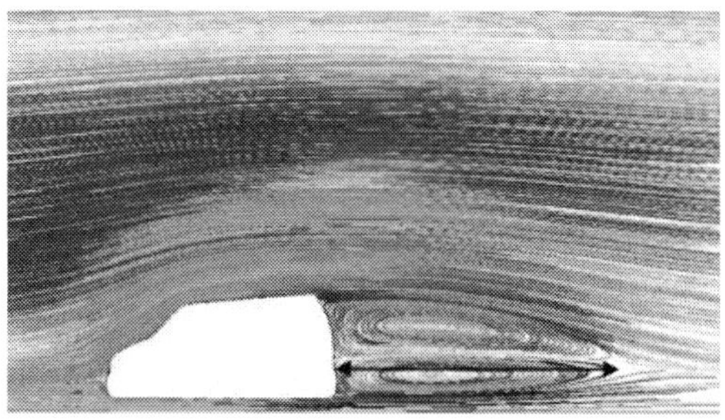

Figure 6 Path line of fluid across Model 1 in 100 km/h

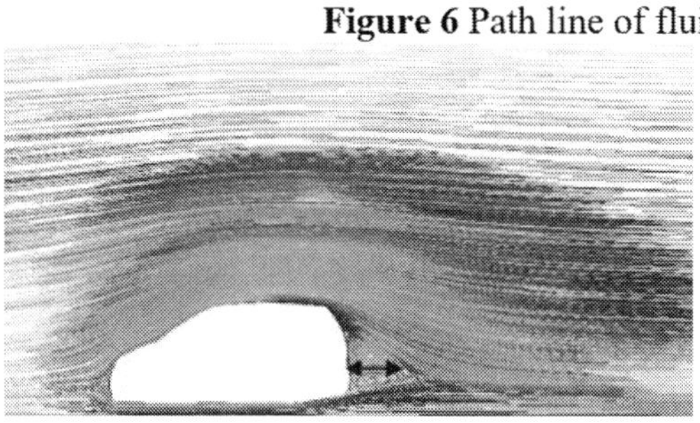

Figure 7. Path line of fluid across Model type 2 in 100 km/h

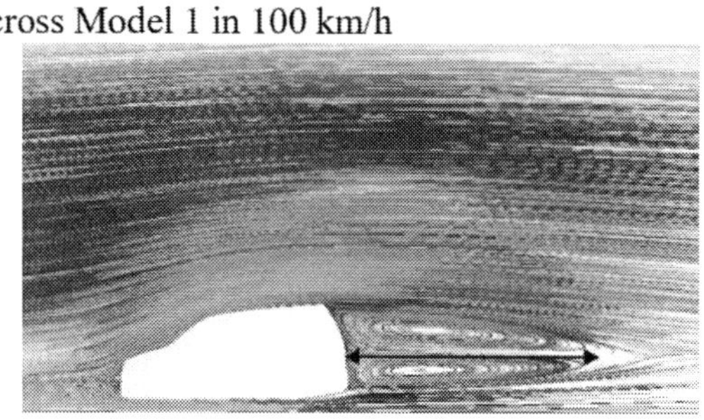

Figure 8. Path line of fluid across Model type 3 in 100 km/h

The ratio of the size of the vortex formed in the wake region is influenced by the velocity of fluid flow at the inlet side. The flow path line can be used to get a clear picture of the size of the vortex behind the car body model. Figures 6, 7 and 8 show the path line of fluids across the Model car body in 100 km/h. In Figure 7 it can be seen that the twin vortex size formed on the Model type 2 has the shortest size compared to the other 2 types. The Model 2 type has a more aerodynamic body shape and has a lower rear. This shape of the car body makes the fluid flow that passes over the car easier to collect again and reduces the wake area that is formed. In addition, the Model type of 2 has a fairly smooth windshield angle. This condition causes fluid stream which flow along the roof of the car has more energy that can be used against the frictional forces and adverse pressure gradient.

In Figures 6 and 8 it can be seen that the rear of the Model type 1 and Model 2 cars have a prominent and taller size. In this section, the fluid flow that is separated leaving the car has a very high energy to fly away from the car body. This condition causes the formation of a wide area of the wake behind the car. The size of the twin of vortex formed behind the car will increase as the wake size increment. In addition, the windshield of the Model 1 and Model 3 car is very sharp angle. The sharper angle of the spacing on the front window of the car, the less energy contained in the fluid flow. This condition causes the fluid flow that flows over the car to have a lower energy to resist the friction force. Comparison of the length of the twin of vortex behind the car part is shown in table 4. Increasingly Long twin of vortices formed on the back of the car indicated the formation of a high enough drag on the car.

Table 4. The length of twin of vortex and coefficient of drag in three generation Model cars

	The length of twin of Vortex	C_D
Model 1	4259 mm	0.410
Model 2	816 mm	0.343
Model 3	4195 mm	0.409

4. Conclusion

Validation has been carried out with previous research and the results show that there are similarities with previous researchers This study showed that Model 2 has the most aerodynamic design than the other type. This is evidenced by the low value of the drag coefficient of Model 2. Moreover, there is no increment of the drag coefficient in Model 2 types when the speed increases above 60 km / h. This condition is caused because Model 2 has the shape of the car body like a bow so that the coefficient of drag is smaller than the Model 1 and Model 3 which has the shape of a box. The Model type 2 has the shortest size compared to the other 2 types. Increasingly long twin of vortices formed on the back of the car indicated the formation of a high enough drag on the car.

5. References

[1] Joost WJ. Reducing Vehicle Weight and Improving U.S. Energy Efficiency Using Integrated Computational Materials Engineering. DOI: 10.1007/s11837-012-0424-z.

[2] Sivaraj G, Parammasivam KM, Suganya G. Reduction of aerodynamic drag force for reducing fuel consumption in road vehicle using basebleed. *Journal of Applied Fluid Mechanics* 2018; 11: 1489–1495.

[3] Norwazan AR, Khalid AJ, Zulkiffli AK, et al. Experimental and numerical analysis of lift and drag force of sedan car spoiler. In: *Applied Mechanics and Materials*. 2012, pp. 43–47.

[4] Venning J, Lo Jacono D, Burton D, et al. The nature of the vortical structures in the near wake of the Ahmed body. *Proceedings of the Institution of Mechanical Engineers, Part D: Journal of Automobile Engineering* 2017; 231: 1239–1244.

[5] Bayraktar S, Bilgili YO. Effects of under body diffuser on the aerodynamics of a generic car. *International Journal of Automotive Engineering and Technologies* 2018; 7: 99–109.

[6] Arya S, Goud P, Sharma V, et al. SIMULATION OF PASSENGER CAR BY ATTACHING ADD-ON DEVICE TO REDUCE DRAG FORCE BY CFD APPROACH. *International Research Journal of Engineering and Technology*.

[7] Saleh ZM, Ali AH. Numerical Investigation of Drag Reduction Techniques in a Car Model. In: *IOP Conference Series: Materials Science and Engineering*. Institute of Physics Publishing, 2020, p. 012160.

[8] Bansal R, Sharma RB. Drag Reduction of Passenger Car Using Add-On Devices. *Journal of Aerodynamics* 2014; 2014: 1–13.

Acknowledgments

The author would like to express their gratitude for the support given by ITATS. The authors are grateful to Prof. Ming Jyh Chern, Prof. Nyoman PA and Prof. Susilo A for their helpful discussions as well.

IOP Conf. Series: Materials Science and Engineering 1010 (2021) 012003 doi:10.1088/1757-899X/1010/1/012003

Characteristics of wall heat transfer from impinging diesel spray flame in low oxygen concentration ambient

R Mahmud[1,2], K Nishida[2], Y Ogata[2], and T Kurisu[2]

[1]Institut Teknologi Adhi Tama Surabaya, East Java, Indonesia
[2]University of Hiroshima, Japan.

rizal@itats.ac.id

Abstract. The objective of this study is to obtain an enhanced understanding of the effect of diesel spray/flame impingement on transient heat flux to the wall. By using a constant volume vessel under engine-like condition, surface heat flux of the wall at spray/flame impingement was measured with three thin film thermocouple heat flux sensors. Fuel was injected using a single-hole injector with a 0.133 mm diameter nozzle. In order to investigate the relation between diesel flame and wall heat loss, two color-method was applied to observe flame temperature distribution. 21 % and 16 % as the various oxygen concentration were investigated at each injection pressure of pressure 80, 120, and 180 MPa. The results showed that the lower oxygen concentration was generated heat loss regardless of injection pressures. The lower flame temperature and flame contact area and period were influence factor on heat loss on the wall. Furthermore, the relation between Nusselt number (*Nu*) and Reynolds number (*Re*) were also investigated to understand the heat transfer phenomena in diesel engines.

1. Introduction

In modern diesel engine, the fuel-air mixture become important factor on the combustion process and exhaust emissions, and then in thermal efficiency. In order to achieve higher efficiency of the diesel engines, it is important to improve the combustion efficiency and decrease the heat transfer through the combustion chamber wall. Therefore, fully understanding of the mechanism of heat transfer by the impinging combustion flame to the wall is required to better thermal efficiency of the diesel engines. Some studies showed that minimizing temperature difference between in-cylinder gas and combustion chamber wall surface was possible by applying insulation material such as ceramic coating on the surface [1][2]. As a result, the heat loss was reduced. Thus, both exhaust energy and piston work increased which, in turn, led to improved thermal efficiency.

Moreover, in order to reduce the emission, exhaust gas recirculation (EGR) is an effective equipment part to reduce nitrogen oxides (N_{Ox}) from diesel engines because it lowers the flame temperature and the oxygen concentration of the working fluid in the combustion chamber [3]. With increasing EGR rate, the oxygen concentration in the fuel/gas mixture decreases and the specific heat of the cylinder increases and leads reaction rate decrease. Further, in-cylinder pressure and combustion decrease [4].

In a simulation study, applying the EGR rate also considered decreasing flame temperature distribution on heat loss. Comparing the single injection results at a different EGR rate, it was confirmed that the heat flux was decreased at a higher EGR rate [5]. It was evident with an experimental study by Kuboyama [7] that the results show that EGR in a diesel engine suggests that it has the effect of reducing wall surface heat loss.

As described by the previous researchers, oxygen concentration has an effect on combustion behavior, which closely relates with lower temperature combustion. Therefore, to clarify an effect of oxygen concentration on decreasing heat loss, variation of oxygen concentration with different injection pressures will be investigated in this study. Furthermore, to investigate the relation between diesel flame and wall heat loss, two color-method was applied to observe local flame temperature distribution. Instantaneous temperature was detected by three Thin Film Thermocouple Heat Flux Sensors (TFTHFSs) which were mounted on the wall surface. By using the measured temperatures as boundary condition, we computed non-steady heat flux through the surface with finite difference method. Additionally, this research also attempts to investigate the heat transfer correlation to understand the effect of flow induced by the spray flame.

2. Experimental Set Up

2.1 High-Pressure and High-Temperature Chamber Vessel

The experiment was done same as the previously conducted one by the authors [7] using high-pressure and high-temperature chamber vessel as shown in Figure 1. It had four side windows and one of them was installed with transparent quartz windows for visualization. Two of them were provided for an injector and a spray impingement flat wall plate facing each other. In order to simulate an engine-like thermodynamic environment, pressurized hot air was supplied through an electric muffle furnace and heated with an electric heater mounted at the bottom of the chamber. The chamber pressure was manually operated, whereas the chamber temperature was controlled by a volt slider for the heater. Four K-type thermocouples were installed around the combustion area. One was fitted at a distance of 5 mm under the nozzle tip, which was placed between nozzle tip and wall, and remaining three were fitted near the wall. Diesel fuel was injected into the chamber when temperature difference of the thermocouples became within 5K. A heat flux measurement was done a little later after hot gas feed for avoiding turbulence in the vessel.

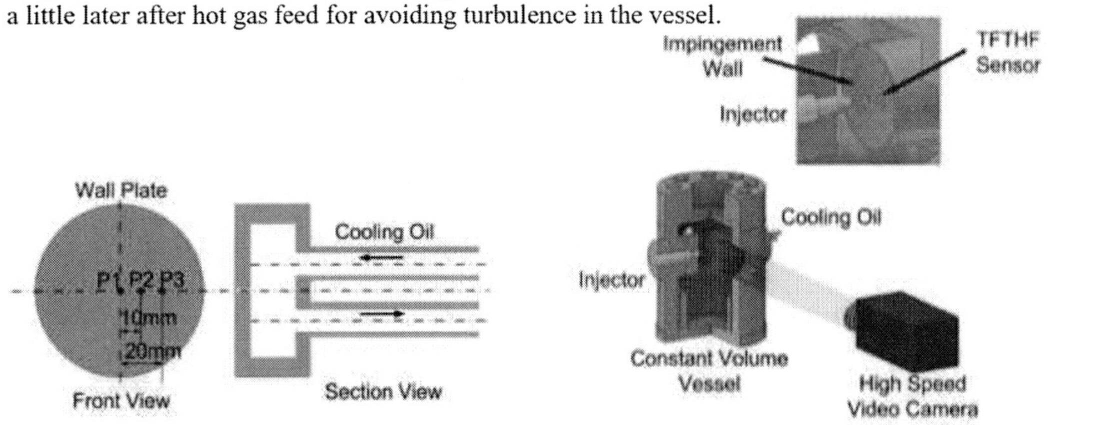

(a). Thin Film Thermocouple Heat Flux Sensor arrangement (b). Optical set-up of high-speed video camera observation

Figure 1. Thin Film Thermocouple Heat Flux Sensor arrangement and optical set-up of high-speed video camera observation

The flat wall was made of stainless steel which has a thickness of 7.3 mm. The back side of the wall plate was cooled by oil in order to form one-directional heat flow across the wall as shown in Figure 1 (a). The cooling oil temperature was kept constant by a thermostatic oil bath. Three micro heat flux sensors were mounted on the impinging surface side of the wall by means of high-temperature adhesive as shown in Figure 1 (b). Fuel, which was supplied from a common rail system, was injected through a single hole nozzle injector controlled by an Electronic Control Unit (ECU). Injection quantity and timing was controlled by a delay pulse generator (Stanford Research Systems DG645). Injection pulse had been calibrated by the injection rate measurement previously.

2.2 Heat Flux and Total Heat Loss Measurement

To obtain the instantaneous surface heat flux on the wall, three Alumel/Chromel TFTHFSs (Medtherm 10702B) were applied as shown in Figure 1 (a). The sensor consists of two thermocouple sensors. One is a surface temperature sensor and the other is an inner temperature sensor. Diameter of the sensors is ø 1.55 mm, and distance between it's thermal contacts (surface and inside) is 3.30 mm. The sensors were radially located on the wall surface with a 10 mm distance each other. Positions of the sensors are designated as Position1 (center of wall impingement), Position2 (10 mm from center) and Position3 (20 mm from center) Instantaneous local heat flux was computed by applying one- dimensional non-steady heat conduction equation using two measured temperatures by the identical sensor as boundary conditions. Steady component of the heat flux, however, was neglected. The equation was explained in the previous report [8].

Table.1 Experimental conditions

Ambient Condition	
Ambient gas	Air (N_2: 79%, O_2: 21%), (N_2: 84%, O_2: 16%)
Ambient pressure (MPa)	4.1
Ambient temperature (K)	873
Ambient density (kg/m³)	16
Injection Condition	
Fuel	Diesel fuel
Injector type	Piezo actuator type
Number of nozzle holes	1
Injection quantity (mm³)	5
Diameter of nozzle hole (mm)	0.133
Injection pressure (MPa)	80, 120, 180
Impingement Condition	
Impingement wall	Flat plate, Stainless steel
Impingement distance (mm)	40
Wall temperature (K)	460 ±10
Cooling method	Oil cooling

Total heat loss was calculated by integrating the heat flux with concentric area and time. In area integration, heat flux affecting areas were defined as concentric circles of 0-5 mm for Area 1, 5-15 mm for Area 2, and 15-25 mm for Area 3, respectively as shown in Figure 2.

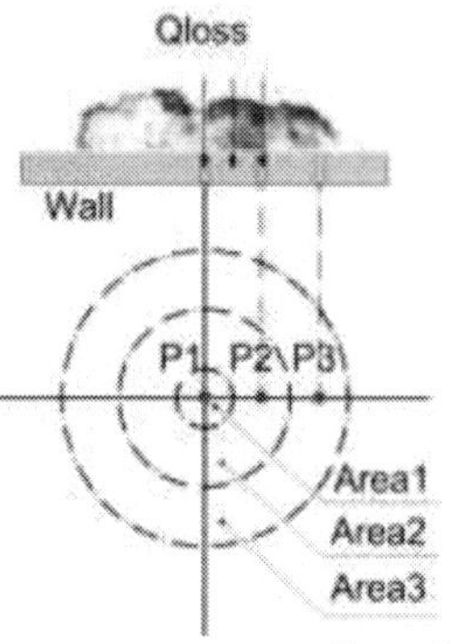

Figure 2. Areas at wall surface

2.3 High-Speed Video Camera Observation

Figure 1 (b) shows optical set-up of high-speed video camera observation and it's arrangement. The camera (Nac Image Technology Inc, HX-3) was utilized to record the flame behavior with an imaging speed of 20,000 fps and resolution of 320 x 448 pixels. Observation of flame was conducted with natural flame luminosity. Moreover, in order to investigate the mechanism of diesel flame and the wall heat loss, the two-color method was used to observe flame temperature distribution from luminous flame. We used a two-color pyrometry software named "Thermias" of Nac Company for analyzing the natural flame color images. A standard light illuminant was used for calibration of it.

2.4. Test Conditions

Experimental measurement conditions are listed in Table 1. It was decided based on actual operation of small size diesel engines. In order to self-ignite the fuel, pressure and temperature were set at those of compression TDC (top dead center) in low load operation for the combustion test. Injection quantity was 5 mm^3 using 0.133 mm nozzle hole diameter. The impinging distance between the nozzle tip to the impingement wall was set at 40 mm. Oxygen concentrations were conducted at 21% and 16% as experimental parameters to investigate the effect of the heat transfer characteristics on the impinging flame on the wall. The all oxygen concentration tests were conducted at three injection pressures, i.e. 80, 120, and 180 MPa. The conditions was set combusting spray at high temperature and high pressure ambient. The gas density is 16 kg/m^3 with 4.1 MPa ambient pressure.

3. Results and Discussion

3.1 Wall Heat Transfer

In this section, effects of the oxygen concentration and injection pressure on wall heat transfer were investigated. Figure 3 presents the combined effect of oxygen concentration/injection pressure on the transferred heat and ratio of heat loss to the total combustion heat. 21 and 16 % as the various oxygen concentration were investigated at each injection pressure of pressure 80, 120, and 180 MPa. The combined effect can be explained as follows:

Firstly, the heat transferred increased gradually with increasing injection pressure at O_2 = 21 %. However, it was not applied to O_2 = 16 %, the heat transferred increased disproportionately even though the injection pressure was increased. It means high velocity by increasing injection pressure was not dominant factor on transferred heat when oxygen concentration under 21 %. Secondly, lower oxygen concentrations consistently reduce heat transfer on the wall in all conditions. This decreased due to the flame temperature was lower. It can be proved by the flame temperature distribution that will be explained in the next section. By higher oxygen concentration as 21 %, it allows more stable combustion with produce high flame temperature then can lead more rate of heat transfer. And then thirdly, the ratio of heat loss through the wall accounts for about 10-15 % of total combustion energy.

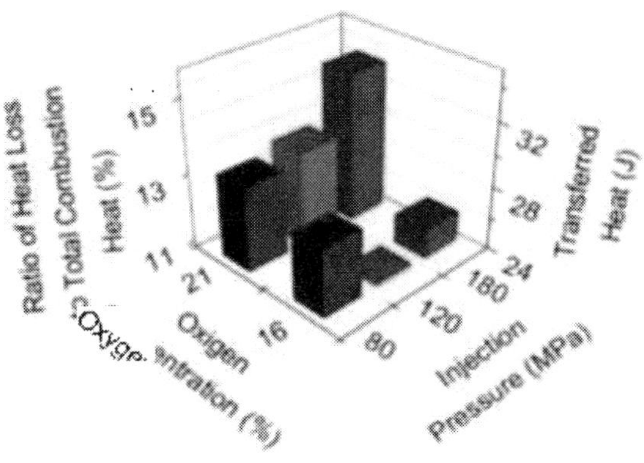

Figure 3. Transferred heat at different oxygen concentration and injection pressure

Considering Figure 3, at lower oxygen concentration, the transferred heat decreased even though injection pressure was increase. It is inversely proportional when oxygen concentration was set O_2 = 21 %. Therefore, the studied will be focus on effect of oxygen concentration, i.e. O_2 = 21 % and O_2 = 16 % at 120 MPa injection pressure. Then, proceed to discuss the following combination of oxygen concentration and injection pressure, i.e. O_2/P_{inj} = 16/80, 16/180 and 21/180 (%/MPa) each. These conditions can represent both comparison different effect of injection pressures and oxygen concentration.

3.2 Heat Flux on the Wall Surface at different Oxygen Concentration

Non-steady local heat fluxes with flame impinging on the wall and integrated luminosity from flame images at different oxygen concentrations are presented in Figure 4 (a-b). The figure shows a significant difference in the ignition timing both conditions. The low temperature due to decreasing oxygen concentration causes unsuitable equivalence ratio for ignition. As a result, local heat flux at Position 1 and 2 started to increase earlier before the ignition occurred. The most surprising is the peak value of local heat flux was dramatically decreased when the ignition occurred later after spray impingement. It is because the both positions increased mainly by convective evaporation spray during the injection duration. Consequently, the rate of heat transfer was lower around area Position1 and 2. However, in case of O_2 = 21 %, the ignition occurred just after impingement in which cover Position1 and 2, as a result Position2 increased simultaneously with ignition and gave substantial amount on local heat flux in that region. The temperature distribution also playing role on this phenomenon.

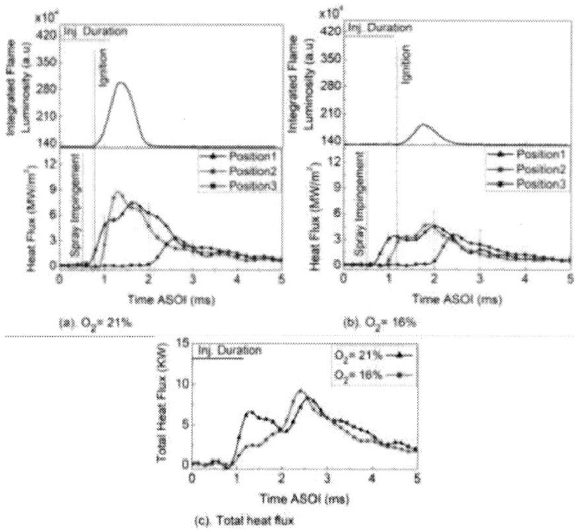

Figure 4. Temporal variation of local heat flux, integrated flame luminosity, and total heat flux at different oxygen concentration

Even though, local heat flux increase and higher peak value was observed at Position1 and 2 under O_2 = 21 %. However, there was no significant difference in Position3 at both conditions. The wall friction and momentum loss due to turbulent mixing took place at Position3 as the far area from the center of impingement.

Figure 4 (c) shows temporal variation of the total heat flux at different oxygen concentrations. It can be seen in the graph, O_2 of 21 % had two peaks waves i.e. there were two stages for the local heat transfer. At first, the local heat flux was increased by turbulence of impingement spray and thereafter when the flame contacts the wall, it develops to a circular area where the flame velocity decreases due to the fact that effects of wall friction takes place, then heat transfer occurs. While O_2 of 16 % accumulated at the second stage for local heat transfer due to significant ignition delay. Consequently, the temperature distribution was appeared far away from the center of impingement in which influence on larger area.

3.3 Flame Natural Luminosity and Flame Temperature

In this section, impinging flame behavior under oxygen concentration are studied. Effects of the oxygen concentrations (O_2), which 21 and 16 % were investigated at impingement distance of 40 mm. Figure 5 shows images of impinging flame natural luminosity at different oxygen concentrations. In this Figure, the luminous flame appeared earlier with O_2 = 21 %. It had started at 0.9 ms ASOI (after start of injection) before the end of injection (EOI) and at 1.25 ms ASOI just after the EOI for O_2 = 16 %. It is found that the reduction of the oxygen concentration in the enclosure mainly affects the reactions in the gas phase and results in changes of the combustion behavior such as ignition timing and flame development. It can be seen at 1.5 to 1.8 ms ASOI, flame natural luminosity start to be decreased at O_2 = 21 %. However, it starts to develop at O_2 = 16 % due to ignition delay was decrease. Consideration the intensity of luminous flame, they were surprisingly different each other. As shown in the figure, lower oxygen concentration generated lower luminous and flame develops quite slowly. Oxygen concentration is therefore required to complete the combustion process since the ignition was not possible occurred without any oxygen.

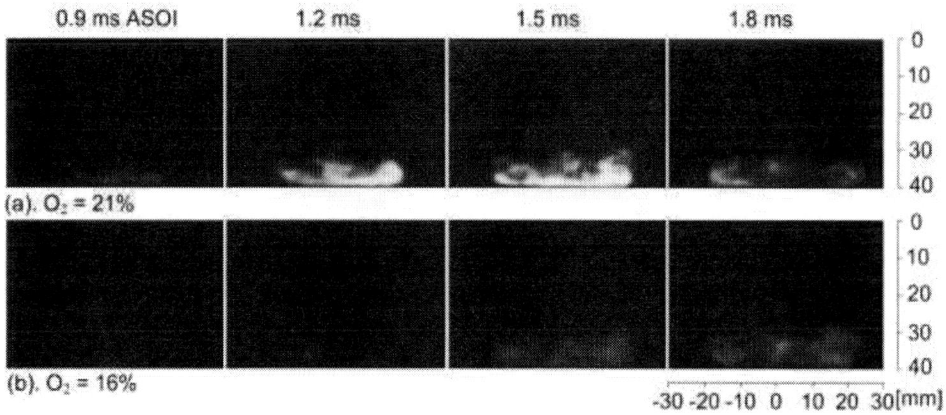

Figure 5. Flame natural luminosity at different oxygen concentration

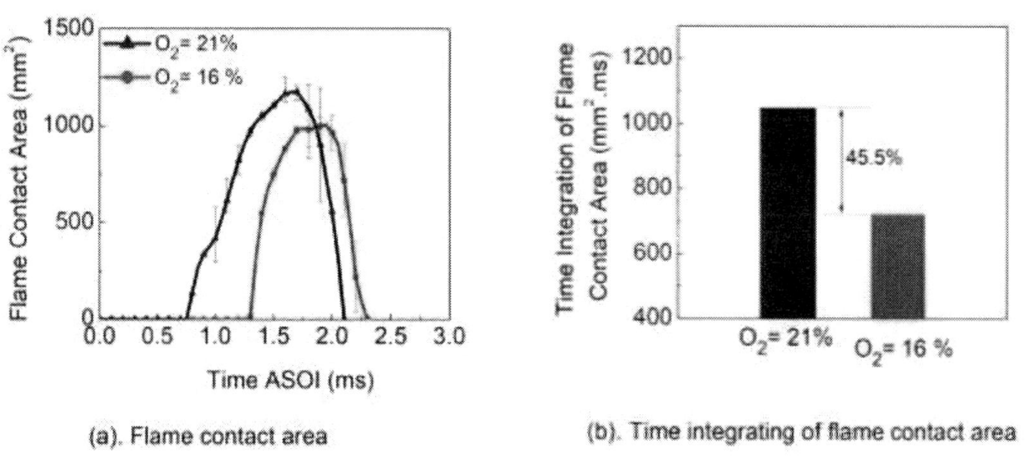

Figure 6. Comparison of flame contact area and time integrating of flame contact area at different oxygen concentration

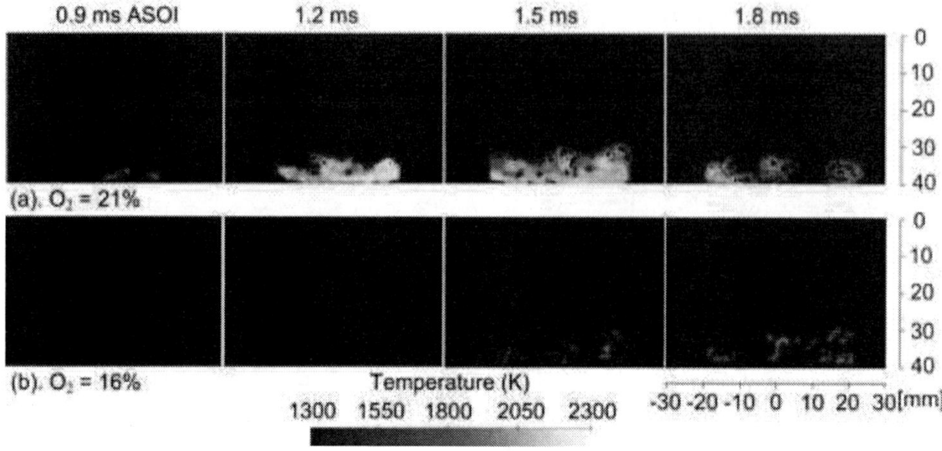

Figure 7. Flame temperature distribution at different oxygen concentration

Effect of flame contact to the wall at different oxygen concentrations are presented in Figure 6. Figure 6 (a) shows that O_2 =16 % had smaller contact area than that O_2 =21 %. It means with oxygen concentration of 21 %, leads to flame contact significantly to spread wide into circumferential area of flat wall. This flame contact area contributes to the heat transfer by conduction. The larger flame contact area causes larger area to be affected to rate of heat transfer. Time integration of the flame contact area can be seen in Figure 6 (b), which shows that flame traveling distance was shorter as the oxygen concentration is decreased.

Figure 7 shows flame temperature distributions at different oxygen concentrations obtained from two-color method analysis, which was performed with flame natural luminosity images. By comparing the temperature distributions with different oxygen concentrations, it can be seen that the temperature distribution was decreased by decreasing oxygen concentration. It can be inferred from Figure 8 that burned gas temperature decreases with decreasing volume fraction of oxygen because the heat capacity of ambient gas increase with decreasing the volume oxygen concentration [8]. When the combustion temperature was low, the temperature difference between the flame and the wall surface were decreased. So that it will have an impact on the reduced energy lost to the wall. Decreasing oxygen concentration was effective parameter for reducing the heat transfer on the wall. Furthermore, soot emission during combustion process also decreased at low temperature as can be seen that the luminous flame significantly decreased. At O_2 = 16 %, the maximum temperature was 1785 K, around 400 K decreased from the peak value of O_2 = 21 %.

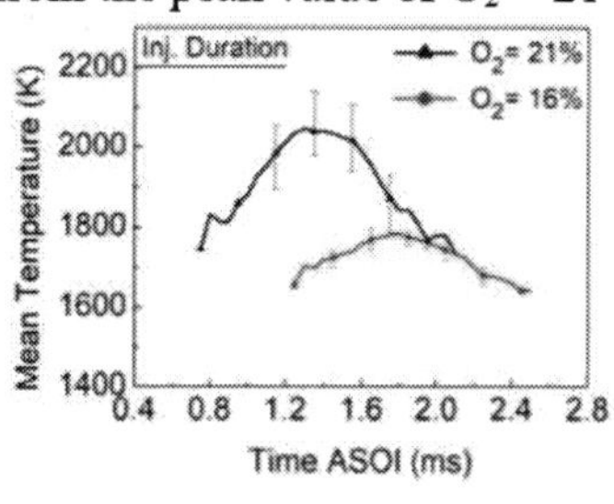

Figure 8. Mean temperature at different oxygen concentration

3.4 Heat Flux on the Wall Surface at Different Oxygen Concentration and Injection Pressure

Non-steady local heat flux and integrated luminosity with the combination of oxygen concentration and injection pressure (O_2/P_{inj}) conditions are presented in Figure 9 (a-c). Figure 9 (a-b) demonstrates the comparison of injection pressure in the combined effect. In the previous report [9], effect of injection pressure was explained in detail. However, since the oxygen concentration of 16 % produced lower flame temperature, it is interesting to clarify the phenomenon in local heat flux history. By comparing the combination of O_2/P_{inj} = 16/80 and 16/180 as shown in the Figure 9 (a-b), local heat flux waveform at all positions started to increase earlier at combination of O_2/P_{inj} = 16/180. It is certainly due to increasing injection pressure related to increasing velocity. On the other hand, local heat flux peak value was almost similar in these combinations. Regarding this phenomenon, increasing injection pressure did not affect more on local heat flux at oxygen concentration of 16 % due to both temperatures are lower. Even though spray mixture increases by higher injection pressure, the time of ignition was not change with pressure, but the combustion process was significant decreased in this condition. At lower injection pressure, combustion duration is slightly longer which causes the flame contact was stayed longer on the wall. As a result, transferred heat to the wall increased as shown in Figure 3.

Figure 9 (b-c) shows a comparison of oxygen concentration in the combined effect. Combination of O_2/P_{inj} = 21/180 had high local heat flux at all positions. With the increased oxygen concentration, increased both the fuel-air mixture and the reaction rate. As a result, the flame temperature increased and this flame gas flow distributed along the wall due to high injection pressure, then lead the increase rate of the heat transfer. However, at lower oxygen concentration, the heat loss was decreased. It is due to the lower flame temperature and the reduced area contact period between the flame and the wall surface as shown in Figure 10 (b), caused by the higher heat capacity and density of the ambient gas [10].

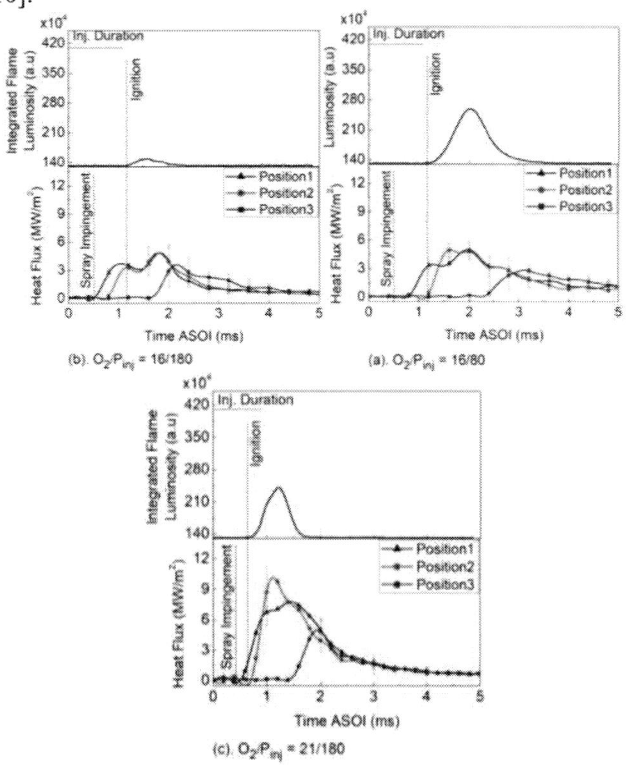

Figure 9. Temporal variation of local heat flux, integrated flame luminosity, and total heat flux at different oxygen concentration and injection pressure

Temporal changes of the flame contact area and time integrating of flame contact area at the different oxygen concentration and injection pressure are shown in Figure 10 (a-b). It shows that combination of O_2/P_{inj} = 16/80 and 16/180 makes flame contact area to increase later comparing O_2/P_{inj}
= 21/180. As mentioned in the previous section that oxygen concentration of 16 % lead ignition occur later after spray impingement on the wall as shown in Figure 5. Comparing three combinations, O_2/P_{inj}
= 16/180 have a smaller flame contact area and shorter resident time flame contact on the wall as shown in Figure 10 (b).

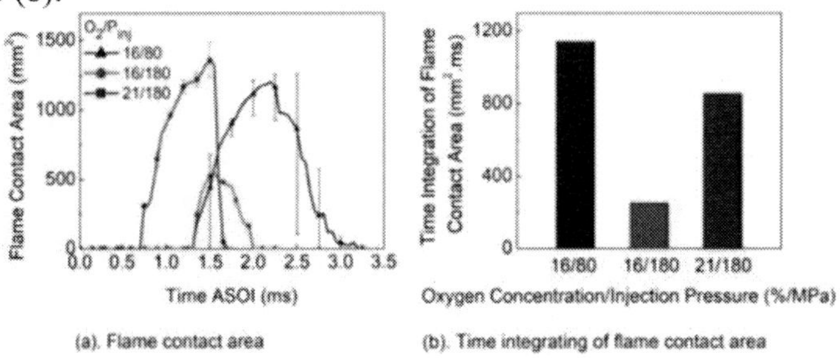

Figure 10. Comparison of flame contact area and time integrating of flame contact area at different oxygen concentration and injection pressure

3.5 Nusselt-Reynold Number Analysis

According to the influence of different injection pressure, the spray flame flow has considerable effect on heat loss through the wall. Then, we studied relation between Reynold number (*Re*) and Nusselt number (*Nu*). The basis of these correlations has relied on dimensional analysis for turbulent flow that corelates the Nusselt (*Nu*), Reynolds (*Re*), and Prandtl (*Pr*) numbers as shown equation (1). The relationship among *Nu, Re,* and *Pr* numbers, which follow those found applied in a turbulent heat transfer equation of the pipe's internal flow or over flat plates [11],[12].

$$Nu = (Const)\ x\ Re^{m} x\ Pr^{n} \tag{1}$$

$$Nu = \frac{h.L}{\lambda} \tag{2}$$

$$Re = \frac{\rho.U.L}{\mu} \tag{3}$$

$$Pr = \frac{\mu.cp}{\lambda} \tag{4}$$

Where *Nu, Re* and *Pr* are difined as the following equation (2), (3), and (4). m and n are defined based on the laminar or turbulent heat transfer. H is heat transfer coefficient, L is characteristic length, λ is thermal conductivity, ρ is density, *U* is characteristic velocity at each position, μ is viscosity and *cp* is specific heat. We used distance between the nozzle tip and the wall as the characteristic length.

As shown in equation (3), Re number requires characteristic flow velocity. As characteristics velocity, mean piston speed is usually used in previously proposed heat transfer equations. In this study, we utilized waveforms of the heat flux as represented characteristics velocity. In order to investigate characteristic flow velocity, flamelet velocity determined according to local combustion flame motion near wall which has direct impact on local heat flux waveform characteristic. It was assumed that the waveform which had clear peaks as a consequence of combustion flame motion. The calculation method of flamelet velocity was explained in detail in the previous report [10].

Figure 11 performing a correlation between *Re* and *Nu* number at difference of oxygen concentrations and injection pressures. For *Re* number calculation, impingement distance is selected as a representative length, i.e. 40 mm. For *Nu* number calculation, thermodynamic properties of air were taken into account. It can be seen from the figure, the *Nu* number was proportional to *Re* number at all conditions. High turbulence flow from combustion which is characterized by high Reynolds number is one of the major factors causing a large amount of heat transfer on the wall.

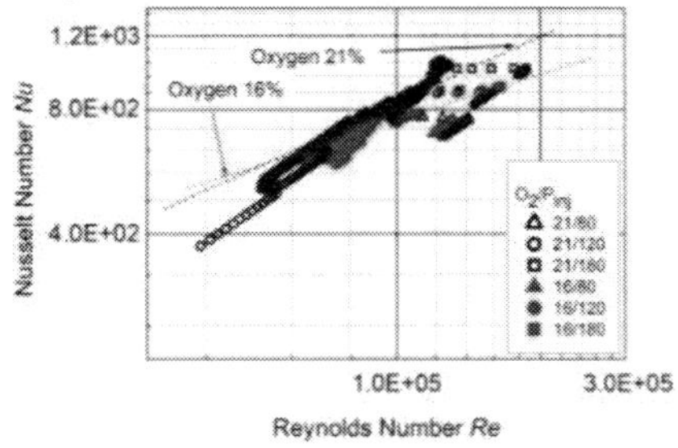

Figure 11. Correlation between Reynolds number and Nusselt number at oxygen concentration/injection pressure.

4. Conclusions

Effects of oxygen concentration and injection pressure on the combustion wall heat transfer under diesel engine-like conditions were investigated. The main conclusions are summarized as follows: By comparing the temperature distributions with different oxygen concentrations, it was found that the temperature distribution and flame contact area were decreased with a decreasing oxygen concentration. When the combustion temperature was low, the temperature difference between the flame and wall surface was smaller. As a result, the local heat flux decreased, and the transferred heat value was also reduced. Under the combined effects of the oxygen concentration and injection pressure, lower oxygen concentrations were consistently reduced the heat transfer on the wall in all conditions. The transferred heat decreased even when the injection pressure was increased. This occurred because the flame temperature was lower compared with the base oxygen concentration of O2 = 21 %. In this study, using the flamelet velocity value, it was found that the *Nu* number was proportional to the *Re* number. Its mean, value of *Nu* number is high when the *Re* number value increased.

Nomenclatures

EGR	: Exhaust Gas Circulation
N_{ox}	: Nitrogen Oxides
TFTHFS	: Thin Film Thermocouple Heat Flux Sensor
ASOI	: After Start of Injection
O_2	: Oxygen Concentration
a.u.	: arbitrary units
Nu	: Nusselt Number
Re	: Reynolds Number
Pr	: Prandtl Number

5. References

[1] T. Kogo, Y. Hamamura, K. Nakatani, T. Toda, A. Kawaguchi, and A. Shoji, —High Efficiency Diesel Engine with Low Heat Loss Combustion Concept - Toyota's Inline 4-Cylinder 2.8-Liter ESTEC 1GD-FTV Engine - Toyota's i,‖ in SAE Technical Papers, Apr. 2016, vol. 2016-April, no. April, doi: 10.4271/2016-01-0658.

[2] K. Fukui, Y. Wakisaka, K. Nishikawa, Y. Hattori, H. Kosaka, and A. Kawaguchi,

—Development of Instantaneous Temperature Measurement Technique for Combustion Chamber Surface and Verification of Temperature Swing Concept,‖ Apr. 2016, doi: 10.4271/2016-01-0675.

[3] M. Zheng, G. T. Reader, and J. G. Hawley, —Diesel engine exhaust gas recirculation - A review on advanced and novel concepts,‖ Energy Convers. Manag., vol. 45, no. 6, pp. 883–900, 2004, doi: 10.1016/S0196-8904(03)00194-8.

[4] X. Cheng, L. Chen, F. Yan, G. Hong, Y. Yin, and H. Liu, —Investigations of Split Injection Strategies for the Improvement of Combustion and Soot Emissions Chraracteristics Baesed on the Two-Color Method in a Heavy-Duty Diesel Engine,‖ SAE Tech. Pap. 2013-01-2523., 2013.

[5] H. Osada, N. Uchida, K. Shimada, and Y. Aoyagi, —Reexamination of Multiple Fuel Injections for Improving the Thermal Efficiency of a Heavy-Duty Diesel Engine,‖ SAE Tech. Pap. 2013- 01-0909, 2013, doi: 10.4271/2013-01-0909.

[6] T. Kuboyama and H. Kosaka, —A Measurement of Heat loss in Combustion Chamber of DI Diesel Engines,‖ Marit. Res. Inst. Technol. Saf., vol. 8, no. 215–221, pp. 89–96, 2008.

[7] R. Mahmud, T. Kurisu, K. Nishida, Y. Ogata, J. Kanzaki, and T. Tadokoro, -Experimental study on flat-wall impinging spray flame and its heat flux on wall under diesel engine–like condition: First report—effect of impingement distance,‖ Proc. Inst. Mech. Eng. Part D J.

Automob. Eng., vol. 233, no. 8, pp. 2187–2202, Jul. 2019, doi: 10.1177/0954407018778153.

[8] T. Kuboyama and Y. Moriyoshi, —Heat Transfer Analysis in a Diesel Engine Based on a Heat Flux Measurement using a Rapid Compression and Expansion Machine,‖ SAE Tech. Pap. 2017-32-0115, p. 6, 2017.

[9] R. Mahmud, T. Kurisu, K. Nishida, Y. Ogata, J. Kanzaki, and O. Akgol, —Effects of injection pressure and impingement distance on flat-wall impinging spray flame and its heat flux under diesel engine-like condition,‖ Adv. Mech. Eng., vol. 11, no. 7, p. 168781401986291, Jul. 2019, doi: 10.1177/1687814019862910.

[10] T. KUBOYAMA, H. KOSAKA, T. AIZAWA, and Y. MATSUI, —A Study on Heat Loss in DI Diesel Engines by Using a Rapid Compression and Expansion Machine (1st Report, The Effects of Oxygen Volume Fraction and Density of Ambient Gas on Heat Loss),‖ Japan Soc. Mech. Eng., vol. 72, no. 721, pp. 2315–2322, 2006.

[11] J. P. Holman et al., -McGraw-Hill Series in Mechanical Engineering INTERNAL COMBUSTION ENGINE Xnderung nur iiber,‖ 1988.

[12] J. Chang et al., —New Heat transfer Correlation for an HCCI Engine Derived from Measurements of Instantaneous Surface Heat Flux,‖ SAE,‖ 2004,. [Online]. Available: https://citeseerx.ist.psu.edu/viewdoc/summary?doi=10.1.1.473.5101.

Acknowledgements
We would like to thank Mazda Motor Corporation for their support with measurement in this study.

IOP Conf. Series: Materials Science and Engineering 1010 (2021) 012004 doi:10.1088/1757-899X/1010/1/012004

Possibility of geothermal offshore in Sangihe archipelago, northern part of Sulawesi, Indonesia

F R Widiatmoko[1,4], D D Dewangga[1], A Gustriandy[2], S Salsabila[2], N Anggraeni[2], M Infithor[2], D Hanifah[2], R A Pratama[2], T H Kusaeri[3] and A Zamroni[5]

[1] Department of Natural Resources and Environmental Studies, National Dong Hwa University (NDHU), Hualien, Taiwan, R.O.C.
[2]Department of Geological Engineering, University of Jenderal Soedirman (UNSOED), Purwokerto, Indonesia.
[3] Department of Geodetic Engineering, Bandung Institute of Technology (ITB), Indonesia.
[4] Department of Mining Engineering, Institut Teknologi Adhi Tama Surabaya (ITATS), Indonesia.
[5] Department of Geological Engineering, Institut Teknologi Nasional Yogyakarta (ITNY), Indonesia.

widiatmokogeo@gmail.com

Abstract. Mostly geothermal power plant placed around volcano on land. Sangihe Archipelago, the northern of Sulawesi Island known as volcano arc that formed by the subduction of Philippines plate beneath Micro-Sunda plate. The geothermal source is located under sea level, this condition needs some approach of remotely applied sciences. The possibility study is conducted by bathymetry, gravity, magnetic, and regional geology. The objective is searching the high value gravity at the high topography as indication of the magmatic product on the seafloor and the low value of magnetic as indication of the active heat resources. A profile section from Philippines oceanic plate to the Celebes Sea by using each data shown a sequence of subduction product, such as trench, outer-arc ridge, and volcano arc. Geological data is used to know the evidence volcano arc and outer-arc ridge. The volcano arc shows that the age is around Quarter which means it is possible for geothermal resources, the outer-arc ridge shows that the rock contents were melange complex and ultramafic rocks. Sangihe area has possibility for further research as geothermal offshores.
Keywords: Geothermal submarine, Gravity, Banua Wuhu, Geomagnetic

1. Introduction

Renewable energy in Indonesia has been developing, since 2010 renewable energy contribution in Indonesia reached 4.42%, meanwhile in 2015 it has increased 6.20% from all of the existing energy contribution coal energy 28.74%, oil energy 43.02%, and gas 22.04% [1]. Indonesian energy production in 2016 was estimated 224.9 kWh, meanwhile 19% of total population without electricity [2], as additional energy demand also occurred from industrial development. The Indonesian government has solution by creating *rencana usaha penyediaan tenaga listrik* (RUPTL) or providing electrical power planning. The last RUPTL period along 2017 to 2026 promote a plan to upgrade electrical power generation about 77.9 GWh, while the upgrading capacity of renewable energy about 21.5 GWh [1]. Geothermal energy contribution for RUPTL is about 6.3 GWh [1], it means the geothermal resources in Indonesia will be used about 36% from a whole geothermal resources in Indonesia.

All of geothermal resources were estimated about 29.5 GWh [1]. But all of them were estimated only on land, meanwhile there are some of geothermal offshore resources. Un-estimated resources are caused by the lack of technology to observe directly. Heretofore, Indonesia has Geomarine ship for marine survey, but the utility purpose did not get in touch to geothermal offshore resources. As another reason geothermal on land resources still as a priority to build geothermal power plant. The lack of technology to observe directly is not the strong obstacle to research about geothermal offshore resources. There are a lot of global geophysical data and geological regional that can be used as an approach for the preliminary observation regarding geothermal offshore resources. The result of preliminary observation will show that the geothermal offshore resource has possibility to be studied further.

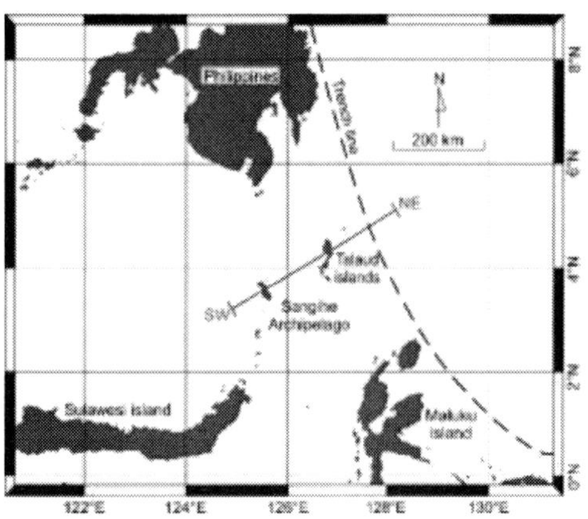

Figure 1. Location of the study area with a trench line as a boundary of each tectonic plates, SW - NE line place as the focus of section study

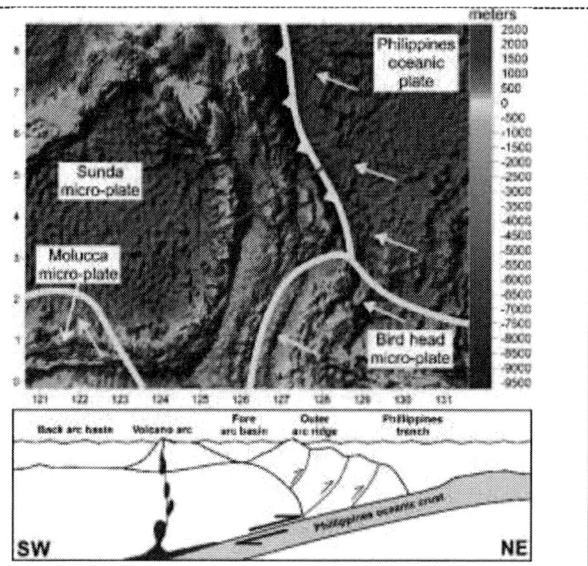

Figure 2. Tectonic settings of the study area by using GEBCO data that is modified from [3] (top), arrows are the direction of plate movement, thick lines are the trench line and boundaries of the tectonic plate, SW-NE line is the profile section line. The SW – NE profile is adopted from Rangin section model [7], [8], and the surface elevation from bathymetry data.

As an example, Sangihe Archipelago, Indonesia (Figure 1) has offshore volcano activity which related to magma activity as well as geothermal resources. Sangare Archipelago is a product of subduction between Micro Sunda continental plate with Philippines oceanic plate [3]. This area becomes interesting to be studied further because of having offshore hot springs complex which is known as Gunung Kawo Barat complex. Hot spring is one of indicator for preliminary geothermal survey due to hot spring represents that heat source is active.

1.1. Geological Settings

Sangihe Archipelago has been formed by Philippines oceanic plate under the subducting Sunda continental micro-plate since the late Miocene [4]. The occurrence is caused by separation plate between Sunda microplate with Philippines plate from the same basin in the late of Oligocene, the movement plate firstly strike-slip followed to be convergent boundaries plate [5]. Sangihe Archipelago has been formed on the Sunda continental micro-plate. The plate thickness was estimated about 6-7 km [6]. The Sangihe Archipelago has been formed by subduction that product magmatic and volcanoes (Figure 2). The basement of Sangihe archipelago is Basaltic [7]. The basement was buried by Quarternary magmatic product (Figure 2) [4]. The SW-NE section (Figure 2) has similarity with the Rangin section model in the outline geology of Nusa Tenggara Timur [8]. The profile section show that Sangihe Archipelago is the subduction product which is included to the volcano arc and Talaud Islands are included to the outer arc ridge [7], [8].

2. Methodology

To determine the possibility of geothermal offshore energy required to do an integrated approach. In this paper, we use bathymetry data, global gravity data, global magnetic data, and regional geological regional maps.

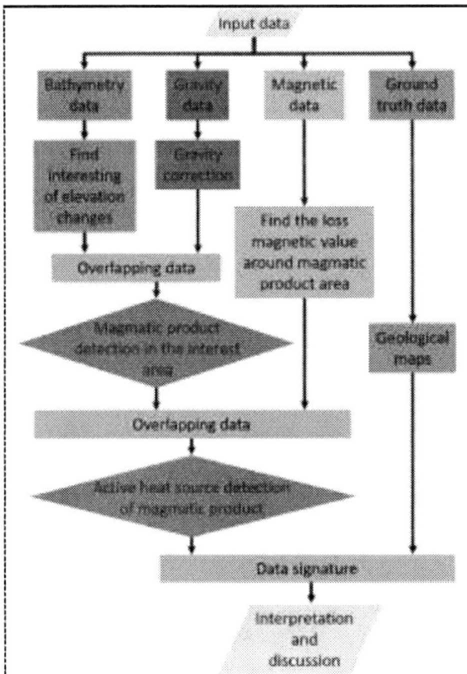

Figure 3. The flow chart of the study, the parallelogram as the first-end, the rectangle as data representative, the diamond shape represent the both of the result.

Bathymetry data is used to determine the elevation in offshore and on land. The elevation changes indicate of different rock content, structural geology, and other geological processes [9]. One of the interest elevation change is used to find the area that has magmatic product (Figure 3). The bathymetry has resolution data 30 arc second that is provided by General Bathymetric Chart of the Oceans (GEBCO) [10].

Gravity data is used to determine the difference of weight force in some areas. The elevation changes on bathymetry data do not have rock content information. One approach for knowing the rock information is using gravity data. The weight force information represents that the area has typical weight of rock (Figure 3). The Gravity data has resolution 60 arc second that is measured by Geosat and ERS 1 satellite altimetry which provided by University of California San Diego [11]. To detect the magma product, the gravity data is required to has complete anomaly bouguer, then we filter the complete anomaly bouguer by using SVD filtering to divide data between regional bouguer and residual bouguer. The residual bouguer map has function to know the shallow gravity value, it is used to know the area has high gravity value [12].

Geomagnetic data is used to determine the susceptibility value on some areas. The high susceptibility value represents that the area containing magnetic elements and some metal elements

(Figure 3). Geomagnetic data has resolution 2 arc minutes that is provided by World Digital Magnetic Anomaly Map (WDMAM) [13].

The integrated approach use bathymetry data, gravity data, and geomagnetic data. The area that has magmatic product is shown by the high elevation changes and having high gravity value (Figure 3). The active magmatic activity is shown by the dropping magnetic value (Figure 3). As our assumption when a magnet is given heat, the magnet will lost the susceptibility force. Magmatic product has high magnetic value, but the active magmatic activity will have the lower magnetic value than the others magmatic product area. As signature, those data is necessary to have ground truth data. Data synchronization is using regional geological maps that were mapped [14] (Figure 3).

3. Results

The processed gravity data are conducted to get the residual anomaly map. The residual anomaly is obtained by separation data from regional anomaly, so the residual anomaly has function to know locally gravity influences [12]. The local gravity was used for interpretation of rock content inside the high elevation.

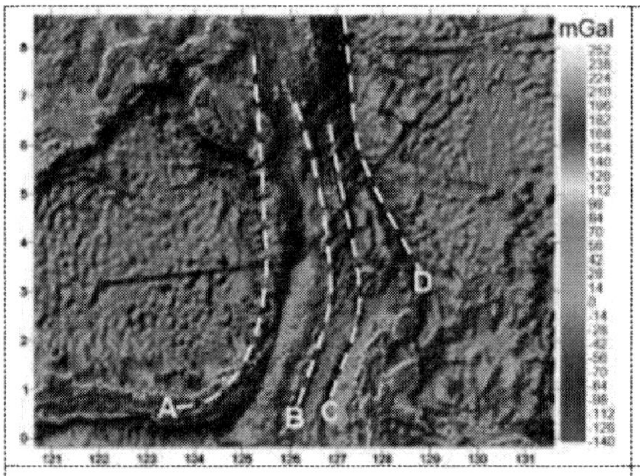

Figure 4. Overlapping residual gravity data in milliGal unit on the bathymetry elevation model from GEBCO, solid lines are the section profile, and dotted lines are the interest line areas.

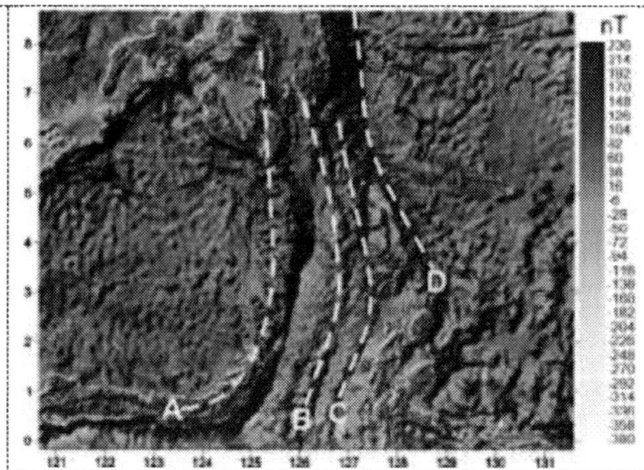

Figure 5. Overlapping geomagnetic data in microteslas or nT unit on the bathymetry elevation model from GEBCO, solid lines are the profile section, and dotted lines are the interest line areas.

In Figure 4 show that sea floor has the same gravity value, it is estimated that both of the sea floor is the same type rock by using gravity approach. Some areas in the middle of map (Figure 4) shows the high gravity values, those areas are estimated have magmatic product, different type rock, structural geology, and deformed formation rocks. Four lines are detected in the middle of residual anomaly map (Figure 4), each of these lines has orientation line to different direction.

In geomagnetic map (Figure 5) show some areas contain high magnetic values, those are estimated containing magnetic and metal sources, it has possibility the rock content is basaltic rock, ultramafic metamorphic rock, and metallic mineral deposits. The low magnetic values are estimated containing the low amount of magnetic and metal sources, it has possibility the rock content are granitic rock, sedimentary rock, non-metallic metamorphic rock, and non-metallic mineral deposits.

4. Discussion

We created profile sections (Figure 6) on both of map (Figure 4 and 5) that across four of those lines. We created those line sections by following the availability of ground truth data and for compare with the section from [7], [8]. The available ground truth data are on the Sangihe Island regional map [4] and Talaud Islands regional map [14].

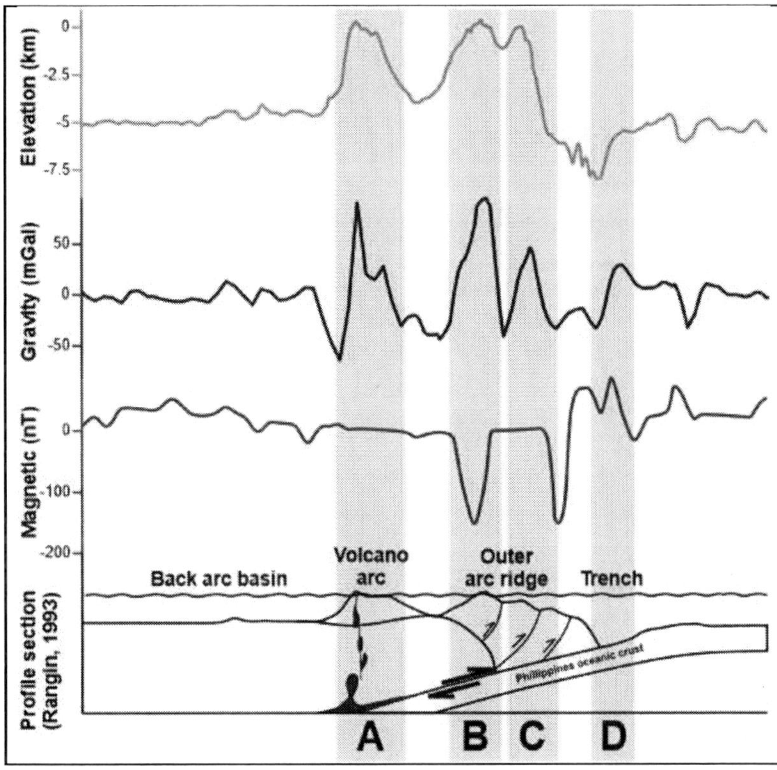

Figure 6. Profile section of Figure 4 and 5. First top curve represents of bathymetry data, second curve represents residual gravity data, third line represent of geomagnetic value, and the rectangles represents each line area in Figure 4 and 5.

The A-line area (Figure 6) has high elevation (-5km to 0.2km), high gravity value (-65mGal to 90mGal), and medium magnetic value (0nT to -20nT). It means the rock content of the A-line area is high density rock that has possibility containing igneous rocks, high density sedimentary rocks, and high density metamorphic rocks. But according to geological settings, this area is included to the volcano arc [7], [8], so it has high possibility containing igneous rock. Meanwhile the magnetic value is about 0nT to -20nT, normally the magnetic value on the magmatic area will be increased, but in this case the gravity curve was flat. We have assumption if magnet was given heat, it will lost the susceptibility force. The A-line area also, if the magmatic inside the A-line area still has active heat source, geomagnetic only detect magnetic value similar with the sea floor magnetic value.

The B-line area (Figure 6) has high elevation (-3.5km to 0.1km), high gravity value (-45mGal to 100mGal), and low magnetic value (-10nT to -150nT). It means the rock content of B-line area is high density rock that has possibility containing igneous rocks, high density sedimentary rocks, and high density metamorphic rocks. But according to the geological settings, is included to the outer arc ridge [7], [8], so it has high possibility containing high density sedimentary rocks and metamorphic rocks. Meanwhile, the magnetic value is about -10nT to -15nT, we estimated the B-line area contains material non-magnetic, such as limestone formation, and high density metamorphic rocks.

The C-line area (Figure 6) has high elevation (-3.5km to 0km), medium gravity value (-40mGal to 50mGal), and flat magmatic value (-5nT to 2nT). According to the geological settings, the C-line area is included to the outer arc ridge, but it has different character with the B line area. Especially the magnetic value is medium (-5nT to 2nT). The rock content in this area has lower density with magnetic similar with sea floor magnetic. We interpreted the C line area was uplifted part of the sea floor that caused of the friction while convergent movement.

The D-line area (Figure 6) has the lowest elevation (-8km to -5.5km), low-medium density (-35mGal to 30mGal), and high magnetic value (-10nT to 100nT). According to the geological settings, the D-line area is included in the trench area. We estimated the D-line area contain basalt sea floor

(Hall et al., 1988), mono-mineral sedimentary rock and some deposit on the base of trench from slump processes.

On the gravity map (Figure 4) show that both of the sea floors have similar gravity value, that means those sea floors are formed from the similar component, according to literature both of those sea floors are basaltic [7]. Previously both of those sea floors were the unity sea floor, then those sea floors got separated by strike slip movement then followed to be convergent boundaries [5]. The convergent movement caused some of basement rocks were uplifted to be land and friction. The friction of those sea floors caused magma generation and produced volcanoes.

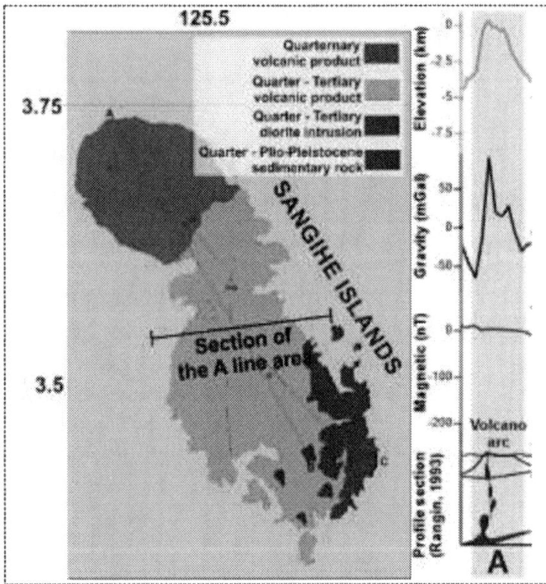

Figure 7. Geological map of Sangihe Islands (left) (Modified from [4]). Profile section of elevation, gravity, magnetic, and geological settings of [7], [8] around the A-line area.

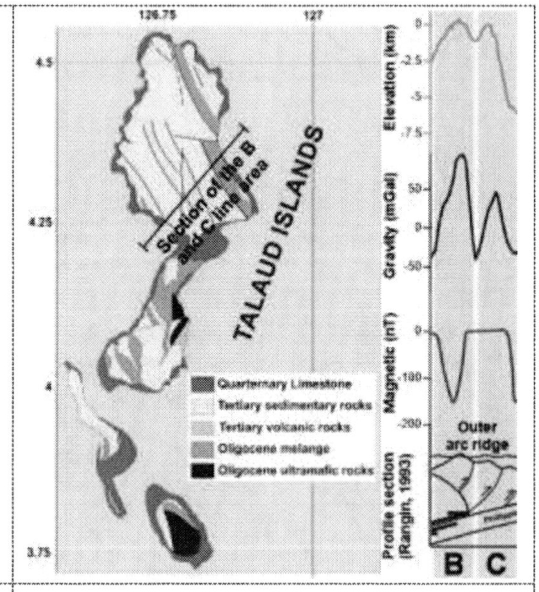

Figure 8. Geological map of Sangihe Islands (left) (Modified from [14]). Profile section of elevation, gravity, magnetic, and geological settings of [7], [8] around the B and C line area..

According to Figure 7. Elevation, gravity, and geomagnetic data show that the active heat source is detected on the A line area. As data signature, we also compare to the ground truth data that was made by [4], [14]. The A-line area is included to Sangihe Island, that area was mapped by [4]. As ground truth data, the A-line area is Quarter – Tertiary volcano product that across the Sahendaruman volcano [4]. We compared with elevation, gravity, and geomagnetic data showing the character of igneous rock, then the lower magnetic than normal value of igneous rock is the active heat source character. 50,000–200,000 year of the volcano formation has high possibility to get geothermal development, that age is about early Pleistocene [15]. Sahendaruman volcano is included to the high possibility of geothermal development. Meanwhile, the younger Awu volcano is on the north of the Sahendaruman volcano, that volcano formed at Quarter [4]. On the south of Sangihe Island there is Karangetang Island and also Ruang Island on the north of Sangihe Island, both of these islands is the young volcano on the Sangihe Archipelago complex.

According to figure 8. Elevation, gravity, and geomagnetic data show that there has two different of rock characters. As data signature, we also compare to the ground truth data that was made by [4], [14]. The B-line and C-line area are included to Sangihe Island, that area was mapped by [14]. As ground truth data, the B-line area is Tertiary sedimentary rocks and on the top has deposited Quarternary limestone [4]. We compared with the elevation, gravity, and geomagnetic data showing the sedimentary rock character that related with the ground truth data. The C-line area is Oligocene melange, Quarternary limestone [14], and interpreted at the Northeastern shoreline there has possibility outcrop of Oligocene ultramafic rocks under the sea level. We compared with these data showing the similarities with the metamorphic rocks.

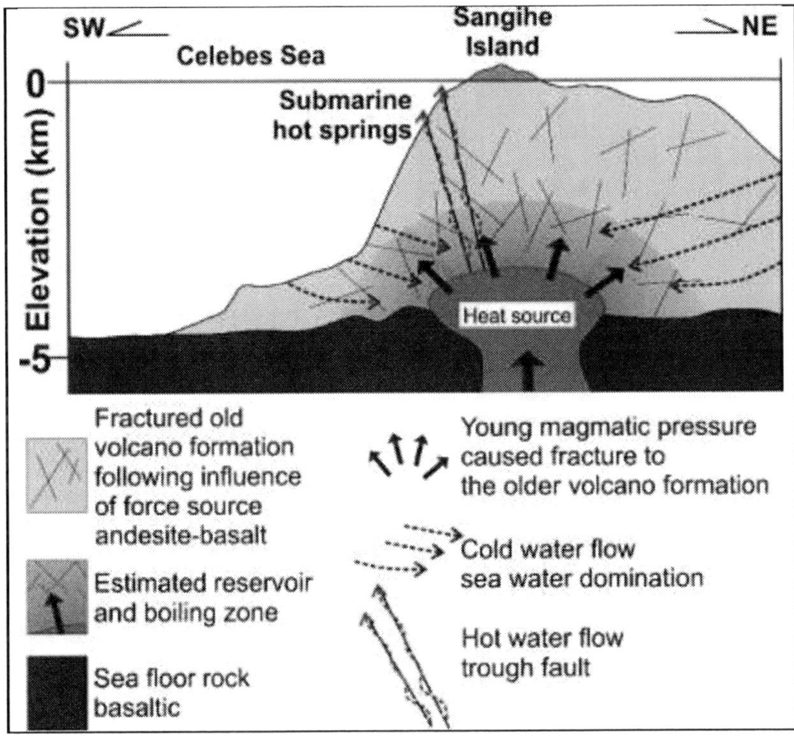

Figure 9. Unscaled conceptual model of geothermal system in the study area, adopted from [16].

We created an unscaled geothermal conceptual model of study area (Figure 9). Basically, this model is adopted from offshore geothermal model that is introduced by Armani and [16]. We modified by adding the estimated reservoir and boiling zone. For the magma positioning followed gravity curve. For the basaltic sea floor rock is used from [5]. We added some fracture that has probability formed that is caused by young magmatic force [17], the fracture direction following influence of the force source by pure shear model [18]. The hot water flow as well flowing through estimated fault [19], and as evidence supports the occurrences of the offshore hot springs [20]. The cold water flow is pushed by hydrostatic pressure of sea level depth, also supported by the circulation water inside the reservoir.

5. Conclusions

As our conclusion, The A-line area especially Sangihe Archipelago as the Tertiary-Quarter volcano belt, Active heat source can be detected by high elevation, high gravity, and low geomagnetic data. The further research about offshore geothermal resources shall be detailed around The A-line area or Sangihe Archipelago volcano belt. For future research about this area is detailed gravity, detailed geomagnetic, and offshore geochemistry survey is necessary.

As possibility study, we suppose this result is enough to be studied further and detail. This method also can reveal a sequence of subduction product, fortunately, we can compare with the available ground truth data as evidence.

6. References

[1] PLN, “Rencana Usaha Penyediaan Tenaga Listrik (RUPTL) PT PLN (Persero),” 2015.
[2] Central Intelligence Agency, “The World Factbook East & Southeast Asia: Malaysia,” 2018.

[3] P. Bird, "An updated digital model of plate boundaries," *Geochemistry, Geophys. Geosystems*, 2003, doi: 10.1029/2001GC000252.
[4] S. H, "Volcanology and Geothermal Energy," Bandung, Indonesia, 1994.
[5] R. Hall, M. G. Audley-Charles, F. T. Banner, S. Hidayat, and S. L. Tobing, "Late Palaeogene-Quaternary geology of Halmahera, eastern Indonesia: initiation of a volcanic island arc.," *J. - Geol. Soc.*, 1988, doi: 10.1144/gsjgs.145.4.0577.
[6] S. Murauchi *et al.*, "Structure of the Sulu Sea and the Celebes Sea," *J. Geophys. Res.*, 1973, doi: 10.1029/jb078i017p03437.
[7] C. Rangin and E. Silver, "Geological Setting of the Celebes and Sulu Seas," in *Proceedings of the Ocean Drilling Program, 124 Initial Reports*, 1990.
[8] H. Darman, "An Outline of The Geology Indonesia," *J. Chem. Inf. Model.*, 2000, doi: 10.1017/CBO9781107415324.004.
[9] T. G. BONNEY, "Theory of the Earth with Proofs and Illustrations," *Nature*, 1899, doi: 10.1038/060220a0.
[10] W. H. F. Smith and D. T. Sandwell, "Global sea floor topography from satellite altimetry and ship depth soundings," *Science (80-.).*, 1997, doi: 10.1126/science.277.5334.1956.
[11] D. T. Sandwell and W. H. F. Smith, "Marine gravity anomaly from Geosat and ERS 1 satellite altimetry," *J. Geophys. Res. B Solid Earth*, 1997, doi: 10.1029/96JB03223.
[12] T. A. Elkins, "The second derivative method of gravity interpretation," *Geophysics*, 1951, doi: 10.1190/1.1437648.
[13] V. Lesur, M. Hamoudi, Y. Choi, J. Dyment, and E. Thébault, "Building the second version of the World Digital Magnetic Anomaly Map (WDMAM) 1. Geomagnetism," *Earth, Planets Sp.*, 2016, doi: 10.1186/s40623-016-0404-6.
[14] S. R and S. N, "Geological Map of Talaud Sheet, Sulawesi," Bandung, Indonesia, 1986.
[15] S. Self, "Volcanology and Geothermal Energy," *J. Volcanol. Geotherm. Res.*, 1994, doi: 10.1016/0377-0273(94)90061-2.
[16] F. B. Armani and D. Paltrinieri, "Perspectives of offshore geothermal energy in Italy," 2013, doi: 10.1051/epjconf/20135402001.
[17] . O., D. A. Ramadhan P, F. R. W, and R. T. A, "Identification of Geothermal Potential Based on Fault Fracture Density (FFD), Geological Mapping and Geochemical Analysis, Case Study : Bantarkawung, Brebes, Central Java," *KnE Energy*, 2015, doi: 10.18502/ken.v2i2.369.
[18] W. R. Buck, F. Martinez, M. S. Steckler, and J. R. Cochran, "Thermal consequences of lithospheric extension: Pure and simple," *Tectonics*, 1988, doi: 10.1029/TC007i002p00213.
[19] F. Pirajno, *Hydrothermal processes and mineral systems*. 2009.
[20] F. R. Widiatmoko, "Pendekatan Analisa Geokimia dengan Multivariate Analysis untuk Mengetahui Tipe Mata Air Panas: Studi Kasus Lapangan Panas Bumi Mapos, Nusa Tenggara Timur," *J. IPTEK*, 2019, doi: 10.31284/j.iptek.2019.v23i2.518.

Acknowledgement
Our thanks presented to Mr. Eko Bayu Purwasatriya who inspired us to bring this topic out. Special thanks to all of colleagues from National Dong Hwa University (NDHU), University of Jenderal Soedirman (UNSOED), Adhi Tama Institute of Technology (ITATS), and Bandung Institute of Technology (ITB). Geological Engineering Department of Jenderal Soedirman University and dr.Bumi Research Group (dBRG) are thanked for supporting this work. Gebco.net are thanked for providing General Bathymetric Chart of the Oceans (GEBCO), to TOPEX University California San Diego for providing gravity data, and also to Geomag.org for providing World digital magnetic anomaly map (WDMAM).

IOP Conf. Series: Materials Science and Engineering 1010 (2021) 012005 doi:10.1088/1757-899X/1010/1/012005

Simplified gasification simulation using zero-dimensional model of empty palm fruit bunch

P N A Nugroho [1] and S Torii [2]

[1]Politeknik Perkapalan Negeri Surabaya, Jl. Teknik Kimia, Kampus ITS Sukolilo, Surabaya 60111, Indonesia.
[2]Department of Advanced Mechanical System Engineering, Kumamoto University, Kurokami, 2-39-1, Kumamoto 860-8555, Japan

priyambodo@ppns.ac.id

Abstract. As one of the most efficient biomass conversions, the gasification process is hard to understand because of its complexity. A simplified gasification process simulation using a zero-dimensional model was designed for an empty fruit bunch considering its potential in Indonesia. A stoichiometric equilibrium model was applied to simulate the whole gasification process. Global gasification reaction, mass, and energy balance with two independent chemical reactions are integrated to perform the model. Some sets of linear and nonlinear equations were solved with the Newton Raphson method in Matlab. The output of the model, like syngas composition and the heating value, was validated with available experimental data published. The zero-dimensional thermodynamic equilibrium model could be used to predict the maximum performance of the gasifier and make the gasification process more feasible.

1. Introduction

Palm oil has been used for centuries ago. Originally from Africa, palm tree *(Elaeis guineensis)* has been naturalized in tropical countries since it grows even better. Palm tree produces crude palm oil (CPO) that could be used in a vast range from cooking oil to biodiesel. CPO become an internationally traded community and is very important for Indonesia as the largest CPO producer.

Along with Malaysia, over 20 million metric tons of CPO produced in the almost 10-million-hectare area. Maitah et al. reported that CPO production also generates a large amount of biomass [1]. For every 1 kg of CPO, about 4 kg of biomass waste. Most of the waste is an oil palm empty fruit bunch; the other is fronds and trunks.

Gasification is the thermochemical process that converts carbonaceous material like coal or biomass into valuable gaseous fuel or chemical substance. The history of Gasification could be traced back to 1659 when Thomas Shirley discovered gas from a coal mine. In 1739 Dean Clayton successfully distilled coal in a closed vessel followed by Robert Gardner patented its first gasification technique in 1788. According to Basu, biomass intended not only energy conversion but also the production of a chemical substance as one of the essential commodities [2].

Similar to Gasification, combustion also changes carbonaceous material into product gases. However, there is distinctness like there is no useful heating value in the combustion product gas, in contrast to Gasification that possesses beneficial heating value. Combustion release energy in its process, while Gasification packs energy into compact chemical bonds. The combustion process occurred in an oxidizing environment and releasing heat, while Gasification happened in inadequate oxygen surrounding and demanding heat.

The gaseous medium of Gasification is involved in the process of converting either fossil or non-fossil fuel into valuable gaseous fuel or other chemicals. Gaseous mediums example is oxygen, air, steam, or a mixture of them. The reason behind transforming potential fuel into another form is to increase the heating value of the fuel by releasing non-combustible matter like nitrogen and water. Another motivation is to remove sulphur and nitrogen from the fuel so that this material will not be released into the atmosphere when burning the fuel. An additional motive is to reduce the proportion of carbon to hydrogen ratio in the fuel.

Generally speaking, the more hydrogen contents in the fuel, the lower the vaporization temperature and increases the likelihood of fuel become in the gaseous phase. Biomass gasification converts solid-state biomass into gaseous state fuel or chemicals. The chemical energy contained in the solid biomass is changed into the thermal and chemical energy of the gas. Hence the chemical energy within the product gas is a function of its chemical configuration. In other words, the quality of the product is driven by chemical composition. Htut et al reported that the outcome of Gasification is called the syngas in which CO and H2 are the main gaseous species [3].

Forecasting performance using thermodynamic computations are commonly used to explain chemical reactions within the gasifier. Performance prediction of a downdraft gasifier has been made by Zainal et al using equilibrium modelling for some biomass materials like wood, paddy husk, paper, and municipal waste [4]. It was found that the heating value of the syngas decrease along with increasing gasifier temperature and also an increase in moisture content.

Wu and Chein have worked on the modelling of biomass gasification with high-temperature inlet as a result of preheating [5]. Combustible species reduction was acknowledged with more than a certain amount of biomass moisture content. Preheated air could enhance gasifier performance; however, the degree of significance was lower than expected.

Align with the great curiosity about the biomass gasification process and make the process more feasible. It is inevitable to model and foresee the gasifier performance before building the actual one. This paper aims to perform syngas composition and heating value analysis composed of the gasifier with a zero-dimensional equilibrium model.

2. Materials and Methods

The gasification process designed in this paper is illustrated in Figure 1. Biomass enters the gasifier at temperature T_b, while air as the gasifying agent has temperature T_a and pressure P_a. Temperature and pressure inside the gasifier are T_g and P_g, respectively. Syngas formed, leaving the gasifier at temperature T_s and pressure P_s.

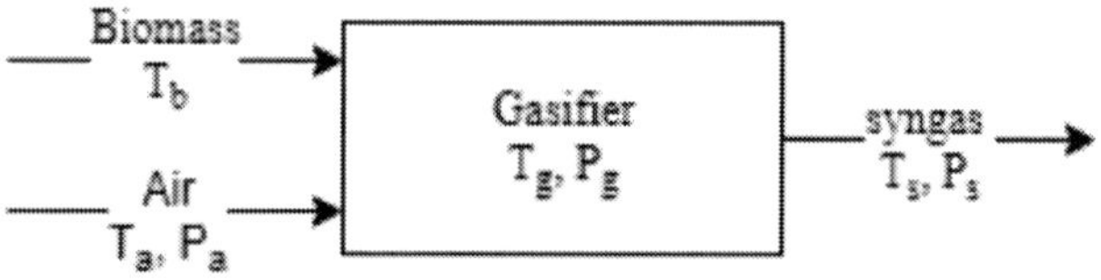

Figure.1 Gasification Process Illustration

Some assumptions, according to Wu and Chein [5] and Htut et al. [3], were taken into consideration to simplify the analysis.

- The thermodynamic equilibrium process and adiabatic (no heat transfer from and to the environment) happens in all gasification reactions.
- Thermodynamics properties like temperature, pressure, and species concentration are uniformly distributed within the gasifier.
- Steady-state condition is already achieved within the gasifier.
- Biomass from palm empty fruit bunch expressed as $CH_aO_bN_c$. a, b, and c are the number of atoms of hydrogen, oxygen, and nitrogen, respectively, derived from the ultimate and proximate analysis in table 1.
- There is no solid carbon remaining after the Gasification. All converted into syngas species in the global gasification model.

Table 1. Empty Fruit Bunch Ultimate and Proximate Analysis [5]

Component	%(Wt)	Element	%(Wt)
Volatile	77.46	C	61.51
Fixed Carbon	17.25	H	10.51
Ash	5.29	O	26.00
		N	1.98

- Syngas produced from the gasification process is consisting of H_2O, N_2, H_2, CO, CO_2, and CH_4. Ideal gas law applied to all of the syngas.

Model input required the following information:

- Biomass chemical composition ($CH_aO_bN_c$), biomass, and initial air temperature derived from the ambient temperature.
- Average gasification equilibrium temperature ranging from 800 to 1000 ^{0}C
- An equilibrium constant (K) used in gasification reaction is from methane forming and water gas shift reaction.
- Thermodynamic data for a chemical substance in gasification reactions like enthalpy for formation (H_of) and specific heat (C_p) taken from Perry [6].

Model output is the syngas composition once the equilibrium state achieved and the syngas heating value calculated from the syngas composition. Syngas heating value could be determined from bomb calorimeter, but also could be estimated with the calculation proposed by Reed [7].

$$\Delta H_c = LHV\ (kJ/kmol) = 0.2326(146.58C + 56.878H - 51.53O - 6.58A + 29.45) \quad (1)$$

Where C, H, O, and A are the mass fractions of carbon, hydrogen, oxygen, and ash, respectively, in the dry biomass.

Global biomass gasification for solid biomass $CH_aO_bN_c$ with air as gasification agent could be written with equation (4) below

$$CH_aO_bN_c + wH_2O + m(O_2 + 3.76N_2) \rightarrow n_1H_2 + n_2CO + n_3CO_2 + n_4H_2O + n_5CH_4 + n_6N_2 \quad (2)$$

Where the amount of water per kmol of biomass is denoted as w, and the amount of oxygen per kmol of biomass is indicated as m. The remaining coefficient of a constituent of products is marked as $n_{1\text{-}6}$. w could be calculated from moisture content per mol of biomass, so w become constant once moisture content was identified.

Mass balance is used from the atomic species balance to express the constraints of the gasifier system

$$C: 1 = n_2 + n_3 + n_5 \tag{3}$$
$$H: 2x + a = 2n_1 + 2n_4 + 4n_5 \tag{4}$$
$$O: b + x + 2y = n_2 + 2n_3 + n_4 \tag{5}$$

Two independent reactions (methane formation and water gas shift reaction) were used to model-independent reactions, the case of reactions without the presence of unconverted carbon in the products.

$$C + 2H_2 \leftrightarrow CH_4 \text{ ; and } K_1 = \frac{x_5}{{x_1}^2} \tag{6}$$
$$CO + H_2O \leftrightarrow CO_2 + H_2\text{; and } K_2 = \frac{x_1 x_3}{x_2 x_4} \tag{7}$$

The equilibrium constant for each reaction is expressed as K_1 and K_2, which is the function of temperature.

The last constraint is the energy balance as one of the essential equations that occurred in the adiabatic process without the transfer of heat or mass of substances between the thermodynamic system and surrounding. Energy is transferred to the surroundings as work only.

$$H^0{}_{f,bio} + w\left(H^0{}_{f,H2O(l)} + H_{(vap)}\right) = n_2 H^0{}_{f,CO} + n_4 H^0{}_{f,H2O(vap)} + n_5 H^0{}_{f,CH4} + \Delta T(n_1 C_{pH2} + n_2 C_{pCO} + n_3 C_{pCO2} + n_4 C_{pH2O} + n_5 C_{pCH4} + 3.76y C_{pN2}) \tag{8}$$

In summary, there are six unknown parameters ($n_{1\text{-}6}$) and six independent relations; three relations from the mass balance reactions (equation 5, 6 and 7), two relations from independent chemical reactions (equation 8 and 9), and one relation from energy balance (equation 10).

Substitutions and eliminations were used on the mass balance, energy balance, and independent reactions to solve the composition of carbon, hydrogen, and oxygen.

After going through the calculation steps, the constant equilibrium K_1 and K_2 for any particular temperature T could be calculated with following general equation for K_1 and K_2

$$lnK_1 = \frac{7082.848}{T} - 6.567\ lnT + \frac{7.466\times10^{-3}}{2}T - \frac{2.164\times10^{-6}}{6}T^2 + \frac{0.701\times10^5}{2T^2} + 32.541 \tag{9}$$

$$lnK_2 = \frac{5872.46}{T} + 1.86lnT - 2.69\times10^{-4}T + \frac{58200}{T^2} - 18.0139 \tag{10}$$

The remaining equations were consisting of linear and nonlinear equations and solved with the Newton Raphson method and fsolve in Matlab.

Biomass gasification performance could be indicated using some indicators. In this paper, syngas composition and LHV are used to indicate the gasification performance. Syngas composition was calculated from equation (11) and (12), while LHV or the Lower Heating Value of biomass could be found from equation (1), reference [4], and [8]. Gasification temperature and ER were varied to achieve the best gasification performance.

3. Result and Discussion

The results were validated with some previously developed models, analytical, numerical, and experimental models. Root Mean Square Error (RMSE) analysis method was employed to verify which model was better.

$$RMSE = \sqrt{\frac{\sum_{i=1}^{N}(Exp_i - Model_i)^2}{N}} \tag{11}$$

Table 2 shows a comparison between the zero-dimensional model (A) and some available data from the literature. As seen in Table 2, The proposed model (A) has lower RMSE compared with other research published (A1), hence in comparison with experimental data; therefore, the zero-dimensional method was relatively close with actual data.

Table 2. Syngas Composition and LHV Comparison

	A1	A	E
H_2 (%)	14.00	14.95	17.50
CO (%)	20.14	20.85	21.30
CH_4 (%)	12.06	16.85	13.30
CO_2 (%)	2.31	2.01	3.10
N_2 (%)	50.79	45.91	44.20
LHV (MJ/m^3)	5.28	5.01	5.69
RMSE	3.44	2.16	

All the models show a good agreement with the experimental data, with syngas composition from the highest to the lowest fraction, which is N_2, CO, CH_4, H_2, and CO_2, respectively. Hence the proposed model could be justified because RMSE evaluates the discrepancy between values provided by the model and the observed value from the experimental analysis.

The goal of gasification modelling is to achieve a maximum heating value from syngas. Syngas further usage following the Gasification should also be considered. According to Htut, syngas for the internal combustion engine usage will require heating value more than 4.2 MJ/m^3 [3]. All of the models and experimental data already meet this requirement.

The effect of gasification temperature on the heating value could be seen in Figure 2. The maximum heating value could be achieved with gasification temperature range 800-900 ^{0}C, or it could be concluded that within the gasification range 800-900 ^{0}C maximum syngas fraction of hydrogen, carbon monoxide, and methane was achieved.

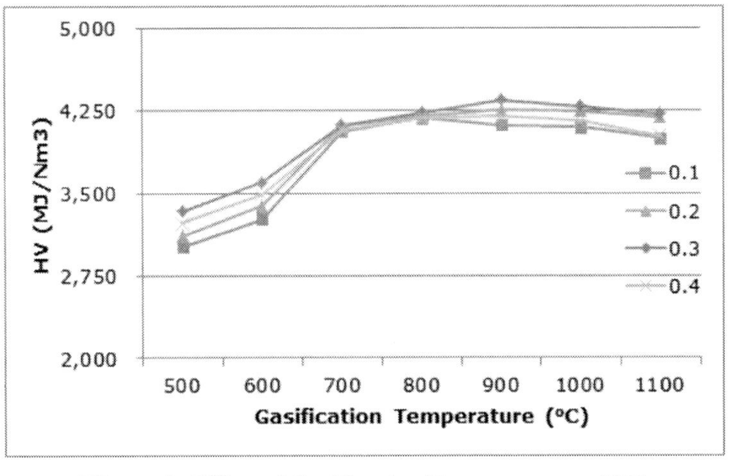

Figure 2. Effect of Gasification Temperature and ER

Equivalence Ratio (ER) variation also affects the Heating Value of syngas produced by the gasifier. Lower ER ratio tends to have lower heating value, an increase from 0.1 to 0.2, and reach the maximum heating value at ER 0.3. Higher ER value resulting in a drop in heating value compared to 0.3

4. Conclusions

Gasification simulation models could be classified based on the spatial dimension from the complicated three-dimensional models to the simple one-dimensional model. A zero-dimensional equilibrium model was established and proven its reliability with the experimental data. Thermodynamic computations are unconstrained with the dimension of the gasifier; hence it is advantageous to analyse the most substantial process parameters.

This model could be used to predict the gasification performance parameter (syngas composition and heating value) with satisfactory precision — the fact that both pyrolysis and syngas products were all need to go through the oxidation zone or the highest temperature in the gasifier. Equilibrium phase could be achieved even in a brief time in the oxidation zone. The advantage of this model is that this model could be able to manage a wide range of biomass sources of materials with different operating conditions and quite simple to operate.

5. References

[1] Maitah, M., Prochazka, P., Pachmann, A., Šrédl, K., Řezbová, H., (2016), Economics of Palm Oil Empty Fruit Bunches Bio Briquettes in Indonesia, International Journal of Energy Economics and Policy, Vol 6(1), pp.35-38.
[2] Basu, P., (2010), Biomass Gasification and Pyrolysis Practical Design and Theory, Academic Press, pp.1-4.
[3] Htut, YM., Khine, MM., Win, MM., (2015), Using a simple modeling and simulation scheme for complicated gasification system, International Journal of Scientific and Research Publication, Vol 5, No 6.
[4] Zainal ZA, Ali R, Lean CH, Seetharamu, KN, (2001), Prediction of performance of a downdraft gasifier using equilibrium modeling for different biomass material, Energy Conversion and Management 42(12):1499-1515.
[5] Wu KT., Chein, RY., (2015), Modelling of biomass gasification with preheated air at high temperatures, Energy Procedia, 75, 214-219.
[6] Robert HP, Don WG., Perry's chemical engineer's handbook. Sixth edition.New York: McGraw Hill,1984
[7] Reed TB, Principles and Technology of Biomass Gasification, Advances in Solar Energy, An Annu of Res and Dev, 1985, 2, 125-74
[8] Saidur R, Boroumand Jazi G, Mekhilef S, Mohammed HA. (2012) A review of the exergy analysis of biomass-based fuels. Renewable and Sustainable Energy Reviews,16:1217– 22.

Acknowledgments

This research was supported by the Indonesian Endowment Fund for Education (LPDP), Ministry of Finance, and the Directorate General of Higher Education (DIKTI), Ministry of Research, Technology and Higher Education, Republic of Indonesia.

Definition, criteria and approaches in designing suspension system with active controls

D A Patriawan[1], H Irawan[1], A Noerpamoengkas[1], B Setyono[1], and A Y Ismail[2]

[1], Mechanical Department, Industrial Faculty, Institut Teknologi Adhi Tama Surabaya, Surabaya, Indonesia.
[2] Mechanical Department, Faculty of Engineering, Kunsan National University

patriawan87@gmail.com

Abstract. Comfort and good handling are always a problem in conventional passive suspension systems. The ideal suspension criterion is to have a short time response, have a small amplitude and eliminate the frequency of causes of motion sickness. The system is only obtained with an active suspension system. The active suspension has an actuator that is able to produce a force to reduce the dynamic force of the vehicle. This actuator will be controlled by a PID active control. The test results showed that the active control was able to reduce the amplitude by 18 mm, passes the Sickness motion frequency and can return to the set point in 0.8 seconds.

1. Introduction

The suspension is a part of a vehicle that affects the dynamics and comfort of the vehicle. Suspension development is always related to vehicle development. Vehicles are getting faster and the distance is getting farther away, making the role of the suspension become increasingly important. So that vehicle designers see the suspension to be one of the important components to produce a vehicle that is fast, comfortable and safe.

The beginning of the vehicle does not use suspension, so the frame of the vehicle is directly connected to the wheel. The history of suspension use is not specifically recorded but is directly related to the development of the vehicle. The possibility of suspension appeared when the vehicle was found with the double pivot steering system in 1893 [1]. When the chain drive system begins to be replaced with an axle system on the rear of the vehicle, it will increase the unsprung mass. Increasing the unsprung mass at the rear of the vehicle requires making better damping.

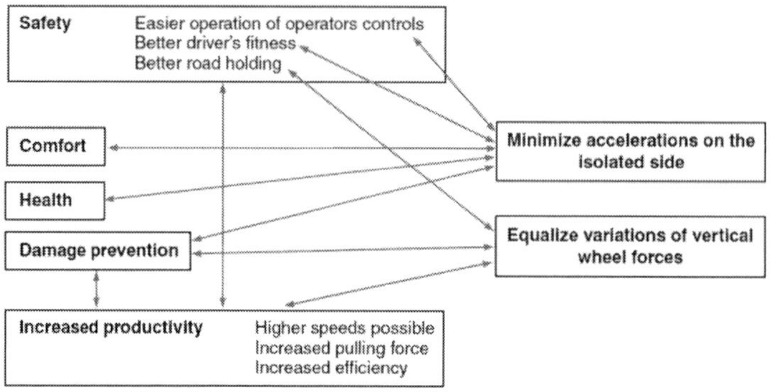

Figure 1. Task and function for suspension system [2].

The suspension system of the vehicle is a collection of mechanical components that are connected to the frame or body of the vehicle. There are three basic components of the suspension, namely linkages, spring and shock absorbers (dampers) [3]. Linkages in vehicles are usually called links because it functions as a liaison between the suspension and the frame or body of the vehicle. The number of links on a vehicle can differ depending on the type of suspension and vehicle. Spring has a different type in each suspension system, including coil, leaf and torsion bar. The type of coil in the form of coil that is most widely used in vehicles, especially small vehicles. The damper on the suspension system usually uses pistons and cylinders with a valve that can be adjusted by hydraulic or pneumatic pressure.

The suspension system has actually become a part of driving life, but many people are not aware of its existence. There are five tasks and functions of the suspension system on the vehicle, including safety, comfort, health, damage prevention and increased productivity [2]. The relationship between the five tasks and functions of the suspension system can be seen in Figure 1. If the five tasks and functions of the suspension system are met, it can be said that the suspension system is the ideal condition of the suspension.

The ideal conditions of suspension cannot be achieved only with conventional suspension systems with a constant damping coefficient [4]. The constant damping coefficient makes conflict occur in determining good handling or driving comfort. If you want to drive comfort to be more important, than handling the vehicle becomes unsteady. The relationship between the two problems can be seen in Figure 2.

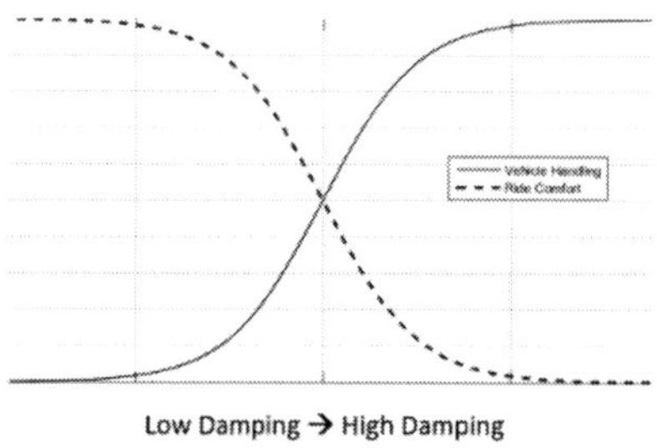

Figure 2. Handling vs comfort with damping ratio.

2. Ideal Suspension Criteria

The first criterion of an ideal suspension is driving stability, this is the first because it involves safety in driving. The role of the suspension in driving stability is to control the vehicle as soon as possible in its reference position after experiencing disturbance. The disturbance can be in the form of road conditions or dynamic motion of the vehicle.

The first criterion cannot be solved only with conventional passive suspension, this is because the damper ratio of the passive suspension is constant, it is can be seen in Figure 3a. In early 1974, Karnopp introduced a semi-active suspension with sky-hook control [5]. Semi-active suspension work system can be seen in Figure 3b. Suspension with an actuator that is able to be fully controlled between sprung and unsprung masses, this suspension is also referred to as an active suspension [6]. Preliminary studies on the performance of active suspensions have been conducted by Hrovat [7].

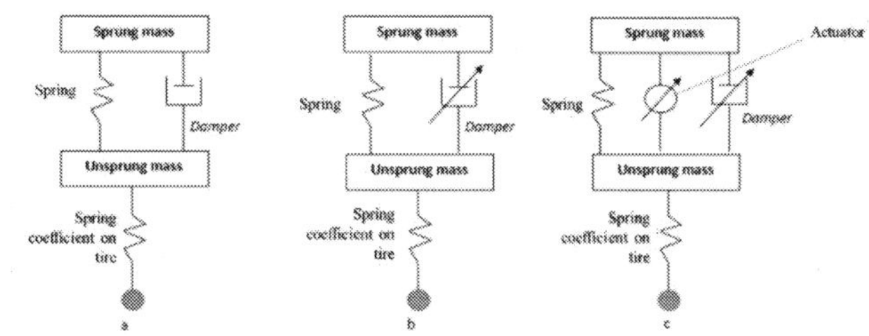

Figure 3. Classification from the suspension system.

The second criterion is health. In addition to the motion sickness described in the introduction, the vehicle can also experience head toss. Head toss can occur when the drive wheel passes through a deep hole. Head toss effect occurs because the receptors of 3 orthogonal oriented channels in each inner ear are activated by the acceleration of the head angle [8]. Head toss occurs if the vehicle enters a frequency of 2-8 Hz. Motion sickness and head toss are produced when the vehicle experiences low frequency. This frequency needs to be eliminated in the vehicle. In addition to frequency, the amplitude generated also needs to be reduced.

The third criterion is driving comfort. These criteria can be obtained by eliminating the high frequency and high amplitude on the vehicle. This high frequency can occur when a vehicle has a suspension with a large damper coefficient. A large damper coefficient makes handling in driving better, but the large frequency it produces makes passengers uncomfortable. The three criteria above are the basis of how to design the ideal suspension so that the vehicle becomes more safe, comfortable and speeds faster.

3. Ideal Suspension Approach

The three ideal suspension criteria can be applied if the damper is not constant and has an actuator to find out the ideal position of the vehicle. These criteria can use active suspension to approach the ideal suspension. Inactive suspension there is an actuator that can produce a force to resist vibrations in the vehicle.

Actuators that are often used as active suspension research subjects are hydraulic and electromagnetic. Both types of actuators have advantages and disadvantages to each. The advantage of a hydraulic suspension system is a large force density, easy to design, reliable and more mature to be commercialized. While the disadvantages of the active hydraulic suspension system include: it requires a continuous compressive system, leakage, and pollution. When compared with a hydraulic suspension system, the advantages of an electromagnet system are to increase dynamic behaviour, increase stability and increase efficiency [9]. Electromagnetic suspension systems have the potential to create an ideal suspension system.

Apart from the actuator, the suspension system must also have a spring and damper. The ideal spring condition is spring without friction, has no mass and moves linearly. Great force is needed at each end to hold the ideal spring when stretching from $|\Delta x|$ given from $F_{applied} \equiv k|\Delta x|$ where k is the spring constant in N/m. So according to Hooke's law, the idea of spring is

$$F_{OS,x} = -kx \quad (1)$$

IOP Conf. Series: Materials Science and Engineering 1010 (2021) 012006 doi:10.1088/1757-899X/1010/1/012006

where $F_{OS,x}$ is 1 vector dimension that is in contact with the force on an object (vehicle). So if the force received by the car and then make the x position change, the value of k will adjust the force. $F_{OS,x}$ can also be obtained from the vertical force on a moving vehicle.

The damper is also one part of the suspension system. Together with the spring, the damper works to reduce shocks or vibrations in the vehicle caused by road conditions or dynamic motion of the vehicle. If the spring and damper are combined into a suspension system, we can get an equation that is in accordance with Newton's second law. The equation is combined with equation (1) to be:

$$F_{OS,x} = -kx - b\frac{dx}{dt} \tag{2}$$

where M is the mass of the vehicle and b is the coefficient of the damping element.

The ideal suspension system of the vehicle can only be obtained with feedback from the sensor to change the damper value and the spring constant. Feedback can be done by adding the accelerometer sensor to the suspension system. Damping values and spring constant changes are obtained from changes in force on the vehicle. The force that changes is read by the sensor in the form of data. Then the data will determine the constant value of the spring and damping. In the mathematical model the equation that occurs in the accelerometer is:

$$M\frac{d^2x}{dt^2} + b\frac{dx}{dt} + kx = M\frac{d^2a}{dt^2} \tag{3}$$

In order to be modeling, equation (3) is converted into Laplace transform, so obtained:

$$Ms^2X(s) + bsX(s) + kX(s) = -MA(s) \tag{4}$$

Where $X(s)$ and $A(s)$ is the Laplace transform of $x(t)$ and d^2y/dt^2, so to get value $X(s)$ is obtained:

$$X(s) = -\frac{MA(s)}{Ms^2+bs+k} \tag{5}$$

4. Research Method

An ideal suspension can be obtained if it meets the requirements if the spring constant and the damping value meet the equation (1) and (2), other than that it must be able to change the spring constant and damping as the force changes in the vehicle. The addition of active control in the suspension system that is able to change the damping value will produce minimal vibration to the vehicle. The suspension system that does not have active controls to change the damping value can be seen in Figure 4. In figure 4 it can be seen that the suspension system has difficulty reaching the reference position.

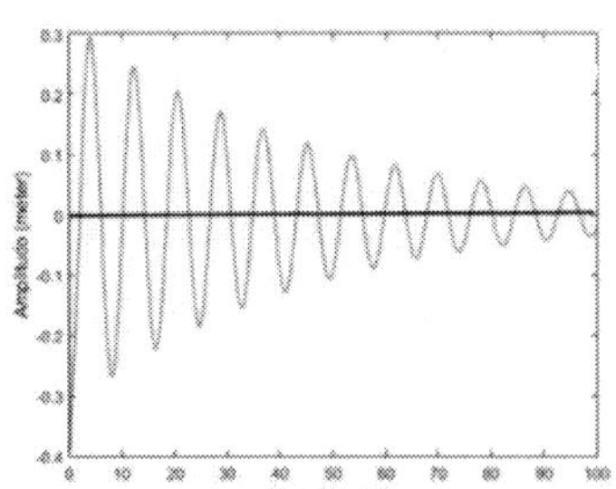

Figure 4. Vehicle position changed due to disturbance [10].

PID controller used in this suspension system. The advantages of using PID controllers include good stability, simple design and easy application on hardware [11]. because of these advantages the PID controller is widely used in the industrial world. PID control also has a disadvantage, one of which is the need to do tuning every change of parameters of the system being modelled.

PID tuning can be done using a variety of methods. The most widely used tuning method is using the Ziegler-Nichols method, the Cohen-Coon method, and the direct synthesis method. Direct synthesis method has better performance than the other two tuning methods [12]. But the direct synthesis method is quite difficult to apply in hardware, so in this paper still uses the Ziegler-Nichols method.

The electromagnetic force suspension system is made with the Maglev effect approach. The Maglev effect is an electromagnetic force that causes objects to float because of the levitator force. The levitator is the source of the electromagnetic force. The free block diagram of maglev can be seen in Figure 5.

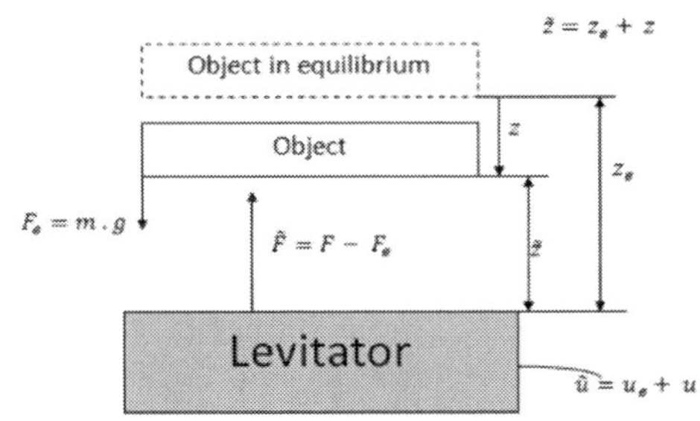

Figure 5. Free block Diagram maglev system.

Preliminary studies of the maglev system with the PID controller produce the specified gap can be achieved immediately even though the maglev system is disturbed by increasing the mass of the raised object [13]. The levitator force that is generated on the Maglev system has two characters, namely attractive and repulsive. Both levitator force characters do have their own advantages and disadvantages, Both levitator style characters do have their own strengths and weaknesses, but if the system wants a fast response then a repulsive force can be a good choice [14].

Maglev system consists of permanent magnets and electromagnets. The electromagnet is a levitator and permanent magnet as a supporting force. The force produced by a permanent magnet can be written with:

$$F_{PM} = BH \times Volume \tag{6}$$

where F_{PM} is the force of the permanent magnet, BH is a magnetic field produced from various types of material from magnets. The permanent material used is neodymium ($Nd_2Fe_{14}B$). The material is able to produce $300kJm^{-3}$.

The electromagnetic force is affected by the magnetic field. Large magnetic fields can be written with Ampere's law approach, where the magnetic field is:

$$B_{EM} = \mu_0 i_0 n \tag{7}$$

where B_{EM} is a large magnetic field, μ_0 is the permeability of air, i_0 is the current through copper wire and n is the number of turns in units of length. With Lorentz's law, equation (7) is processed to become an electromagnetic force into:

$$\vec{F}_{EM} = q\vec{v} \times \vec{B} \tag{8}$$

q is a positive charge and $\vec{v}$ is the vector value of speed. The direction of the force in equation (8) can be obtained by the right-hand rule, so the equation of force becomes:

$$\vec{F}_{ME} = \vec{I}L \times \vec{B} \tag{9}$$

Where $\vec{I}$ is the current passing through the length of the wire from L. The results of this equation are then made into a block diagram from the suspension system. Figure 6 can be seen a block diagram of the suspension system. Whereas in table 1 are the parameters of modelling a quarter of a car.

Table 1. The parameters of the quarter car suspension modelling

Parameter	Value	Description
K_1	25 kN/m	Spring stiffness
K_2	220 kN/m	Tire Stiffness
M_1	375 kg	¼ vehicle mass
M_2	25 kg	Unsprung mass
b_1	1000 Ns/M	Damping passive suspension
b_2	10 kNs/M	Tire passive damping

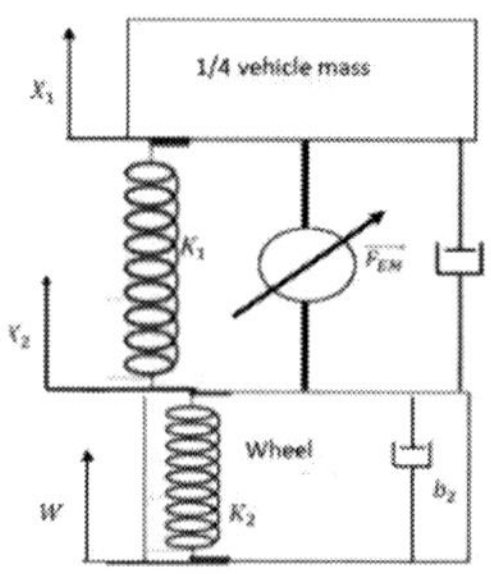

Figure 6. Block diagram of quarter of a vehicle.

One disadvantage of using electromagnets is that a large amount of power is needed. Large power can overload the vehicle engine performance. The hybrid suspension system can be used as a solution [15]. Future research will add air suspension to the active suspension system with electromagnets.

5. Result and Discussion

The first modelling is to observe how much vibration occurs in the vehicle without interference and active control. This modelling needs to be done to determine the response of the suspension system from the force generated from the vehicle itself. The results of modelling and simulation can be seen in Figure 7. The results shown in Figure 7 show that vehicles naturally have vibrations caused by dynamic motion. But the vibration produced is very small, the largest amplitude only reaches 0.0605 mm, this is not felt in driving. The frequency that occurs is also quite small around 1.21 Hz. These frequencies can cause passengers to experience motion sickness.

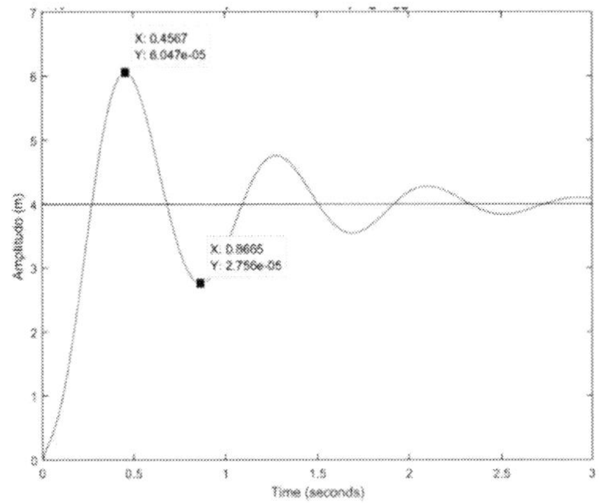

Figure 7. Vehicle response without disturbance

The second test is to see how the response of the suspension system that gets disturbance. Disturbances used are speed limiting devices adjusted to the Minister of Transportation regulation No. 31/1994 Article 5. The article reads: "Vehicle speed limit must be made with a maximum height of 12 cm, a minimum width of 15 cm and a sloping side with a maximum slope of 15°. Disturbances received by the suspension system result in changes in position. Changes in position increased from 0.0605 mm to 109.8 mm. The resulting frequency also experienced a slight change that is to 1.23 Hz. Also in 3 seconds, the suspension system has not been able to return to the reference position (steady-state). The results can be seen in Figure 8.

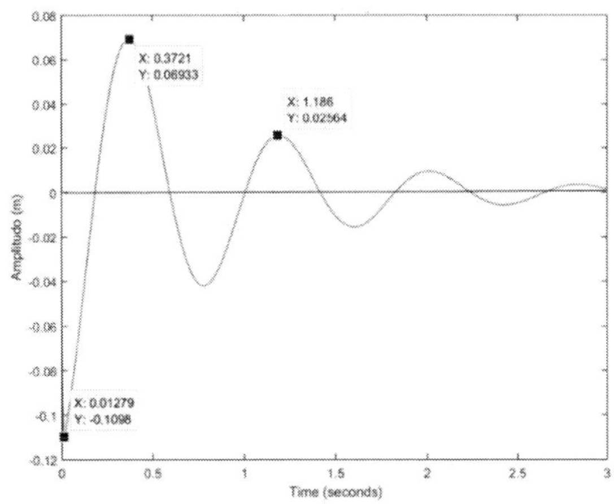

Figure 8. Vehicle response with disturbance.

In the third test is to add active control to the suspension system. The controller used is the PID controller, this is because it is easily applied to hardware. The tuning process uses the Ziegler-Nichols method which then continues to improve the response results. The results of testing added with active control are able to produce a much better response. The change in the position shown from the amplitude is only 18 mm adrift with a suspension system without active control. However, the response time required for the suspension system to return to its original position is 0.8 seconds. The results of this response are much faster than the suspension system without active control, this makes the handling of the active suspension system better. The initial frequency produced is around 4.5 Hz, so the suspension is not at the motion sickness frequency. The results of the suspension response with active control can be seen in Figure 9.

IOP Conf. Series: Materials Science and Engineering 1010 (2021) 012006 doi:10.1088/1757-899X/1010/1/012006

Frequency becomes an interesting problem to be discussed next. High frequency can make passengers uncomfortable when driving long distances. Frequency and amplitude can be removed if the suspension system and the vehicle are not directly connected directly or by using the maglev system.

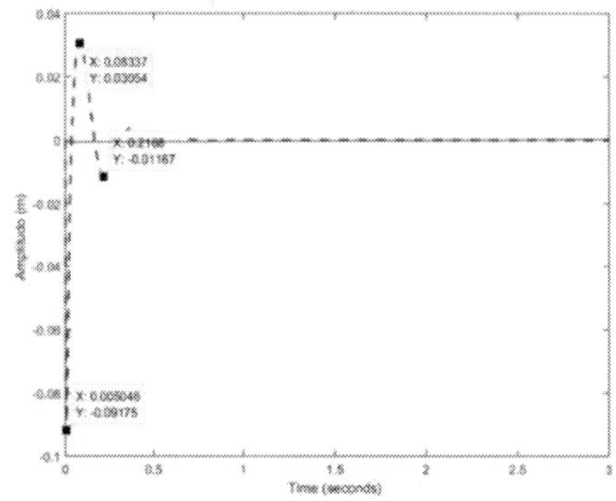

Figure 9. Suspension response with disturbance and active control.

6. Conclusions

The results of the suspension system without disturbance still slightly produces vibrations that make changes in the position of the vehicle. The highest change in position is 0.06 mm with an initial frequency of 1.21 Hz. Within 3 seconds, the suspension system has not been able to reach the reference position.

In the second test, the suspension system was given a disturbance in the form of a speed limit with a height of 120 mm. The results of the test obtained changes in the height of the suspension position is 109.8 mm with an initial frequency of 1.23 Hz. For 3 seconds the position of the vehicle is not able to return to its original position.

The third test by adding active control makes the suspension work better. While the frequency rises to 4 Hz. This frequency does not enter the frequency in motion sickness. The position of the vehicle is also very fast to return to the reference position, this makes handling of the vehicle better.

7. References

[1] Singh E P and Sharma E M, 2017 Suspension Systems : A Review *Int. Res. J. Eng. Technol.* **4**, 4.

[2] Wolfgang Bauer, 2011 *Hydropneumatic Suspension Systems* 1 st Weinheim: Springer.

[3] Tata R P, 2012, Automotive Suspension Systems Credit: 2 PDH, 877, .

[4] Patil S A and Joshi S G, 2014 Experimental analysis of 2 DOF quarter-car passive and hydraulic active suspension systems for ride comfort *Syst. Sci. Control Eng.* **2**, 1 p. 621–631.

[5] Karnopp D Crosby M J and Harwood R A, 1973 Vibration Control Using Semi-Active Force Generators. *ASME Pap* 73-DET-122 p. 619–626.

[6] Omar M El-kassaby M M and Abdelghaffar W, 2017 A universal suspension test rig for electrohydraulic active and passive automotive suspension system *Alexandria Eng. J.* **56**, 4 p. 359–370.

[7] Hrovat D, 1997 Survey of Advanced Suspension Developments and Related Optimal Control Applications *Automatica* **33**, 10 p. 1781–1817.

[8] Gahlinger P M, 1999 Motion sickness: How to help your patients avoid travel travail *Postgrad. Med.* **106**, 4 p. 177–184.

[9] Gysen B L J Paulides J J H Janssen J L G and Lomonova E A, 2010 Active electromagnetic suspension system for improved vehicle dynamics *IEEE Trans. Veh. Technol.* **59**, 3 p. 1156–1163.

[10] Patriawan D A Wahyu E Widodo R and Hariyanto I, 2015 Pemodelan Suspensi Aktif dengan Elektromagnet untuk Menghasilkan Kenyamanan dan Manuver yang Lebih Baik dalam

Berkendara *Semin. Nas. Sains dan Teknol. Terap. III 2015* p. 219–224.
[11] Åström K J and Hägglund T, 2001 The future of PID control *Control Eng. Pract.* **9**, 11 p. 1163–1175.
[12] Handy Wicaksono, 2004 Analisa Performansi dan Robustness Beberapa Metode Tuning Kontroler PID pada Motor DC *J. Tek. Elektro* **4**, 2 p. 70–78.
[13] Patriawan D A Pramujati B and Nurhadi H, 2014 Preliminary study on magnetic levitation modeling using PID control *Appl. Mech. Mater.* **493** p. 517–522.
[14] Pramujati B Nurhadi H and Patriawan D A, 2015 A Study on the Effect of an Attractive and a Repulsive Forces with Feedback Control *Int. J. Sci. Eng. Res.* **6**, 6 p. 56–60.
[15] Hu Q Lu W and Jiang J, 2019 Dynamic modeling and adjustable damping layered control of air suspension hybrid system *Aust. J. Mech. Eng.* **0**, 0 p. 1–13.

Acknowledgments
The authors thank you very much for the support of LPPM ITATS and Ristekdikti (Ministries of Research, Technology and Higher Education) grant.

IOP Conf. Series: Materials Science and Engineering 1010 (2021) 012007 doi:10.1088/1757-899X/1010/1/012007

Lateral bearing capacity analysis of pile foundation using a spring modelling system

J Propika[1], L L Lestari[1], and A D Puspasari[1,2]

[1]Department of Civil Engineering, Institut Teknologi Adhi Tama Surabaya, Surabaya 60117, Indonesia
[2]Department of Civil Engineering, Chung Yuan Christian University, Zhongli District, Taoyuan City, Taiwan 320

jakapropika@itats.ac.id, laraslaila.lestari@itats.ac.id, aisyahdwipus@gmail.com

Abstract. Pile foundations are often required to resist lateral loads and moments in addition to their primary use as axially loaded members. The interaction between material strength of axial, moment and shear against load capacity and soil interaction is the most important factors in the pile analysis. This study aims to obtain the lateral bearing capacity of pile foundation using a spring modelling system which will be calculated manually for the K_v spring constant in the vertical direction and the k_h coefficient in the horizontal direction using the Nakazawa method and SAP 2000 v14.2.5 program. The result shows that the maximum shear load and the maximum axial tensile load received by the pile is still within the allowable shear load and allowable axial tensile load capacity resulting from the manual calculation. However, the maximum axial compression load and maximum moment is exceeded its each allowable capacity. Hence, it is necessary to re-modelling by adding the depth or number of piles.

1. Introduction

Pile foundations are common foundation used in the construction, especially in the pillar bridge construction. Piles are often required to resist lateral loads and moments in addition to their primary use as axially loaded members. The goals are to determine deflections and stresses in the selected soil-pile system in order that they may be controlled within tolerable limits [1]. Nowadays, most of pile foundation's planning separate the strength calculation of axial bearing capacity and lateral bearing capacity. The calculation analysis is done without involving interaction between material strength of axial, moment and shear against load capacity and soil interaction. In fact, soil-structure interaction is one of the most important factors in the analysis [2].

Soil as a damper cannot be assumed to be the whole fixed. In subgrade reaction approach, it is assumed that soil acts as a series of independent linear elastic springs [2]. Pile is divided into several segments (mass) where each segment has a center of mass in the middle and the ground is considered as a linear spring [3]. The discrepancy between the modelling of pile foundation at this time with the actual conditions in the construction project field will affect to the lateral bearing capacity. So that we need an analysis of the calculation of the lateral bearing capacity of pile foundations in the construction of bridge piers using a spring modelling system. Soil that is modelled as an elastic has the ability to resist the load depending on the estimated spring constant Kv in the vertical direction and the estimated coefficient of kh from the subsurface soil reaction in the horizontal direction [4].

The structural data analysis as same as with the previous research that is taken from the construction of the bridge's pile of Surabaya's Outer West Ring Road (planning project) STA 0+400, Surabaya, Indonesia [5].

The purpose of this paper is to obtain the lateral bearing capacity of pile foundation using a spring modelling system which will be calculated manually for the Kv spring constant in the vertical direction and the kh coefficient in the horizontal direction using the Nakazawa method and SAP 2000 v14.2.5 program. By conducting these methods, it is expected that the results are closer to the actual conditions in the construction project field.

1.1 Literature Review

1.1.1. Lateral Load. Piles are always required to be designed to with stand the lateral loads in addition to the compression and tension loads [6]. Pile foundations often receive inclined loads which are generally the resultant of dead loads and horizontal loads due to wind, earthquakes, water pressure, ground pressure, etc. The horizontal component of the inclined load is usually relatively small compared to the vertical component. However, for certain cases such as retaining walls, bridge pillars and piers, the horizontal force acting is quite large and is a fixed load, which means that the load works throughout the life of the building [7].

The allowable lateral loads on pile is determined from the following two criteria; 1. Allowable lateral load is obtained by dividing the ultimate (failure) load by an adequate factor of safety; 2. Allowable lateral load is corresponding to an acceptable lateral deflection. The smaller of the two above values is the one actually adopted as the design lateral load. Method of calculating lateral resistance of vertical piles can be broadly divided into two categories:

- Methods of calculating ultimate lateral resistance
- Methods of calculating acceptable deflection at working lateral load.

According to Prakash and Sharma (1990), methods of calculating lateral resistance of vertical piles among others Brinch Hansen's method (1961) and Broms' method (1964). Both of them are based on earth pressure theory. Brinch Hansen's method applicable for c-Ø soils and layered system, but this method only can applied for short piles and requires trial-and-error solution to locate point of rotation. Otherwise, Broms' method applicable for short and long piles, considers purely cohesive and cohensionless soils, and considers free-head and fixed-head piles that can be 2nalysed separately, but this method cannot applied to layered system and does not consider c-Ø soils.

Methods of calculating acceptable deflection at working load among others Modulus of subgrade reaction approach (Reese and Matlock, 1956) and Elastic approach (Poulos, 1971). Modulus of subgrade reaction approach is assumed that soil acts as a series of independent linearly elastic springs. This method is relatively simple and can incorporate factors such as nonlinearity, variation of subgrade reaction with depth, and layered systems. Moreover, this method has been used in the practice for a long time so a considerable amount of experience has been gained in applying the theory to practical problems. However, this method has some disadvantages such as it ignores contuity of the soil and modulus of subgrade reaction is not a unique soil property but depends on the foundation size and deflections.

Furthermore, the elastic approach, in this method the soil is assumed as an ideal elastic continuum. It is based on a theoretically more realistic approach and it can give solutions for varying modulus with depth and layered system. But this method is difficult to determine appropriate strains in a field problem and the corresponding soil module. It needs more field verification by applying theory to practical problems [3].

Different from the methods above, this is using a calculation method adapted from technical standard in Japan called Nakazawa method [4]. As we know, Japan is an earthquake-prone country dominated by soft soil. This is similar with the condition in Indonesia [5]. In this method, the lateral bearing capacity of pile foundation is obtained by the manual calculation of spring constant Kv in the vertical direction and the coefficient of k_h from the subsurface soil reaction in the horizontal direction at each depth of soil foundation.

2. Research Methodology

2.1 Calculation of Pile Structural Capacity.

The pile foundation used in the construction planning is strong, if it is able to withstand the loads that act on the pile. This can be determined by comparing the allowable stress of the pile material and the loads that act on the pile. The allowable stress of the pile material is the force strength per unit area based on the pile material used to obtain the allowable axial compression capacity, the allowable axial tensile capacity, the allowable shear capacity, and the allowable moment capacity. This value is the maximum strength of the pile that is allowed to withstand loads acting on the pile by considering the strength of the material from the pile itself which is formulated by SNI T-12-2004 [8] as follows:

Allowable axial compression capacity

$$\begin{aligned} Pu &= \emptyset \,.\, P_n \text{ compression} \\ &= 0.7 \{(0{,}60 \,.\, fc' .\, Ag) - (0{,}70 \,.\, fpu \,.\, Asp \,.\, n)\} \end{aligned} \tag{1}$$

Allowable axial tensile capacity

$$\begin{aligned} Pu &= \emptyset \,.\, P_n \text{ tensile} \\ &= 0{,}8 \left\{(0{,}50 \sqrt{fc'} .\, A_g) + 0.85\,(0{,}70 \,.\, fpu \,.\, Asp \,.\, n)\right\} \end{aligned} \tag{2}$$

Allowable shear capacity

$$\begin{aligned} V_u &= \emptyset \,.\, V_n \\ &= \emptyset\,(V_c + V_s) \\ &= 0{,}75 \left[\left(1+\frac{N_u}{14A_g}\right)\left(\frac{\sqrt{f_c'}}{6}\right) b_w d\right] + \left[\frac{A_v\, f_y\, d}{s}\right] \end{aligned} \tag{3}$$

Information :

P_n compression	= The nominal compressive axial limit strength of the structural components (N)
P_n tensile	= The nominal tensile axial limit strength of the structural components (N)
V_n	= The nominal limit of shear strength of structural components (N)
V_c	= Shear strength of concrete (N)
V_s	= Shear strength of shear reinforcement (N)
N_u	= Ultimate axial load (N)
Ø	= Reduction factor
fc'	= Compressive strength of concrete (Mpa)
f_y	= Yield stress of non-prestressed reinforcement (Mpa)
fpu	= Tensile strength of pc bar (Mpa)
A_g	= The cross-sectional area of the pile (mm^2)
A_{sp}	= The cross-sectional area of PC bar (mm^2)
A_v	= The area of the shear reinforcement in the range s (mm^2)
n	= Number of reinforcement
b_w	= Diameter (mm)
d	= Distance from the outer compression area to the center of the tensile reinforcement (mm)
s	= Spacing of shear reinforcement parallel to longitudinal reinforcement (mm)

Besides getting the allowable stress of the pile material by manual calculation, this value can also be obtained from the Wika Beton pile brochure specifications, 2017 [9]. **Table 1** shows the specification of Wika Beton production piles for a diameter of 600 mm

2.2 Calculation of Lateral Bearing Capacity.
In this study, the spring constant Kv and the coefficient of kh are calculated manually by using Nakazawa method then inputed into SAP 2000 program as a spring stiffness and generate maximum shear force or called lateral bearing capacity.

Estimated Kv (spring constant) in the vertical force of the pile is an elastic constant which is expressed as a force in the vertical direction and can cause displacement on the masthead. Kv is used in the calculation of the size of the pole head reaction or elastic decline that occurs [10]. The following is the equation for estimating Kv constants [4].

$$K_v = a \cdot \frac{A_p \cdot E_p}{l} \tag{4}$$

For steel pipes	$a = 0{,}027\ (l/D) + 0{,}2$
For prestressed concrete	$a = 0{,}041\ (l/D) - 0{,}27$
For cast in situ	$a = 0{,}022\ (l/D) - 0{,}05$

Information:
a = Multiplier factor based on pile material
A_p = Net cross-sectional area of pile (cm2)
E_p = Modulus of elasticity of pile (cm2)
l = Pile length (cm)
D = Pile diameter (cm)
The formula above is used when $l/D \geq 10$

Table 1. Pile specification

Size (mm)	Class	Bending moment Crack (ton.m)	Bending moment Break (ton.m)	Allowable compression (ton)	Decompression tension (ton)
600	A1	17.00	25.50	252.70	70.52
	A2	19.00	28.50	249.00	77.68
	A3	22.00	33.00	243.20	104.94
	B	25.00	45.00	238.30	131.10
	C	29.00	58.00	229.50	163.67

Source: Wika Beton Brochure, 2017

The following is the equation for estimating the coefficient of soil reaction below the surface in a horizontal force [4].

$$k_h = k_0 \cdot y^{-1/2} \tag{5}$$

$$k_0 = 0{,}2 \cdot E_0 \cdot D^{-3/4} \tag{6}$$

Information:
k_0 = Value k if the shift on the surface is made 1 cm (kg / cm2)
y = Value of deformation (cm)
E_0 = Modulus of foundation soil deformation, usually estimated from E0 = 28 N using N value from standard penetration experiments (standard penetration test)
D = Pile diameter (cm)

3. Data and Analysis Result

The structural data analysis is taken from the construction of the bridge's pile of Surabaya's Outer West Ring Road (planning project) STA 0+400, Surabaya, Indonesia.

Spun pile specification

Pile diameter, *D*	: 0.6 m
Pile class	: C
Pile material	: Prestressed concrete
Concrete quality, *fc'*	: 52 MPa
Concrete volume weight, *Wc*	: 2500 kg/m^3
Number of piles in group, *n*	: 25
Pile depth, *l*	: 16 m

3.1 Allowable axial compression capacity

$$P_u = \emptyset \,.\, P_n \text{ compression}$$
$$= 0.7\,\{(0{,}60 \,.\, f_c' \,.\, A_g) - (0{,}70 \,.\, f_{pu} \,.\, A_{sp} \,.\, n)\}$$
$$= 0.7\{(0{,}60 \,.\, 52 \text{ MPa} \,.\, 157.080 \text{ mm}^2) - (0{,}70 \,.\, 1030 \text{ MPa} \,.\, 89.92 \text{ mm}^2 \,.\, 26)\}$$
$$= 0.7\,\{4.900.896 \text{ N} - 1.685.640{,}32 \text{ N}\}$$
$$= 2.250.678{,}98 \text{ N}$$
$$= 229{,}51 \text{ ton}$$

3.2 Allowable axial tensile capacity

$$P_u = \emptyset \,.\, P_n \text{ tensile}$$
$$= 0.8\,\{(0{,}50\,\sqrt{f_c'} \,.\, A_g) + 0.85\,(0{,}70 \,.\, f_{pu} \,.\, A_{sp} \,.\, n)\}$$
$$= 0.8\,\{(0{,}50 \,.\, \sqrt{52} \text{ MPa} \,.\, 157.080 \text{ mm}^2) + (0{,}85 \,.\, (0{,}70 \,.\, 1030 \text{ MPa} \,.\, 89.92 \text{ mm}^2 \,.\, 26)\}$$
$$= 0.8\,\{566.359{,}99 \text{ N} + 1.432.794{,}27 \text{ N}\}$$
$$= 1.599.323{,}41 \text{ N}$$
$$= 163{,}09 \text{ ton}$$

3.3 Allowable shear capacity

$$V_u = \emptyset \,.\, V_n$$
$$= \emptyset\,(V_c + V_s)$$
$$= 0.75\left\{\left[\left(1+\frac{N_u}{14A_g}\right)\left(\frac{\sqrt{f_c'}}{6}\right) b_w d\right] + \left[\frac{A_v\, f_y\, d}{s}\right]\right\}$$
$$= 0.75\left[\left(1+\frac{1.821.595 \text{ N}}{14 \,.\, 157.080 \text{ mm}^2}\right)\left(\frac{\sqrt{52}}{6}\right) 600 \text{ mm} \,.\, 560{,}65 \text{ mm}\right] + \left[\frac{12{,}57 \text{ mm}^2 \,.\, 40 \text{ MPa} \,.\, 560{,}65 \text{ mm}}{100 \text{ mm}}\right]$$
$$= 0.75\,\{739.176{,}00 \text{ N} + 2.818{,}13 \text{ N}\}$$
$$= 556.495{,}60 \text{ N}$$
$$= 56{,}75 \text{ ton}$$

3.4 Spring constant Kv (z-axis)

The spring constant Kv at each depth is obtained by first calculating the *a* value according to the material of the pile, which is made of prestressed concrete by using equation 1. The results of the recapitulation of the spring constant Kv at each depth are shown in **Table 2** below.

IOP Conf. Series: Materials Science and Engineering 1010 (2021) 012007 doi:10.1088/1757-899X/1010/1/012007

Table 2. Recapitulation of the spring constant K_v

Depth (m)	K_v (kN/m)	Depth (m)	K_v (kN/m)	Depth (m)	K_v (kN/m)
1	0.00	11	419609.99	21	531616.61
2	0.00	12	439211.15	22	537216.94
3	0.00	13	455796.74	23	542330.29
4	7985.66	14	470012.97	24	547017.52
5	137353.30	15	482333.70	25	551329.78
6	223598.40	16	493114.33	26	555310.32
7	285202.04	17	502626.66	27	558996.01
8	331404.78	18	511082.06	28	562418.43
9	367340.23	19	518647.42	29	565604.83
10	396088.60	20	525456.25	30	568578.80

Source : Calculation result, 2019

3.5 Coefficient k_h (x and y-axis)

The soil reaction coefficient below the surface in a horizontal direction at each depth is obtained by first calculate the value of k_0 (eq 6.) and the value of k_h (eq 5.) whilst the shift on the surface is made of 1 cm (kg / cm^3). To find the value of the k_h coefficient per unit length, it must be multiplied by the diameter of the pile and the length of the segment. The results of the recapitulation of the coefficient k_h at each depth are shown in Table 3 below.

Table 3. Recapitulation of the spring constant k_h

Depth (m)	k_h (kN/m)	Depth (m)	k_h (kN/m)	Depth (m)	k_h (kN/m)
1	0.00	11	39739.27	21	47381.43
2	0.00	12	39739.27	22	47381.43
3	15284.33	13	42796.13	23	47381.43
4	18341.20	14	47381.43	24	51966.74
5	22926.50	15	51966.74	25	55023.60
6	29040.23	16	50438.30	26	55023.60
7	30568.67	17	45853.00	27	51966.74
8	35153.97	18	42796.13	28	48909.87
9	36682.40	19	45853.00	29	61137.34
10	38210.83	20	47381.43	30	68779.50

Source: Calculation result, 2019

3.6 SAP 2000 v14.2.5

The value of the spring constant K_v and the coefficient k_h at each depth obtained by using the manual calculation of the Nakazawa method are then input into the SAP 2000 v14.2.5 program to produce lateral bearing capacity. The modelling of bridge pillar construction with 25 pile foundations without springs is shown in **Figure 1**.

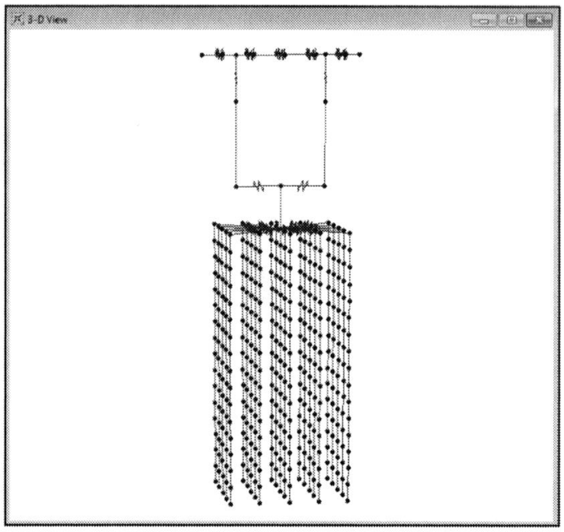

Figure 1. Modelling of 25 piles (16m depth)
Source: SAP 2000 v14.2.5, 2019

Then, input the spring constant value K_v for the z-axis in Translation 3 and the k_h coefficient for the x and y-axis in Translation 1 and Translation 2 at each depth from the manual calculation of the Nakazawa method shown in **Table 2** and **Table 3** into Joint Springs setup through *Assign - Joint - Springs* command (**Figure 2.**)

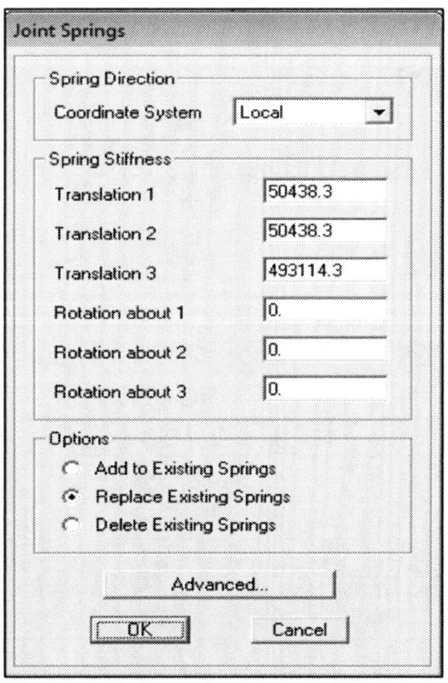

Figure 2. Joint springs setup
Source: SAP 2000 v14.2.5, 2019

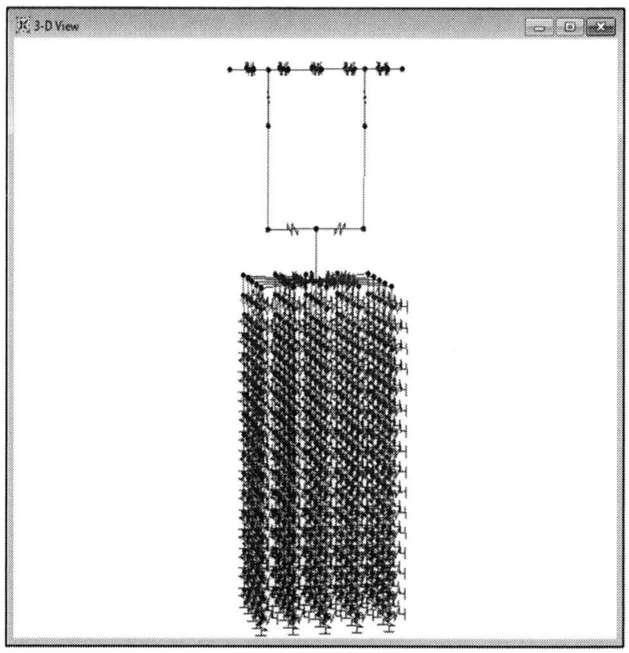

Figure 3. Modelling of 25 piles (16m depth) with springs
Source: SAP 2000 v14.2.5, 2019

Figure 3 shows a modelling of bridge pillar construction with 25 pile foundations with springs. The springs modelling is inputted per 1 meter as explained. The following table is the force generated by the piles in spring modelling using SAP 2000 v14.2.5.

IOP Conf. Series: Materials Science and Engineering 1010 (2021) 012007 doi:10.1088/1757-899X/1010/1/012007

Table 4. Recapitulation of the force generated by piles in the spring modelling

Output Case	Step Type	P	Vx	Vy	Mx	My
		kN	kN	kN	kN-m	kN-m
SERVICE 1D	MAX	-14.98	1.07	2.27	3.45	3.72
SERVICE 1D	MIN	-900.81	-8.38	-0.29	-1.00	-12.65
SERVICE 1T	MAX	-13.64	1.08	2.27	3.45	3.75
SERVICE 1T	MIN	-843.47	-8.38	-0.29	-1.00	-12.55
SERVICE 2D	MAX	-15.94	1.10	0.00	0.00	3.81
SERVICE 2D	MIN	-923.55	-8.60	0.00	0.00	-12.99
SERVICE 2T	MAX	-14.21	1.11	0.00	0.00	3.85
SERVICE 2T	MIN	-849.01	-8.60	0.00	0.00	-12.87
EXTREME XD (1)	MAX	1159.09	**340.55**	119.28	185.05	526.88
EXTREME XD (1)	MIN	-2612.36	-340.55	-119.28	-185.05	-526.88
EXTREME XT (1)	MAX	1177.48	340.55	119.28	185.05	**526.90**
EXTREME XT (1)	MIN	-2595.05	-340.55	-119.28	-185.05	-526.85
EXTREME YD (1)	MAX	1319.63	102.21	397.32	616.36	158.13
EXTREME YD (1)	MIN	**-2772.95**	-102.21	**-397.32**	**-616.36**	-158.13
EXTREME YT(1)	MAX	1340.07	102.21	397.32	616.36	158.16
EXTREME YT(1)	MIN	-2755.64	-102.21	-397.32	-616.36	-158.10
KUAT 1 D	MAX	-21.05	1.97	0.00	0.00	6.87
KUAT 1 D	MIN	-1273.45	-15.48	0.00	0.00	-23.39
KUAT 2 T	MAX	-18.65	1.99	0.00	0.00	6.92
KUAT 2 T	MIN	-1170.24	-15.48	0.00	0.00	-23.21
KUAT III	MAX	-17.34	1.03	10.60	16.08	1.61
KUAT III	MIN	-1021.02	-0.13	-1.34	-4.68	-0.44

Source: Calculation result, 2019

Based on the pile load generated from spring modelling with a total of 25 piles and a length of 16 m, the maximum material capacity and pile loads/forces could be seen in **Table 5**. The conclusions shows the comparison between material capacity and loads received by piles. The design analysis is valid when material capacity is bigger than pile loads.

Table 5. Recapitulation of the material capacity and pile loads/forces

No.	Type of load	Material capacity (ton)	Pile load (ton)	Conclusions
1	Axial compression	229,51	282.76	Exceeds
2	Axial tensile	163,09	136.66	OK
3	Shear	56,75	40.52	OK
4	Moment	58,00	62.85	Exceeds

Source: Calculation result, 2019

The maximum shear load received by the pile is 397.92 kN or 40.52 ton. This value is still within the allowable shear load capacity resulting from the manual calculation of 56.75 ton. However, the maximum axial compression load received by the pile is 2,772.95 kN or 282.76 ton that is exceeds the allowable axial compression load capacity resulting from the manual calculation of 229.51 ton which is similar to the value in Wika Beton brochure (Table 1.)

The maximum axial tensile load received by the pile is 1340.07 kN or 136.66 ton, this value is still within the allowable axial tensile load capacity resulting from the manual calculation of 163,09 ton which is equivalent with the value in Table 1. While the maximum moment received by the pile is 616.36 kN-m or 62.85 ton-m, this value also exceeds the allowable moment capacity of 58 ton-m from Pile specification in Table 1. Hence, it is necessary to re-modelling by adding the depth or number of piles for the further study.

4. Conclussions

The analysis result shows that the lateral bearing capacity of pile foundation with a spring modelling system by manual calculation for the K_v spring constant in the vertical direction and the k_h coefficient in the horizontal direction using the Nakazawa method and SAP 2000 v14.2.5 program. The pile load generated with a total of 25 piles and a length of 16 m, the maximum shear load and the maximum axial tensile load received by the pile is still within the allowable shear load and allowable axial tensile load capacity resulting from the manual calculation. However, the maximum axial compression load and maximum moment is exceeded its each allowable capacity. Hence, it is necessary to re-modelling by adding the depth or number of piles for the further study.

5. References

[1] DAVISSON MT, "Lateral Load Capacity of Piles," *Highw Res Rec*, no. 33, pp. 104–112, 1970.

[2] O. Taheri, R. Z. Moayed, and M. Nozari, "Lateral Soil-Pile Stiffness Subjected to Vertical and Lateral Loading Lateral," *J. Geotech. Transp. Eng.*, vol. 1, no. 2, pp. 30–37, 2015.

[3] S. Prakash and H. D. Sharma, *Pile Foundations in Engineering*. 1990.

[4] S. Sosrodarsono and K. Nakazawa, *MEKANIKA TANAH & TEKNIK PONDASI*. JAKARTA: PT. PRADNYA PRAMITA, 2000.

[5] L. L. Lestari, J. Propika, and A. D. Puspasari, "Axial Bearing Capacity Analysis of Pile Foundation using Nakazawa Method," *J. IPTEK*, vol. 24, no. 1, pp. 45–52, 2020.

[6] P. K. Jayasree, K. V. Arun, R. Oormila, and H. Sreelakshmi, "Lateral Load Capacity of Piles: A Comparative Study Between Indian Standards and Theoretical Approach," *J. Inst. Eng. Ser. A*, vol. 99, no. 3, pp. 587–593, 2018.

[7] J. T. Hatmoko, *Dinamika Tanah Dan Liquefaction*. Yogyakarta: Cahaya Atma Pustaka, 2016.

[8] Badan Standardisasi Nasional Indonesia, "SNI T-12-2004 Perencanaan Struktur Beton untuk Jembatan," pp. 1–142, 2004.

[9] WIKA, "Brochure The Precast Concrete Manufacturer," 2017.

[10] D. K. Fitriyah, J. Propika, L. L. Lestari, H. Istiono, D. Pertiwi, and R. Sekartadji, "Pile Foundation Analysis on High - Rise Building using Finite Element-Spring Method on Sandy Clay Soil," *IOP Conf. Ser. Mater. Sci. Eng.*, vol. 462, no. 1, 2019.

ERP compatibility on business performance through the inventory system and internal integration

P Sebayang[1], Z J H Tarigan[2*], T W S Panjaitan[3]

[1] Tama Jaga Karsa University, Jl. TB Simatupang No.152, Tj. Bar., Kec. Jagakarsa, Kota Jakarta Selatan, Special Capital Region of Jakarta 12530,
[2] Lecturers with department of Master Management, Faculty of Business and Economics, Petra Christian University, Jl. Siwalankerto 121–131, Surabaya 60236, Indonesia.
[3] School of Earth and Environmental Sciences, Queensland University, St. Lucia QLD 4057, Australia

* zeplin@petra.ac.id

Abstract. The information technology used by companies today is ERP (Enterprise Resources Planning) technology. This study aims to examine the role of ERP compatibility in improving the performance of MM-2100 manufacturing companies. Data is collected at manufacturing companies that have made ERP system adjustments. Data collection was carried out through questionnaires distributed to manufacturing companies, and as many as 78 questionnaires were obtained, which could be further processed. Retrieval of data using questionnaires submitted to the relevant section that has access to company inventory data. Data analysis was performed using the SEM-PLS technique using smart PLS software version 2.0. The results of the study show that ERP compatibility can increase internal information integration and inventory system continuously. However, ERP compatibility has no direct effect on manufacturing performance. Internal integration affects inventory systems and manufacturing performance with coefficients of 0.346 and 0.378, respectively. Internal integration allows other functions to be able to access inventory in the company. The company's inventory system that is controlled can perform efficiency and effectiveness for the company so that it has an impact on manufacturing performance with a coefficient of 0.766.

1. Introduction

Globalization is indeed not a new phenomenon because globalization has developed in recent years, but lately, it has experienced very rapid progress and is affecting life in various aspects. The global competition enables companies to reach international markets with business models, less capital intensive. This globalization increases the risks and challenges for various countries in formulating new policies to keep abreast of this global phenomenon. Increasing competition has forced companies to improve their business performance operations so they can survive [1]. Operation performance is one of the measurements used in assessing the performance of an organization [2]. The measure includes reliability, inventory turnover, lead time, and cycle time. Operation business performance is an essential aspect because it affects customer satisfaction and market share. The company is necessary to implement supply chain management practices to maintain and improve business performance operations [3]. Another factor is that Supply chain management also is critical to help company performance more effectively and efficiently. The company can maintain the stability of the procurement of raw materials by building communication and collaboration integrated with suppliers on an ongoing basis. Then, the company carries out an internal integration of processes with integration between functions within the company [4]. Furthermore, the company serves the demands of customers by building customer relationship management. The company carries out supply chain practices by managing the process from beginning to end by involving various functions in the company. Controlling internal and external processes can be carried out in an integrated manner through the implementation of ERP (enterprise resources planning) technology. ERP implementation enables companies to produce efficient and effective processes and fast coordination [5]. ERP is an integrated technology between functions in an organization which functions automatically and in real-time. ERP can integrate the main business functions and support of manufacturing companies well. The main business functions of manufacturing consist of marketing, procurement, production, warehouse functions. Supporting functions consist of administrative, human resources, accounting, and finance functions [6]. ERP technology can build transparency in the company so that all department data can be accessed by personal employees who have access rights. ERP is also able to integrate data and be accessed in a multinational company. ERP systems build company competitiveness by integrating processes and data with partners and optimizing the use of company resources. ERP capability is related to how ERP systems can run correctly and according to the needs of manufacturing companies. ERP adoption is related to ICT infrastructure, Technical availability, perceived compatibility, perceived value, and ERP system security. ERP success is affected by several factors. Those factors are associated with ERP compatibility, which includes such as user involvement, business process reengineering, business plan and vision, knowledge integration, information technology legacy systems, data accuracy, software testing, and development and customization. ERP system can provide increased business performance for companies [7].

The ability of ERP to integrate business functions includes sharing data inside and outside the company, sharing information, and collaborating with external parties, and accessing data information in real-time. ERP systems that are in line with company conditions can provide significant added value in carrying out business practices and supporting management decision making [8]. ERP practice does not have a significant impact on the operational performance of manufacturing companies in Turkey. The reason is due to the utility and module in the ERP system are customized made to the company and does not directly have an impact on the size of the company's inventory level and on-time delivery. But it needs real action in the company's operations [2]. An exemplary ERP implementation provides good performance improvement and will be different if the company runs the implementation incorrectly. Inappropriate ERP interferes with data integration in the company and results in data in the form of reports generated by ERP that cannot help management in making decisions. ERP compatibility in manufacturing companies can be upgraded continuously to be able to adjust to the needs of the company [9]. Adjustment of ERP with company conditions in the form of software and hardware needs to be done continuously under internal processes in the company. The company's ERP system can build all business processes into a single unit in the form of an integrated database between functions. Internal integration can create effective and efficient systems for companies. ERP systems enable companies to integrate with suppliers and customer integration and integrate within the company.

Internal integration allows the company to establish good cooperation between functions in the company, reducing conflicts related to differences in data or reports between functions so that coordination between the activities of operations is well established [4]. ERP companies can build an inventory system that can be controlled precisely in determining inventory levels. ERP system can be set to determine the minimum stock level and maximum stock level and can calculate the optimal stock and adjust the warehouse capacity. The inventory system is an essential function for a company regardless of the size of the business. The more inventory the company stores in the warehouse, the higher the costs the company will incur. Therefore, a company needs to be able to formulate a policy so that inventory investment can be optimal and can reduce costs. Companies that can control inventory levels can maintain a balance between inventory cost risk and stock out inventory risk. The use of information technology can reduce the company's level of inventory efficiency [10]. Internal integration enables companies to integrate business functions so that relevant departments can access the company's inventory data. Thus, other functions can monitor and control business processes by adjusting inventory availability in the warehouse. Internal integration between marketing functions with the inventory section can help marketing flexibly meet customer demands according to inventory availability. The inventory level mediates the link between the material procurement department, and the company's shipping department increases the level of efficiency in transportation costs [11]. ERP implementation can improve company business performance. ERP implementation is determined by top management to be able to provide improved performance for the company hardware [9]. Research conducted by States that ERP can improve business performance on an ongoing basis. Based on the explanation above, the ERP compatibility set by the company and updated on an ongoing basis will be able to build a system of integration between functions within the company properly. This condition results in all functions being able to access inventory data properly and accurately to make the right decisions according to the external conditions of the company in order to improve performance.

1.1. Enterprise Resources Planning

Enterprise resource planning is a software package that can integrate functions within the company and become an operating system used by the company. ERP is also able to provide condition reports in real-time to top management when decisions are needed [6]. ERP resources in the company are investments made by companies related to application software, application hardware, and skills from the information technology department that focuses on the company's performance on an ongoing basis. ERP can describe the company's processes for all functions in the company from the process of procurement of materials to the delivery of products in support of the company's business [12]. ERP system implementation can bring added value to companies through best business practices, which are provided through various functional applications that are supported by comprehensive processes and data integration between functions in the company [8]. Indicators measure ERP system upgrades in manufacturing companies to maintain compatibility in the form of data integration, ERP system stability, ERP system continuous upgrades, and ERP technology utilization [9].

The ERP stage has been implemented, called the post-implementation, where users develop ERP features following company needs. ERP users' perceptions related to ERP capability, ERP value, ERP timing, and ERP acceptance. ERP capability with indicators measuring the sustainability of the current ERP system, data can be accessed at any time, accurate data, ease of using ERP systems, and ease to do customization. The ERP provides an added value which is measured by the indicator of ease of moving the old system to the ERP system, the benefits of ERP are more significant than the costs incurred, understanding the value of the ERP system, able to provide coordination in company activities. The ERP system can support my career in the company. ERP implementation can give support to supplier capability, and measurements made in ERP implementation are data integration, data configuration following company needs, ERP systems can adapt to user training according to ERP implementation needs [21, 22]. This study uses indicators in measuring ERP compatibility is software according to the needs of corporate organizations, hardware according to the needs of ERP users, easy to customize ERP and ERP capability.

1.2. The Internal Integration

Company has integrated data as a whole with a single database, but previously the company used data integration only departmentally. Previous data integration was still limited to the integration of two or three departmental functions, but after the company used ERP, data integration between functions could be done throughout the company. Internal integration aims at synchronizing between processes and carrying out procedures for each internal function [4]. Supply chain integration is carried out between the company's internal and external functions. Internal integration can be measured by an integrated inventory management system, an integrated logistics system, sharing data between functions, forming cross-functional teams. Besides, internal integration also aims to improve processes, periodically meet functions, connections in real-time, and inventory level information. Internal integration is related to operation integration within the company, which is sharing information between departments, joint activities, interface activities, working together in achieving company goals [5]. In addition, internal integration also includes the determination of production plans and collaborative decision making between functions in the company. Internal integration can be achieved well if the position is willing and able to commit and cooperate in integrating business functions in the company's internal [5]. This research stipulates that internal integration is an integration between business functions within a company, starting from the process of procuring raw materials of the company to the operation of shipping the company's finished products. The indicators used in measuring research variables are the existence of data sharing between company functions, data integration in real-time, intensive coordination between tasks in the company, and joint decision making within the company

1.3. Inventory System

Inventory manufacturing companies can be classified in the form of corporate raw materials, the company supporting materials, office stationery materials, company spare part materials, machete work in process, materials in the form of components, and company finish products. Material acceptance from suppliers and delivery of products to customers in the company is often determined by intermediaries in the company in the form of inventory and work in process, so it needs to be controlled appropriately by directly integrating with the production and transportation system [11]. Demand with product variance and product volume increases have an impact on increasing the level of company inventory rising company inventory costs. This condition will weaken the company's competitiveness because it has a high level of stock, resulting in inefficient companies. In contrast, companies are required to be efficient to be more flexible and able to be adaptive [13]. Inventory level reduction can be made by using computer technology in providing real-time inventory levels, repeated order approaches by shortening the lead time value, and the reorder point value. Hence lowering or eliminate the company's safety stock.

ERP system can provide facilities related to implementation in real-time and can be set according to company needs [11]. Internal integration in the company can provide fast information to the warehouse in the procurement of materials, and relevant departments can access the availability of the material in real-time according to the needs of other departments. The company's ability to utilize ERP will be able to provide efficiency and reduce inventory levels in the company [12]. The company's inventory has an essential role in the company and has implications for the company's shareholders. The ability to use information technology can increase the efficiency of a company's inventory and have an impact on the company's stock price and restore the stock market increase [10]. The indicators used to measure the inventory system are the determination of lead time of material orders, determination of quantity reorder points, determination of safety stock, and material input-output procedures that are easy to implement.

1.4. Manufacturing Performance

Companies always try to produce products that can compete. The company always strives to improve its operational performance to be able to make these products. Manufacturing performance is the result achieved by the company in a certain period. Company performance results obtained from secondary data in the form of data recorded in company reports that are objective related to the confidentiality of the company compared to data conducted subjectively related to the perception of the company represented by user. Business performance shows the results of manufacturing performance as a whole when associated with the object of the company that produces the product. Performance measurement of manufacturing industry management that is non-financial performance is on-time delivery, work productivity, and customer satisfaction. The indicators used in the measurement of manufacturing operational performance in manufacturing industry companies in Turkey are on-time delivery, accuracy in forecasting, lead time production, service after sales, and the value of inventory levels [2].

Research conducted by Truong et al. [1] states that operational performance measured by the management cost, lead time, order time, damaged material, and late delivery. Tarigan Research [14] states that operational performance in the small and medium business industry in Surabaya is measured by meeting demand management, delivery speed according to the promise of product delivery flexibility, and flexibility in changing order quantities. Indicator,s used to measure Jordanian manufacturing firm performance are market share performance and financial performance, while indicators for measurement supply chain performance are flexibility, integration, customer responsiveness and supplier performance [3]. The indicators used to measure manufacturing performance in this research are on-time delivery, the accuracy of predicting demand, lead time production, and flexibility product delivery.

2. Methodology

This study aims to examine the effect of ERP compatibility on manufacturing performance with internal integration and inventory systems as intervening variables in manufacturing companies located at MM 2100. This research was conducted by examining related literature, making research measurement tools through questionnaires, and distributing questionnaires to manufacturing companies in the region. Based on the results of the literature obtained research model Figure 1. The study population is a manufacturing company engaged in producing products in the MM 2100 area located in the industrial estate town area Cikarang Bekasi area totaling 178 manufacturing industries based on the data list. This study took 78 manufacturing companies that have implemented ERP and have made ERP system adjustments. The sampling technique used judgmental sampling which has the criteria is to have upgraded ERP software or hardware as an adjustment to the company's system, the company produces finished products in the form of pure goods or core goods, respondents are in the middle management up position. The distribution of questionnaires was collected by field officers and submitted to researchers during the 3-month data collection process.

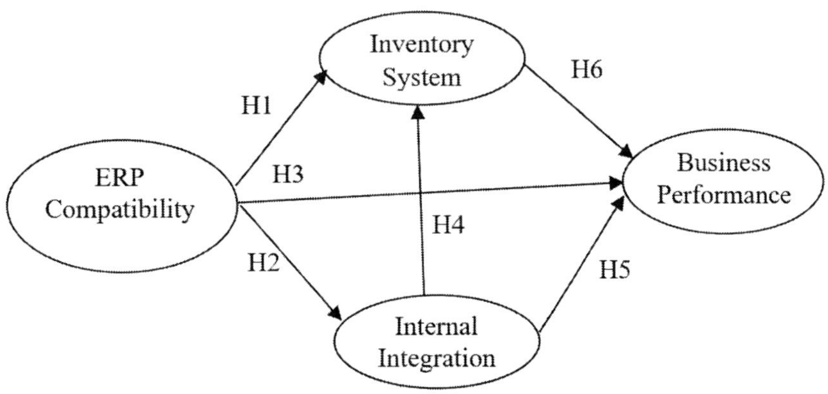

Figure 1. ERP Compatibility to Business Performance Manufacture

Based on Figure 1, it can be determined six research hypotheses are as follows:

(H1): ERP compatibility affects the inventory system
(H2): ERP compatibility influences internal integration
(H3): ERP compatibility influences business performance
(H4): Internal integration influences inventory system
(H5): Internal integration influences business performance.
(H6): Inventory system influences business performance.

The process of data retrieval with the questionnaire technique asks questions to the respondents of manufacturing industry practitioners. Data collection used a closed questionnaire namely questions made in such a way that respondents were limited in giving answers to only a few alternatives or just one answer [15]. Analisa PLS is to predict the effect of variable and describes the theoretical relationship between these two variables; this study uses PLS software. Assessment of the measuring instrument used to test the validity of research with composite discriminant validity and reliability.

3. Research Analysis and Discussion

The distribution of questionnaires conducted to companies in MM-2100 was 88 questionnaires, and researchers submitted all questionnaires, then conducted the first category by paying attention to the face validity of the requirements that had been set to fill the questionnaire. Of the 88 questionnaires, four questionnaires were filled out by company staff; three questionnaires were not completely filled in. Researchers can process further data by 78 questionnaires using PLS. Assessment of the research instruments in terms of validity and the reliability was conducted. A validity test is performed by assessing if the factor loading of each indicator greater than 0.5 as the acceptable minimum value. The reliability test is conducted by assessing if the reliability value greater than 0.7 as the acceptable minimum value (Table 1), shows the result of the assessment in terms of the values of factor loading and reliability. Factor loading assessed the validity of those indicators.

Table 1. Result of Validity and Reliability Test

Variables / Indicators Research	Value of Validity / Reliability	Decision
ERP compatibility	0.863	Reliable
Software according to the needs of the organization	0.782	Valid
Hardware according to the needs of ERP users,	0.768	Valid
Easy to customize ERP	0.854	Valid
ERP capability	0.721	Valid
Internal Integration	0.835	Reliable
Data sharing between company functions	0.800	Valid
Data integration in real-time	0.756	Valid
Intensive coordination between functions	0.727	Valid
Joint decisions in the company's internal	0.513	Valid
Inventory system	0.797	Reliable
Determination of lead time material orders	0.790	Valid
Determination of quantity reorder points	0.717	Valid
Determination of safety stock	0.852	Valid
Material procedures are easy to implement	0.620	Valid
Manufacturing performance	0.894	Reliable
Timeliness of delivery	0.773	Valid
accuracy of predicting demand	0.880	Valid
Lead time production	0.794	Valid
The flexibility of shipping product	0.845	Valid

Table 1 that the lowest value of factor loading from all the indicators is 0.513 is respect to one of the indicators of the inventory system. This lowest value is greater than 0.500 as the acceptable minimum value of factor loading. This result means that those indicators of the four variables are considered valid. The next assessment is regarding the reliability of the block indicators of each variable. As it can be seen in Table 1, the lowest value of reliability is 0.797 in respect to the inventory system. The result means that those indicators of each variable are considered reliable. Based on this finding, further analysis can be pursued. The research hypothesis was obtained by conducting the PLS test in Table 3.

Table 2. Sixth Hypothesis of the Research Test

The hypothesis of Research	Original sample estimate	Mean of subsamples	Standard deviation	t-statistic
ERP Compatibility -> Internal Integration	0.404	0.429	0.138	2.931
ERP Compatibility -> Inventory System	0.248	0.284	0.131	1.988
ERP Compatibility -> Manufacturing performance	0.034	0.049	0.495	0.294
Internal Integration -> Inventory System	0.346	0.370	0.137	2.023
Inventory System -> Manufacturing performance	0.766	0.775	0.066	11.647
Internal Integration -> Manufacturing performance	0.378	0.395	0.123	2.069

Note: Significant if t-statistic ≥ 1.96

Based on the first hypothesis test that ERP compatibility has an effect on the internal integration of 0.404, the standard deviation value is 0.138, and the t-statistic is 2.931. This shows that the first hypothesis is accepted, and it can be concluded that ERP compatibility in manufacturing companies in MM-2100 has an impact on internal integration. Companies that have implemented ERP well in a manufacturing company can provide data integration between right departments, report integration that can be used by related departments in decision making and hardware that is used according to the needs of the company represented by the user. ERP technology can build transparency in the company so that all department data can be accessed. The data can be assessed by the employees who have access rights and can be accessed multinationally and can also integrate data in multinational companies. The second hypothesis is that ERP compatibility influences the inventory system by 0.248, the standard deviation value is 0.131, and the t-statistic is 1.988. This condition illustrates that the second hypothesis can be accepted, that ERP compatibility affects the inventory system in MM-2100 manufacturing companies. Companies that have implemented ERP and can manage the level of company inventory, due to the company being able to use existing software and hardware and ease of customization so that it can be implemented in the company's warehouse. ERP that has been implemented can automatically report data in the warehouse regarding the amount of the company's inventory level, the lead time of the company's material procurement, and the safety stock of each product in the warehouse. This situation has an impact on the relevant parts of material control. This research is in line with a study conducted by Mishra et al. [10], which states the use of integrated information technology (ERP) can reduce the level of company inventory.

The third hypothesis that ERP compatibility influences manufacturing performance are 0.034, the standard deviation value is 0.115, and the t-statistic value is 0.294. It can be said that ERP compatibility has no direct effect on manufacturing performance. Software and hardware that is updated in manufacturing companies and ease of customization are not able to give effect to the increased flexibility in shipping products to customers. The company's ability to control the level of inventory in the company will have an impact on shipping flexibility. ERP practice does not have a significant impact on the operational performance of manufacturing companies in Turkey [2]. The fourth hypothesis is that internal integration has an impact on the inventory system of 0.346, the value of standard deviation 0.137, and the t-statistic value of 2.023. It can be said that internal integration has a positive impact on the inventory system, because with the integration of data between the relevant departments able to provide better inventory level control. Related departments are able to adjust work rhythms and by paying attention to data sharing and integration within the company. This condition can control the level of safety stock, and control lead time better for manufacturing companies. Companies in MM-2100 can manage inventory levels better when there is integration between departments so that inventory costs decrease. Internal integration as measured by an integrated inventory management system, integrated logistics system capable of real-time connections between functions in the company, and real-time in obtaining inventory levels. The Fifth Hypothesis obtained internal integration in the company can provide better manufacturing performance. This condition is seen in the t-statistic value of 2,069, which gives the conclusion that the fifth hypothesis is accepted. Internal integration has an effect of 0.378 on manufacturing performance significantly. The company's ability to control the internal system will be able to determine predictions and direction and the company's strategy going forward. The increase in the inventory system given to manufacturing performance was 0.766. The company's ability to control safety stock and inventory data sharing to other functions in the company can deliver on-time delivery and accuracy in predicting demand. Manufacturing companies always pay attention to the level of company inventory so that inventory costs can be controlled. The manufacturing industry still tries to determine the maximum-minimum inventory so that there is no excess inventory, and there is no shortage of inventory, so there is no stock out.

4. Conclusions

ERP adjustments for manufacturing companies need to be done on an ongoing basis to produce a potential performance of manufacturing companies in a sustainable manner. Based on the results of data analysis, the first conclusion can be that ERP compatibility can provide an increase in internal integration that increases so that there is cross-functional between departments, especially in data sharing companies. ERP compatibility can give control of the company's inventory level because it is easy to automate the determination of the company's lead time and the determination of safety stock. However, ERP compatibility does not directly affect the performance of manufacturing companies. Internal integration built by the company allows related departments to be able to access company inventory data making it easier to control company inventory. This condition causes internal integration to improve inventory system control. Internal integration in the company has an impact on the performance of the manufacturing company because it can provide data quickly so that the MM-2100 manufacturing company can improve on-time delivery and accuracy in predicting demand.

5. References

[1] Truong, H. Q., Sameiro, M., Fernandes, A. C., Sampaio, P., Duong, B. A. T., Duong, H. H., and Vilhenac, E. (2017). Supply chain management practices and firms' operational performance. *International Journal of Quality & Reliability Management*, 34 (2), 176-193.

[2] Acar, M. F., Zaim, S., Isik, M., and Calisir, F. (2017). Relationships among ERP, supply chain orientation and operational performance: An analysis of structural equation modelling. *Benchmarking: An International Journal*, 24 (5), 1291-1308.

[3] AL-Shboul, M. A. R., Barber, K. D., Garza-Reyes, J. A., Kumar, V., and Abdi, M. R. (2017). The effect of supply chain management practices on supply chain and manufacturing firms' performance. *Journal of Manufacturing Technology Management*, 28 (5), 577-609.

[4] Zhao, X., Huo, B., Selen, W. and Yeung, J. H. Y. (2011). The impact of internal integration and relationship commitment on external integration. *Journal of Operations Management*, 29, (1-2), 17-32. https://doi.org/10.1016/j.jom.2010.04.004

[5] Tarigan, Z. J. H., Siagian, H., and Jie, F. (2020). The role of top management commitment to enhancing the competitive advantage through ERP integration and purchasing strategy, *International Journal of Enterprise Information Systems*, 16 (1), DOI: 10.4018/IJEIS.202001010

[6] Tarigan, Z. J. H., Basana, S.R. and Suprapto, W. (2018). Enterprise Resources Planning Project Manager Competency on Improving Organizational Performance through Process Design and Quality Performance. *Proceedings of the 2nd International Conference on E-Education, E-Business and E-Technology*, 153-157, https://doi.org/10.1145/3241748.3241777

[7] Awa, H. O., and Ojiabo, O. U., (2016). A model of adoption determinants of ERP within TOE framework. *Information Technology & People*, 29 (4), 901-930.

[8] Jagoda, K., and Samaranayake, P. (2017). An integrated framework for ERP system implementation. *International Journal of Accounting & Information Management*, 25 (1), 91-109, https://doi.org/10.1108/IJAIM-04-2016-0038

[9] Tarigan, Z. J. H., Lianto, and Basana, S. R., (2019). The impact of organizational commitment on upgrading ERP for maintaining the quality of information and the ERP performance, *IOP Conf. Series: Materials Science and Engineering*, 473, 012051.
[10] Mishra, S., Modi, S. B., and Animesh, A. (2013). The relationship between information technology capability, inventory efficiency, and shareholder wealth: A firm-level empirical analysis. *Journal of Operations Management*, 31 (6), 298-312.
[11] Wang, D. Y., Grunder, O., & Moudni, A. E. (2015). Integrated scheduling of production and distribution operations: A review. *International Journal of Industrial & Systems Engineering*, 19 (1), 94–122, https://doi.org/10.1504/IJISE.2015.065949
[12] Badewi, A., Shehab, E., Zeng, J., and Mohamad, M. (2018). ERP benefits capability framework: orchestration theory perspective. *Business Process Management Journal*, 24 (1), 266-294.
[13] Torkul, O., Yilmaz, R., Selvi, I. H., and Cesur, M. R., (2016). A real-time inventory model to manage variance of demand for decreasing inventory holding cost. *Computers & Industrial Engineering,* 102, 435–439, http://dx.doi.org/10.1016/j.cie.2016.04.020
[14] Tarigan, Z. J. H. (2018). The impact of organizational commitment to the process and product innovation in improving operational performance. *International Journal of Business and Society*, 19 (2), 335-346
[15] Sekaran, U., and Bougie, R. (2019). *Research Methods for Business, (6th Ed.)* Jakarta: Salemba Empat

Information system planning process design based on clause 8 iso 20000-1:2018 using sysml language

R Aurachman[1,*], Y Sudianto[2] and N S Utomo[1]

[1] Telkom University, Bandung, Indonesia
[2] Institut Teknologi Telkom Surabaya, Indonesia

*rio_aurachman@yahoo.com

Abstract. information systems have become inseparable from current products and services. Improving information systems will add value to the quality of products and services. Some references can be used to produce a good information system. One reference that can be used is ISO 20000-1: 2018. Because of the new standards, there are not many studies that guide the implementation. This paper aims to contribute in providing references for application in aspects of information system planning based on clause 8 ISO 20000-1: 2018. The modelling process uses SysML modelling language. We find the guidance to implementation of this standard and the example of Model Based System Engineering using Block Definition Diagram.

1. Introduction

The organization has a strategy to achieve the organization's vision, mission and goals. The process of preparing the strategy through the process of analyzing the external and internal conditions of the organization [1] [2]. Strategies can be divided into macro level strategies to micro level strategies. The BCG matrix, for example, is a tool for choosing which business portfolio is superior [3]. Even at the technical level there is a strategy to maintain the services of the organization. one form of its strategy is to manage services and certify those services so that quality is maintained.

ISO 20000-1:2018 is one form of certification to information systems services. The certification is carried out to provide information to consumers that the organization is committed to maintaining service quality. The certification also proves that the external organization, which is the certification body, is also involved in maintaining the quality through the control and audit process.

Some organizations are at risk of experiencing difficulties in implementing ISO 20000-1: 2018 standards. This paper aims to provide guidance to organizations in implementing the implementation of ISO 20000-1: 2018. The guidance provided is mainly in the aspects of planning and implementing information system services. The guide that was built refers mainly to clause 8. This is because in the concept of high level structure, besides clause 8 it does not focus on the main theme of the standard. Other than clause 8, it does not specifically regulate the theme of ISO 20000-1: 2018, which is information system services.

Some previous research is a guide and reference. Previous research is not only related to ISO 20000-1: 2018 but also for the 2011 series. There is research that also examines the understanding and interpretation of ISO 20000-1:2018 [4]. Other research discusses the proposed development of the 2011 series [5]. Other research attempts to apply in several fields, selling system [6] and project management [7]. There is also research that examines the latest series of ISO 20000-1:2018. The amount of research is not much. Some existing research is trying to synchronize ISO 20000-1: 2018 with other management practices; as an inspiration for risk management [8], as a reference for digital transformation [9], and as a reference for handling information systems security [10]. There is also research that offers procedures for implementing ISO 20000-1:2018 [11].

2. Resource and Method

Systems thinking approach is an alternative form of analytical thinking that breaks down problems into the smallest components to be solved [12]. Some of our previous research was also enriched by system thinking, especially in planning mathematical models [13] [14] [15], distribution planning [16]. Even on a larger scale system can utilize the system approach to solve problems and understand the system [17] [18].

In this research, a system of thinking and the use of a system approach are also used in modeling the information system service system. SysML is a modeling language used to model this information system. Each standard sentence is analyzed and analyzed in order to get a picture of the process map and system design.

3. Result and Discussion

In the PDCA cycle, the first process is Plan and Do. That means planning the process and implementing the process. In managing information system services, there are also Plan and Do. phases. ISO 20000-1: 2018 clause 8 governs the operation of the service. Clause 8 can be a guidance or guide in the planning and implementation of information system services. The following will explain the form of implementation of that clause 8 for planning and implementation. The System map, using Block definition Diagram are shown in Figure 1.

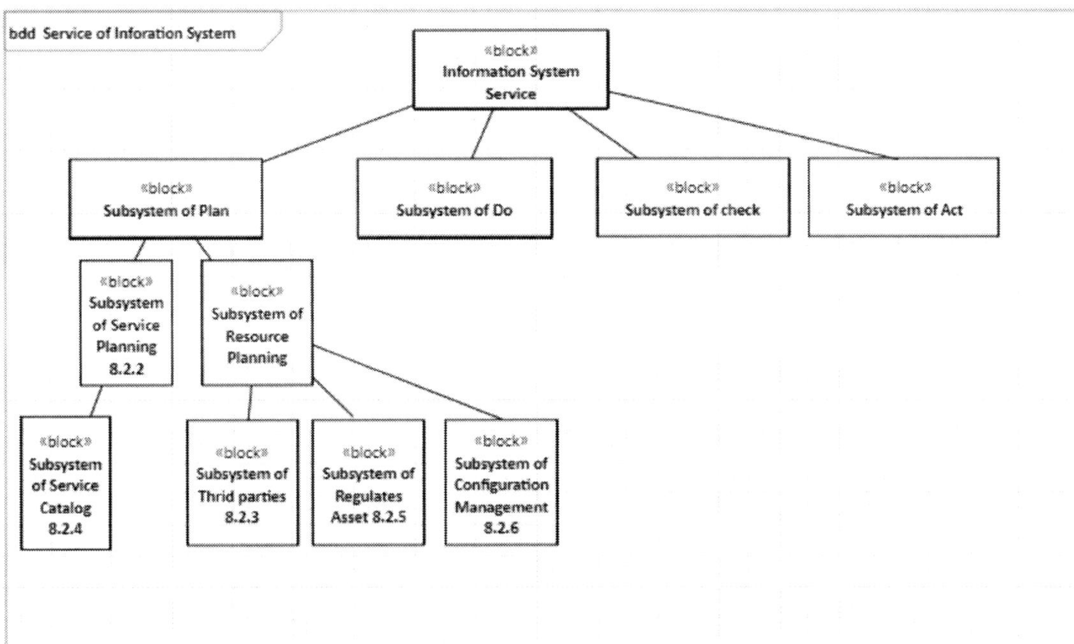

Figure 1. Block Definition Diagram of Currency Prediction

Clause 8.2.1 instructs that the organization must coordinate and carry out processes so that services can be realized. Clause 8.2.2 is about service planning. This clause shows that organizations need to explore what the requirements of the service will be. In addition, in the plan, it is necessary to see how the priority of the service is and the relationship between one service and other services. Changes must also be managed properly. Change needs to be prioritized when it relates to the interests of the organization and business.

Clause 8.2.3 Regulates the control of third parties involved. Clause 8.2.3.1 explains some rules about controlling the responsibility of the process. Criteria need to be applied in selecting suppliers. Some services provided from other parties must be documented what services are provided, service components, or processes provided to third parties. Even though the process is carried out by a third party, it still needs to be integrated and maintained harmony between processes that are controlled internally and externally. Clause 8.2.3.2 Explains the organization's necessity to determine the procedures for controlling external parties. There needs to be a clear measure in evaluating performance.

Clause 8.2.4 regulates the management of service catalogs. The service catalog contains some information contained in standard documents. The catalog also needs to be organized so that it can be accessed by anyone. Service catalogs can be posted on the website so they can be accessed by customers and prospective customers. The service catalog needs to be agreed on in the 2011 version. While the 2018 version doesn't need to be agreed with the customer. Whereas SLAs need to be agreed with the customer. Even for each customer can have different SLA requirements.

Clause 8.2.5 regulates asset management. Every asset that supports service delivery, must be managed properly so that it meets the requirements and guidelines that exist. For example the license used must meet existing regulations. Does not violate intellectual property regulations. Such a list of assets can be made and their compliance with applicable regulations.

Clause 8.2.6 deals with configuration management. Configuration Item (CI) must be defined and each service classified as CI. CI is an element that needs to be controlled to ensure that services can be provided. CI can be interpreted as an asset to realize services. Documents and equipment can be considered as CI's. SLA can also be considered as CI. If every entity already has a CI then there is no need to be given a new CI identification. For example, human resources already have an employee ID number, no need to be numbered.

In the context of configuration management, links between other CIs need to be identified. This distinguishes between Configuration management and asset management. The benefit of this configuration is that it illustrates how the impact of changes on other CIs. As an example, a renewal of the UPS will be carried out, then identification will be approximately connected to any CI. The status of CI also needs to be explained so that it can be monitored whether it is still active or not. Need to verify configuration information.

The CI must contain several information attributes. As with other information, CIs must be properly tracked and stored. CI must be updated and monitored for compliance. This CI also needs to be accessible by every related service.

Clause 8.3 contains relations management and agreement management. Clause 8.3.1 explains what can be delegated to the supplier. Suppliers can carry out services, or parts of services, and can carry out processes that support services.

Clause 8.3.2 regulates business relationship management. This clause further regulates the process after planning and implementation, namely the process of controlling and repairing.

Clause 8.3.3 regulates service level management. The organization and customer need to agree on the service level. Service requirements need to be clearly documented. And this is the basis of SLA. SLAs also need to include tagets, workloads, and exceptions. For example, internet services can work with the assumption that there are so many users. Also for example displayed exceptions for example at certain times there will be a reduction in service quality.

There needs to be a process of monitoring, reviewing, and reporting on service level agreements. The process needs to include comparison of SLAs and their realization. There should also be a comparison between workloads that actually exist and those written on the SLA. If it turns out that the two indicators cannot be met, then an effort is needed to identify the improvement plan and analyze the cause.

Clause 8.3.4 Regulates supplier management. Clause 8.3.4.1 regulates the management of external suppliers. While clause 8.3.4.2 regulates the management of internal suppliers.

In clause 8.3.4.1, it is written that, Like the customer, the supplier must be monitored by at least one responsible individual. For each external supplier, there must be a contract document governing the collaboration between the organization and the external party. The contract needs to contain some information as stated in the standard documents from points a) to d). It is necessary to study the relationship between agreements with external parties and compliance with SLAs and risks.

Relationships with external parties also require well-defined and clear media. As with services to customers, the performance of external parties also needs to be monitored and followed up if not achieved. Contracts that have been made need to be evaluated and updated as needed. If there is disputes between the organization and external parties, there needs to be handling and resolution.

Clause 8.3.4.2 regulates internal suppliers and customers who act as suppliers. As with external parties, internal suppliers need to establish agreements with organizations. Also if there is a failure to reach an agreement, it must be followed up and resolved.

Clause 8.4 Regulates the management of supply and demand. This clause governs budgeting and accounting for services, demand management, capacity management. Clause 8.4.1 regulates the budgeting and accounting of services. The budgeting process is indeed closely related to planning. But the accounting process is closely related to the control and follow-up process. So this is not explained in this paper. Clause 8.4.2 deals with demand management. More related to the control process and follow-up. Based on the history of capacity usage, it can be followed up with capacity planning in the coming period.

Clause 8.4.3 regulates capacity management. The capacity of each resource needs to be measured and its relation calculated in supporting services and service level services. The organization needs to provide sufficient capacity and conduct evaluations and follow up on its realization.

Clause 8.5 Includes the process of making services, their development, and transitions or changes. This clause covers change management, service design and transition, and release and deployment. Some processes include planning and implementation. Some other processes include control and follow-up. Clause 8.5.1 contains information about change management. Then this clause is not explained in detail in this paper.

Clause 8.5.2 regulates service design and transition. It includes service plans, service designs, and regulates builds and transitions. Clause 8.5.2.1 Plan new or changed services, is to set plans before live services. This plan includes several things contained in standard documents ranging from signs a) to signs h). This clause also regulates services that undergo deletion and transfer. Ci also needs to be considered

Clause 8.5.2.2 Design, Set about how to design our services after live. The design must include points a) to g). Clause 8.5.2.3 regulates the build and transition process. Services that are built need to be made and go through the stages of testing, verification, and meeting the requirements. The service also needs to be in accordance with documented design. If it turns out there are requirements that have not been met, there needs to be a follow-up agreement from the relevant parties. Release and deployment need to be used when transitioning in a live environment. This transition process needs to be continually communicated to related parties. Clause 8.5.3 explains about release and deployment management. Each release needs to be classified into several classifications.

Clause 8.6 Explains resolution and fulfillment. Clause 8.6 covers incident management, service request management, and problem management. clause 8.7 describes service assurance, which is how to follow up and maintain the service process. Both of these clauses also cover the process of control and follow-up so that they are not explained in this paper.

4. Conclusions

After reviewing the ISO 20000-1: 2018 clause 8 standard. A best practice of information system planning has been obtained. From these best practices, organizations can implement standards in their respective organizations. Block Definition diagrams have also been obtained. The Block Definition Diagram is needed if the implementation process will be carried out with a System Engineering approach, especially the Model Based System Engineering.

Some further research can be done. Research can be done on the procedure for implementing other clauses standard. In addition, it can be analyzed several other models of SysML in the context of implementing information systems.

5. Reference

[1] E. Chumaidiyah, W. Tripiawan and R. Aurachman, "Exploring the internal and external constraint of it business start up in Bandung, Indonesia," *International Journal of Innovative Technology and Exploring Engineering,* 2019.

[2] E. Chumaidiyah, R. Aurachman and B. H. Sagita, "Strategy for Capability Development of Knitting Small Medium Enterprises Using SWOT Analysis," in *Proceedings of the International Conference on Industrial Engineering and Operations Management*, Bandung, 2018.

[3] S. Guntara, B. Yogaswara and R. Aurachman, "Strategi Transformasi Pt Greeneration Indonesia Menuju Perusahaan Ramah Lingkungan Yang Unggul Dengan Pendekatan Matriks Boston Consulting Group," in *eProceedings of Engineering 4, no. 3*, Bandung, 2017.

[4] G. Disterer, "Iso 20000 for IT," *Business & Information Systems Engineering,* vol. 1, p. 463, 2009.

[5] A. Tanović and I. S. Marjanovic, "Development of a new improved model of ISO 20000 standard based on recommendations from ISO 27001 standard," in *2019 42nd International Convention on Information and Communication Technology, Electronics and Microelectronics (MIPRO)*, 2019.

[6] S. Romadini, A. F. Santoso and I. Santosa, "Perancangan Sistem Manajamen Layanan Teknologi Informasi Pada Layanan Reseller Dan Dropship Bandros Menggunakan Iso 20000-1: 2011 Area General Requirements Dan Design And Transition Of New Or Changed Services (studi Kasus: Cv Kabita Informatika)," *eProceedings of Engineering,* vol. 5, 2018.

[7] B. Federowicz, "The project service management system according to ISO 20000-1 in a company providing services to the construction industry," 2019.

[8] B. Barafort, "Integrated risk management process improvement framework in it settings based on ISO standards," 2019.

[9] F. M. Siddiqui, "Digital Transformation of Modern Airports by Exploiting Fog as a Service Model," in *2019 Integrated Communications, Navigation and Surveillance Conference (ICNS)*, 2019.

[10] T. Tagarev and D. Polimirova, "Main Considerations in Elaborating Organizational Information Security Policies," in *Proceedings of the 20th International Conference on Computer Systems and Technologies*, 2019.

[11] D. Haven, "A Guide to ISO/IEC 20000-1: 2018 Service Management," 2018.

[12] H. D. D. McNickle and S. Dye, Management science: decision-making through systems thinking, Palgrave macmillan, 2012.

[13] E. G. B. Kuncoro, R. Aurachman and B. Santosa, "Inventory policy for relining roll spare parts to minimize total cost of inventory with periodic review (R, s, Q) and periodic review (R, S)(Case study: PT. Z)," in *IOP Conference Series: Materials Science and Engineering*, Jakarta, 2018.

[14] R. Aurachman and A. Y. Ridwan, "Perancangan Model Optimasi Alokasi Jumlah Server untuk Meminimalkan Total Antrean pada Sistem Antrean Dua Arah pada Gerbang Tol," *JRSI (Jurnal Rekayasa Sistem dan Industri),* vol. 3, no. 2, pp. 25-30, 2016.

[15] A. Desiana, A. Y. Ridwan and R. Aurachman, "Penyelesaian Vehicle Routing Problem (vrp) Untuk Minimasi Total Biaya Transportasi Pada Pt Xyz Dengan Metode Algoritma Genetika," in *eProceedings of Engineering*, Bandung, 2016.

[16] B. M. Muttaqin, S. Martini and R. Aurachman, "Perancangan Dan Penjadwalan Aktivitas Distribusi Household Product Menggunakan Metode Distribusi Requirement Planning (DRP) Di PT. XYZ Untuk Menyelaraskan Pengiriman Produk Ke Ritel," *JRSI (Jurnal Rekayasa Sistem dan Industri),* pp. 56-61, 2017.

[17] R. Aurachman, "Model Matematika Dampak Industri 4.0 terhadap Ketenagakerjaan Menggunakan Pendekatan Sistem," *Jurnal Optimasi Sistem Industri,* pp. 14-24, 2019.

[18] R. Aurachman, "Perancangan Influence Diagram Perhitungan Dampak Dari Revolusi Industri 4.0 Terhadap Pengangguran Kerja," *Jurnal Teknologi dan Manajemen Industri,* vol. 4, no. 2, pp. 7-12, 2018.

IOP Conf. Series: Materials Science and Engineering 1010 (2021) 012010 doi:10.1088/1757-899X/1010/1/012010

Vehicle routing problem with simulated annealing using python programming

R. Aurachman[1], D B Baskara[2], J Habibie[1]

[1]Telkom University, Bandung, West Java, Indonesia
[2]Institut Teknologi Telkom Surabaya. East Java, Indonesia.

Abstract. Vehicle Routing Problems (VRP) are a type of linear programming problem. This paper tries to explain the completion of VRP using Python Programming with the Simulated Annealing algorithm. Python Programming is used as a tool by utilizing the wealth of packages in python. Python is a popular programming language. Python is easy to learn for beginners. However, there is still no example and guide for solving VRP using simulated annealing in python programming with a process paradigm. This design can be a reference and guide for beginners in dealing with similar problems or the development of existing problems.

rio_aurachman@yahoo.com

1. Introduction

The transportation system is one of the important systems in human life. The transportation system connects one location to another. Between locations there are entities that move using the Transportation system. Transportation is a waste. As much as possible the movement is minimized. However, transportation is often very difficult to avoid. So humans try to find ways so that transportation activities can be minimized.

One form of transportation is the delivery of goods in a network. One vehicle must take many items to many places in turns. The simple model of moving goods is the Traveling Salesman Problem (TSP) Several studies have examined the mathematical modelling of the TSP. Other research studies the method of its solution using certain algorithms [1] [2]

The form of development of the Traveling Salesman Problem is the VRP. In VRP, there are more than one vehicle involved in the transportation process. Each vehicle has a capacity. VRP has developed rapidly and one of the studies has mapped various variations of the VRP model. Some of our research has also discussed Transportation problem, especially VRP [3] [4] [5] [6].

On the other hand, programming is growing and advancing rapidly. Academics have begun to explore several alternative programming languages that can be used. Among students in Indonesia and also many researcher around the world, many solve transportation problems using MATLAB [7] [8] [9] [10] [11] . But not enough literature has tried to solve the problem of the use of open source programming software such as Python, Julia, and others. Therefore, this research tries to contribute the programming design using Python Programming to solve VRP Heterogenous Fleet. It is hoped that the results of this research can be a reference for academics and students to solve the transportation problems they face.

IOP Conf. Series: Materials Science and Engineering 1010 (2021) 012010 doi:10.1088/1757-899X/1010/1/012010

2. Method

The research process goes through several stages. The framework used is mathematical modelling. The process of mathematical modelling begins with understanding the problem. After that, an interpretation of the variables representing the problem elements is carried out. After that the relationship between variables is examined. The relationship between these variables is modelled in the form of mathematical equations. After the mathematical model is formed, the next step is to validate and verify the mathematical model.

In designing source code and algorithms, mathematical equations are converted into functions in programming. The paradigm we use in this study is a functional approach. These functions are interrelated with one another to form systems and sub-systems.

The last step is to test the mathematical model and test the software that has been designed. Then evaluate the efficiency and computational time.

3. Result and Discussion

Here is the designed programming code

```
import functools,operator,os,pandas,math,random, numpy, copy, matplotlib.pyplot as plt
xls =pandas.ExcelFile('data node VRP.xls')
sheet1 = pandas.read_excel(xls,'Sheet1')
sheet2 = pandas.read_excel(xls, 'Sheet2')
nodes=sheet1.as_matrix()
depot=nodes[len(nodes)-1]
cities=nodes[:(len(nodes)-1)]
vehicle=sheet2.as_matrix()
tour=[0,1,2,3,4,5,6,7,8,9,10,11,12,13,14,15,16,17,18,19,20]
n=len(cities)

def distance(x1,y1,x2,y2):
  return math.sqrt((x1-x2)**2+(y1-y2)**2)

def totaldistancetour(tour):
  d=0
  for i in range(1,len(tour)):
    x1=cities[tour[i-1]][0]
    y1=cities[tour[i-1]][1]
    x2=cities[tour[i]][0]
    y2=cities[tour[i]][1]
    d=d+distance(x1,y1,x2,y2)
  x1=cities[tour[len(tour)-1]][0]
  y1=cities[tour[len(tour)-1]][1]
  x2=depot[0]
  y2=depot[1]
  d=d+distance(x1,y1,x2,y2)
  x1=cities[tour[0]][0]
  y1=cities[tour[0]][1]
  x2=depot[0]
  y2=depot[1]
  d=d+distance(x1,y1,x2,y2)
  return d
```

Programming begins with import packages or libraries that will help the programming process become more efficient and faster. After that, in the second row import the spreadsheet file which is used as reference data. The data used is dummy data. Line 3 and line 4 are the specific import process sheets containing data to be processed. The fifth to seventh lines are instructions for defining the point that is the depot and the point that is the destination. This programming imports the coordinate data and not the distance matrix.

The distance function is to measure the distance between two Cartesian coordinates. Whereas the totaldistancetour function is a function whose task is to measure the total distance at which the tour becomes an input. Example tour is located in line 9. Tour is an array that contains the destination city numbers ranging from 1 to 20.

```
def subtour(slice,tour):def subtourslice(tour,vehicle):
  capacityused=numpy.zeros(len(vehicle))
  k=0
  slice=[]
  mass=[]
  for i in range(len(vehicle)):
    while capacityused[i] <= vehicle[i][1] and k<=(len(tour)-1):
      capacityused[i]=capacityused[i]+cities[tour[k]][2]
      if capacityused[i] > vehicle[i][1]:
        capacityused[i]=capacityused[i]-cities[tour[k]][2]
        k=k-1
        slice.append(k)
        k=k+1
        break
      k=k+1
  sub=[]
  sub.append(tour[:(slice[0]+1)])
  for i in range(0,len(slice)-1):
    sub.append(tour[(slice[i]+1):(slice[i+1]+1)])
  return sub
```

In code above there are two functions, namely subtourslice and subtour. The function of the subtourslice is to cut a long tour into a small tour. Each small tour will be assigned to a different vehicle. Then each deduction is based on the capacity of each vehicle. If for example between tours 1,2,3,4,5,6 have a total weight of goods that must be transported beyond the capacity of vehicle 1, then this algorithm will cut off so that vehicle one only goes to subtour 1,2,3. Whereas the function of subotour is to combine all the subtours that have been cut by the subtourslice function.

In code below, it starts with the allvehicledistance function. This function is used to measure the total distance of all vehicles that have each subtour. While the tourtodistance function is a function whose task is to summarize the function of the previous line, so that when inputting the initial tour, and the vehicle capacity, this function will automatically generate the total distance travelled by all vehicles.

Lines 80 to 90 contain the iteration process that operates functions from the previous line. Iteration is performed using the Simulated Annealing Algorithm [12].

Figure 1 shows the final appearance of the programming that has been designed. The system can generate routes from many vehicles. Each circle shows one vehicle route. The system tries to find solutions that produce optimum performance, which is the minimum mileage. Each point shows the coordinates of the Destination city. while the point in the centre of the picture is the depot where the vehicle will start departing and will return after delivering to several points.

```
def allvehicledistance(sub):
   alldistance=functools.reduce(operator.add,(totaldistancetour(x) for x in sub),0)
   return alldistance

def tourtodistance(tour,vehicle):
   u=subtourslice(tour,vehicle)
   v=subtour(u[0],tour)
   total=allvehicledistance(v)
   return total

for temp in numpy.logspace(0,5,num=100000)[::-1]:
   existingDistances=tourtodistance(tour,vehicle)
   [i,j] = sorted(random.sample(range(n),2))
   updateTour =tour[:i] + tour[j:j+1] +  tour[i+1:j] + tour[i:i+1] + tour[j+1:]
   updateDistances=tourtodistance(updateTour,vehicle)
   if math.exp((existingDistances - updateDistances)/temp)>random.random():
     tour =copy.copy(udpateTour)
     betterDistances=copy.copy(updateDistances)
betterslice=subtourslice(tour,vehicle)
bettersubtour=subtour(betterslice[0],tour)

combinesubtour =numpy.concatenate([x+[21] for x in bestsubtour])
combinesubtour = numpy.append([21],combinesubtour)
plt.plot([nodes[combinesubtour[i]][0]                    for                    i                    in
range(len(combinesubtour))],[nodes[combinesubtour[i]][1] for i in range(len(combinesubtour))], 'xb-')
for i in range(len(combinesubtour)):
   plt.annotate(combinesubtour[i],xy=(nodes[combinesubtour[i]][0],nodes[combinesubtour[i]][1]))
plt.show()
```

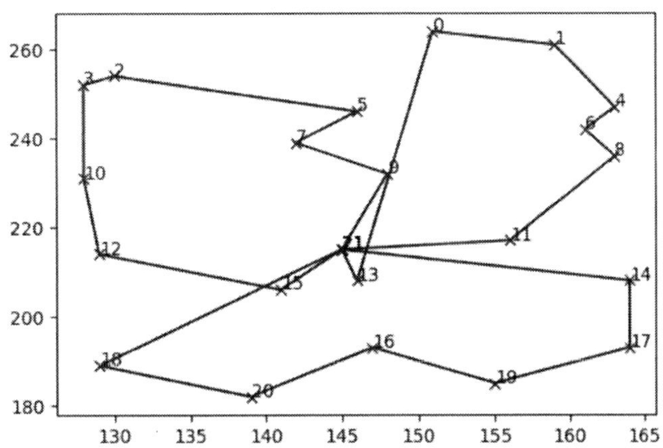

Figure 1. Result of Optimum Solution

4. Conclusion

Based on the research, Python programming code has been obtained to solve VRP using the Simulated Annealing algorithm. The function uses 6 functions. Programming is also assisted by 9 libraries; functools, operators, os, pandas, math, random, numpy, copy, matplotlib. It is hoped that this programming will become a reference for academics to solve VRP problems using open source software.

Python is a preferrable programming language for Meta heuristic Optimization, both for learning and for professional practice.

The next research can continue the programming design using the Genetic Algorithm algorithm or other algorithms. You can also use alternative open source programming code like Julia.

References

[1] S. Kaewman and R. Akararungruangkul, "Heuristics Algorithms for a Heterogeneous Fleets VRP with Excessive Demand for the Vehicle at the Pickup Points, and the Longest Traveling Time Constraint: A Case Study in Prasitsuksa Songkloe, Ubonratchathani Thailand," *Logistics,* vol. 2, p. 15, 2018.

[2] L. Zheng, Z. He and W. Liang, "VRP Problem Solving Based on Adaptive Dynamic Search Ant Colony Algorithm".

[3] R. Aurachman, "Optimization model of the number of one-way servers to minimize cost and constraint," in *Understanding Digital Industry: Proceedings of the Conference on Managing Digital Industry, Technology and Entrepreneurship (CoMDITE 2019), July 10-11, 2019, Bandung, Indonesia*, 2020.

[4] R. Aurachman and A. Y. Ridwan, "Perancangan Model Optimasi Alokasi Jumlah Server untuk Meminimalkan Total Antrean pada Sistem Antrean Dua Arah pada Gerbang Tol," *JRSI (Jurnal Rekayasa Sistem dan Industri),* vol. 3, no. 2, pp. 25-30, 2016.

[5] A. Desiana, A. Y. Ridwan and R. Aurachman, "Penyelesaian Vehicle Routing Problem (vrp) Untuk Minimasi Total Biaya Transportasi Pada Pt Xyz Dengan Metode Algoritma Genetika," in *eProceedings of Engineering*, Bandung, 2016.

[6] B. M. Muttaqin, S. Martini and R. Aurachman, "Perancangan Dan Penjadwalan Aktivitas Distribusi Household Product Menggunakan Metode Distribusi Requirement Planning (DRP) Di PT. XYZ Untuk Menyelaraskan Pengiriman Produk Ke Ritel," *JRSI (Jurnal Rekayasa Sistem dan Industri),* pp. 56-61, 2017.

[7] E. Harahap, A. Harahap, A. Suryadi, D. Darmawan and R. Ceha, "LINTAS: Sistem simulasi lalu lintas menggunakan SimEvents MATLAB," 2018.

[8] M. K. A. Al Mufti, W. Wahyudi and B. Setiyono, "Simulasi Lalu Lintas Simpang Empat Menggunakan Cellular Automata Dengan Sistem Kontrol Berbasis Artificial Neural Network pada Matlab," *TRANSIENT,* vol. 7, pp. 508-514, 2018.

[9] G. O. N. G. Jin-li and Z.-m. CHENG, "The Realization of Genetic Algorithm of VRP Based on the Matlab [J]," *Logistics Sci-Tech,* vol. 7, 2006.

[10] N. Rathore, P. K. Jain and M. Parida, "A MATLAB-Based Application to Solve Vehicle Routing Problem Using GA," in *Advances in Simulation, Product Design and Development*, Springer, 2020, pp. 285-298.

[11] Z. Li, "Improved ant colony optimization for emergency rescue VRP based on matlab," in *2010 International Conference on Computer Application and System Modeling (ICCASM 2010)*, 2010.

[12] F. Mar'i,, W. F. Mahmudy and P. B. Santoso, "An Improved Simulated Annealing For The Capacitated Vehicle Routing Problem (Cvrp)," *Jurnal Ilmiah Kursor,* vol. 9, no. 3, pp. 119-128, 2018.

IOP Conf. Series: Materials Science and Engineering 1010 (2021) 012011 doi:10.1088/1757-899X/1010/1/012011

Design and integration of portable health sensors

M U H A Rasyid, He Y Martono, B N D Ariyadi and I N Nasution

Politeknik Elektronika Negeri Surabaya (PENS), Surabaya, Indonesia

udinharun@pens.ac.id

Abstract. Body Area Network is a measurement concept using sensors with a pair directly attached to the body to monitor the condition directly. The benefits in the world of health are enormous plus remote monitoring capabilities. Currently the use of medical devices as a means of detecting heart conditions, blood pressure, and so forth is still considered less efficient, because in every health tool required several sensors and different platforms. Thus, the idea for integrating these sensors is to improve service quality and tool efficiency and time, to meet the Body Area Network concept, this research proposes a new approach to direct health monitoring. The purpose of this research is to integrate some medical devices such as Temperature, Electrocardiogram (ECG), Electromyogram (EMG), Galvanic Skin Response (GSR) and SPO2 Pulseoxymeter into one platform portable health sensor equipped with two delivery media for data storage that is by using Bluetooth and network Wi-Fi so that in a health check only use one tool. In the application of the results of the use of GSR, ECG and EMG sensors in the form of graphs that can be seen fluctuations based on measurements in the body, and on the use of temperature sensors and SPO2 Pulseoxymeter results from visualization on the sensor in accordance with the measuring value range in normal body.

1. Introduction

Health is an important aspect of human rights where everyone has the right to an adequate standard of living for health. Public awareness to carry out health checks is very influential in preventing early health problems. By carrying out regular checks, the community can also get information about health conditions at that time. The problem is that not all devices for detecting health are integrated into one platform. The majority of medical devices are separate and only owned by the hospital. Not all people have health devices in their homes that can be connected to hospitals.

With the development of Internet of Things (IoT) and Wireless Sensor Network (WSN) technologies that form a sensor network, each health device unit with a unique ID can collect health data. The medical device unit can be owned by everyone and can be taken anywhere. A person can see the development of his health condition and send the health condition status to the doctor at the hospital from his home[1].

The objectives of this research is to make portable Electronic Health (E-Health) device to monitor real time patient health condition with Integration of five sensors (Temperature, Electrocardiogram (ECG), Electromyogram (EMG), Galvanic Skin Response (GSR) and SPO2 Pulseoxymeter) in one platform supported by wireless networks used as data transmission medium for later medical analysis without requiring the patient to face to face with the doctor and facilitate the patient to conduct health checks anytime and anywhere independently[2].

Several studies have been conducted by several researchers relating to monitoring health conditions by utilizing sensors and wireless network. Sandeep et al. [3] developed an Internet of medical things (IoMT) framework for organizing, monitoring and managing health applications. The IoMT Framework consists of 3 layers, namely the data collection layer of the health device, the data transmission protocol layer, and the remote health monitoring layer with an efficient health device. To overcome the problem of limited energy availability, they proposed an energy efficient ON-OFF algorithm (EEOOA) method for sending health data from the source node to the destination node in the medical system. The experimental results show that the energy efficiency using the EEOOA method is better than the traditional method on IoMT. Soumya et al. [4] proposed an IoT system in the health field that provides information on the patient's current condition. The system will provide information on the location of the nearest hospital and information on the availability of health services in the hospital.

Pace et al. [5] proposed an IoT-based health monitoring service called INTER-Health. INTER-Health provides information about body weight, blood pressure, and physical activity carried out by the patient. The data is sent via sensors and analysed on cloud servers. Alpana et al. [6] propose the IoT framework model for health monitoring for hospitals. IoT in the field of health monitoring can be applied in several models: smart devices, digital health knowledge resources, remote health monitoring, remote weighting, electronic health data, elderly care, personal health applications, health information systems and hospitals. Udin et al. [7] developed a health device to detect fever symptoms in children to avoid the condition of febrile seizure. A temperature sensor is installed in the child's clothes which will sense the child's body temperature and send the sensor data to the cloud and his parents' mobile devices.

In this portable health research have differences with previous researches: There are 5 sensors used, namely SPO2 Pulse Oximeter, Galvanic Skin Response, Electrocardiogram, Electromyogram, and Temperature. The 5 sensors are integrated in one tool that can be used interchangeably. There is a direct display of the tools that have been made, so that the results of the inspection can be seen directly. There are several options available to see the results of the examination, namely through mobile phones by utilizing Bluetooth communication and through the web, where the examination data is stored in the cloud first, then the data will be processed so that the patient can see the results every examination on the web.

2. System Architecture

2.1. Design System of Portable Health platform

This portable health implementation is a combination or integration of several health measuring instruments as shown in figure 1 the design system of portable health platform.

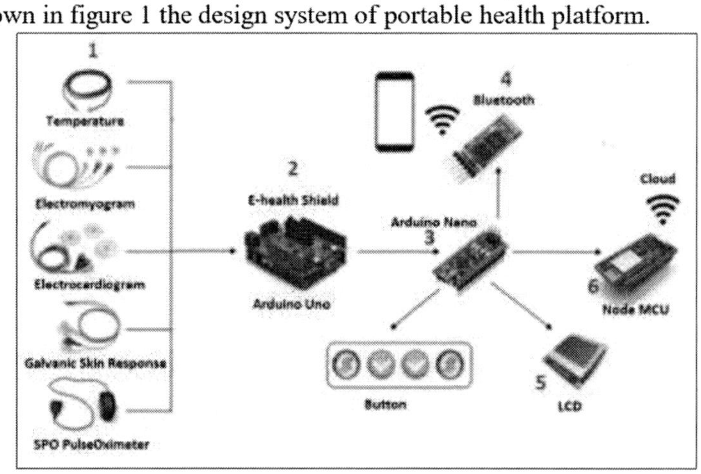

Figure 1. Design system of portable health platform.

The functions in the section above are:

1. On the left side there are 5 sensors that will be used for measuring or knowing the condition of the patient, the 5 sensors are Body Position, Electrocardiogarm, Electromyogram, Galvanic Skin Response, SPO2 Pulseoximeter, Temperature and Blood Pressure.
2. Arduino Uno and E-health Platform, act as data readers and process measurement data that is carried out with 5 sensors which then the measurement data will be forwarded to Arduino Nano.
3. Arduino Nano is the main brain that functions how this tool runs, on Arduino Nano is connected with a button that can be used to choose which sensor to read, besides Arduino Nano can manage what is displayed and the data delivery media using Bluetooth or forwarded to NodeMCU to be sent to the cloud provided.
4. Bluetooth HC - 05 functions for communication lines between smart u - health with mobile phones wirelessly.
5. There is an LCD Nokia 5110 which functions as an interface with patients to see first-hand the data of the examination.
6. NodeMCU ESP8266 is a Wi-Fi module that functions as a sender of data to the cloud via the internet network.

Figure 2 shows the E-Health platform, Arduino Uno, NodeMCU, Arduino Nano, and Bluetooth have been combined into one. The device is used to read all sensors connected to Arduino Uno and E-Health platform.

(a) (b)

Figure 2. (a) Sensing device, (b) Processing device.

Arduino Uno and the E-health platform are used to read the results of the measurement to the 5 sensors used. There is a connector that is used to receive commands obtained from the device, in addition there is a data path used for sending data from the measurement. Then, if the E-Health platform for measuring EMG and ECG has to move jumpers located at the top of the module, the switch has provided a switch that can be used without having to move the jumper to retrieve data.
On this device there are several modules installed at once such as Arduino Nano, Nodemcu, step-down and Bluetooth. Through this layout all devices can be connected, both from buttons, LCD, Arduino Uno and also MCU Nodes. This tool is working at a voltage of 5V so that a step-down module is needed to reduce the voltage so that it corresponds to the working voltage on this device.

There are 4 buttons provided to select the menu on this tool. Then provided 3 pairs of switches that can be used to select the sensor read EMG or ECG, also provided a switch to turn off and turn on Bluetooth, so that when unused Bluetooth can be turned off alone to save power of the battery.

Figure 3 is a box packaging design of smart E-Health prototype that will be made, in the design has been provided a place for LCD, button, power and place to connect the sensor that will be used for measurement. The prototype of smart E-Health has been assembled and connected all in this tool. The patient can perform a health check using the 5 sensors. There is a LCD with a resolution of 84x48 pixels provided to display the measurement data, 4 buttons used for menus, and also 6 switches.

IOP Conf. Series: Materials Science and Engineering 1010 (2021) 012011 doi:10.1088/1757-899X/1010/1/012011

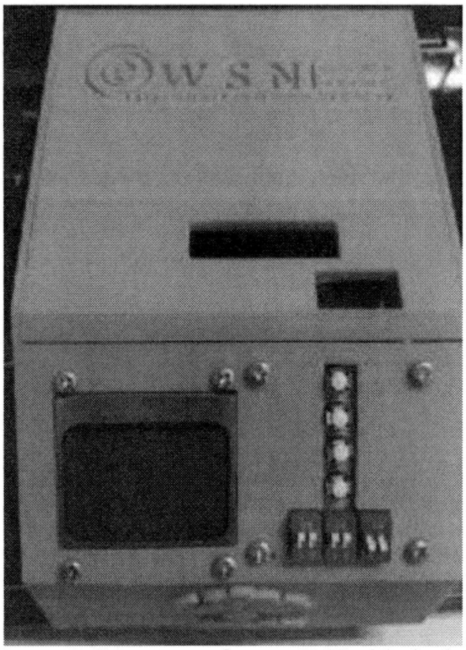

Figure 3. Prototype of smart E-Health device.

Figure 4 is a source code that will be uploaded to Arduino Uno for reading of GSR, ECG, EMG, Temperature and SPO2 Pulseoxymeter sensors. Sensor readings on e-health occur on the line where there is source eHealth. Get Skin Resistance (), eHealth. getBPM(), eHealth. Get Oxygen Saturation(), and the other that serves to read sensor readings that are then stored into variables that have been provided. The program is embedded in Arduino Uno. The similar code is also done for Arduino Nano and NodeMCU.

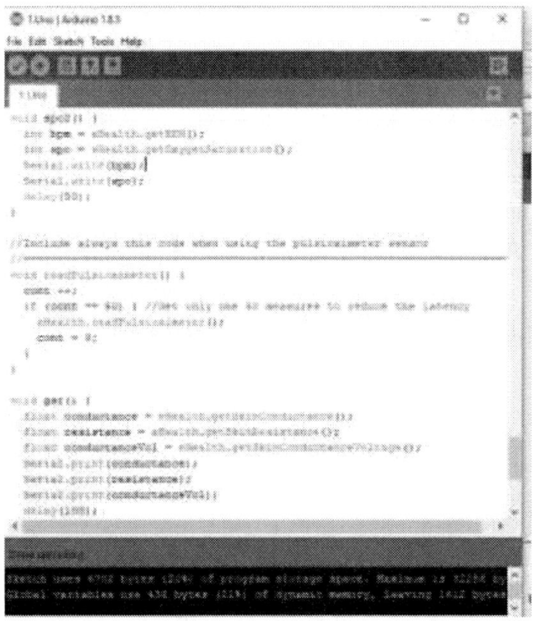

Figure 4. Source code program software for reading sensors.

Figure 5 describes the sensor data transmission flow in Arduino Uno where Arduino Uno will read sensor data according to the command received from Arduino Nano which then the data is stored into

IOP Conf. Series: Materials Science and Engineering 1010 (2021) 012011 doi:10.1088/1757-899X/1010/1/012011

the corresponding variable with the sensor data respectively, then the data is sent to Arduino Uno via the serial line. After the user chooses the sensor button, Arduino Uno will command the E-Health sensor to send the sensor data. Once the data is received, the data will be from the 5 sensors has been read. The data saved into the variable is proceed and displayed on the LCD that has been provided.

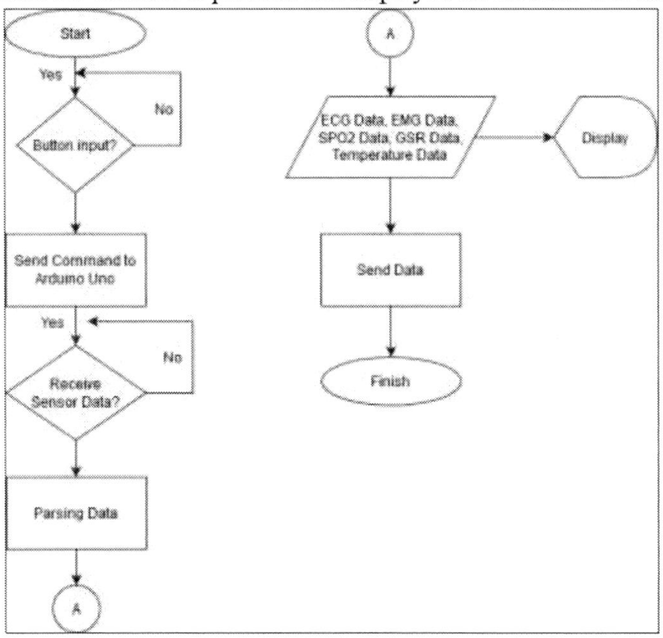

Figure 5. Sensor data transmission flow.

2.2. Design of E-Health Cloud System

The E-Health system was also developed by adding cloud computing as a data processing and storage centre as shown in figure 6.

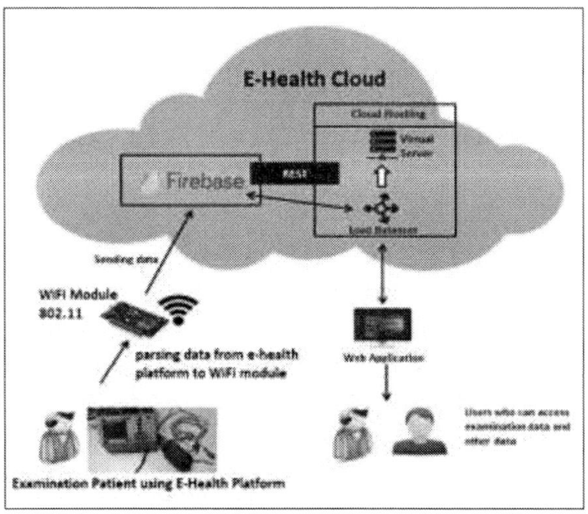

Figure 6. E-Health cloud system.

The system starts from patient biometric data collection. Then the health sensor data is sent to the cloud after the Wi-Fi module gets an internet connection. The cloud section contains Firebase as a data storage centre and cloud hosting subscription as a web server. Data from sensors in real time is stored in Firebase database which has a structure like the JSON data format. Every single data from the sensor that is successfully sent will be saved into one database tree.

The web server will forward data requests by the user to Firebase using REST. The web server used is cloud hosting subscription. It provides its own virtual network infrastructure that is isolated from each other. So if one of the servers goes down then the other web will have no effect. Network management is managed by more professional third parties.

The user interface is available through the E-Health web application. Users can find out the results of their health checks through a web browser connected to the internet by accessing the E-Health Cloud URL. The web page is created using PHP 7 and the Yii framework.

3. Result and Discussion

3.1. Testing Temperature Sensor

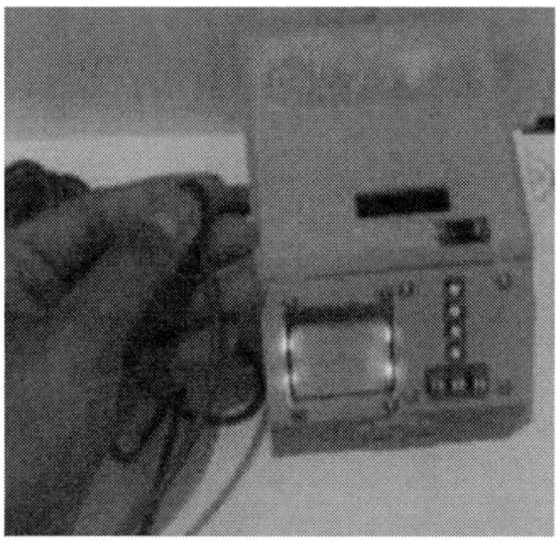

Figure 7. Installation temperature sensor.

The first stage is the installation of a temperature sensor. The temperature sensor is placed on the fingers as shown in figure 7 to begin measuring. There are several menus on the device that have been made, so the user can choose the temperature sensor. A variable temperature with the float data type used to store sensor readings with an eHeatlh. Get Temperature function found in the eHealth.h library. After the data is stored then it is displayed on the serial monitor. In this experiment, data collection was taken every 200 ms. The results of measurements with temperature sensors are displayed on the serial monitor that has been provided by Arduino IDE with the condition of normal body temperature is in the range of 35 C - 37.5 C.

3.2. Testing SPO2 Pulseoxymeter Sensor

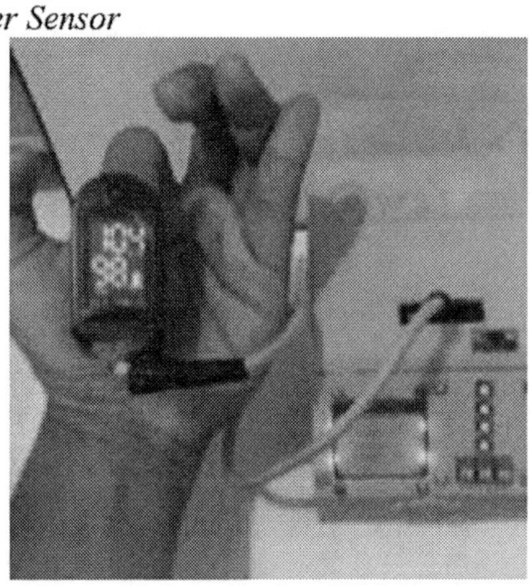

Figure 8. Installation SPO2 sensor.

The SPO2 sensor is attached to the finger as shown in figure 8 to start measuring. In the Pulseoxymeter SPO2 reading, there are two stored values, pulse rate bits per minute (bpm) and oxygen in the blood (SPO2) which is the value of reading the percentage of hemoglobin molecules in the arterial blood. The results above show that the measurement of SPO2 in the blood is 98% where the value is still in the normal range measurements using SPO2, the normal hemoglobin level in blood is between 95% - 100%.

The data retrieval of Pulseoxymeter SPO2 use digital pin 6 to set the interrupt of sensor reading so that when interrupt send interrupt on digital pin 6 function of read Pulsioximeter will be executed which then will update data stored on private variable in program. The eHeatlh.getOxygenSaturation () and eHeatlh.getBPM () functions are started on the program i, to retrieve the heart rate data per minute and hemoglobin level in the blood. In this experiment the data retrieval is taken every 500 ms then displayed to see the conformity of the measurement data on a device with processed data displayed on the monitor serial.

3.3. Testing Galvanic Skin Response (GSR)Sensor

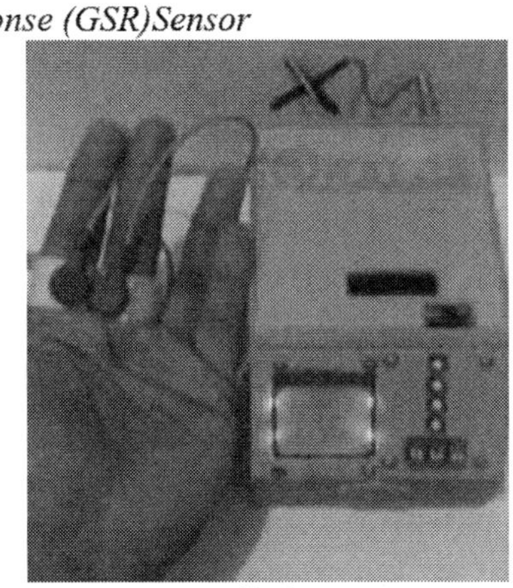

Figure 9. Installation GSR sensor.

The installation of GSR sensor is mounted on both fingers as shown in figure 9. The patient must be in a state of relaxation in this examination, because the sensor is sensitive to sweat. If the patient is emotional or in a condition that is not relaxed will make the sweat glands secrete more sweat so that will affect the results of the measurement. The function in GRS sensor is used to read the conductance, resistance, and voltage of the conductance resulting from the reading done. The measurement results are stored in the variable conductance, resistance, conductance Vol in the float data type, then adds the Serial.print () command to retrieve the results of the measurement. The results of the measurements can be seen in the monitor serial feature provided by the Arduino IDE.

After the trial shows the sensor data to serial monitor successfully then the author tries to develop to take the value on serial monitor and process it into a desktop application with C# language as shown in figure 10.

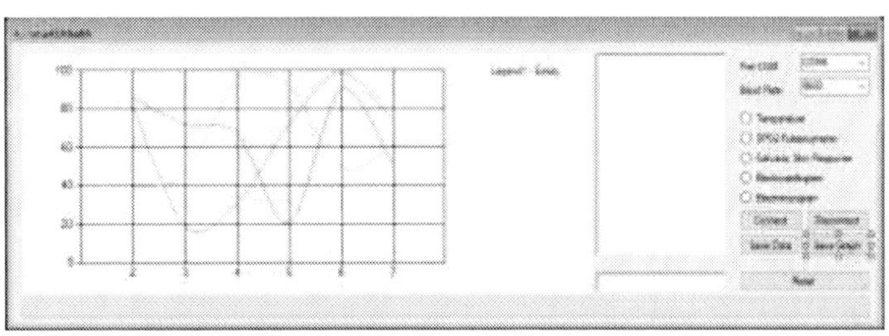

Figure 10. Desktop application monitoring.

This application has several features such as that can store the results data examination of numbers and graphs as follow.

1. Connect button: This button serves to start taking data and display data in the form of graphics and numbers from the E-Health smart prototype.
2. Disconnect button: The button functions to Disconnect from application with tool.
3. Save Data button: This button is used to store data from the measurement results in the form of numbers.
4. Save Graph Button: This button is used to store data from the measurement result of the graph.
5. Combo box baud rate: function to determine the baud rate on the port to be read. This baud rate must be the same as the one set in the Arduino.
6. Combo box Port Com: function to determine which device will be done serial reading.
7. Reset Button: Works to restore applications at initial values where the graphs and existing data are emptied.
8. Radio Button: On the radio button there is a choice of 5 sensors that will be read, namely Temperature, SPO2 Pulseoxuymeter, GSR, Electrocardiogram (ECG), and EMG (Electromyogram).

3.4. Implementation of Web Applications

Sending data to the database can be done by setting Wi-Fi delivery on the device. The MCU node on the E-Health platform can send data to Firebase as long as the device is connected to the internet. To connect to the internet, you need to initiate an SSID and access point password that is used. If the connection status is successful, the IP address will be displayed. After that the connection request to the Firebase account is done. This is done with the help of FirebaseArduino and ArduinoJSON libraries.

Before sending data, the patient's name must be inputted first on the E-Health Cloud web page. After that, the patient begins to do data checking and parsing. Patient name, date of examination, and patient status data will be asked for NodeMCU for the location of the data tree to be sent. The parsed data is made into JSON format and then sent using the Firebase.push() method.

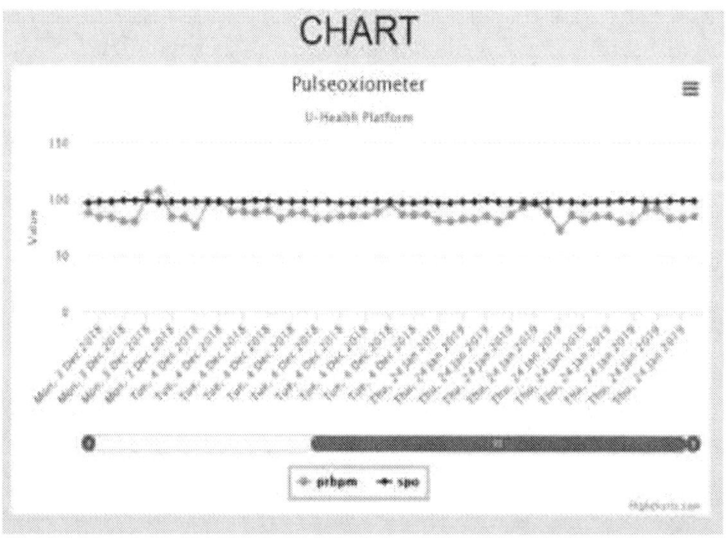

Figure 11. Pulseoximeter sensor graphics in the web.

Figure 11 shows the results of inspection data using a pulseoxiometer sensor. There is a pulse rate bit per minute (prbpm) value that shows the number of pulses, the SPO2 value for oxygen saturation in the blood, and the data category that is normal.

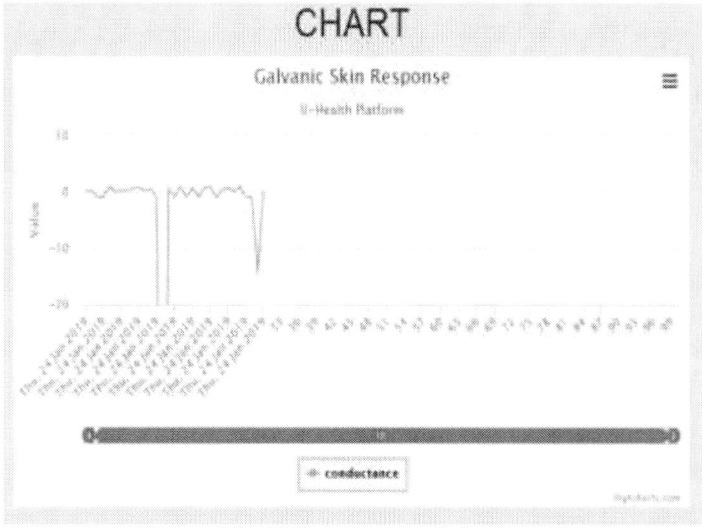

Figure 12. GSR sensor graph in the web.

Figure 12 shows the results of the inspection data using the GSR sensor. There is a conductance value from the user's sweat. Data graphs are taken from measurements in two conditions. On the graph, the first condition is to the left of the red line. The graph line looks smoother which shows the psychological condition of the user tends to be good. Changes occur in the line to the right of the line from the measurement results in the second condition. Visible lines experienced sharp fluctuations. In this condition the psychological user is feeling not good.

4. Conclusion

This research designed and implemented a portable Health to detect Temperature, Electrocardiogram (ECG), Electromyogram (EMG), Galvanic Skin Response (GSR) and SPO2 Pulseoxymeter sensors into one device. The data sensors will send to the cloud by using Bluetooth and network Wi-Fi. The experimental results show that Desktop applications for monitoring are used successfully. The measurement results display using the SPO2 Pulseoximeter sensor from the visualization of the sensor with the data displayed after processing accordingly. Measurement with temperature sensors in normal body conditions according to the temperature range contained in the sensor description. With the implementation of this research, the data of each sensor can be collected and monitored. The results of collected biometric data that can be monitored in the form of fluctuations in graph movements obtained from the measurement results. The web application for monitoring is successfully used. In future research, this device can be added with other several vital sign monitoring sensors.

5. References

[1] H. Habibzadeh, K. Dinesh, O. Rajabi Shishvan, A. Boggio-Dandry, G. Sharma, and T. Soyata, "A Survey of Healthcare Internet of Things (HIoT): A Clinical Perspective," *IEEE Internet of Things Journal*, vol. 7, no. 1. Institute of Electrical and Electronics Engineers Inc., pp. 53–71, Jan. 01, 2020, doi: 10.1109/JIOT.2019.2946359.

[2] M. U. Harun Al Rasyid, F. Setiawan, I. U. Nadhori, A. Sudarsonc, and N. Tamami, "Anomalous data detection in WBAN measurements," in *International Electronics Symposium on Knowledge Creation and Intelligent Computing, IES-KCIC 2018 - Proceedings*, Jan. 2019, pp. 303–309, doi: 10.1109/KCIC.2018.8628522.

[3] S. Pirbhulal, W. Wu, S. C. Mukhopadhyay, and G. Li, "A Medical-IoT based Framework for eHealth Care," Nov. 2018, doi: 10.1109/ISSI.2018.8538031.

[4] S. Yattinahalli and R. M. Savithramma, "A Personal Healthcare IoT System model using Raspberry Pi 3," in *Proceedings of the International Conference on Inventive Communication*

and Computational Technologies, ICICCT 2018, Sep. 2018, pp. 569–573, doi: 10.1109/ICICCT.2018.8473184.

[5] P. Pace *et al.*, "INTER-Health: An Interoperable IoT Solution for Active and Assisted Living Healthcare Services," in *IEEE 5th World Forum on Internet of Things, WF-IoT 2019 - Conference Proceedings*, Apr. 2019, pp. 81–86, doi: 10.1109/WF-IoT.2019.8767332.

[6] K. Alpana and S. Shaurya, "An iot equipped hospital model: A new approach for e-governance healthcare framework," *Indian Journals*, vol. 8, no. 3, pp. 36–42, 2019, Accessed: Oct. 27, 2020. [Online]. Available: https://www.indianjournals.com/ijor.aspx?target=ijor:ijmrhs&volume=8&issue=3&article=006 .

[7] M. U. Harun Al Rasyid, I. K. Wibowo, and M. A. Windhy Saputra, "Mobile Monitoring of Toddler's Body Temperature for Early Effort of Febrile Seizure Prevention," in *Proceedings - 2019 International Seminar on Application for Technology of Information and Communication: Industry 4.0: Retrospect, Prospect, and Challenges, iSemantic 2019*, Sep. 2019, pp. 313–318, doi: 10.1109/ISEMANTIC.2019.8884294.

Acknowledgments
This research was supported in part by Ministry of Research, Technology and Higher Education of Indonesia, under scheme 'Penelitian Terapan Unggulan Perguruan Tinggi (PTUPT)' 2020.

Carte server implementation for improving data quality management application performance in profiling module

K F Salmawati[1], T F Kusumasari[1], E N Alam[1]

[1] Information System Department, School of Industrial and System Engineering, Telkom University, Bandung, Indonesia

karinafarizki20@gmail.com

Abstract. Data is a critical component in the management of systems that support business in an organization. The application of proper data quality management can help organizations in making policies and maintaining data by reducing inconsistent data. The initial stage in the data quality management process begins with the profiling process. The use of job trigger executors in DQM applications affects the application performance level. This paper aims to make changes to the executor used by using a carte server to improve the performance of the DQM application that has been made. Carte server is a web server that allows running files remotely. The results of this study will compare application performance from the use of different executors namely job trigger (pan.bat) and carte server to find out which is superior. With that knowledge can be obtained about the importance of performance in the application.

1. Introduction

The application of technology in business is one of the benefits that benefit the organization. Data is an important component of technology. The use of data is not only used for operations but is used at a strategic level [1]. Within a few years, there was a rapid increase in data [2]. Diverse data sources with different database designs can't be ascertained that the data is in good quality, such as the amount of data lost and different standards [3]. To produce useful and reliable information, good quality data is needed [3]. The data management process includes a series of concepts, roles, and responsibilities [4]. Data quality management consists of planning, implementing, and controlling supported by methodologies, tools to measure, assess, improve, and ensure data quality [5]. Data quality management has several phases, namely data profiling, data cleaning, data quality assessment, and data quality monitoring [6]. Data profiling is a process to detect inconsistent, wrong, lost, and duplicate data in a data set that will be repaired [7]. Several application tools can be used to help process data profiling [8]. Some of the tools available are paid tools. Therefore it can switch to open source tools [8]. Often open-source tools differ from paid tools such as functionality and application performance.

Application performance is one of the problems that often appear in application implementation [9]. Often users perceive the application used has a problem so that the impact of the application will not be reused [10]. Based on physiological measurements, poor application performance can affect human reactions to decision making [11]. For that application performance is one of the factors that need to be considered in application implementation. Appraisal performance application uses a predetermined time limit so that it can be used as a reference in the assessment. In the current era, applications become one of the most widely used tools to help in business. For this reason, application performance has an essential role in business continuity. If the level of application efficiency increases, it results in higher user satisfaction and retention [11]. Studies show a delay of response for 1 second can have an impact on user conversions by 7% [12]. This makes application performance a challenge in developing sustainable applications.

In previous studies, a data quality management application has been designed to assist data processing in a government organization. In the application, there are features of cleansing, profiling, and monitoring. The profiling feature in the DQM application still has a low level of performance. This problem is of particular concern so that further application development is needed. After analyzing existing applications, it is known that the cause of the low level of DQM application performance on the profiling feature is the use of job triggers as executors. By looking at this condition, it is necessary to optimize application performance on the profiling feature so that it becomes more optimal and can reduce the resulting impact. The development of application performance requires several steps including the replacement of the job trigger executor by using a carte server. This study was made for carte server implementation to improve the performance of DQM applications in case studies of government agencies.

Carte server is implemented to improve performance in data quality management applications, especially in the profiling feature. The writing of the paper consists of 4 parts. Section 2 explains the related theories. Section 3 introduces the method used. Part 4 is the implementation process and part 5 is the conclusion.

1.1. Theory and Related Work

1.1.1. Data Quality Management. Data quality has an essential role in the decision making and planning process [13]. There are several data quality criteria, including correctness, consistency, completeness, accuracy, no redundancy, etc. [6]. Data quality management (DQM) is a series of planning, implementation, and control activities that are supported by data quality methodologies and applying appropriate quality management techniques (DQT) and tools (DQt) to measure, assess, improve and ensure data quality [4]. In general, there are four phases in data quality management, namely data profiling, data cleansing, data quality assessment, and data quality monitoring [6]. One phase that has an essential role in managing data quality is data profiling [14]. Data profiling is a process to detect inconsistent, wrong, lost, and duplicate data in a data set that will be repaired [7]. Information data obtained from the results of profiling can be used to conduct analysis and references in cleansing. The process of cleansing in data cleansing takes the form of erasing data permanently and repairing data. Data deletion is done only if needed. In data quality management after going through the process of data profiling and data cleansing, monitoring of the data will be carried out. Monitoring is carried out to ensure that data is always in good quality. The results of the profiling process will be monitored to remain by organizational rules [15].

1.1.2 Data Quality Tools. To support the process in data quality management several applications can be used, paid apps and open source applications. Here are some tools that have the main features in data quality processing, such as Pentaho Kettle, Talend Open Studio, and Data Cleaner. Pentaho kettle is a data integration tool that includes the profiling process using ETL (extract, transform, load) techniques [16]. This tool supports various input and output formats. Formats that can be used are text files, datasheets, and database engines [16]. Talend Open Studio is an open-source tool used for the profiling process with the ability to monitor repository metadata [16]. This tool can facilitate access to databases, applications, and accept input with different formats [16]. Data Cleaner is another open-source tool that can be used for data profiling [16]. This tool uses drag and drops components in use.

Table 1. Core function data profiling [24].

No.	Data Quality Tools	Pattern Discovery	Table Analysis	Domain Analysis
1	Pentaho Kettle	X	X	X
2	Talend Open Studio	X	✓	X
3	DataCleaner	✓	X	✓

1.1.2. Application Performance. Application performance is a criterion for evaluating the system [17]. In physiological measurement, examine the human body's reaction to performance in the decision making process [18]. In physiological measurements are useful in providing optimal levels of application performance. Assessment based on a survey, evaluation process based on emotional reaction, frustration level, or satisfaction level based on the performance function of the application [18]. The results of the survey evaluation in the form of frustration and satisfaction levels which are emotional reactions can affect user perceptions about the credibility and quality of the system. In general waiting time standards that can be used as a basis for measuring performance are 1 - 30 seconds [18].

Table 2. Response time performance category [18].

Kategori	**Target Response Time**	**Maximum Response Time**
Basic Operation	2 *s*	2 *s*
Complex or Ambiguos Search or Save Operations	5 *s*	5 *s*
Integration or Major Calculation	5 *s*	15 *s*
Heavyweight Operation	10 *s*	30 *s*

Application performance testing can be done using the performance testing method. Performance testing is a type of testing to determine the level of response, throughput, and reliability or scalability of the application [19]. Performance testing is not intended to find bugs in the app but to eliminate bottlenecks [20]. The purpose of performance testing is to know the stability and maximum load limit of the application. Performing performance testing can use mathematical principles such as standard deviations. If the value produced by the calculation of the standard deviation has smaller, then the application is more consistent. Following is the standard deviation formula used [21] :

$$(\sigma) = \sqrt{\frac{1}{n-1}\sum_{i=1}^{n}(x_i - \bar{x})^2} \quad (1)$$

Explanation :

σ = standard deviation
n = calculation of response time
x = value of response time per iteration
$\bar{x}$ = average response time

2. Proposed Methodology

In conducting research, researchers analyze and make adjustments to data quality management applications, especially on the profiling feature. The research method is divided into four phases, namely the analysis phase, the research phase, the implementation phase, and the testing phase.

Figure 1. Implementation methodology.

The first phase carried out was the analysis phase. In this phase, analyzing the existing DQM applications. An analysis is done on the performance of the profiling function. In this phase, the review of existing applications will be used as a reference in determining the proposed solution. From this phase, it can also be seen that the use of job trigger executors is a performance problem from the DQM application. The second phase is the research phase, in this phase the researcher collects the literature relating to research to be used as a reference for implementation. In the third phase, in this phase, the carte server will be implemented in the DQM application to improve the performance of the profiling feature. While the last phase is testing, this phase measures the stability and performance of the application based on response time.

3. Result of Experiment

The conditions that occur in government organizations today are the application of data quality management with cleansing, profiling, and monitoring features that can facilitate the processing of data quality. However, existing applications, especially in profiling features still have a performance with a low level of performance. The low level of performance in apps used by organizations can harm the running business processes. By looking at the condition of government organizations like that, a review of the performance of the data quality management application on the profiling feature is needed. This research will focus on reviewing the executor used in the data quality management profiling feature that uses the job trigger. In data quality management applications, the process of data profiling that is run in applications is designed using Pentaho Data Integration. Profiling features available in the app are divided into several functions, namely pattern, value distribution, data completeness, show null, and clustering.

3.1. Analysis of Existing Applications

Researchers will use one of the profiling functions, namely the show null profiling function in analyzing job triggers as executors used in carrying out functions in applications. Before conducting a review of the executor used, the researcher measures the time required to carry out the null profiling show process. When making measurements, it is known that the time needed by the function still exceeds the predetermined time standard of 1 - 30 seconds.

Table 3. Application performance before compliance

	Performance Function				
Profiling Function	Iteration 1	Iteration 2	Iteration 3	Iteration 4	Iteration 5
Show Null	51 *s*	43 *s*	49 *s*	46 *s*	44 *s*

The show null profiling function is used to find cells in the selected column that have a null value. In this function, the resulting output is to return all values from a null column. The purpose of the function is to be able to find out the number of cells in the column that have null values so that it can assist in the cleansing process. Profiling with the show null function in a data quality management application is designed in a transformation file in Pentaho.

In this case study, the executor used is pan.bat because the executable file has an extension (.ktr) or transformation file. The use of the executor affects the performance of the null show profiling function. Before running the file, the pan.bat executor will initiate it to begin the execution process.

These conditions cause the file does not immediately execute so it requires quite a long time. The time needed for the executor to initiate before executing the file is 33 seconds. While the time required if directly performing the Pentaho file is 4 seconds. So that when added up the time needed to run the show null function is 37 seconds, which means it exceeds the standard time of 1 - 30 seconds. To that end, changes were made to the use of the pan.bat trigger job executor with the carte server.

3.2. Carte Server Implementation

The use of the executor adjusts to the type of file to be executed. For pan.bat used in files with extensions (.ktr) while kitchen.bat for files with extensions (.kjb). The results of the analysis conducted, the use of the executor affect the performance of the DQM application. Executor replacement is intended to improve the performance of the show null profiling function in data quality management applications. The replacement executor used is a carte server. Carte server is a simple web server that allows running transformations or jobs remotely [20]. The use of a carte server makes it possible to monitor remotely or run files without the need to use an executor in general and open the pentaho application.

To run a carte server, two configurations can be used, using configuration by default and performing the manual setup. The settings that are used for carte servers by default do not need to do a specific configuration. When running the server, you only need to use carte.bat then insert the hostname and port used. Make sure the hostname and port used are unused. Naming the hostname for default settings usually uses "localhost" while the port used starts with 80 "80xx". As for setting up a carte server manually by creating a configuration file with the XML file extension. The XML file contains settings to define the name of the carte server, hostname, and port. The naming and port used in the manual configuration carte serve are what the user wants. But it should be noted, the hostname and port used must be in unused condition.

When running a carte server, simply use the command:

Carte.bat {hostname} {port} → for default settings

Carte.bat {configuration_file_name}.xml → for manual settings

In this case study, the researcher implements a carte server with a default configuration. When executing a transformation file using the carte server, simply use the URL:

http://{hostname}:{port}/kettle/executeTrans/?trans={filelocation}

To find out the status of the process being executed or monitor the running process can be controlled through a web page by visiting http://{hostname}:{port}. On that page, we can know the name of the executable file, id, status, and log. Status indicators that usually appear are in the process, finished, and finished (with error).

Next, adjustments are made to the code used in the DQM application. In the DQM application, the website used is built based on the PHP laravel framework. For this reason, it is necessary to make adjustments in the form of replacing the executors that have been installed on the web to be used in the DQM application. Following is the syntax used in existing applications :

```
$exec = 'D:/pentaho_location/pan.bat /file:"D:/pentaho_location/param';
```

Other changes made to make adjustments are to add variables to the env file, create a file that is used to configure the URL to be executed, add code to the controller file, and add notifications for processes that run like success or failure. The .env file contains the configuration of the laravel project created, all configurations included in the file. The purpose of saving the URL into a .env file is to make it easier when there are changes that must be made. Adding variables to the .env file is as follows:

```
URL_PENTAHO="http://localhost:8095/kettle/executeTrans/";

LOCATION_PENTAHO="D:/file_location/";

CREDENTIAL_PENTAHO="cluster:cluster";
```

The `URL_PENTAHO` variable is used to store the URL that is used to execute the transformation file which saves the pentaho logic of the profiling functions. For the `LOCATION_PENTAHO` variable, it is used to store the address or storage location of the transformation file to be executed according to its function in the profiling feature. When using carte serve when running, you will be asked to enter a username and password, for that username and password you use are stored in one variable with the name `CREDENTIAL_PENTAHO`.

In this study, the researchers used the guzzle as the URL executor. Guzzle is an HTTP PHP client that is used to make it easy to send HTTP POST requests to be integrated with web services. In the following code, 'Authorization' has been inserted to authorize when accessing the URL used to execute the pentaho transformation file.

```
<?php
use Illuminate\Support\Facades\Session;
use Zttp\Zttp;
function requestToPentaho(array $queryParams)
{
    $credentials = base64_encode(env('CREDENTIAL_PENTAHO'));
    $res = Zttp::withHeaders([
      'Authorization' => ['Basic '. $credentials],
    ])->get(env('URL_PENTAHO'), $queryParams);
    if($res->isOK()) {
        Session::flash('success', 'Success');
      }else {
        Session::flash('error', 'Error');
      }
}
```

The file controller is used to control and connect between models and views. The controller functions to take requests, parse, call the model, and then take the response that will be sent to the view. The following is the code that is selected for making adjustments:

```
  ...
  $filename = 'filename.ktr';
        $queryParams = [
          'trans' => env('LOCATION_PENTAHO') . $filename,
          'host' => $host,
          'db_name' => $db_name,
          'db_username' => $db_username,
          'port' => $port,
          'col' => $column,
          'tab' => $table

  ...
```

The addition of notification is one of the adjustments made because in previous studies conducted, a log of the process will be displayed and the process will automatically finish. The status of the success or failure of the process is in the log. Whereas in this study, the log is not displayed because it is in a different source from the carte server.

3.3. Application Testing

After the executor changes that are used to be a carte server, the researcher tests the DQM application by using the performance testing method. The test was carried out using factory asrot tables, column names, and the amount of data in 4483. The results (Table 5) showed significant changes in terms of the time taken in each iteration when running the show null profiling function, the response time given was below the standard performance value 1 - 30 seconds. This value shows a significant increase in performance.

Table 4. Application performance after adjustment.

Profiling Function	Performance Function				
	Iteration 1	Iteration 2	Iteration 3	Iteration 4	Iteration 5
Show Null	7 s	8 s	7 s	4 s	6 s

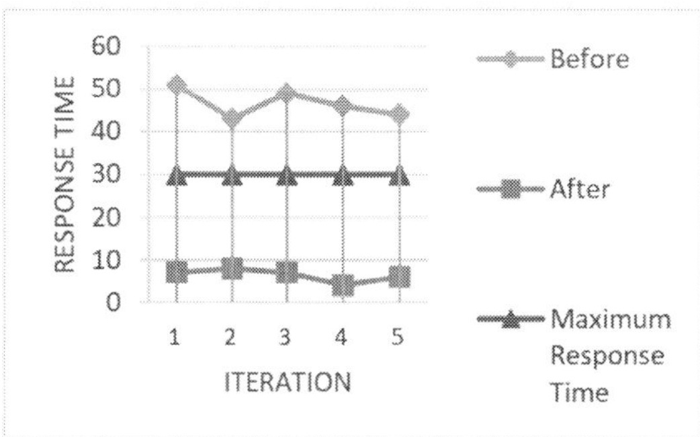

Figure 2. The response time of the show null profiling function.

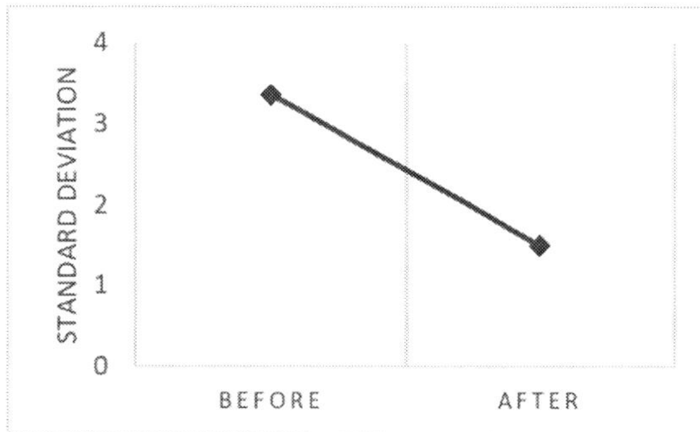

Figure 3. Transformation the standard deviation value of the show profiling function to null.

Table 5. Performance testing.

Profiling Function	Average Response Time		Standard Deviation	
	Before	After	Before	After
Show Null	46.6	6.4	3.36	1.5

The test results show the calculation with standard deviation (Equation 1) of the show null profiling function after the executor changes used has a smaller value than before the change was made. This shows the reliability and consistent level of the function increases.

4. Conclusions

Poor level of performance in applications can cause various consequences for users, especially in organizations such as obstruction of running business processes (decision making processes). For this reason, the accuracy of the selection of the executor in the application affects the performance of the application. The use of job triggers in this case study makes DQM application performance worse. This is indicated by the resulting standard deviation is 3.36. Implementing a carte server in the application can improve the performance of the application. This is evidenced by performing performance testing, resulting in standard deviation is 1.5. This figure shows the difference that is quite far from the standard deviation value when the application uses the executor.

5. References

[1] R. Sabtiana, S. B. Yudhoatmojo, A. N. Hidayanto, 2018, "Data Quality Management Maturity Model : A Case Study in BPS-Statistics of Kaur Regency, Bengkulu Province, 2017", The 6th International Conference on Cyber and IT Service Management (CITSM 2018).

[2] C. Jin, S. Liu, C. Yang, L. Wu, L. Pan, X. Meng, 2016 "Piecewise Linear Representation Method Based on Importance Data Points for Time Series Data", IEEE 20th International Conference on Computer Supported Cooperative Work in Design.

[3] M. H. Tekieh, B. Raahemi, 2016, "Importance of Data Mining in Healthcare: A Survey", IEEE/ ACM International Conference on Advances in Social Networks Analysis and Mining.

[4] L. Jiang, J. Zhao, 2012, "An empirical study on risk data quality management", 2012 International Conference on Information Management, Innovation Management and Industrial Engineering.

[5] The Data Management Asociation, 2009, *The DAMA Guide to The Data Management Body of Knowledge First Edition*, Bradley Beach: Technics Publications, LLC.

[6] D. Apel, Datenqualität erfolgreich steuern: Praxislösungen für Business-Intelligence-Projekte, Heidelberg: dpunkt.verlag, 2015.

[7] A. Munzberg, J. Sauer, A. Hein, N. Rosch, 2018, "The use of ETL and data profiling to integrate data and improve quality in food database", Sixth International Workshop on e-Health Pervasive Wireless Application and Service 2018.

[8] T. F. Kusumasari, Fitria, 2016. "Data Profiling for Data Quality Improvement with Openrefine", 2016 International Conference on Information Technology System and Innovation (ICITSI).

[9] S. Yang, J. Chen, 2007 "A Study of Security and Performance Issue in Designing Web-based Applications", IEEE International Conference on e-Business Engineering.

[10] A. Redouane, 2003, "Expressing Performance Issues in Web Application Design", The Second IEEE International Conference on Cognitive Informatics.

[11] Nishitha, Why is application performance management important?. Retrived April 27, 2020, from WitttySparks : https://wittysparks.com/why-is-application-performance-management-important/

[12] M. Sharma. 2015. The Ever-Increasing Importance of Application Performance. Retrived April 27, 2020, from TechWell Insight : https://www.techwell.com/techwell-insights/2015/08/ever-increasing-importance-application-performance

[13] D. Rao, V. N. Gudivada, V. V. Raghavan, 2015, "Data Quality Issues in Big Data", IEEE International Conference on Big Data.

[14] M. R. Effendy, T. F. Kusumasari, M. A. Hasibuan, 2019, "Star Schema Implementation For Monitoring in Data Quality Management Tool (A Case Study at A Goverment Agency", Fourth International Conference of Informatics and Computing (ICIC).

[15] B. Vani, B. Suriya, R. Deepalakshmi, 2014, "MANAGING PERFORMANCE OF WEB BASED APPLICATION THROUGH AGILE APPROACH AND IMPROVING ROI", India, ICICES 2014.

[16] V. S. V. Pulla, C. Varol, M. Al, "Open Source Data Quality Tools: Revisited", 2011.

[17] T. Z. Tan, R. S. M. Goh, V. March, S. See, 2009. "Data Mining Analysis to Validate Performance Tuning Practices for HPL", Singapore, IEEE International Conference on Cluster Computing and Workshop.

[18] Meier, J.D., Farre, C., Bansode, P., Barber, S. and Rea, D. 2007. Performance Testing Guidance for Web Applications, Microsoft Patterns & Practices (Chapter 15 – Key Mathematic Principles for Performance Testers). Retrieved August 8, 2014, from MSDN: http://msdn.microsoft.com/enus/library/bb924370.aspx

[19] K. Z. J. F. Y. Li, 2010, "Research the performance testing and performance improvement strategy in web application", 2nd International Conference on Education Technology and Computer (ICETC).
[20] Hitachi Vantara, Use Carte Cluster. Retrived April 19, 2020, from Pentaho Documentation: https://help.pentaho.com/Documentation/8.0/Products/Data_Integration/Carte_Clusters
[21] A. Dajan, "Pengantar Metode Statistika Jilid I", LP3S, 2007.

Binary Particles Swarm Optimization for Power Plant Schedule by Considering "Take or Pay" Contract

T Wati[1,*], I Masfufiah[1], T Suheta[1], S Muharom[1], N E Setyawati[1] and S Triwijaya[2]

[1] Institut Teknologi Adhi Tama Surabaya , Indonesia
[2] Politeknik Perkeretapian Indonesia

*trisnaw@itats.ac.id

Abstract. This study discuss 24-hours load scheduling to get the minimum generation cost. Fuel cost in load scheduling need to be considered, because fuel and power suppliers are bound by take or pay schedule. In addition load scheduling considering heat rate, start-up cost, minimum up time and minimum down time. Load scheduling calculate by priority list method that is completed using artificial intelligence, Binary Particle Swarm Optimization (BPSO). The data used in this study is IEEE 30 bus and Java-Bali 500 kV.

1. Introduction

PT PLN (Persero) is the largest electricity producer in Indonesia. PT PLN cooperates with private electricity producers, and is bound by a power purchase contract. In the power purchase contract, PT PLN bears a significant risk because it is bound by Take or Pay. The Take or Pay contract requires PLN to absorb electricity from private producers in a minimum amount of the total capacity of the power plant. For example, the capacity of 150 MW, 80% availability factor, the PLN must buy a minimum of 120 MW. If PLN is unable to absorb 120 MW, PLN must pay a Take or Pay penalty to the private electricity producer. In this case, it is necessary to schedule a generator in order to meet the load requirements and not be subject to a Take or Pay penalty.

Generator scheduling aims to obtain economical costs by considering various constraints such as rotating reserves, heat rates, minimum up / down time, and minimum operational costs. Generating scheduling is more complex when compared to economic dispatch so a method is needed to solve this problem if the system is used on a large scale. Based on the function of generating scheduling costs are divided into non-convex and convex costs. To schedule a generator can be solved mathematically or with artificial intelligence. The priority list and Lagrange can only complete the convex cost function [1]. Artificial intelligence or heuristic approach methods can solve non-convex cost functions and can solve in large systems[2]

Load scheduling can be mathematically solved by the Quadratic Programming and Unit Decommission (QPUD) method, this method can schedule generating units precisely and convergently and is able to complete scheduling by considering minimum up / down time, backup operations, and load balances [3]. Conventional methods such as the priority list are compared to the lagrange method [4]. With proper initialization, the lagrange method gets an economical scheduling value. However, when compared to the priority list method, the lagrange method has the disadvantage of having a limited state so that it can only schedule generators on small systems. To complete the scheduling of more complex plants and systems used on a large scale, the artificial intelligence method is one of the recommended methods such as the Artificial Bee Colony Algorithm (ABC) [5]. Load scheduling in addition to considering minimum up / down time, backup operations, and load balance, can also consider contingencies [6] Binary Particle Swarm Optimization (BPSO) [7]. Artificial intelligence is one of the methods most developed by researchers today. Artificial intelligence is divided into three namely neural networks, fuzzy systems, and evolutionary computing [8]

In this research, artificial intelligence used is Binary Particle Swarm Optimization (BPSO). In scheduling the generator using the priority list method and finished with BPSO. Load scheduling for 24 hours by considering take or pay as well as minimum up / down time limits, backup operations, and load balance. The system used is the Java-Bali 500 kV system. Figure 1 shows the workflow scheduling diagram using BPSO. Meanwhile, to determine Take or Pay using the gamma search method shown in Figure 2 [9].

The cost function in general generator scheduling can be written in the following equation:

$$F_H = \sum_{h=1}^{H}\sum_{n=1}^{N}[F_{nh}(P_{nh}) + STC_{nh}(1 - U_{n(h-1)})]U_{nh} + DC_{nh}(u_n(h-1)^{-0}) \tag{1}$$

with

$$F_{nh}(P_{nh}) = a_n(P_{nh})^2 + a_n(P_{nh}) + c_n\left|e_n \sin(f_h x(P_h min - P_n^t))\right| \tag{2}$$

STC_{nh} is

$$HSC_n\ if\ MDT_n \leq T_n^{off} \leq MDT_n + CSH_n \tag{3}$$

$$CSC_n\ if\ T_n^{off} > MDT_n + CSH_n \tag{4}$$

with

$$\sum_{n=1}^{N} P_{nh} = D_h \tag{5}$$

$$\sum_{n=1}^{N} P_{nh}(max) \geq D_h + R_h = D_h \tag{6}$$

$$U_{nh}P_n(max) \geq P_{nh} \geq U_{nh}P_n(min) \tag{7}$$

$$T_n^{off} \geq MDT_n \tag{8}$$

$$T_n^{on} \geq MUT_n \tag{9}$$

Where:

F_H	Operation cost every h
N	Number of each generating unit
H	Hourly number for each scheduling period
n	Generator index per unit
h	Hourly number
STC_{nh}	*Start Up Cost* unit n every hour h
HSC_n	*Hot Start Up Cost* unit n
CSC_n	*Cold Start Up Cost* unit n
U_{nh}	ON / OFF each unit n at h
DC_{nh}	*Shut Down Cost* unit n
a_n, b_n, c_n, e_n, f_n	Cost function of each unit n
P_{nh}	Generating each unit n at h hours
$P_n(\max)$	Maximum generation each unit n
$P_n(\min)$	Minimum pembangkit each unit n
D_h	Peak Load at h

R_h	*Spinning reserve* at h
T_{on}/T_{off}	On or OFF time
MUT_n	*Minimum Up Time* each unit n
MDT_n	*Minimum Down Time* each unit n

BPSO is the development of Particle Swarm Optimization (PSO) [10], the difference of the two methods is the update and position speed. Following of the BPSO steps are:

1. Population Initialization

   The parameters used for initialisation consist of, the number of particles represented as the number of generators, the number of iterations, constant variables and manipulation variables. Generating initial initial population in binary form. To determine the initial initialization can be written as follows [10]
   Initial Position X_i = (x_i1, x_i2, ..., x_iD) (10)
   for i = 1,2,3 ...., m and velocity V_i for i = 1,2, ..., m with all particles randomly set.
2. Update speed and position
   For each article i, with dimension d,

   V_id (t) = w * V_id (t-1) + c_1 * r_1 * (〖pbest〗_id- x_id (t-1)) + c_2 * r_2 * (〖gbest〗_d-x_d (t-1)) (11)

   X_id (t) = 1, if r_2 <1 / (1+ 〖exp〗 ^ (- vid (t-1))) (12)

   Where w is the weighting of inertia, c_1 and c_2 are constant speeds, r_1 and r_2 are generated randomly with a range [0,1]
3. Fitness function evaluation.
   Evaluate the fitness function using equation (1). If it violates the limitation of the fitness function value, a penalty factor will be imposed so that the particle candidate will not be selected as a good solution candidate. After evaluating the fitness function, update the particle position:

   if f (x_i)> f (〖pbest〗_i) then update 〖pbest〗_i = x_i, for k is the area of x_i if f (x_k)> f (〖gbest〗_i), so the value 〖gbest〗_i = x_k , where f () is an evaluation of the fitness function
4. Termination of iteration

   Repeat steps 2 and 3, until you find the position and speed that has been determined by the fitness function evaluation

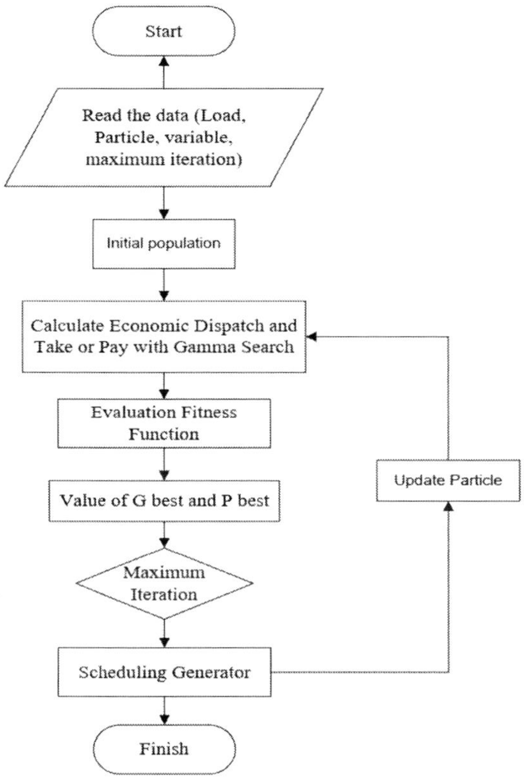

Figure 1. Flow Chart of Load Scheduling Using BPSO

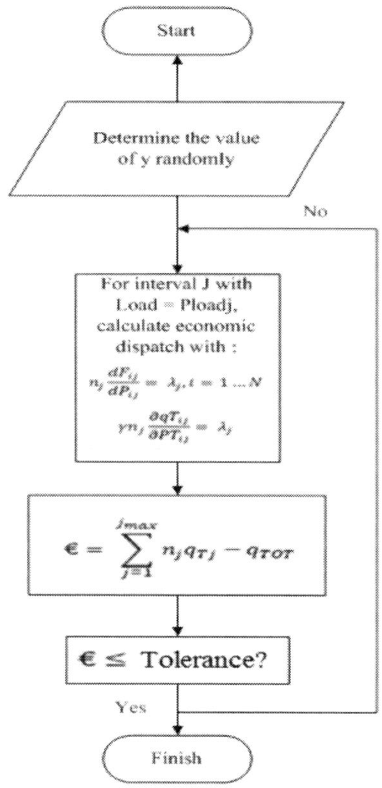

Figure 2. Take or Pay Calculation Flow Chart with Gamma Search.

2. Results And Discussion

Table 1. 500 kV Java-Bali generator data

Unit	Generator Cost Coefficient (Rp/hour)			P min	P max	Fuel Cost
	a	b	c	(MW)	(MW)	(Rp/MBtu)
Suralaya	-0.0015	34.2155	275.1741	1600	3400	0.41
Muaratawar	-0.0512	178.862	624.9346	600	1400	1.25
Tanjung Jati	-0.0019	27.2541	348.5491	1200	2100	0.85
Gresik	-0.0003	52.8313	538.3751	900	2100	0.26
Paiton	-0.0083	151.780	-212.5033	1800	4300	1.74
Grati	-0.0061	45.6393	-119.3939	290	800	0.38

Java-Bali electricity generation daily load data on April 20, 2016, and made into 4 periods [11]. The Heat Rate value is obtained from the factory specifications. The Java-Bali plant's Heat Rate data is shown in Table 2. PT PLN is bound by a contract with a private power plant of 206,000 MW per day, so the load scheduling is carried out taking into account the Take or Pay contract which must fulfill the contract within a day so as not to be subject to penalties or fine. Optimal load scheduling aims to reduce fuel costs to a minimum. The results of the Java-Bali power plant scheduling simulation using Lagrange are shown in Table 3 while using BPSO are shown in Table 4. The simulation results using lagrange are obtained by generating power by meeting the needs with a total power of 228,319 MW and a total cost of Rp 379,318,526,529.78 while BPSO generates a total power of 205,860 MW and a total cost of Rp 1,192,518,224,448.21 to a difference of Rp 260,066,704,081.57, making it cheaper to use the BPSO method compared to the lagrange method. If using a Lagrange the total power exceeds the Take or Pay contract that has been specified. While the total power using the BPSO method the total power is less than the Take or Pay contract with an error of 0.068%.

Table 2. Heat Rate of the Java-Bali generator 500 kV

Unit	Power Generated (MW)				Heat Rate (Btu/kWh			
	1	2	3	4	1	2	3	4
Suralaya	1703	2221	2561	3247	76492.24	74493.38	73454.29	71796.5
Muara Tawar	666	826	993	1140	112582.8	112253.6	100783.9	98182.3
Tanjung Jati	1227	1525	1812.8	1982.8	28800.93	28483.89	28186.52	27978.62
Gresik	1141	1382	1649	1973	191161.2	189915.8	189237.6	188630.8
Paiton	2071.5	2792.5	3358.75	4005	76161.72	73013.27	70840.3	68897.35
Grati	320	400	560	795.6	124583.9	111932.4	108890.5	106665.5

Table 3. 500 kV Java-Bali Power Plant Scheduling Simulation Results by Using Lagrange

Hour	load (MW)	Power (MW)	Suralaya	Muara tawar	Tanjung Jati	Gresik	Paiton	Grati	Cost (Rp/hour)
1	8350	8553	2090	0	1980	789	2928	0	2.634.764.974.61
2	8197	8263	2068	0	1980	621	2928	0	2.458.304.681.75
3	8138	8318	2063	0	1980	582	2928	0	2.418.151.802.24
4	8138	8318	2063	0	1980	582	2928	0	2.418.151.802.24
5	8304	8494	2087	0	1980	738	2928	0	2.582.626.627.24
6	8218	8405	2075	0	1980	657	2928	0	2.496.523.853.49
7	8138	8319	2063	0	1980	582	2928	0	2.418.332.057.67
8	9256	9551	2236	0	1980	945	2928	697	4.635.032.343.2
9	9237	9543	2252	0	1980	945	2928	673	4.580.321.011.58
10	9337	9475	2090	1324	1980	238	2928	150	7.132.355.710.59
11	9320	9459	2098	1308	1980	238	2928	150	7.063.192.603.35
12	9318	9457	2098	1306	1980	238	2928	150	7.054.609.225.58
13	10361	10508	2158	1980	1980	697	2928	0	10.314.418.140.81
14	10454	10609	2171	1980	1980	785	2928	0	10.408.568.440.43
15	10401	10552	2164	1980	1980	735	2928	0	10.355.202.199.74
16	10401	10552	2164	1980	1980	735	2928	0	10.355.202.199.74
17	10454	10609	2171	1980	1980	785	2928	0	10.408.591.565.09
18	11292	11526	2299	1980	1980	945	2928	629	12.217.691.353.57
19	11301	11536	2300	1980	1980	945	2928	638	12.242.560.960.50
20	10734	10912	2211	1980	1980	898	2928	150	10.948.570.166.60
21	10454	10609	2171	1980	1980	785	2928	0	10.408.591.565.09
22	8885	9142	2183	0	1980	945	2928	341	3.687.352.069.37
23	7997	8167	2044	0	1980	449	2928	0	2.284.277.122.60
24	7196	7338	1665	0	1980	0	2928	0	1.651.698.915.30

Table 4. 500 kV Java-Bali Power Plant Scheduling Simulation Results Using BPSO

Hour	Load (MW)	Power (MW)	Suralaya	Muara Tawar	Tanjung Jati	Gresik	Paiton	Grati	Cost (Rp/hour)
1	8350	8553	2095	0	1816	980	3089	0	2.334.714 .834.36
2	8197	8263	2072	0	1972	986	3098	0	2.400.216 .681.05
3	8138	8318	2072	0	1937	933	2468	0	2.211.709 .192.12
4	8138	8318	2082	0	1367	764	3578	0	2.211.709 .192.12
5	8304	8494	2090	0	1346	874	3209	0	2.362.232 .138.98
6	8218	8405	2077	0	1904	846	3289	0	2.186.233 .954.24
7	8138	8319	2069	0	1368	978	3227	0	2.110.332 .497.19
8	9256	9551	2240	0	1579	864	3213	897	3.120.032 .943.21
9	9237	9543	2261	0	1487	987	3097	978	4.128.128 .011.58
10	9337	9475	2190	1355	1478	892	2487	638	7.110.138 .130.59
11	9320	9459	2198	1209	1238	792	3198	293	6.139.198 .133.35
12	9318	9457	2198	1431	1323	927	3221	348	6.019.109 .139.58
13	10361	10508	2160	1978	1083	872	3218	983	8.135.418 .140.81
14	10454	10609	2215	1967	1479	928	3098	0	10.193.56 8.440.43
15	10401	10552	2181	987	1585	627	3218	0	7.921.202 .199.74
16	10401	10552	2176	923	1974	821	3127	0	9.013.202 .199.74
17	10454	10609	2341	977	1390	921	3189	0	6.103.591 .565.09
18	11292	11526	2344	1815	789	917	3812	782	8.019.691 .353.57
19	11301	11536	2211	1753	987	934	2341	812	8.139.560 .960.50
20	10734	10912	2173	1920	908	819	3109	267	5.418.139 .166.60
21	10454	10609	2191	1867	907	712	3179	0	6.920.161 .565.09
22	8885	9142	2051	0	1139	983	3109	539	3.130.130 .902.37
23	7997	8167	1670	0	1940	791	3501	0	2.371.277 .122.60
24	7196	7338	651	0	1497	0	3417	0	1.552.123 .983.30

3. Conclusion

With the BPSO method the total cost of generation is more minimum compared to the lagrange method. But in the BPSO method it still has a big error when considering Take or Pay. While the lagrange method, the total power compared to Take or Pay results exceeds the specified contract. So in this case the BPSO method is better when compared with the lagrange method. For further research it is expected that the Take or Pay error is close to 0% so as not to harm the PLN

4. References

[1] Raj VM, Chanana S. Analysis of Unit Commitment Problem Through Lagrange Relaxation and Priority Listing Method. 6.

[2] Wati T, Suheta T, Masfufiah I. Pengoptimalan Pembebanan Ekonomis dengan Mempertimbangkan Emission Constraint menggunakan Hybrid CPSO. *Jurnal EECCIS* 2019; 13: 46–50.

[3] Sawa T, Furukawa K. Unit commitment using quadratic programming and unit decommitment. In: *2012 IEEE Power and Energy Society General Meeting*. San Diego, CA: IEEE, pp. 1–6.

[4] Govardhan M, Roy R. Profit based unit commitment using Gbest Artificial Bee Colony algorithm. In: *2013 3rd International Conference on Electric Power and Energy Conversion Systems*. Istanbul, Turkey: IEEE, pp. 1–6.

[5] Sartika N, Abdullah AG, Hakim DL. Schedulling Economical Thermal Power Plant 500 KV Java-Bali System Using Lagrange Multiplier. *IOP Conf Ser: Mater Sci Eng* 2017; 180: 012073.

[6] Bigdeli M, Karimpour A. Optimal reserve requirements and units schedule in contingency Constrained Unit Commitment. In: *2014 14th International Conference on Environment and Electrical Engineering*. Krakow: IEEE, pp. 443–448.

[7] Wibowo RS, Utama FF, Putra DFU, et al. Unit commitment with non-smooth generation cost function using binary particle swarm optimization. In: *2016 International Seminar on Intelligent Technology and Its Applications (ISITIA)*. 2016, pp. 571–576.

[8] Nath RP, Balaji VN. Artificial Intelligence in Power Systems. 7.

[9] Wood AJ, Wollenberg BF, Shebl&amp eacute, et al. Power Generation, Operation and Control. 658.

[10] Chakraborty B. Binary Particle Swarm Optimization Based Algorithm for Feature Subset Selection. In: *2009 Seventh International Conference on Advances in Pattern Recognition*. 2009, pp. 145–148.

[11] Sartika N, Abdullah AG, Hakim DL. Schedulling Economical Thermal Power Plant 500 KV Java-Bali System Using Lagrange Multiplier. *IOP Conf Ser: Mater Sci Eng* 2017; 180: 012073.

IOP Conf. Series: Materials Science and Engineering 1010 (2021) 012014 doi:10.1088/1757-899X/1010/1/012014

Performance analysis of two layer leach algorithm based on area partition (tl-leach-p) for wsn

M U H A Rasyid[1*], D I Permatasari[1], D J Karim[1]

[1]Politeknik Elektronika Negeri Surabaya (PENS), Surabaya, Indonesia

Email address: udinharun@pens.ac.id

Abstract. Currently, much research has been done to develop routing algorithms to improve the lifetime of sensors from Wireless Sensor Network (WSN). Routing protocols based on the network structure can be classified into two categories: flat protocols and hierarchical protocols. One of the most popular hierarchical protocols is the Low Energy Adaptive Clustering Hierarchy (LEACH) algorithm. However, LEACH routing has a high packet loss in a busy network. This research propose Two Layer LEACH based on area Partition (TL-LEACH-P) routing algorithm based on the LEACH routing algorithm where the cave layer is done to prioritize the node that can become Cluster Head to decrease energy consumption. LEACH and TL-LEACH-P algorithms are simulated using MATLAB. The results show that the TL-LEACH-P algorithm has a better performance of 60% reaching LEACH in terms of network lifetime. The TL-LEACH-P algorithm has a 50% increase in the round where 50% of the nodes are dead from LEACH. Best TL-LEACH-P algorithm performance is when the k cluster value = 5. And best performance for LEACH algorithm is when the k cluster value = 15.

1. Introduction

Wireless Sensor Network (WSN) has a very important role in various fields of life. WSN supports remote monitoring of a sensor without having to be in that location. In its development, WSN is widely used for monitoring the civilian environment, such as buildings, bridges, reservoirs, and in the military supervision system under certain very dangerous conditions. In addition, WSN is also used in part monitoring systems and marine monitoring for the detection of natural disasters.

WSN is an infrastructure network that consists of a set of sensor nodes that are spread over a sensor area. Each sensor node has the ability to collect data around it and redistribute it to the sink node through intensive radio transmission. The collected data can be in the form of temperature, pressure, movement of an object, humidity, and so on. In general, WSN consists of two components, namely sensor nodes and sinks. Sensor nodes are a unitary component of a network that can produce information, can be a sensor or a structured actuator in the entire operation. In general, sensors are spread with high volume and density. Sink is an integrated process of gathering information from sensor nodes so that further information processing can be done [1].

In its implementation, the main problem of WSN is energy consumption. This is caused by the power supply on the sensor node only supplied by the battery for its operation, so it has limited energy reserves. If one node is off or not active, it will change network performance in terms of topology and routing. On the other hand, obstacles will arise if you have to conserve energy repeatedly because it will increase costs and disrupt network performance. So that energy consumption is the most important factor to determine the lifetime of a network, so that the energy used must be as efficient as possible in order to produce maximum performance. Therefore, the routing algorithm for determining routes that use more efficient energy is a top priority in designing WSN. Hierarchical routing approaches increase the lifetime of network performance by cutting the whole network into clusters which are led by cluster head. Machine learning and computation intelligence bring out the optimization of clustering algorithms for WSN [2].

One of the popular routing algorithms for WSN is the LEACH algorithm. LEACH algorithm is an extension of Low-energy Adaptive Clustering Hierarchy which is a TDMA-based MAC protocol that is integrated with clustering and simple routing protocols at WSN. The purpose of the LEACH algorithm is to reduce the energy consumption needed to create and maintain clusters to increase the lifetime of WSN [3].

Problem of WSN is the network has very limited energy; therefore a protocol communication algorithm is needed to increase the node lifetime. The purpose of this research is to propose Two Layer LEACH Based on Area Partition (TL-LEACH-P) to conduct a performance comparison analysis on the TL-LEACH-P and LEACH (standard) algorithms which cover rounds where the first node die (first node dies), half node die (50 nodes dead), and all die nodes (100 dead nodes). So we will know which is the best algorithm should be implemented based on the conditions needed.

Several research works have already done by researcher in order to optimize the LEACH algorithm. Mohammed et al. [4] proposed Kangaroo method (KM) using the optimal path adaptive routing to reduce sensor node energy dissipation. The KM algorithm is inspired by the simulated annealing method. The data transmission of the data from the cluster head to the base station is done by muti-hop mode. The Matlab simulation results show that the proposed KM method better regarding residual energy and extend the lifetime of networks. Sweety et al. [5] proposed Limited Communication-LEACH-Mobile Fuzzy (LC-LEACH-MF) to utilize the allocation of energy to sensor nodes in efficient way. The proposed method is developed based on LEACH algorithm, and LEACH-Mobile (LEACH-M) using fuzzy inference system LEACH-Mobile Fuzzy (LEACH-MF). The simulation result show that the proposed method better than the LEACH method and other traditional techniques in terms of packet delivery ratio, last node dead, half node dead, first node dead, and energy consumption.

Maurizka et al [6] proposed delay tolerant network (DTN) to LEACH algorithm to reduce the packet loss in a busy network. The proposed system adds bundle layer to save the data temporarily. The simulation results show that the proposed system decrease the packet loss in changing the number of node and improve the packet delivery ratio by changing the capacity of buffer. Muhammad et al. [7] proposed enhanced three layer hybrid clustering mechanism (ETLHCM) to limits the exchange of control packets between nodes when select lower layer heads of cluster. The higher layer heads of cluster is selected by base station to reduce backward transmission of packet data in network. Simulation results show the proposed method increase the lifetime of network.

2. System Design

In this section, we discuss our proposed system for experiment and analysis preparation. This contains overall system workflow, simulation model attributes used, and flowchart of each algorithm.

2.1. Overall System Workflow

The general overview of the system design process is shown in figure 1.

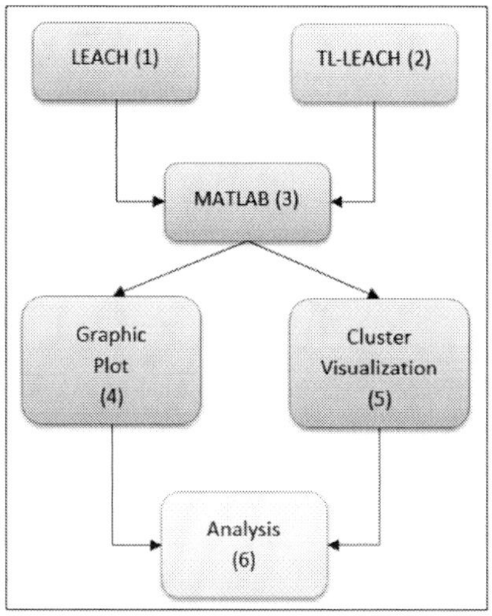

Figure 1. General system design.

Figure 1 shows the general system design that will be implement and analysis. First, the parameters needed in each algorithm both for LEACH and TL-LEACH-P algorithm files to be compiled in MATLAB are determined. For parameters, it will be explained on the next section about simulation model attributes.

After all parameters have been successfully determined, then an algorithm file is formed which will be compiled and obtained data to be processed. Implement the source code in MATLAB for the two algorithms tried, namely LEACH and TL-LEACH-P. After the source code has been completed, it will run to get the data visualization of the node's location when the two algorithms are running and compare directly in graphical form.

2.2. Simulation model attributes

The simulation mode attributes and parameters we used are shown below in table 1.

Table 1. Simulation model attributes used and parameter value.

Parameter	Value
Total node	100
Node location	Random
Node Area	100 x 100
K cluster	5, 10, 15,20
Base Station location	50,175
Initial Energy of each node	0.5 Joule
Simulation Time	3000 rounds

2.3. LEACH and TL-LEACH-P

After the node has been raised and spread, the next is to determine the Cluster Head and Cluster Member to determine the cluster so that the delivery of information is known to its origin and destination. In this case the determination of the cluster head and cluster members is still random or the cluster head does not choose the closest cluster member but the cluster member determination is random.

Determination of cluster head randomly according to which leach algorithm when the energy of all nodes is the same then the cluster node is chosen randomly, only then when the energy starts to fall then the node with the greatest energy is chosen, in this case the number of cluster nodes cannot be determined it depends from the energy that exists at each node, whereas for node switching is done simultaneously with other cluster heads. The formation of cluster head is the process after spreading the node and determining the parameters for the simulation. Furthermore, to determine the cluster head itself, it is determined from residual energy (E) greater than the threshold that is done, the node will become a cluster head and the one that is not suitable will become a cluster member.

In LEACH algorithm, the node performs the settings made by the node or the node stands alone without having to have commands from the Base station or cluster head. All cluster members send data to the cluster head, which the cluster head receives data from all the cluster members. LEACH divides activities into several rounds. Round is a time session needed for cluster formation until data transmission is complete. Each round produces a number and arrangement of different clusters. This is because the determination of cluster head is done randomly.

After we design how LEACH determines the cluster head, then we design how the TL-LEACH-P done the same thing on the same area and the same node location. In the process of determining the Cluster Head and Cluster Members have similarities between the LEACH and TL-LEACH-P. However, there are some differences. The first is the minimum area value of the layer. Determine the minimum value of this layer based on the number of clusters specified. By default the width of the area will be divided into two at 50 in the y axis for area partition. The node distribution factor is proven when the use of partition topologies where the number of nodes is spread evenly is better than the random topology at the same environment size [8]. Furthermore, if there are no nodes above the layer boundary, then the layer is no longer needed. Also, the preferred node to become a cluster head is a node with the y position above the layer boundary. In this case, we will set the limit to half the area becomes 50. The flowchart of TL-LEACH-P is shown in figure 2.

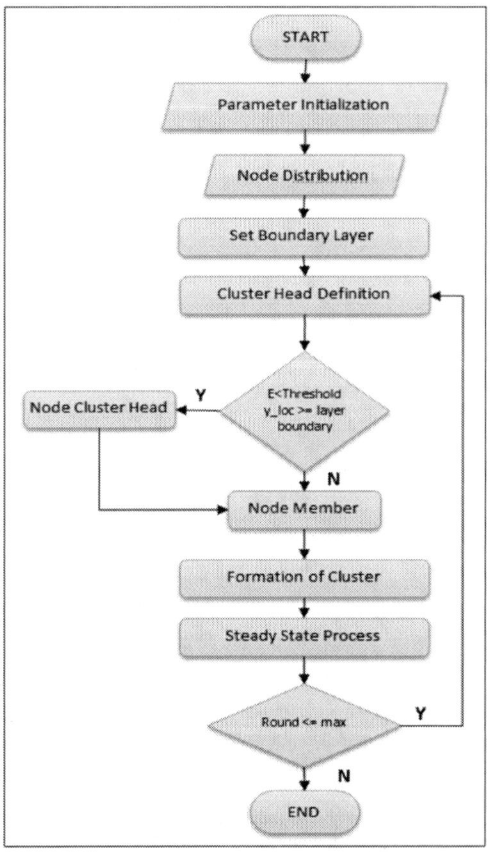

Figure 2. Flowchart of TL-LEACH-P algorithm.

Based on the figure 2, the TL-LEACH-P topology illustration how the data flow will be sent to the base station is shown in figure 3. The difference between TL-LEACH-P algorithm and the standard LEACH lies in the selection process of the cluster head. Where in the setup state, this algorithm regulates the number of cluster heads to be measured using the average silhouette method [9] to determine the optimal number of cluster heads when selecting cluster heads in that round. After the optimum value k is obtained. Furthermore, the area of the node will be divided into 2 layers. So that what will be prioritized to become a cluster head is a node with high energy and on the layer closest to the Base Station.

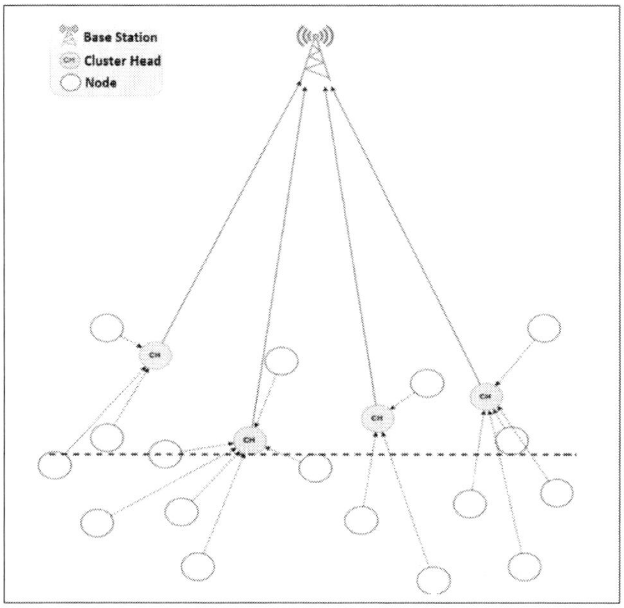

Figure 3. TL-LEACH-P topology illustration.

3. Experimental Results

In this section will be testing process of implementation LEACH and TL-LEACH-P algorithm on performance analysis using MATLAB. This test will compare the performance of node alive, energy average, first node die (FND), half node die (HND), and all node die (AND) for each algorithm.

3.1. Total Remaining Energy

Energy has become an important concern in WSN research. While many energy-saving protocols have been proposed to extend network lifetime, various network lifetime definitions have also been used for different scenarios and protocols. We conduct the experiment to compare the total remaining energy (J) results in the LEACH and TL-LEACH-P algorithms for each cluster k, namely 5, 10, 15, and 20 clusters. After we run the simulation on each algorithm and k clusters obtained graph results as shown in figure 4. Figure 4 shows that the total remaining energy in TL-LEACH-P algorithm is better than the total remaining energy in the LEACH algorithm.

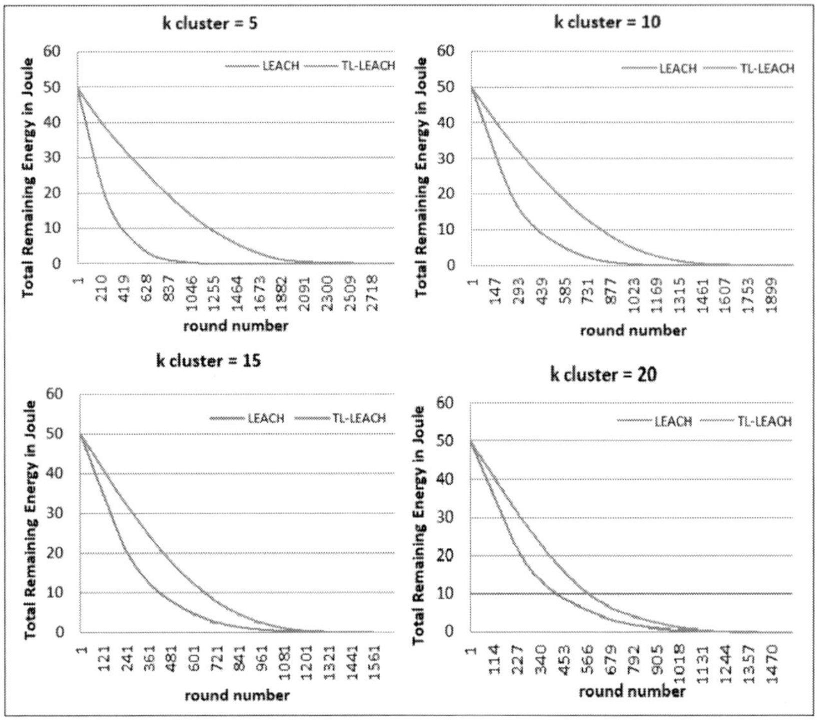

Figure 4. Total remaining energy with k cluster = 5, 10, 15, and 20.

3.2. Average Remaining Energy

In this test, the calculation of the average energy of each node in each round is carried out. So, in each round will be calculated the total energy owned by all nodes divided by the number of nodes in the round. This calculation is performed on each algorithm, namely leach and TL-LEACH-P. After we do the simulation running on each algorithm and cluster k, the graph results are as shown in figure 5. From the results of the graph above from 5, it can be seen that the remaining energy in the leach algorithm has completely exhausted in the round to 1300s. This is directly proportional to the number of nodes that are still alive in the round. If in that round the living nodes are still many, then the average remaining energy in the round will be high. Likewise, if the number of nodes that live in the round is getting smaller, the average total energy in the round will be the smaller. It can be seen that the TL-LEACH-P algorithm has more efficient energy utilization compared to the leach algorithm.

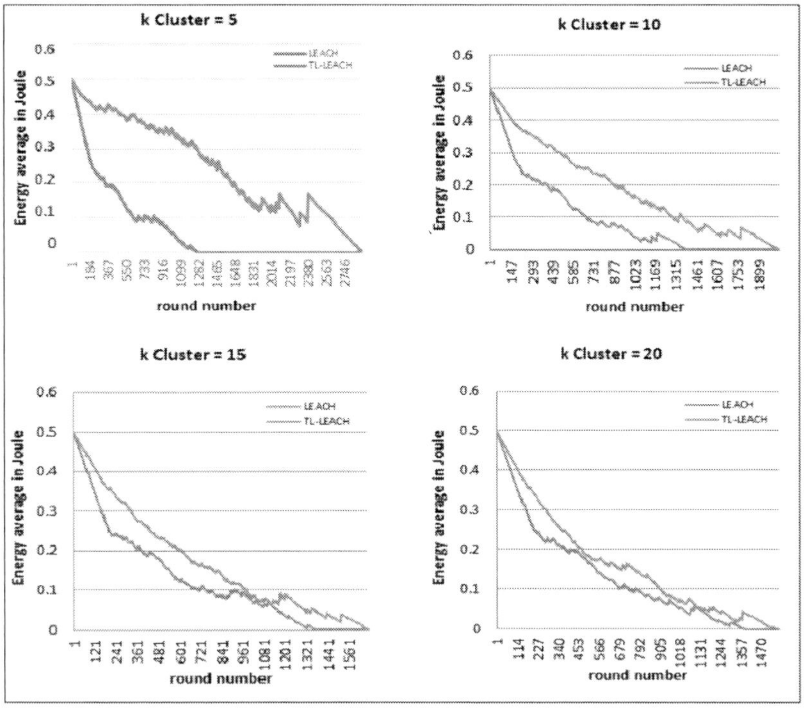

Figure 5. Average remaining energy with k cluster = 5, 10, 15, and 20.

3.3. *First Node Die (FND), Half Node Die (HND), All Node Die (AND)*

In this test we do a comparison on the number of nodes that have the same value. In this case it is when the first node dies or the First Node Die (FND), 50% of the total nodes die or Half Nodes Die (HND), and all nodes die or All Nodes Die (ALND) on each algorithm and each cluster k. In this comparison we will appear in the form of a separate table showing the comparison of the number of dead nodes compared to the round. Here are the results of the comparison of FND, HND and AND on the LEACH algorithm.

Table 2. Comparison of FND, HND and AND on LEACH algorithm.

State	LEACH (round)			
	5 CH	**10 CH**	**15 CH**	**20 CH**
First Node Die (FND)	140	176	160	150
Half Node Die (HND)	410	436	428	417
All Node Die (AND)	1266	1389	1371	1380

Table 2 shows the result of FND, HND and AND obtained from the search in what rounds occurred in situations where First Node Die (FND), Half Node Die (HND), and All Die Nodes (AND) on each cluster k for LEACH algorithm. It can be concluded that the LEACH algorithm is the best First Node Die (FND) value of k cluster with a value of 10. Similarly, for the Half Node Die (HND) and All Node Die (AND) values the best value is in k cluster with a value of 10.

Table 3 shows that the TL-LEACH –P algorithm has a good performance for First Node Die (FND) on k clusters with a value of 15, which in k = 15 the round value in the FND situation is 197 rounds. As for performance in Half Node Die and All Node Die situations, the best performance is in the 5 cluster with Half Node Die being in the 890th round and All Node Die in the 2919th round. From table

2 and 3, it can be shown that the network lifetime of TL-LEACH-P is more that the network lifetime of LEACH.

Table 3. Comparison of FND, HND and AND on TL-LEACH-P algorithm.

State	TL-LEACH-P (round)			
	5 CH	10 CH	15 CH	20 CH
First Node Die (FND)	107	140	197	175
Half Node Die (HND)	890	758	705	626
All Node Die (AND)	2919	2042	1668	1569

4. Conclusions

This paper proposes Two Layer LEACH based on area Partition (TL-LEACH-P) routing algorithm in order to decrease the energy consumption of network nodes for WSN. TL-LEACH-P determines the first minimum area value of the layer based on the number of clusters specified for area partition. After carrying out various stages starting from the design stage, making the system which is then continued at the testing and analysis stage, it can be concluded that the TL-LEACH-P Algorithm has advantages compared to LEACH in terms of node remaining energy, network lifetime, first node die (FND), half node die (HND), and all node die (AND).

5. References

[1] Aidil S K, M. Udin H A R and Iwan S 2019 Efficient Energy for Cluster Head Selection using New LEACH-based routing protocol in Wireless Sensor Network *International Electronics Symposium (IES)* (Surabaya:Indonesia) p 70

[2] Damien W S, Blaise O Y, Anna F and Paul D 2019 Optimized Clustering Algorithms for Large Wireless Sensor Networks: A Review *MDPI Sensors Journal* **19** 322

[3] M. Udin H A R, Nur M and Jeni R 2019 Performance Analysis LEACH Based Genetic Algoritm In Wireless Sensor Network *International Seminar on Application for Technology of Information and Communication (iSemantic)* (Semarang:Indonesia) p 394

[4] Mohammed K, Khelifa B and Mohammed O 2017 LEACH-KANG: A New Routing Protocol for WSN based on Leach Protocol and Kangaroo Method *International Conference on Mathematics and information Technology* (Adrar: Algeria) p 273

[5] Sweety S and Nitin M 2018 An Improved LEACH-MF Protocol to Prolong Lifetime of Wireless Sensor Networks *IEEE 8th International Advance Computing Conference (IACC)* (Greater Noida:India) p 174

[6] Maurizka A R, Leanna V Y and Ratna M 2018 Energy Consumption and Packet Loss Analysis of LEACH Routing Protocol on WSN over DTN *International Conference on Wireless and Telematics (ICWT)* (Nusa Dua:Indonesia)

[7] Muhammad F U, Junaid I and Khawaja Q M 2019 Enhanced Three Layer Hybrid Clustering Mechanism for Energy Efficient Routing in IoT *MDPI Sensors Journal* **19** 829

[8] M. Udin H A R, Bih-Hwang L, Iwan S and Mokhammad M A 2018 LEACH Partition Topology for Wireless Sensor Network *IEEE International Conference on Consumer Electronics – Taiwan (ICCE-TW)* (Taichung: Taiwan)

[9] Silhoutte (clustering) [online] https://scikit-learn.org/stable/auto_examples/cluster/plot_kmeans_silhouette_analysis.html

Acknowledgments

This research was supported in part by Ministry of Research, Technology and Higher Education of Indonesia, under scheme 'Penelitian Dasar' 2020.

An experiment on different type of muffler on spark Ignition engine 110 cc performance

G Setyono[1], M Ulum[2], and Z Lillahulhaq[3]

[1]Institut Teknologi Adhi Tama Surabaya, East Java, Indonesia

gatot_mesin@itats.ac.id

Abstract. The design of the exhaust system affects many parameters such as reducing the noise level, control velocity, temperature, back-pressure of exhaust gas and reduce pollution production. The exhaust system has many parts which consist of an exhaust manifold, resonator, catalytic converter, muffler, and connecting pipes. Muffler has the main function as a noise absorber. Furthermore, the temperature distribution on mufflers is also affected by fuel consumption by reducing back-pressure effect. This study has conducted an experiment to determine the effect of modified muffler shape in a motorcycle which consists of standard, scorpion, and 3v3 type. The effects of mufflers' performance on the motorcycle are described on torque and power graphics. Furthermore, this study explains the muffler geometry effect on noise reduction from the combustion process. The installation of 3v3 muffler decreases torque and power in a motorcycle. Moreover, the installation of 3v3 muffler type raises noise pollution level which generates by the engine combustion process.

1. Introduction

1.1. Exhaust system on reducing pollution particle

The automotive vehicle uses is very effective to transport goods and passengers immediately. However the use of automotive vehicle enhancement pollution to our environment such as noise, emission gas, and increase the global warming effect. Emission gas flows through the exhaust system which consists of piping lines from the combustion chamber to the surrounding air. The amount of pollution product can be reduce by using biofuel, ethanol, catalyst pollution control, and modified exhaust design. Ethanol (C_2H_5OH) is a low price renewable energy which can be regenerated by fermentation process. Ethanol-gasoline is a suitable candidate as an alternative engine fuel with 25-45% efficiency range [1]. Biofuel reduce air pollution such as smoke, carbon monoxide (CO) and hydrocarbon (HC). Meanwhile using biofuel in combustion process resulted cold flow properties, high viscosity, poor oxidation stability and high nitrogen oxides (NO_x) emissions[2]. Emission production in combustion process can be control by using catalyst pollution control. The commercial vehicles reduce the amount of particles production, but produce very fine and reactive fullerene-like soot. The fine particles size potentially absorb to human lung membranes. A particulate matter catalyst system (PMKAT) consisting of a platinum oxidation catalyst and metallic substrates with open cells for storage and oxidation of soot. This system can be reducing the amount of soot particle spreading through the environment[3]. Exhaust system is modified to reduce noise and air pollution for example by using asbestos fibre material[4]. Combustion engine modifying can be done in order to reduce emission product in combustion process for example by modifying the material that used in spark plug. Iridium spark plug reduce HC and CO emission on high speed motor cycle effectively[5]. Moreover, the use of spark plugs from iridium material can also improve engine performance which is indicated by an increase in engine power output[6].

The design of exhaust system affects many parameters such as reducing the noise level, control velocity, temperature, back pressure of exhaust gas and reduce pollution production[7][8]. Back pressure is defined as gas emission which is flowing back to the combustion chamber and causes nonstoichiometric combustion[9]. This condition is affected by the high pressure of inlet air which used in the combustion process is mixed with emission gas and reduces the amount of oxygen. The part of the exhaust system consists of exhaust manifold, resonator, catalytic converter, muffler, and connecting pipes.

1.2. The modified of exhaust manifold on reducing pollution particle

Exhaust manifold is commonly modified to create optimum by considering parameters include pipe diameter, length, and geometry. The exhaust manifold is installed to prevent exhaust gases from entering the combustion chamber. Manifold tube Intake and exhaust diameter aim to maintain exhaust gas backflow. The increment diameter of exhaust manifold hole can reduce exhaust gas flow to the combustion chamber[10]. Moreover the angle of manifold elbow also affected on back pressure forming process. Exhaust manifold with 90° bend angle reduce amount of emission gas flows back to the combustion chamber[11]. A Reactive muffler has resonator chamber to reduce low-frequency noise generated by combustion process. The Reactive muffler can be modifying as U-shaped corrugated pipes which reduce noise pollution significantly above the upper cut-off frequency range[12]. Furthermore perforation area inside the muffler aim to reduce back pressure and emission gas especially NO_x[13]. The study on the combustion process in an automotive vehicle can be done through a simulation process by software[7] [8] [13][14]. The results of the simulation process can be described through qualitative data such as velocity vector, temperature, and pressure couture. Simulation study should be accompanied by a validation process to get relevant data that close to the real condition[15].

Muffler can be classified as absorptive and reactive based on its function. The reactive muffler dissipates noise energy in specific geometry. Circular muffler absorb in the frequency range less than 200 Hz. Meanwhile rectangular muffler has better performance to absorb noise more than 200 Hz [16]. Although the main function of muffler is to reduce noise pollution, muffler temperature distribution also affected on fuel consumption by reducing back pressure effect [17]. This study is conducted an experiment to determine the effect of different geometry on modified muffler shape which used on 4 strokes motorcycle, Jupiter Z 110 cc. The effects of mufflers on motorcycle performance are described on torque and power graphics. Furthermore, this study explains the muffler geometry effect on noise reduction from combustion process.

2. Methodology

This experiment conducted on 3 types of mufflers which consist of standard, 3v3, and scorpion type as shown in figure 1, figure 2, and figure 3. The performance of motorcycle measures by chassis dynamometer 3.2. The data of motorcycle performance which obtained dynamometer test is torque and power on specific angular velocity (RPM). Tachometer is used ensure the speed of motorcycle is at certain rpm in testing process. Muffler noise measures with digital noise analyser.

Experiment procedure begins with turning on and running the machine without intercooler at idle condition. Then set up the carburettor in stoichiometry condition to prevent too rich and lean air fuel ratio in combustion process. Engine performance tests with dynamometer between 2000 - 10000 rpm with increment of 1000 rpm in each data retrieval.

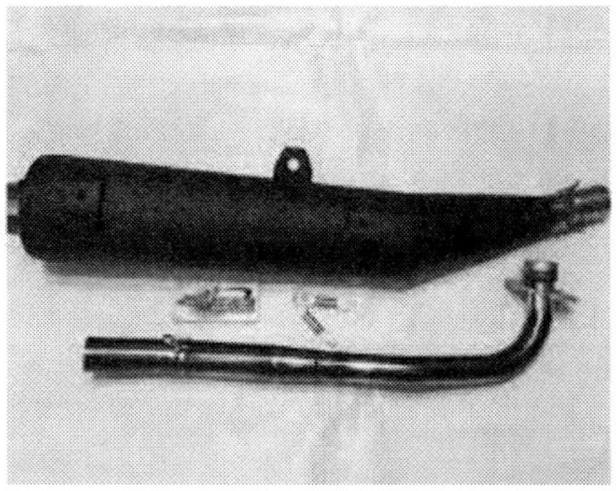

Figure 1. Muffler Jupiter Z 110 cc Standard type

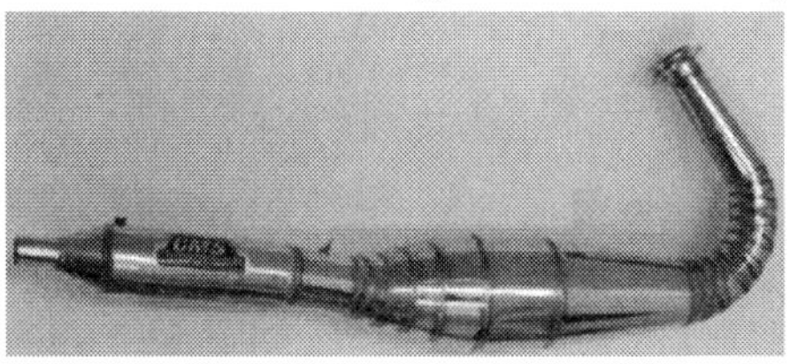

Figure 2. Muffler Jupiter Z 110 cc type 3v3

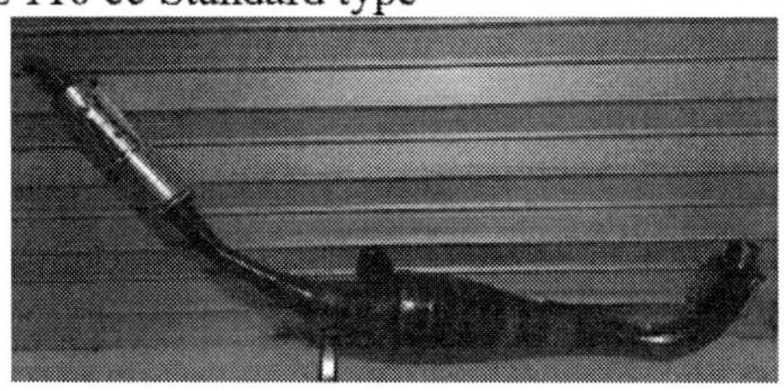

Figure 3. Muffler Jupiter Z 110 cc type Scorpion

3. Equations and Mathematic

The motorcycle performances are described on torque, power, and brake specific fuel consumption (B_{SFC}). There are several equations that can be used in this study including:

$$B_{sfc} = \frac{\dot{m}_f}{\dot{W}_b} \tag{1}$$

$$\dot{W}_b = \eta_m . \dot{W}_i \tag{2}$$

$$\dot{W}_i = W_{net} . N \tag{3}$$

$$W_{net} = \eta_t . Q_{in} \tag{4}$$

With:

With:	B_{sfc}	: Brake specific fuel consumption	(kg/hp.hr)
	$\dot{m}_f$	: Mass fuel consumption	(kg/hr)
	$\dot{W}_b$	: Brake horse power	(hp)
	$\dot{W}_i$	: Indicative horse power	(hp)
	W_{net}	: Mean horse power	(hp)
	Q_{in}	: Inlet Energy	(kJ)
	N	: Angular velocity	(rpm)
	η_m	: Mechanical efficiency	
	η_t	: Thermal efficiency	

In this study the power and torque data from the motorcycle were obtained from the result of dynamometer test. Meanwhile the mass flow fuel consumption can be calculated using Pyrex and a digital stopwatch.

4. Result and Discussion

4.1. The difference of geometry muffler on motorcycle performance

The effect of different muffler geometry on motorcycle performance is discussed in this section. The performance of motorcycle which analysed in this section is power and torque from dynamometer test. Torque is described as the force to pull vehicle and causes rotation. A higher torque level at low engine RPM indicates the motorcycle pulls some load without much effort and speed losses. The torque resulted by dynamometer test is shown in figure 4. Torque produced by the engine increases along with motorcycle RPM increment. Increased speed on the motorcycle causes increased fuel consumption and produces high torque on the engine. "In figure 4 shows that a motorcycle with scorpion muffler type produces higher torque compared to standard muffler. The 3v3 muffler produces the lowest torque compared with the other type. Moreover different types of mufflers also affected on the maximum rpm that can be achieved by the engine. Standard muffler has the shortest range of angular velocity in 7500 rpm. Meanwhile the modified mufflers have longer range where the maximum range of 3v3 muffler is 9000 rpm and scorpion muffler 9500 rpm.

Power is the result of multiplication between torque and angular velocity (RPM). The higher torque and angular velocity (RPM) that produced by the engine resulted higher the power output. The power resulted by dynamometer test is shown in figure 5. The graphics of torque and power have similar trend line. The installation of 3v3 mufflers resulted unsatisfied performance on a 110 cc Jupiter Z motorcycle. It has lowest torque trend line which gain less power among another muffler. High rpm and torque on scorpion muffler resulted higher engine power in motorcycle. Higher engine power has an impact on shorter acceleration time.

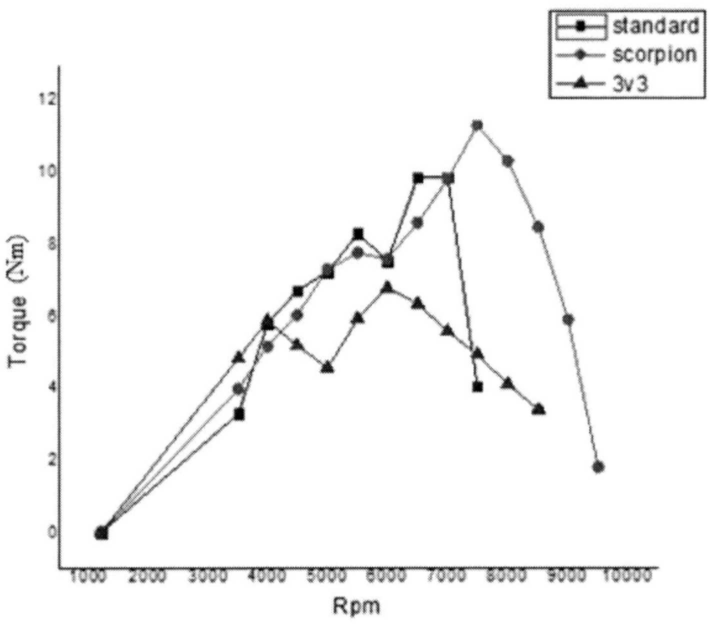

Figure 4. Torque as a function of angular velocity of dynamometer test result in various mufflers type without intercooler

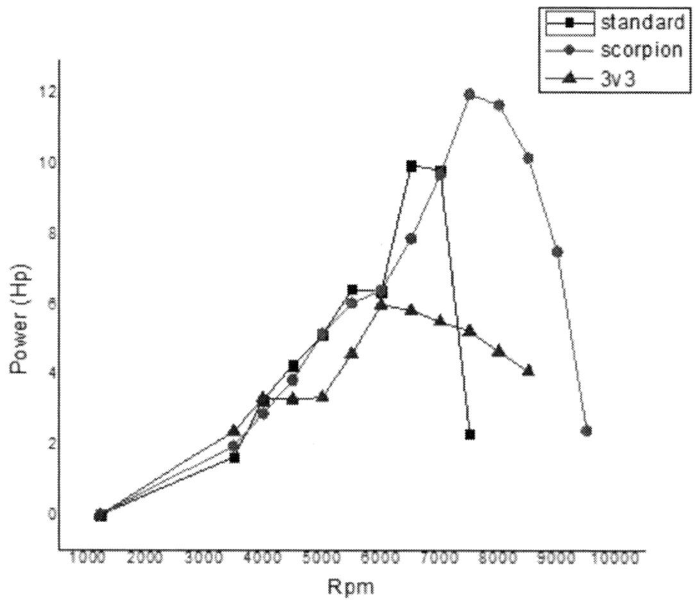

Figure 5. Power as a function of angular velocity of dynamometer test result in various mufflers type without intercooler

At the highest point of rpm, engine throttle is fully open rapidly and the combustion process occurs in rich conditions. In this point the fuel combust incompletely and result lower power. This combustion process produce high amount of harmful pollution such as CO, C, CO_2 and unburned fuel. These exhaust gas can flow back into the engine and ruin combustion process. When a motorcycle with 3v3 muffler reaches maximum rpm, it has higher power compared to another muffler. This condition may be affected by the effectiveness of 3v3 muffler in reducing back pressure.

4.2. The difference of geometry muffler on noise reduction

The main function of the muffler installation is to reduce the noise which generated by the combustion process. Noise reduction which affected by muffler installation is shown in figure 6. Standard mufflers produce sounds with intensity of 101.7 db. This noise level can damage human hearing system in 2 hour. Scorpion muffler is effective in reducing noise in the motorcycle. Scorpion mufflers are equipped with a resonator chamber, perforated pipe and long exhaust pipe. This installation absorbs noise pollution generated by the engine effectively until 99.4 db. Meanwhile the 3v3 muffler produces higher noise pollution than standard muffler with intensity of 102.2 db. Exhaust pipe in 3v3 muffler has a shorter size than the scorpion muffler. The shorter size of exhaust pipe causes back pressure condition on exhaust gas.

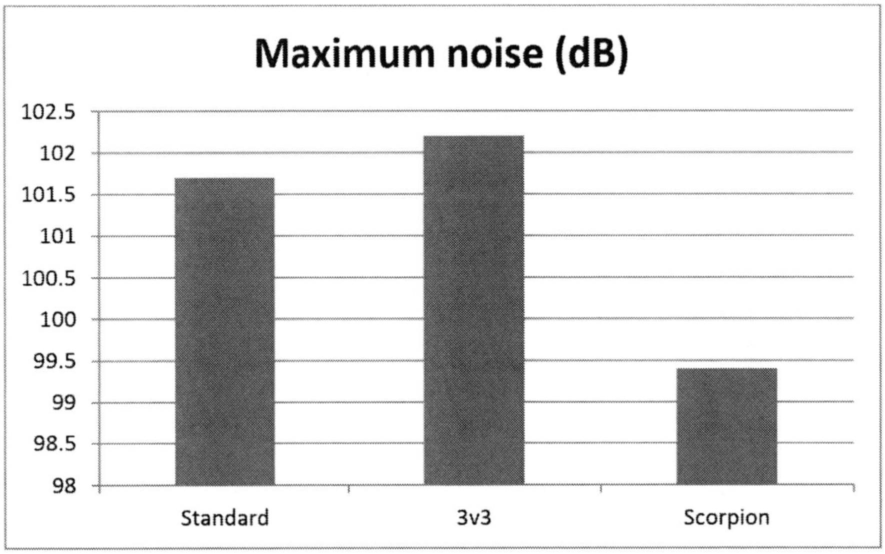

Figure 6. Noise pollutions level produce by various types of muffler.

5. Conclusions

The scorpion muffler is effective to reduce noise pollution and resulted high power on Jupiter z 100 cc. This type of muffler is suitable to increase the motor cycle performance. The installation of 3v3 muffler decrease torque and power in motorcycle. Moreover the installation of 3v3 muffler type raises noise pollution level which generate by engine combustion process.

6. References

[1] Setyono G, Arifin AA. EFFECT OF ETHANOL-GASOLINE MIXES ON PERFORMANCES IN LAST GENERATION SPARK-IGNITION ENGINES WITHIN THE SPARK-PLUG NO GROUND-ELECTRODES TYPE. *Mek J Tek Mesin* 2019; 5: 19–26.

[2] Mahalingam A, Dinesh, Munuswamy B, et al. INVESTIGATION ON THE EMISSION REDUCTION TECHNIQUE IN ACETONE-BIODIESEL ASPIRATED DIESEL ENGINE. DOI: 10.21894/jopr.2018.0020.

[3] Su DS, Müller JO, Jentoft RE, et al. Fullerene-like soot from EuroIV diesel engine: Consequences for catalytic automotive pollution control. In: *Topics in Catalysis*. Springer, 2004, pp. 241–246.

[4] Kathiravan Bahir R. *An Experiment on Noise and Emission Control in a Modified Exhaust System Energy system technology for power electronics View project Natural Fibers and its composites View project An Experiment on Noise and Emission Control in a Modified Exhaust System*, http://www.ripublication.com (2018, accessed 2 June 2020).

[5] Setyono G, Kawano DS. PENGARUH PENGGUNAAN BUSI BERELEKTRODA NIKEL, PLATINUM DAN IRIDIUM TERHADAP PERFORMA MOTOR BENSIN TORAK SPARK IGNITION ENGINE (SIE) 4 LANGKAH 1 SILINDER.

[6] Setyono G, Kawono DS. Pengaruh Penggunaan Variasi Elektroda Busi terhadap Performa Motor Bensin Torak 4 Langkah. *J SAINTEK*; 11, www.kopertis7.go.id, (2014, accessed 4 June 2020).

[7] Obodeh O, Ogbor AD. Improving the performance of two-stroke motorcycle with tuned adjustable exhaust pipe. *Res J Appl Sci Eng Technol* 2009; 1: 59–65.

[8] Mangukiya S. *Optimizing the Design of Intake & Exhaust System of a Single Cylinder Motorcycle Engine for Formula Student Vehicle*, http://www.ripublication.com (2018, accessed 6 May 2020).

[9] Sharma T, Khan Z, George J. INQUISITIVE ANALYTICS OF DIVERSE EXHAUST SYSTEM CONTINGENT TO POLLUTION. *Int Res J Eng Technol* 2008; 1226.

[10] Rojan MA, Alias Z, Bakar SA, et al. Enhancement of MODENAS CT115'S motorcycle muffler performance by optimization of backpressure. DOI: 10.1088/1757-899X/670/1/012063.
[11] Tkaczyk M. CFD Tests of the Exhaust System of a Sports Motorcycle. *J KONES* ; 25, https://content.sciendo.com/view/journals/kones/25/4/article-p429.xml (2019, accessed 6 May 2020).
[12] Xue F, Sun B. Experimental study on the comprehensive performance of the application of U-shaped corrugated pipes into reactive mufflers. *Appl Acoust* 2018; 141: 362–370.
[13] Mishra PC, Kar SK, Mishra H. Effect of perforation on exhaust performance of a turbo pipe type muffler using methanol and gasoline blended fuel: A step to NOx control. *J Clean Prod* 2018; 183: 869–879.
[14] Meena R, Ramachandra P, Dube A. A 3D-CFD Study of Flow Dynamics on Mixture Preparation for Fuel Injected Motorcycles.
[15] Lillahulhaq Z, Maulana HS. Pengaruh Model Turbulensi Aliran Terhadap Simulasi Numerik Aircurtain. *Mek J Tek Mesin* 2019; 5: 27–42.
[16] Ranjbar M, Arslan H, Dalkılıç B, et al. *On Muffler Design for Transmitted Noise Reduction Multidisciplinary Design Optimization of Auxetic Structures View project Review on applications of the AUXETIC beams View project On Muffler Design for Transmitted Noise Reduction*, https://www.researchgate.net/publication/320895026 (2017, accessed 10 May 2020).
[17] Mishra PC, Kar SK, Mishra H, et al. Modeling for combined effect of muffler geometry modification and blended fuel use on exhaust performance of a four stroke engine: A computational fluid dynamics approach. *Appl Therm Eng* 2016; 108: 1105–1118.

Physics and chemical activation to produce activated carbon from empty palm oil bunches waste

A Budianto[1*], E Kusdarini[2], N H Amrullah[1], E Ningsih[1], K Udyani[1] and A Aidawiyah[1]

[1]Department of Chemical Engineering, Institut Teknologi Adhi Tama Surabaya
[2]Department of Mining Engineering, Institut Teknologi Adhi Tama Surabaya

budichemical@itats.ac.id

Abstract. This research was conducted by laboratory-scale trials to make activated carbon from empty oil palm bunches. Empty bunches were palm oil processing industrial waste. This research was an effort to optimize waste management. Empty palm oil bunches were studied to become carbon active. Activated carbon is needed to meet the needs of the food industry and non-food industries. This research aimed to determine the appropriate method, optimal operating condition, and get the specification of standard carbon active. Making activated carbon was carried out by multiple activations; chemical and physical. The feed carbonation process was conducted at 400° C for 4 hours with N_2 gas flow. After the charcoal is formed, chemical activation is carried out using an H3PO4 activator with a concentration of 9-15% wt. Physical activity was conducted by heating the activated carbon method in 700°C for an hour. The result of the study showed that multiple activations produced activated carbon, and the quality filled the standard. H_3PO_4 concentration of 13% was the most appropriate concentration to provide activated carbon with the best quality. The best quality activated carbon, which has produced of this research has complied with Indonesian National Standards 06-3730-1995 with parameters: 3% moisture content, 3.92% ash content, 0.7% volatile matter, 95.38% bonded carbon, iodine number 945.47 mg / g, and surface area of 272.9 m_2/g.

1. Introduction

Activated carbon is activated charcoal or hollow reliable, which had an amorphous characteristic, and it is a non-graphite carbon compound [1]. Industrial needs activated carbon, both food-industrial and non-food industrial. Activated carbon can adsorb heavy metals, like Cr (VI), Pb (II), Mn (II), and Bi (III) [2,3]. The activated carbon needs for industrial are increasing. In 2018, Indonesia needed an activated carbon more than 200 tons per month, and 47% was still imported from China [4]. There are many characteristics of activated carbon varieties on the market. Generally, activated carbon has a high surface [5,6] developing pores [7,8] and good adsorption [9]. The different aspects of activated carbon are influenced by materials, activities, and production of the operating condition [10]. One of the materials of activated carbon, which often used is agricultural and plantation waste, like orange peel [11], sawdust of *Sapelli* wood [12], mangrove wood charcoal [13], tree bark [3], and palm oil waste [10]. Regarding the use of agricultural and plantation waste as a material of activated carbon, oil palm empty bunches is a potential oil palm industry in Indonesia. Oil palm empty bunches is a waste oil palm industrial which is existing in Indonesia. It is because Indonesia is the biggest producer of oil palm in Indonesia [14].

According to the United States Department of Agriculture (USDA), in 2018, Indonesia produced palm oil in 55% of the world's total palm oil production. Besides Indonesia, several countries in Southeast Asia and some countries in West Africa also produced palm oil in a large amount. Malaysia produces 28%, Thailand was 4%, Colombia was 2%, Nigeria produced 1% and 10% from other countries [15]. It showed that the potential of empty oil palm bunches was able to contribute to activated carbon in several states that produced oil palm.

The empty palm oil bunches are the most massive solid waste which is produced by the palm oil industry. Every ton of processing of fresh fruit bunches would produce empty oil palm bunches about 22 – 23% or 220 – 230 kg [16]. Nowadays, empty oil palm bunches are only utilized as manure or put it in the garbage. It is happened in many countries, especially in Indonesia. If the people can process empty palm oil bunches became activated carbon, Indonesia has an excellent opportunity to no longer import activated carbon, and otherwise, Indonesia can export it. Activated carbon is produced to meet quality standards, appropriate producing must be chosen, and the operating condition must optimal in the production process. Several scientists have learned the choosing of the processing method and operating conditions in the production process. Guo et al. explained that activated carbon was produced by a plant, *phragmites australis*, which was activated in physics, and chemist using H_3PO_4 reagent had iodine number 1017.5 mg / g and pore surface area of 337 m^2 / g [17]. Budianto et al. defined that activation in physics and chemical used H_3PO_4 reagent; it was able to activate the mangrove charcoal. So, it could be activated carbon, which had an iodine number of 1019.87 mg /g and a pore surface area of 354,977 m_2 / g [13]. Another researcher described that H_3PO_4 reagent combined with NH_4HCO_3 was able to activate sub-bituminous coal in well. So, it produced activated carbon with an iodine number of 1238,544 mg / g and a pore surface area of 86,213 m^2/g [18]. Udayani et al. explained that physics activation used a microwave and chemical activation used H_3PO_4 reagent were able to activate mangrove charcoal into activated carbon and the large of pore surface area of 936,221 m^2/g [19]. Activation of activated carbon from cassava skin has been studied using the help of a microwave with a phosphate acid chemical activator. The results show that the amount of iodine produced is 3173 mg/g [20]. Research on making activated carbon from the skin of Cocoa has also been carried out. Cocoa skin is carbonized and chemically activated with phosphoric acid and physical activation by heating. The results showed that the specific surface area of activated carbon was 211 m^2/g, and the iodine number was 1194 mg/g [21].

The research showed that physics and chemical activation using the H_3P0_4 reagent was a good enough method to produce activated carbon. But, this method has not been proven useful if it was used to activation of empty palm oil bunches. Chosen of the appropriate processing method could save the cost, and the product result met the specification. This research studied and tested multiple activation methods using a phosphoric acid chemical activator and getting activated carbon specification, which was produced from empty palm oil bunches. The result of activated carbon characteristics compared to the Indonesian National Standard (SNI) 06-3730-1995.

2. Method

2.1 Material and Tools Preparation.

There were few materials that must prepare in making activated carbon. They were empty palm oil bunches, Technical H_3PO_4 85%, technical I2, technical starch, and technical $Na_2S_2O_3.5H_2O$. The tools were porcelain saucer, oven, drop pipette, desiccator, volumetric flask, pH paper, furnace, funnel, burette, porcelain crucible, analytic balance, filter paper, beaker glass, aluminum foil, measuring cup, 60 mesh sieve, and Erlenmeyer.

2.2 *General Procedure.*

The research procedure was started by material reparation, which was heating of empty palm oil bunches inside of the furnace, and the temperature was 400^{0}C in 4 hours and flowed by N_2 gas. Then, the charcoal that has been formed was cooled down inside of the desiccator. After that, activated carbon was physics activated inside of furnace in 700°C in an hour. The next procedure was the process of charcoal cooling divided into four parts. Those four parts were chemically activated by soaking activated carbon using H_3PO_4 liquid with a concentration of 9%, 11%, 13%, and 15% for 22 hours. The last, activated carbon was washed using aqua dest until the pH of the solution becomes neutral, and then it must be dried in the oven for 2 hours.

2.3 *Analysis.*

Activated carbon from empty palm oil bunches was analyzed continuously to obtain the specification of activated carbon. The analysis was water content, ash content, volatile matter, bound carbon, iodine number, and surface area. Analysis of water content, ash content, and the volatile matter was using the ASTM method. Fixed carbon calculated based on equation (1).

Fixed Carbon = 100% - (water content + volatile matter + ash content) (1)

Activated carbon with the best specification results was measured its surface area using the BET method.

3 Result and Discussion

There were four types of activated carbons that were produced from the result of the study. It consisted of the concentration of H_3PO_4 activators. The parameters were water content, ash content, volatile matter, bound carbon, and iodine number. Those four activated carbons were presented in Fig.1 until Fig. 5. The measured parameters from activated carbon were compared to the Indonesian National Standard (SNI) 06-3730-1995.

3.1 Water Content

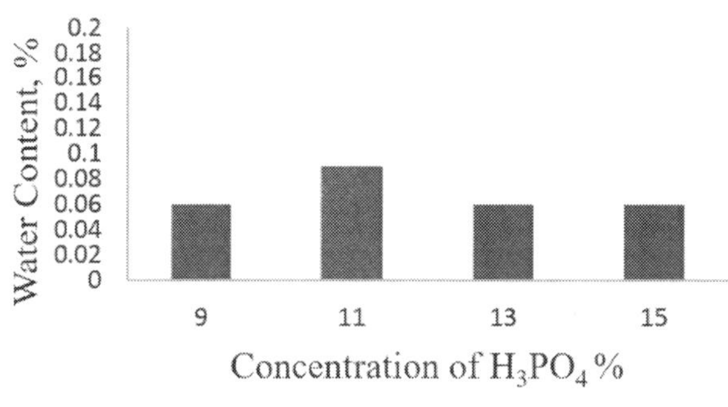

Figure 1. Water content from activated carbon which was activated with four types of H_3PO_4 concentration

Figure 1. showed the amount of content water which was contained in activated carbon from empty palm oil bunches. The water content was 0,06 - 0.09 %. It presented that the process chosen was beneficial to decrease water content, especially in the physics activation process. In this study, water content which was contained in activated carbon met the quality standard of activated charcoal-based on SNI. 06-3730-1995, which was a maximum of 15% to activated charcoal in the form of powdered [22]. Overall, the water content in activated carbon was better than activated carbon, which was produced by Budianto et al. in their research. The materials used were mangrove and activating agent of H_3PO_4 with content water of 0.11%. Low water content showed that free water content and bound water contained inside of the materials have evaporated during the carbonation and physics activation process. High water content was caused by several factors, like well-bound water, which had forms both steam and liquid trapped into the charcoal molecule, and they did not go out in the heating process at low temperature. At low temperatures, the C atom bond in charcoal has not yet been broken down by heat. So, the water vapor was still trapped inside of the C atom molecule bond, which was between C atom and another atom. Conversely, the higher activation temperature and activation duration were, the more water was existing in the cavity of the C atom bond that came out and evaporated. It was because the C atom bond was opened inside of the charcoal when it got a high temperature, and it made the micropore formation became larger [23].

3.2 Ash Content

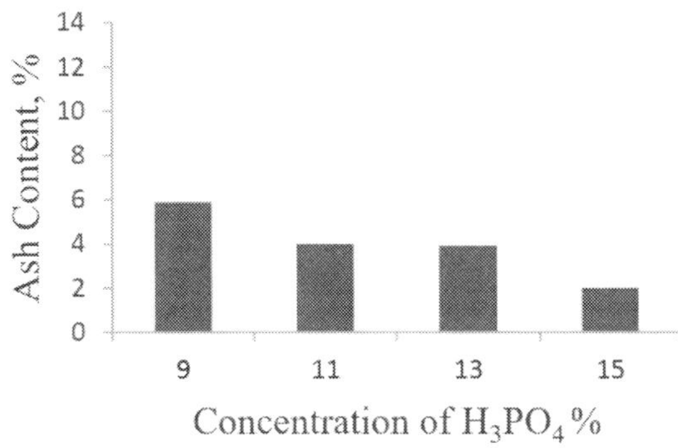

Figure 2. Ash content from activated carbon was activated by four types of H_3PO_4 concentration

Figure 2. showed that ash content which was contained inside of activated carbon from empty palm oil bunches charcoal. Ash content of activated carbon was about 2% - 5.88%. Ash content was only appeared to have a little different. It explained that the concentration different used in chemical activation using H_3PO_4 agent activation was not really giving influence to the ash content. Ash content is the rest of the combustion, which has no carbon element and calorific value. Ash content value describes the number of residue from the end of the combustion process in the forms of minerals that are not lost during the combustion process. Carbon active makers can reduce ash carbon if they use higher temperatures while they make it. In this research, ash carbon met the activated carbon charcoal based on SNI. 06-3730-1995, which maximally 10% for powder active charcoal [22]. In this study, the smallest ash carbon contained in activated carbon was 2%, and it was better than activated carbon, which made from mangrove material, 6.8% [13].

3.3 Volatile Matter Level

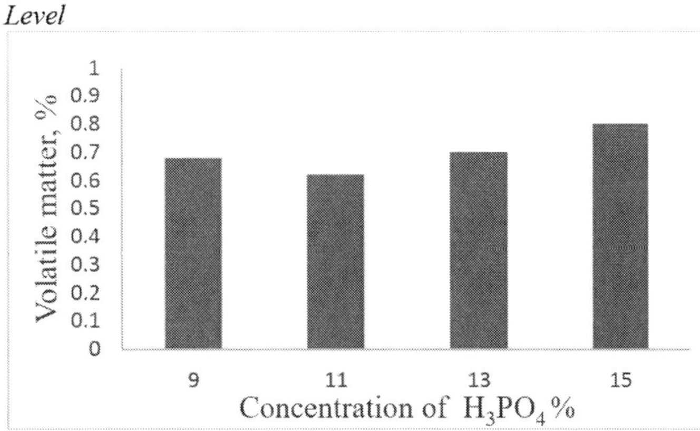

Figure 3. Volatile matter level from activated carbon from four types of H_3PO_4 concentration

Figure 3. described the volatile matter level from four samples with the variation of H_3PO_4 concentration were between 0.62-0.8%. Activating agent concentration used was not really influence the volatile matter level, which was contained inside of activated carbon. The volatile matter level provided inside of activated carbon has met activated carbon quality standards based on SNI. 06-3730-1995, which was maximally 25% for powder activated carbon [22]. In this research, volatile matter level in the activated carbon produced smaller than activated carbon, which was activated by the same reagent of sub-bituminous coal with volatile matter level 39.1-45% as the material [18] and mangrove charcoal with volatile matter 19.8-23.4% [13]. The amount of volatile that is determined by the total amount of organic delegated during the pyrolysis process [23]. If the pyrolysis process took a long time and the temperature was increased, it would make more volatile matter wasted. So that it would get lower the level of volatile matter. Increasing the temperature of carbonization would evaporate volatile compounds that were still left, especially *tar*-. It would cause the number of pores formed to increase. Carbon with this condition might be used as activated carbon with a surface that was not covered by polar compounds, so it had the ability to adsorb [24].

3.4 Fixed Carbon

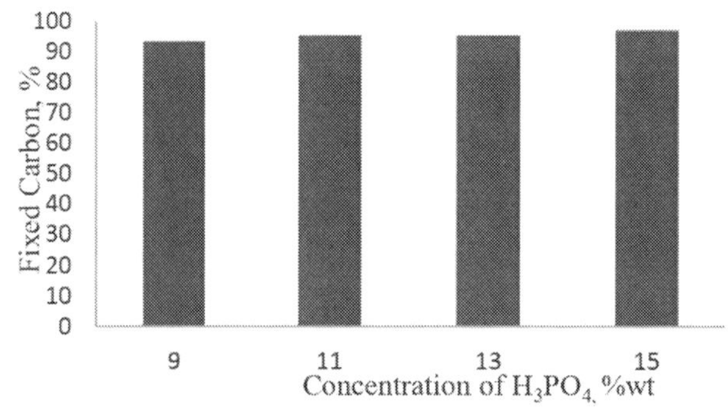

Figure 4. Fixed carbon level from activated carbon which was activated by four types of H_3PO_4 concentration

Figure 4. showed that the lowest fixed carbon level was 84.64% in H_3PO_4 strength nine %wt, and the highest fixed carbon was 97,20% in H_3PO_4 concentration 15%. If fixed carbon was referred to the quality standard requirement for activated carbon quality SNI. 06-3730-1995, in this research, bound activated carbon, which was produced by activated carbon, has met the standard, which is minimally 65% [22]. High and low bound carbon level is influenced by the content of volatile matter level and ash matter. The lower matter volatile level and ash matter get, the higher bound carbon level gets. Fixed carbon was influenced by cellulose and lignin content, and it could be converted to the carbon atom [24].

3.5 Iodine Number

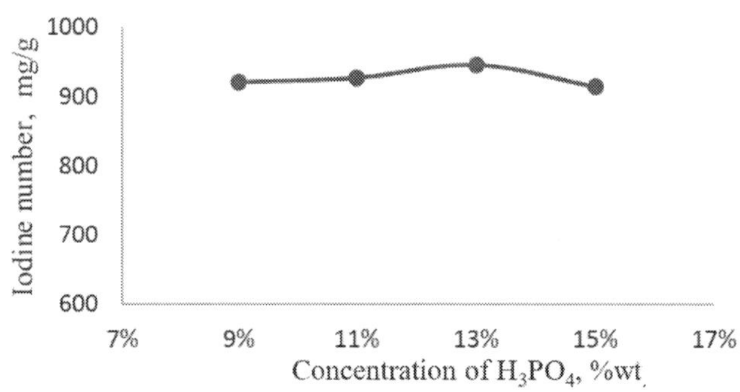

Figure 5. Iodine adsorption of activated carbon which was activated from four types of H_3PO_4 concentrations

Figure 5 explained that I2 adsorption (iodine number) was about 913.8-945.6 mg/g. Iodine number tended to increase with increasing concentration of H_3PO_4 solution. The use of 9%, 11%, 13%, and 15% H_3PO_4 solution produced activated carbon with an iodine number that has met SNI. 06-3730-1995, which was a minimum of 750 mg/g [22]. The research showed that in the concentration of 9%, 11%, 13%, and 15% H_3PO_4 solution increasing of iodine number was associated with H_3PO_4 concentration. This was because the higher of H_3PO_4 concentration raised, the stronger chemical solution bounded tar- compounds residual of carbonation to pass pores micro from palm oil bunches. Increased adsorption is possible because there are so many carbon atoms that form hexagonal crystallites. It caused pores formed between the layers of crystallites to become greater [25]. The decrease of iodine number at 15% H_3PO_4 concentration indicated the saturation in the H_3PO_4 solution in the process of activating the pores from activated carbon. This saturation caused the formed of the P_2H_5 compound, and it was the result of excess H_3PO_4 decomposition, and it was not trapped into activated carbon. P_2H_5 compound is a micropore and mesoporous structure in the inner structure [25]. Excess P_2H_5 compound, which was not trapped in activated carbon, caused the adsorption of iodine from activated carbon to decrease.

3.6 Surface Area

Physics and chemical activation could help the formation of micropore and mesoporous structures in the surface of activated carbon, so it could make larger activated surface larger. Adding of H_3PO_4 chemical activator made the activated charcoal sample more transparent or thinner, so the carbon contact power would be more significant. The surface area of activated carbon with the best iodine number, which was activated with 13% H_3PO_4, was 272.9 m^2/g. It showed that volatile matter and *tar*-were increasingly detached from activated carbon because of the use of the H_3PO_4 solution. H_3PO_4 solution was caused mesopore structure to have a surface area, and pore volume became bigger [26].

4 Conclusion

Empty palm oil bunches were proved to produce became activated carbon, and the quality met SNI. 06-3730-1995. The appropriate method and produce the best-activated carbon was physics and chemical multiple activated methods. Physics activated with charcoal heating was having 700^0c for the temperature for an hour while chemical activated with phosphate acid marinade was having 9-15%. All of the operation conditions produced activated carbon, which met SNI. The best-activated carbon, which was produced, had an iodin number up to 945,6 mg/g, and the pore surface area was until 272.9 m_2/g.

5 References

[1] Mayyas M and Sahajwalla V 2019 Carbon nano-sponge with enhanced electrochemical properties: A new understanding of carbon activation *Chem. Eng. J.* **358** 980–91
[2] Rao H J, King P and Kumar Y P 2016 EXPERIMENTAL INVESTIGATION ON ADSORPTION OF LEAD FROM AQUEOUS SOLUTION USING ACTIVATED CARBON FROM THE WASTE RUBBER TIRE:OPTIMIZATION OF PROCESS PARAMETERS USING CENTRAL COMPOSITE DESIGN *Rasayan J. Chem.* **9** 254–77
[3] Wanjari A K and Chaudhari U E 2017 REMOVAL OF Cr(VI), Pb(II), Mn(II) and Bi(III) FROM AQUEOUS SOLUTIONS USING GRANULAR ACTIVATED CHARCOAL PREPARED FROM Cordia Macleodii TREE BARK *Rasayan J. Chem.* **10** 82–5
[4] Hermansyah 2018 *Mini Plant Karbon Aktif Kapasitas 1 Ton per Hari Dibangun di Samarinda*
[5] Khan J H, Marpaung F, Young C, Lin J, Islam M T, Alsheri S M, Ahamad T, Alhokbany N, Ariga K, Shrestha L K, Yamauchi Y, Wu K C-W, Hossain M S A and Kim J 2019 Jute-derived microporous/mesoporous carbon with ultra-high surface area using a chemical activation process *Microporous Mesoporous Mater.* **274** 251–6
[6] Zhang Y L, Li S Y, Tang Z S, Song Z xing and Sun J 2019 Xanthoceras sorbifolia seed coats

derived porous carbon with unique architecture for high rate performance supercapacitors *Diam. Relat. Mater.*

[7] Kesavan T, Partheeban T, Vivekanantha M, Kundu M, Maduraiveeran G and Sasidharan M 2019 Hierarchical nanoporous activated carbon as potential electrode materials for high performance electrochemical supercapacitor *Microporous Mesoporous Mater.* **274** 236–44

[8] Sharma M, Joshi M, Nigam S, Shree S, Avasthi D K, Adelung R, Srivastava S K and Kumar Mishra Y 2019 ZnO tetrapods and activated carbon based hybrid composite: Adsorbents for enhanced decontamination of hexavalent chromium from aqueous solution *Chem. Eng. J.* **358** 540–51

[9] Altintig E, Onaran M, Sarı A, Altundag H and Tuzen M 2018 Preparation, characterization and evaluation of bio-based magnetic activated carbon for effective adsorption of malachite green from aqueous solution *Mater. Chem. Phys.* **220** 313–21

[10] Ayinla R T, Dennis J O, Zaid H M, Sanusi Y K, Usman F and Adebayo L L 2019 A review of technical advances of recent palm bio-waste conversion to activated carbon for energy storage *J. Clean. Prod.* **229** 1427–42

[11] Lam S S, Liew R K, Wong Y M, Yek P N Y, Ma N L, Lee C L and Chase H A 2017 Microwave-assisted pyrolysis with chemical activation, an innovative method to convert orange peel into activated carbon with improved properties as dye adsorbent *J. Clean. Prod.* **162** 1376–87

[12] Thue P S, dos Reis G S, Lima E C, Sieliechi J M, Dotto G L, Wamba A G N, Dias S L P and Pavan F A 2017 Activated carbon obtained from sapelli wood sawdust by microwave heating for o-cresol adsorption *Res. Chem. Intermed.* **43** 1063–87

[13] Budianto A, Kusdarini E, Effendi S and Aziz M 2019 The Production of Activated Carbon from Indonesian Mangrove Charcoal *Mater. Sci. Eng.* **462** 1–8

[14] Santika T, Wilson K A, Budiharta S, Law E A, Poh T M, Ancrenaz M, Struebig M J and Meijaard E 2019 Does oil palm agriculture help alleviate poverty? A multidimensional counterfactual assessment of oil palm development in Indonesia *World Dev.* **120** 105–17

[15] USD.A 2019 Palm Oil Production by Country in 1000 MT [WWW Document]

[16] Fuadi A M and Pranoto H 2016 Pemanfaatan Limbah Tandan Kosong Kelapa Sawit Sebagai Bahan Baku Pembuatan Glukosa *Chemica* **3** 1–5

[17] Guo Z, Zhang A, Zhang J, Liu H, Kang Y and Zhang C 2017 An ammoniation-activation method to prepare activated carbon with enhanced porosity and functionality *powder Technol.* **309** 74–8

[18] Kusdarini E, Budianto A and Ghafarunnisa D 2017 Produksi Karbon Aktif dari Batubara Bituminus dengan Aktivasi Tunggal H3PO4, Kombinasi H3PO4-NH4HCO3, dan Termal *Reaktor* **17** 74–80

[19] Udyani K, Purwaningsih D Y, Setiawan R and Yahya K 2019 Pembuatan Karbon Aktif Dari Arang Bakau Menggunakan Gabungan Aktivasi Kimia dan Fisika Dengan Microwave *J. IPTEK* **23** 39–46

[20] Purwaningsih D Y, Budianto A, Ningrum A A and Kosagi B T 2019 Produksi Karbon Aktif Dari Kulit Singkong Dengan Aktivasi Kimia Fisika Menggunakan Gelombang Mikro *Seminar Nasional Sains dan Teknologi Terapan VII* ed LPPM ITATS (Surabaya: LPPM ITATS) pp 663–70

[21] Budianto A, Romiarto R and Fitrianingtyas F 2016 PEMANFAATAAN LIMBAH KAKAO (Theobroma cacao L) SEBAGAI KARBON AKTIF DENGAN AKTIFATOR TERMAL DAN KIMIA (Surabaya: LPPM ITATS) pp 1–8

[22] BSN 1995 Standar Nasional Indonesia untuk Karbon Aktif Teknis SNI 06-3730-1995

[23] Zhang Y, Cong Y, Zhang J, Li X, Li Y, Dong Z, Yuan G, Zhang J and Cui Z 2018 Effects of activation temperatures on the surface structures and supercapacitive performances of porous carbon fibers *Surf. Coatings Technol.* **349** 384–91

[24] Pari G 2004 *Disertasi, Kajian Struktur Arang Aktif dari Serbuk Gergaji Kayu sebagai*

Adsorben Emisi Formaldehida Kayu Lapis
[25] Yue Z, Economy J and Mangun C L 2003 Preparation of fibrous porous materials by chemical activation 2. H3PO4 activation of polymer coated fibers *Carbon N. Y.* **41** 1809–17
[26] Baquero M C, Giraldo L, Moreno J C, Suárez-Garcı́a F, Martı́nez-Alonso A and Tascón J M D 2003 Activated carbons by pyrolysis of coffee bean husks in presence of phosphoric acid *J. Anal. Appl. Pyrolysis* **70** 779–84

Acknowledgment
The researchers say thanks to LLDIKTI. Region VII and DPRM Ministry of Research and Technology of the Republic of Indonesia, which helped fund this research. We are also thankful to all of the leadership of YPTS-ITATS, which supported us in doing this research. The thankful is also we deliver to LPPM and department of chemical engineering, which had supported and gave us motivation for the success of this research.

IOP Conf. Series: Materials Science and Engineering 1010 (2021) 012017 doi:10.1088/1757-899X/1010/1/012017

Monitoring of road damage detection systems using image processing methods and Google Map

R Sulistyowati[1], A Suryowinoto[1], H A Sujono[1] and I Iswahyudi[1]

[1]Institut Teknologi Adhi Tama Surabaya, East Java, Indonesia

riny.971073@itats.ac.id

Abstrak. In the traffic environment, accidents often occur due to road conditions with holes. This can be fatal for motorcycle riders and car drivers. Therefore, this paper aims to provide a road contour damage information system by detecting potholes on the highway and reporting damage on Google's maps. So road users, especially vehicle drivers, can be careful when they want to pass that road. The method used in this paper, which uses Werner D. Streidt algorithm threshold values and edge detection in digital image processing to detect holes in the road, as well as mark the coordinates of hole locations and uploads them on Google's maps, using the Raspberry Pi two embedded devices as a management center data, with a 6M NEO U-box Global Positioning System (GPS) sensor, as geotagging location and CSI camera interface. A A raspberry pi 2 for taking pictures of road surface contours. System testing is carried out in several stages at different times. In daytime conditions, in sunny weather, the system's success rate in detecting potholes is 83.2%. Whereas with tree barriers, the success rate is 67%, with detection failure 33%. In conclusion, this system succeeded in showing all points of the hole location by giving a mark on the Google Map display application.

1. Introduction

Damage to roads is not a strange thing for Indonesians. Most of the Indonesian people have often encountered damage to a road segment both on alternative roads and main roads. The types of damage are very diverse, generally in the form of cracks (cracking), in the form of waves (corrugation), in the form of grooves / depressions in the direction of extending the road around the track of the vehicle wheels (rutting), in the form of asphalt puddles on the road surface (bleeding), and some are in the form of holes (pothole) [1]. The type of damage that most motorists cause accidents is in the form of a hole (pothole).

Hole detection used by related parties is currently relatively simple, which still relies on information from the public and it is not uncommon for this information to be obtained after an accident occurs at the location of the hole. With these problems, it is necessary to create a road condition information system where this system can segment the location and real conditions in the form of images of holes on the highway. To perform the data segmentation process in the form of images, a special method such asage processing is needed.

From these problems, a research is proposed with the title, "Monitoring Of Road Damage detection systems using image processing methods and Google Map.

IOP Conf. Series: Materials Science and Engineering 1010 (2021) 012017 doi:10.1088/1757-899X/1010/1/012017

1.1 Literature Review

Road damage is caused by many things. Among other things such as cracked roads, bumpy, hollow road [1]. One of the road damages that caused an accident for motorists and drivers was a hole. One of them is by doing early detection of the presence of holes in the road. In 2015, a research was conducted on the application of image processing to the detection of holes on the highway by T. N. R. Kumar [2]. In this study, the detection of holes was carried out using the K-Mean clustering method where it was concluded that the method used had successfully detected holes but did not mention the effectiveness of its success. In 2015, S. Nienaber et al. [3] also carried out a research on the detection of holes on the highway. In this study, the convex hull algorithm method was used where it was concluded that the success of this system was 81.8%. Many studies that discuss information, among others, by SMS gateway aims to develop information systems that can determine the state of conditions or conditions, among others [4-6]. In a study on image processing applications for hole detection on the highway by Himanshu Punjabi, et al. [7]. In this study, the effectiveness of the application of image pixel counting methods, first-order moments, moments or seconds, and cross-correlation where of the four methods the most effective is the current method or the first and second moments with each method. has a strong effect. 100% but not an accurate tool testing method. The five studies above are used as reference and evaluation materials in designing a road damage information system application using image processing in this study. In this study, an information system was developed to detect potholes and location coordinates, and to classify road images. In addition, it can be identified damage to potholes in using the drawing process, as well as marking data, at the location of the existence of potholes, in a form that also marks the Google Maps application. The benefits that can be given are providing preliminary information to road users, so users can take precautions by reducing vehicle speed and increasing vigilance when driving through the road. An important part of this research is the Google Maps *Application Programming Interface* (API)[8] special Google's server map [9] which includes a database of satellite imagery, road views, altitude profiles, directions, maps with a touch of style, demographics, analysis, and large databases and can be equipped with various specific attributes according to the desire of the programmer, to build areal time application to estimate the camera pose[10]. In the information system to detect information on hole damage on the highway, among others. Figure 1 shows a map of the location that is integrated with the image display.

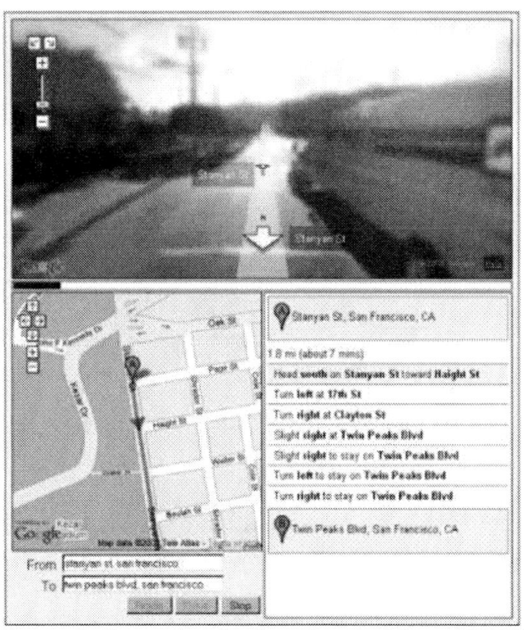

Figure 1. Example of using Google Maps API [6]

1.1.1 Global Positioning System (GPS)

The navigation system that provides position and time information using satellites. The GPS[8] receiver has data output related to the information above. This data was issued by the association of marine electronic device manufacturers in America, NMEA (National Marine Electronics Association) under the name NMEA-0183 with the header $ GPGGA [10] decoding explanation in Table 1. In this study, using the GPS module NEO 6 GPS block, as shown in figure 2, with a high-performance position tracking machine, with programming application interface [11-13] using serial TTL (RX / TX), which can be accessed from a microcontroller with UART Function and the default baud speed is set at 9600 bps. Process up to 50 signal channels quickly with Cold TTFF (Cold Start Time-To-First-Fix

Tabel 1. The GPS code report of NMEA-0183 with header $GPGGA [10]

Position	Content	Example	Explanation
1	Header	SGPGGA	GP is for GPS receiver, GGA – NMEA sentence type
2	UTC time in hours, minutes, and seconds	090507	Time for fix is 5 min 7s after 9 am (GMT)
3	Latitude	3540.468	35° 40.4689° (decimal min not s)
4	N or S	N	North
5	Longitude	13931.896	1.39° 31.8960°
6	E or W	E	East
7	GPS quality indicator (0=invalid; 1= standalone; 2= Diff, fix)	1	Single positioning fix
8	Number of satellites in use	8	Not all satellite in view
9	HDOP	1.42	Horizontal dilution of position
10	Antenna altitude	87.672	Above or below geoid (mean sea level)
11	Antenna heigh unit	M	Meter
12	Geoidal separation	29.1	Difference between WGS-84 ellipsoid and geoid
13	Units of geoidal separation	M	Meter
14	DGPS update age	10	Time in second since last update from reference station
15	Reference station ID number	0515	
16	checksum	*5B	Always begin *

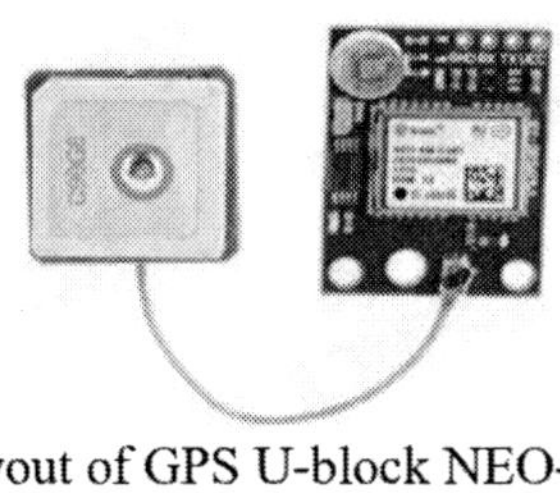

Figure 2. Layout of GPS U-block NEO-6m board [11]

1.1.2. Determine the Image Model

Known as 'image' can be interpreted as a representation or similarity of an object. Images are divided into two types namely analog images and digital images. Digital images are divided into three types based on their intensity, including, color images, gray scale images and binary images. Image contrast and brightness settings are the operating point for each pixel of the image. To make images brighter or darker, is by changing the brightness of the image by adding (or reducing) constant b to each pixel, I (x, y) in the image. Meanwhile, to increase the image contrast multiplying each pixel value, I (x, y) with a constant can be determined in equation 1.

$$I'(x, y) = a \times I(x, y) + b \tag{1}$$

To determine the contrast and brightness change algorithm used by Werner D. Streidt can do clipping on the image. Mathematically, the algorithm can be divided into two conditions namely if the contrast value is more than 0 and the contrast value is less than 1. The equation used can be seen in equations 2 and 3.

$$\begin{aligned} delta &= \frac{127 \times contrast}{100} \\ a &= \frac{255}{(255 - delta \times 2)} \\ b &= a \times (brightness - delta) \end{aligned} \tag{2}$$

Whereas contrast values of less than 1 indicate the equation 3

$$\begin{aligned} delta &= \frac{-128 \times contrast}{100} \\ a &= \frac{(256 - delta \times 2)}{255} \\ b &= a \times brightness + delta \end{aligned} \tag{3}$$

1.1.3 Comma Separated Value (CSV)

Separated Value is a file storage format in Excel programs where data or values contained in CSV files are separated by comma (,) characters. If you change the row, it is considered new data. One reason information is stored in CSV format compared to standard format (XLS) or XLSX is that files can be exchanged with many people who do not use Excel or for export / import data into database software.

2. Research methodology

In this section, we explain how a hole in a highway is detected using the image method, and the Google map can be seen in Figure 3.

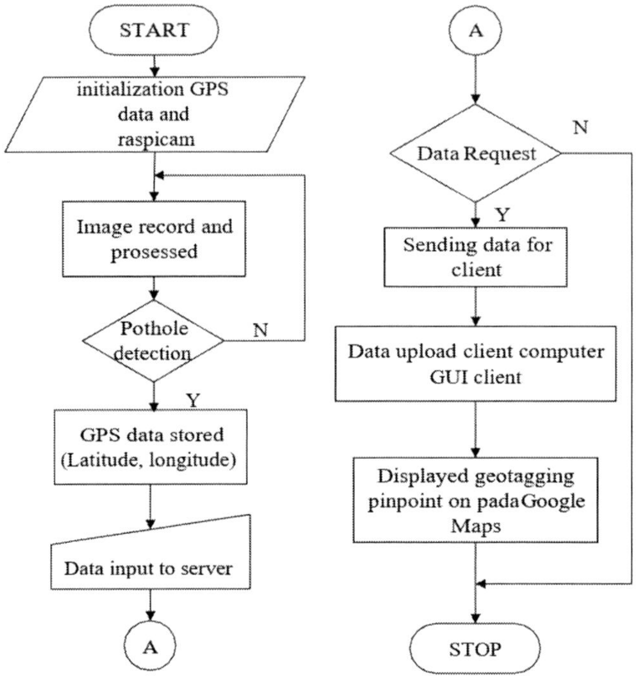

Figure 3 . Flowchart system

From the flowchart system flow in Figure 2 can be explained as follows.

The first step is to determine the location with the detection of the highway hole. Furthermore, the second step is detected by using an algorithm on unique hole characteristics in the form of changes in the color contrast between the area around the hole with bright colors and the edge area of the hole with intensities that tend to be dark. The third step is to take live video, then determine the ROI with a size of 640 x 480 pixels, convert it to a gray scale using the Werner D Striedt algorithm, then detect the edge of the hole using the Canny method. The fourth step determines the average pixel value of the image with an average value (threshold) ranging from 0.65 to 130. The next step is to save the image requesting location data from the GPS U-block NEO-6m, and save the location data in a Comma Separated Value (CSV) file. Extracting latitude and longitude data that links access to the Google map API script in GMLib, then creates pinpoint geotagging locations. The ublox NEO-6m GPS module serves as a means to determine the location of the hole's existence with data output in the form of degrees of longitude and latitude. The GPS module is connected to Arduino via the transmitter pin (Tx) of the GPS which is connected to the Arduino 2 pins. The Receiver (Rx) pin is connected to the Arduino 3 pins. To change the function of pin 2 and pin 3 so that it functions as Tx and Rx, serial library software used. Next, the final step is a microSD card breakout board that functions as a data storage location for holes in the form of CSV files. The process of identifying location data requests is run when the system detects a hole in the image. The request is in the form of sending 'get' character data from raspberries to Arduino using the command. Terminal access on UNIX uses the command "struct termios serial; specified characters to be sent, char * str = (char *) "get". The interconnection process between Raspi Cam, Raspberry Pi 2 and Arduino, and the NEO-6m U-block GPS module can be shown in Figure 4.

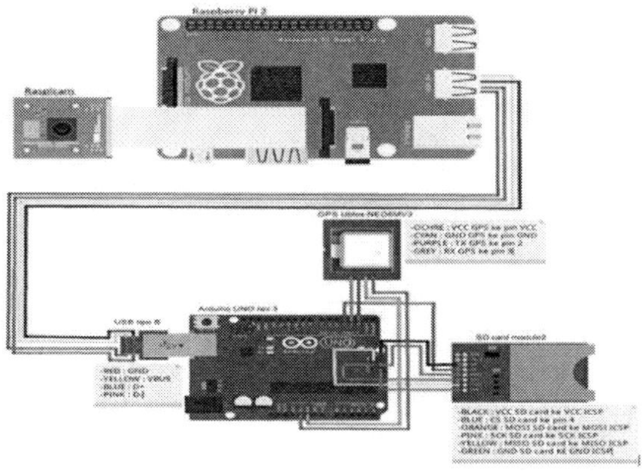

Figure 4. The system interconnection between RaspiCam, Raspberry Pi 2 with Arduino, and GPS U-block NEO-6m module

3. Result and discussion

The results of system testing, from experiments conducted on two road objects with criteria, namely the first test road without lines and signs with lots of tree shadows, and the second test road without lines and signs with few tree shadows. Where each test has five subject periods is due. The testing time starts from 09.00 until 14.00 WIB (Western Indonesia Time). It is around 08.00 hours from GMT. The vehicle's average speed is 20 km/h to 40 km/h when testing the system can be shown in the Table 2. The two type of the road we where test, is to determine the accuracy of potholes detection based on type of the road, and difference vehicle speed, when the system capture the image for image processing, detection and geo tagging can be displayed in the Table 3.

Table 2. Testing on road 1, vehicle speed on average 20k m/h, road without lines and signs with lots of tree shadows

Number of testing	Number of potholes	Potholes detected	Potholes not detected	Miss detect
I (09.00-10.00)	12	6	5	1
II (10.00-11.00)	12	10	2	0
III (11.00-12.00)	12	11	1	0
IV (12.00-13.00)	12	11	1	1
V (13.00-14.00)	12	10	2	0
Percentage		80 %	20 %	3,2 %

Table 3. Testing on the 2nd road, the average vehicle speed of 40 km/h, the road without lines and signs with little shadow.

Number of testing	Number of potholes	Potholes detected	Potholes not detected	Miss detect
I (09.00-10.00)	9	9	0	0
II (10.00-11.00)	9	9	0	0
III (11.00-12.00)	9	9	0	0
IV (12.00-13.00)	9	9	0	0
V (13.00-14.00)	9	9	0	0
Percentage		100 %	0%	0%

As the result of twice system testing on Table 2 and Figure 5, Table 3 and Figure 6 the system performs better with little error on the road without lines and signs with little shadow. From the Werner D. Streidt algorithm (as on equation 2 and 3 the values of a and b are used in several experiments on a number of different hole image samples. From these experiments, it is known that the contrast and brightness values that generally can show the characteristics of a maximum aperture are 200 and 100, as shown in the following Table 4.

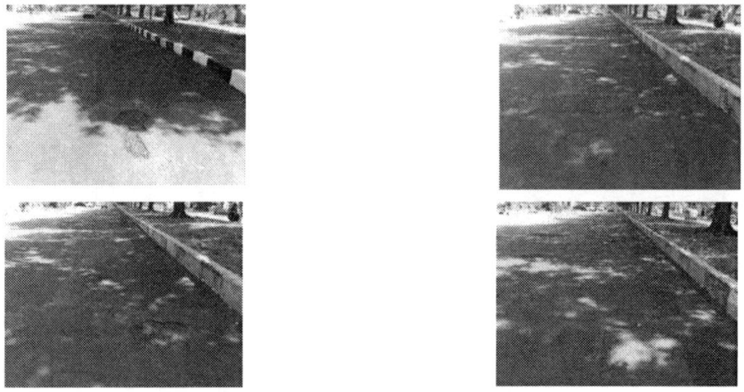

Figure 5. Image data of test results on a road without markers with lots of shadows

Figure 6 Image of test results on an unmarked road with a little shadow

Table 4. Testing parameter of Werner D. Streidt algorithm for contrast and brightness road characteristics

Brightness, contras	**image sampling 1**	**Image sampling 2**	**Image sampling 3**
0,0	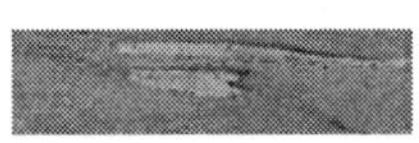		
50,100			
100,150	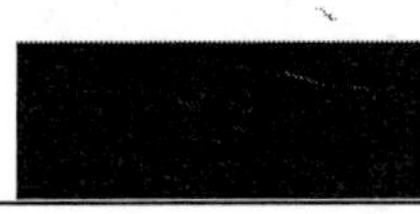		

Brightness, contras	image sampling 1	Image sampling 2	Image sampling 3
100,200			
150,200			

In testing the system, the data will be shown the suitability of the detection of pothole by requesting location from GPS u-boxNEO-6m, through the Google Maps GUI application. The results of the test, the data is displayed in a GUI application such as in Figure 7 that has its parameters stored, in CSV format, complete with satellite view of the road.

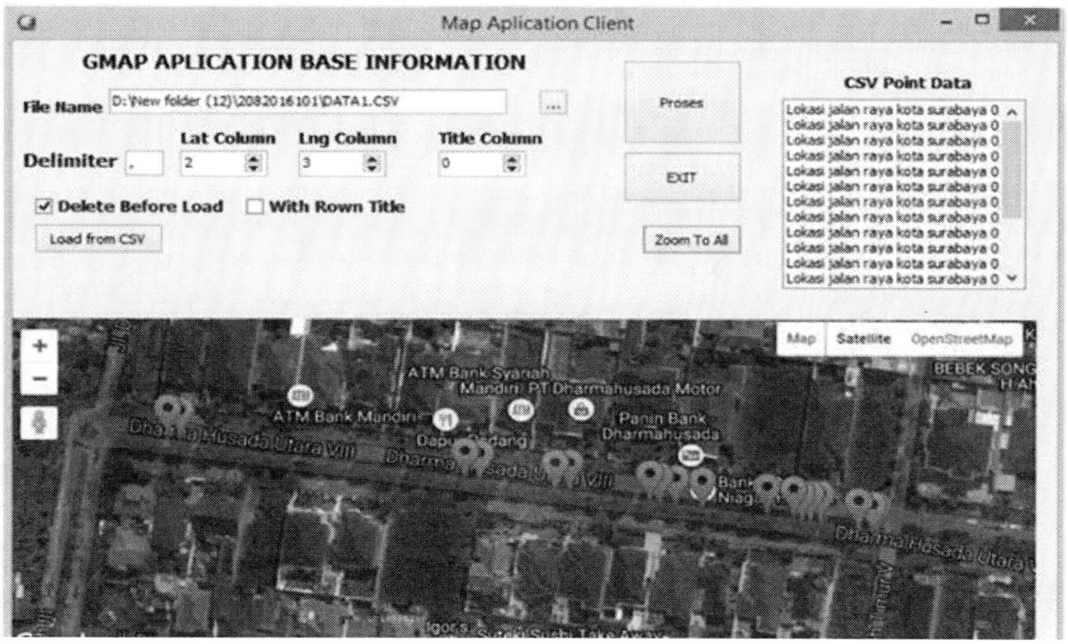

Figure.7 Data storage and geotagging on CSV point data

4. Conclusions

System testing is carried out in several stages at different times and type of roads. In daytime conditions, in sunny weather, the system's success rate in detecting potholes is 83.2%. Whereas with tree barriers, the success rate is 67%, with detection failure 33%. In conclusion, this system succeeded in showing all points of the pothole location by giving a mark on the Google Map display application.

5. References

[1] Hall, Kathleen T. dkk, Rehabilitation Strategies for Highway Pavement, National Cooperative Highway Research Program, Amerika Serikat, 2001.
[2] Kumar, T.N.R., A Real Time Approach for Indian Road Analysis using Image Processing and Computer Vision, Department of Computer Science M.S.Ramaiah Institute of Technology Bangalore, India, 2015.
[3] Nienaber, S. dkk, Detecting Potholes using Simple Image Processing Techniques and Real-World Footage, Department of E&E Engineering Stellenbosch University, Afrika Selatan, 2015.
[4] R. Sulistyowati, A. Suryowinoto, A. Fahruzi, and M. Faisal, "Prototype of the Monitoring System

and Prevention of River Water Pollution Based on Android," IOP Conf. Ser. Mater. Sci. Eng., vol. 462, no. 1, 2019, doi: 10.1088/1757-899X/462/1/012028.

[5] R Sulistyowati, HA Sujono, AK Musthofa "A river water level monitoring system using android-based wireless sensor networks for a flood early warning system". Proceedings of Second International Conference on Electrical Systems Proceedings of Second International Conference on Electrical Systems, Technology and Information, 401-408 , ICESTI.2015,Springer.

[6] R Sulistyowati, HA Sujono, AK Musthofa, " Design and field test equipment of river water level detection based on ultrasonic sensor and SMS gateway as flood early warning ,R Sulistyowati, HA Sujono, AK Musthofa, AIP conference proceedings 1855 050003, https://doi.org/10.1063/1.4985517, 2017.

[7] Buza, Emir dkk, Pothole Detection with Image Processing and Spectral Clustering, Faculty of Electrical Engineering Department for Computer Science and Informatics Zmaja od Bosne bb. Kampus Univerziteta, Bosnia Herzegovina, 2013

[8] Punjabi, Himanshu dkk, Intelligent Pothole Detection System, Department Thadomal Shahani Engineering College, India, 2014

[9]. Bray, M., "Application Programming Interface", The Software Engineering Institute, 1997.

[10] M. Pérez, I. Strictest, C. Confidence, P. One, I. Strictest, and C. Confidence, Mastering OpenCV, no. January. 2003.

[11] G. R. Bradski and A. Kaehler, Learning OpenCV - computer vision with the OpenCV library: software that sees. 2008.

[12] I. G. Petrovski, "GPS, GLONASS, galileo, and beidou for mobile devices," GPS, GLONASS, Galileo, BeiDou Mob. Devices, vol. 9781107035843, pp. 1–312, 2012, doi: 10.4324/9781139565455.

IOP Conf. Series: Materials Science and Engineering 1010 (2021) 012018 doi:10.1088/1757-899X/1010/1/012018

Effect of welding methods for different carbon content of ss304 and ss304l materials on the mechanical properties and microstructure

VA Setyowati[1*], Suheni[2], F Abdul[3], and S Ariyadi[4]

[1,2,4] Mechanical Engineering Department, Institut Teknologi Adhi Tama Surabaya, Indonesia
[3] Materials and Metallurgical Engineering Department, Institut Teknologi Sepuluh Nopember, Surabaya, Indonesia
Corresponding Author*, E-mail address: vuri@itats.ac.id

Abstract. Austenitic stainless steel types of SS304 and SS304L are most widely used in the industrial manufacturing because it has excellent weldability and corrosion resistance. It is undeniable that welding is one of the common steps in the manufacturing process and repairing damaged equipment. In this study, three different welding methods including TIG (Tungsten Inert Gas), SMAW (Shielded Metal Arc Welding), and MIG (Metal Inert Gas) are described to compare mechanical properties and morphological analysis. The characterizations were conducted by tensile test, XRD, and Metallography. The results of this study indicate that MIG causes higher tensile strength compared to TIG and SMAW, the highest tensile stress value occurs in SS304 material with MIG welding of 679.64 MPa, while the lowest tensile stress value in SS304L material with SMAW welding of 604.89 MPa. Based on microstructure analysis, Both materials present austenite, ferrite, and Cr carbide phases.

1. Introduction

Stainless steel is an alloy steel with high alloy content (high alloy steel) and has suitable properties for corrosion and high-temperature resistance. Corrosion-resistant properties are caused by the stable oxide layer (especially Chrome) that adheres to the surface and protects the steel against a corrosive environment. Stainless steels are commonly used for the manufacture of vacuum tubes, plasma storage components, supporting structures, and various other subsystems [1]. SS304 alloy steel is a type of stainless steel austenitic stainless steel that composition of 0.042% C, 1.19% Mn, 0.034% P, 0.006% S, 0.58% Si, 18.24% Cr, 8.49% Ni, and the rest Fe. Some mechanical properties owned by type SS304 stainless steel include tensile strength of 646 MPa, yield strength of 270 MPa, 50% elongation, hardness of 82 HRB.

SS304L is used as a defense material and nuclear field because of its corrosion resistance in the excellent seawater environment. Such properties are due to having molybdenum content to prevent corrosion due to the element chloride. Besides, low carbon content can increase the resistance to grain boundary corrosion (intergranular corrosion) [2]. SS304L alloy steel is a type of stainless steel austenitic stainless steel which has a composition of 0.027% C, 0.39% Si, 1.61% Mn, 0.025% P, 0.011% S, 8.01% Ni, 18.27% Cr, 69.24% Fe. The low carbon content of 0.03 wt% in SS 304L causes a small amount of carbide precipitation to form. Therefore, SS304L is suitable for welding and used as a power plant application [3]. The chemical composition of SS304 and SS304L alloys is the same except the carbon content. SS304 has a higher carbon content of 0.8% due to high carbides content. It is also applied as pressure vessel reactor materials [4].

SS application for industrial equipment cannot be separated from the welding process. Welding is an attempt to connect the material both the same material and different materials using heat energy and affect both mechanical properties, metallurgical structure, deformation, and thermal stress. Welding techniques need to be considered because they affect the mechanical properties of the material and the performance of the equipment produced. When welding occurs, there are several problems involving changes in stress on the materials caused by changes in the microstructure in the area around the weld. Such a phenomenon will result in a decrease in material strength due to residual stress and making cracking easier [5].

The two types of austenitic SS304 and SS304L are used as power plant materials, one of which is a pressure vessel. In the power plant, there are reactor pressure vessels and SS pipes. Therefore, the welding process should be carried out. In this type of SS welding, many failures occur in the heat-affected zone (HAZ) and the welded zones [6]. Some failure factors for SS welding need to be known both in terms of welding implementation and base metals. Based on previous research, mechanical properties and microstructure analyzes have been carried out separately for welding SS304 or SS304L material (7,8,9,10). Both materials have different compositions in carbon content. Therefore, this study will compare the mechanical properties of SS304 and 304L materials. This research aims to analyze the effect of carbon levels of SS304 and SS304L which are carried out by various of welding methods on mechanical properties and microstructure analysis.

2. Experimental Method

The experimental procedures of this research are divided into three processes such as specimen preparation, welding process, and characterization. In the specimen preparation process, the materials used in this study are SS304 and SS304L. Both materials were prepared by cutting and making V groove for 60° of groove angle. Furthermore, welding is carried out using several types of welding which are TIG, MIG, and SMAW with a current of 100 amperes and voltage of 220 volts. The protective gas used in TIG, MIG welding is argon 99.9%. The fillers of TIG and MIG are ER308L in diameter of 1.6, and the SMAW uses electrodes of type E308-16 in diameter of 2.6. Welded materials were characterized by tensile test, metallography, and X-Ray Diffractometer (XRD). Tensile test is carried out to determine the tensile strength, yield strength, and elongation materials. The schematic of tensile test specimens was shown by Figure 1. Macro observation determined the three zones in welding: The Heat Affected Zone (HAZ), the base metal, and the weld metal area. The sample surface was grinded using SiC paper from grade 200 to 2000 and polished by alumina powder. The etched surface should be smooth and continued by etching process to clarify the grain boundary. An optical microscope was used to show the microstructures of samples and XRD analysis determined the phases formed after the welding process. XRD analysis is carried out from a diffraction angle of 20-100°.

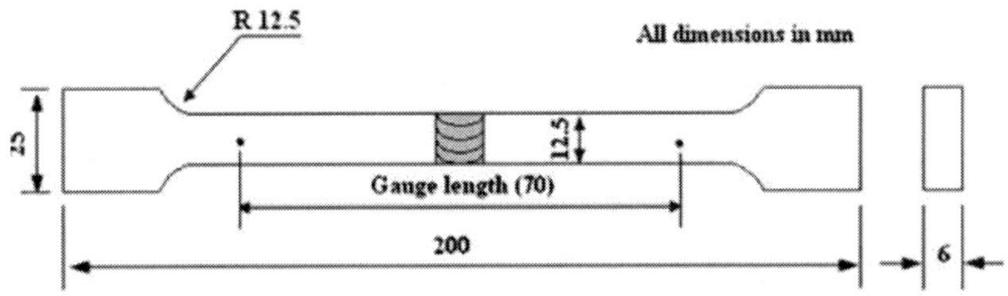

Figure 1. Schematic of welded samples for tensile test [11]

3. Results and Discussion

Figure 2 shows the yield strength of welded SS304 and SS304L material using various welding methods. The highest yield strength of SS304 material in MIG has a yield strength of 523.3 MPa. The lowest yield strength of SS304 materials under SMAW method has a yield strength of 468.02 MPa. In SS304L materials, the highest yield strength of TIG is 445.66 MPa. The lowest yield strength of MIG welding

SS304L material is 432.3 MPa. The yield value of SS304 materials is higher than SS304L caused by the SS304 materials increases the yield strength of filler metal. Otherwise, the yield strength of SS304L material is lower than SS304 due to the yield strength of filler metal decreases after welding process.

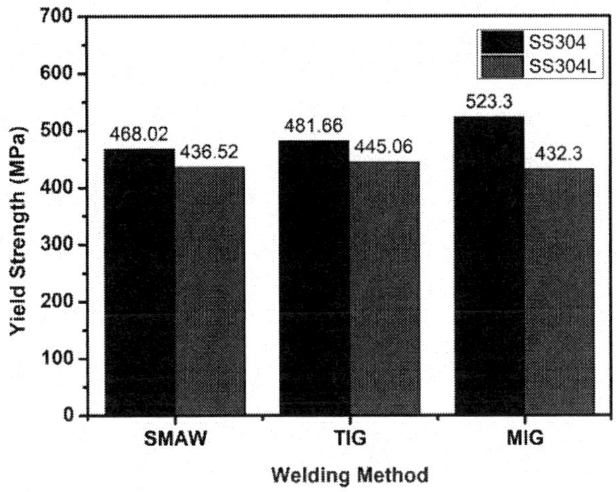

Figure 2. Yield Strength of materials using variations of welding methods

Based on Figure 3, the highest tensile stress value occurs in SS304 material with MIG welding because the mechanical strength of ER308L as a filler metal have increased in MIG welding method. The tensile strength of ER308L filler is 550 MPa and increases for 19.07% to 679.64 MPa after welded. The lowest tensile stress value on welded 304L belongs to SMAW welding. Tensile stress values in SS304L material are slightly different between SMAW, TIG, and MIG welding. Otherwise, SS304 materials using TIG and MIG welding have a significant difference in tensile stress elucidating that welding methods affects the tensile strength of materials. The fracture area of the tensile test specimen for SS304 occurs on the weld metal.

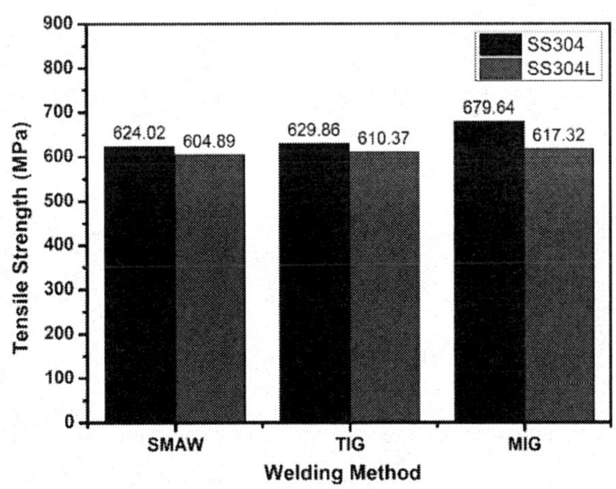

Figure 3. Tensile Strength of materials using variations of welding methods

The tensile strength of SS304 material showed the highest tensile strength results for all welding methods because SS304 has Si element content while SS304L has less Si element content. Si element metal in metals has the role of improving the tensile strength of the materials [12]. The lowest tensile strength value of the test sample occurs when the SMAW welding method is used due to the mechanical strength of filler materials decreases after the SMAW welding with E308-16 electrodes.

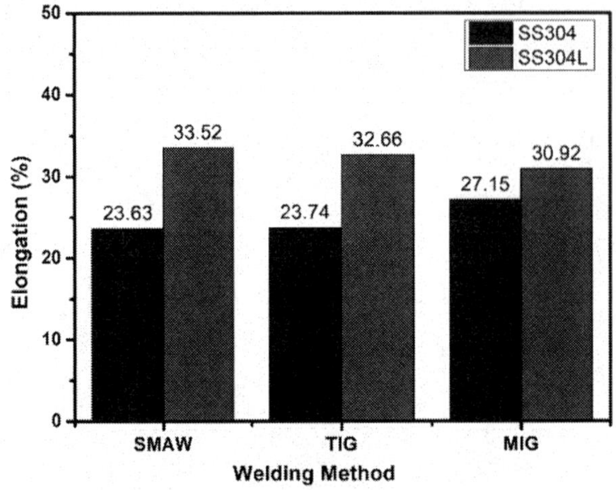

Figure 4. Elongations of materials using variations of welding methods

The elongations properties of SS304 and SS304L welded using different methods were shown in Figure 4. SS304L material has inversely proportional between yield strength and elongations. SMAW has the lowest tensile strength, while elongation has the highest value of 33.52%, compared to TIG welding of 32.66% and at MIG of 30.92%. Welded SS304 material in SMAW method has the lowest elongation value of 23.63 % followed by TIG and MIG of 23.74% and 27.15%, respectively. The highest elongation value at SS 304 welded joints occurs in MIG welding that has an elongation of 27.15%.

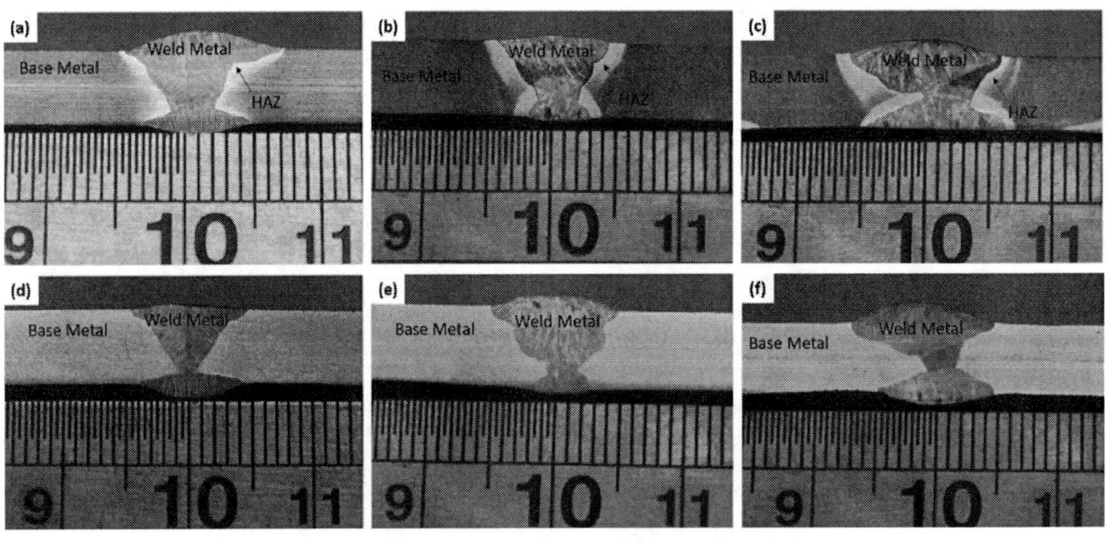

Figure 5. Macro structures for (a) SMAW (b) TIG (c) MIG for SS304 and (d), (e), (f) for SS304L, respectively

Macrostructure analysis confirms three zone of welded materials including base metal, HAZ, and welded metal. This analysis was conducted by ASTM E08. Figure 5 presents the macro image of SS304L material with TIG, MIG, and SMAW welding methods. HAZ zone of SS304 appears clearly and is shown by white spot but HAZ of SS304L is not visible. Low carbon content of SS304L causes a narrow area of HAZ [13]. The weld metal distribution on the 304L SS welding by the TIG method was slightly widened and also in MIG welding. SS304L welded by SMAW method shows clearly in V groove shape.

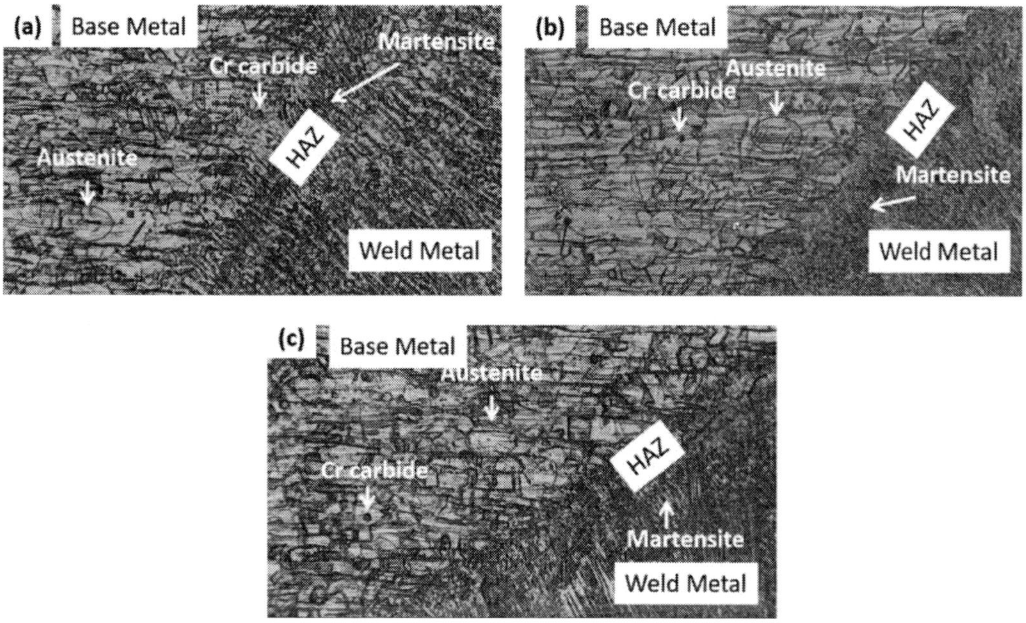

Figure 6. Microstructures for SS304 (a) SMAW (b) TIG (c) MIG

Regarding macrostructure images, further observations was conducted by microstructure analysis under optical microscope. Based on Figure 6, it can confirm the difference between three welding zones that have significant difference of grain structure. The morphological structure of HAZ describes the boundary of base metal and weld metal zone. It has grain structure among other zones because of the rapid cooling in the base metal. Therefore, HAZ has the highest hardness value. There are visible phases on microstructure images which consist of austenite, martensite, and Cr (chromium) carbide with fine granules (black spots) [14]. The weld and HAZ regions for specimens with TIG welding produce the most delicate grain structures dominated by austenite dendrite structures [15]. Moreover, the microstructure of the welding area and the HAZ of MIG welding produce a grain structure. The large austenitic grains using SMAW welding can be ascertained the strength and toughness from the tensile test. The mechanical properties of SMAW welding are the lowest value that other various welding methods.

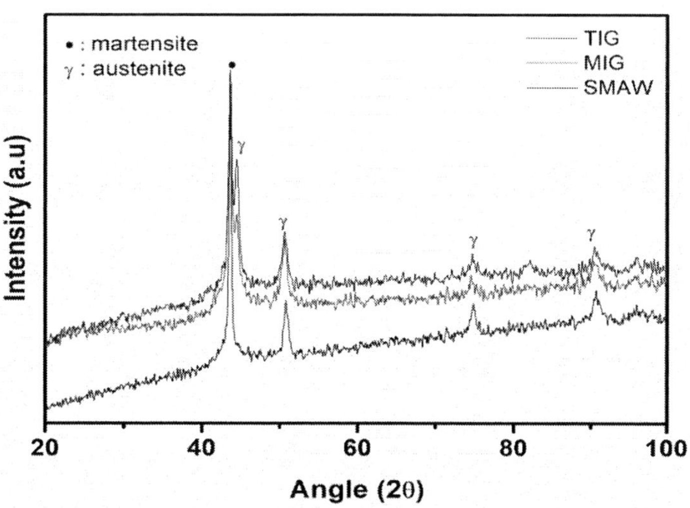

Figure 7. XRD patterns of SS304 materials using variations of welding methods

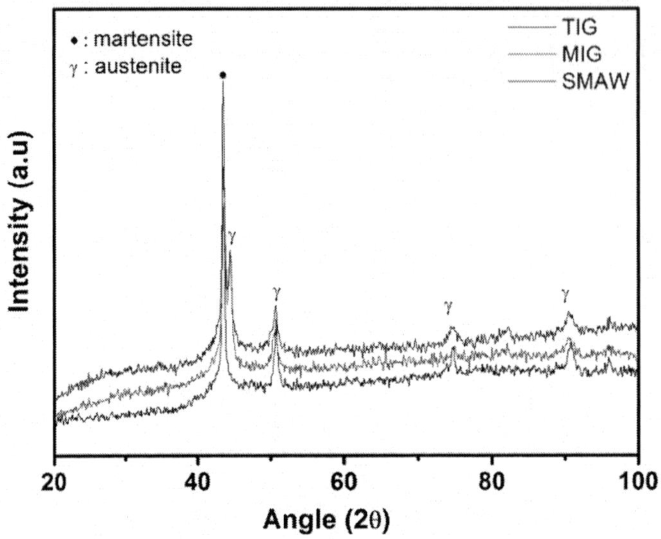

Figure 8. XRD patterns of SS304L materials using variations of welding methods

XRD analysis used to analyze the phases contained in material SS304 and SS304L after welding with different methods shown by Figure 7 and figure 8. XRD is tested at the angle range of 20-100°. Both SS304 and SS304L materials have a similar XRD pattern and the samples have a predominant phase in the form of austenite and martensite [16]. The presence of austenite is more than martensite phase in both SS304 and SS304L materials because SS304 materials have more austenite phase than other types of stainless. Cr carbide and ferrite cannot be identified by XRD pattern because of low amount phases [17]. A clear difference in this XRD analysis is that the SMAW welding carried out on both SS304 and SS304L samples has a different peak position 2Θ of 44.5°. This angle shows the peak austenite that is not owned by samples welded using TIG and MIG. The existence of this peak causes the low mechanical properties of the material welded using the SMAW method for both SS304 and SS304L.

4. Conclusions

Based on those data observations, it can be concluded that the TIG, MIG, and SMAW welding techniques affect each material, including the results of the tensile test, microstructure test, and macro. SS304 welded by MIG method has the highest tensile strength of 679.64 MPa. Otherwise, the SS304L material has the highest tensile strength of 617.32 MPa resulting from MIG welds. It can be concluded that MIG method affects the highest strength of both materials. In the macro photo test, there is a difference in HAZ width caused by elemental content of stainless steel. Moreover, there are obvious phases are austenite, Ferrite, and Cr (chrome) carbides on SS304 using various welding methods. It is confirmed by XRD pattern which austenite exists in SS304 and SS304L materials.

5. References

[1] Buddu R Chauhan N Design P R-F E and and 2014 undefined, Mechanical properties and microstructural investigations of TIG welded 40 mm and 60 mm thick SS 316L samples for fusion reactor vacuum vessel applications *Elsevier*.

[2] Desu R K Nitin Krishnamurthy H Balu A Gupta A K and Singh S K, Jan. 2016 Mechanical properties of Austenitic Stainless Steel 304L and 316L at elevated temperatures *J. Mater. Res. Technol.* **5**, 1 p. 13–20.

[3] Simms H G, 2011, OXIDATION BEHAVIOUR OF AUSTENITIC STAINLESS STEELS AT HIGH TEMPERATURE IN SUPERCRITICAL PLANT.

[4] Arivazhagan N Singh S Prakash S and Reddy G M, May 2011 Investigation on AISI 304 austenitic stainless steel to AISI 4140 low alloy steel dissimilar joints by gas tungsten arc,

electron beam and friction welding *Mater. Des.* **32**, 5 p. 3036–3050.
[5] Luder D *et al.*, Jan. 2016 Failure analysis and metallurgical transitions in SS 304L air pipe caused by local overheating *Eng. Fail. Anal.* **59** p. 292–303.
[6] Singh P J Guha B and Achar D R G, Dec. 2003 Fatigue life prediction for stainless steel welded plate CCT geometry based on Lawrence's local-stress approach *Eng. Fail. Anal.* **10**, 6 p. 655–665.
[7] Unnikrishnan R *et al.*, 2014 Effect of heat input on the microstructure, residual stresses and corrosion resistance of 304l austenitic stainless steel weldments *Mater. Charact.* **93** p. 10–23.
[8] Ullah M Wu C S and Shah M, Jan. 2016 In situ delta ferrite estimation and their effects on FCPR at different orientations of multipass shielded metal arc welded SS304L *J. Manuf. Process.* **21** p. 107–123.
[9] Kristianto H Arie A A Susanti R F Halim M and Lee J K, 2016 The effect of activated carbon support surface modification on characteristics of carbon nanospheres prepared by deposition precipitation of Fe-catalyst *IOP Conf. Ser. Mater. Sci. Eng.* **162**, 1.
[10] Agrawal B P Chauhan A K Kumar R Anant R and Kumar S, Nov. 2017 GTA pulsed current welding of thin sheets of SS304 producing superior quality of joint at high welding speed *J. Brazilian Soc. Mech. Sci. Eng.* **39**, 11 p. 4667–4675.
[11] Kumar S and Shahi A S, Jun. 2011 Effect of heat input on the microstructure and mechanical properties of gas tungsten arc welded AISI 304 stainless steel joints *Mater. Des.* **32**, 6 p. 3617–3623.
[12] 2006 *Steel Heat Treatment: Metallurgy and Technologies (Steel Heat Treatment Handbook, Second Edition)* CRC Press.
[13] Widodo E W R Setyowati V A Suheni and Hardianto A R, 2019 Influences of groove angles and filler metals on 304L stainless steel to AISI 1040 carbon steel dissimilar joint by gas tungsten arc welding *E3S Web Conf.* **130**.
[14] Setiawan A and Ir. Pramuko Ilmu Proboputro M, 2016 Penelitian Stainless Steel 304 Terhadap Pengaruh Pengelasan Gas Tungsten Arc Welding (GTAW) Untuk Variasi Arus 50 A,100 A dan 160 A Dengan Uji Komposisi Kimia, Uji Struktur Mikro, Uji Kekerasan Dan Uji Impact.
[15] Yakub Y (Yunus) and Nofri M (Media), 2018 Variasi Arus Listrik Terhadap Sifat Mekanik Mikro Sambungan Las Baja Tahan Karat Aisi 304 *Widya Eksakta* **1**, 1 p. 249172.
[16] Enhanced Pitting Corrosion Resistance of 304 SS in 3.5 wt% NaCl by Ultrasonic Nanocrystal Surface Modification.
[17] Setyowati V A Suheni Widodo E W R and Hermanto S A, 2019 Normalising of 316L Stainless Steel using Temperature and Holding Time Variations in *IOP Conference Series: Materials Science and Engineering* **462**, 1 p. 012012.

IOP Conf. Series: Materials Science and Engineering 1010 (2021) 012019 doi:10.1088/1757-899X/1010/1/012019

Maritime industry - ports and supporting activities: literature review

Juvinal Lucas Monteiro[1], Lukmandono[2], Pramudya Imawan Santoso[3] and Rony Prabowo[4]

[1]Postgraduate Student, Master ofIndustrial Engineering Department, InstitutTeknologiAdhi Tama Surabaya, Indonesia
[2,3,4]Lecturer, Master of Industrial Engineering Department, InstitutTeknologiAdhi Tama Surabaya, Indonesia

lukmandono@itats.ac.id

Abstracts- Queues are waiting for lines from customers to get service. The queue is caused by the needs of consumers to be served beyond the ability of service facilities, so consumers who come cannot immediately get service. Queue problems at ports often occur every day. Ship loading and unloading services is a queuing phenomenon in daily life. This queue can be caused by damage to the equipment that supports loading and unloading, labor, availability of warehouses, and the limited capacity of berths at berths and others. In maintaining optimal port services, it is necessary to avoid long waiting times and low queuing system utilities. Queues at the mooring service system occur due to mismatch between planning and realization. Waiting time for vessels inside and outside the port area is calculated using the simulation method in queuing theory so that it becomes a queuing model. Until finally, it produces the optimum berthing service system. Based on the simulation results obtained value, and which is much shorter than the conditions on realization. The potential causes of these conditions are divided into three factors, namely, internal factors, external factors, and weather factors.

1. Introduction

As an archipelagic nation, Indonesia holds great potential, especially its strategic location because it is located at a crossing of international trade routes. Shipping has become an important sector in supporting social, economic, government, defense, security, cultural and other life to unite separate islands and the vast ocean. Seeing the development of port infrastructure and facilities in Indonesia, it is not impossible that the port will become a sector capable of boosting national competitiveness and economy. As stated in the 2018 Global Competitiveness Report, if Indonesia maintains and improves performance and competitiveness it will be able to increase higher and sustainable income in the future.

The central pillar of the logistics system is to guarantee effectively and efficiently, which is reflected in low logistics costs, and responsive and satisfying service. Efficient and effective logistics management will help businesses to be more competitive in the competition through the creation of higher added value.Harbor is a part of its waters protected from storms, safe and well or suitable for accommodation for ships shelter, refueling, inventory, repair and loading, and unloading of goods[1]. Port is a protected harbor with marine terminal facilities consisting of mooring or dock for loading and unloading of goods from ships, warehouses, transit, and other accumulation to store products in the short term or term-long[2].

Ports can also be interpreted as terminals and areas where ships loading or unloading at the dock, at the landing site, in the buoy-bui or the like and include waters were waiting for their turn to get service[3]. Based on the notions stated above, then the port can be interpreted as a place for ships anchored, maneuvering, and berthing to do activities securely raising and or lowering passengers and goods and safe. There are six assessment indicators for maritime logistics success: customs, infrastructure, ease of managing international shipping, competence logistics from local operators and service providers, domestic logistics costs, and delivery times[1, 9].How many things need to be considered in analyzing the ship queue time include the average waiting time of the ship at the port, the characteristics of the queue of the ship seen from the level of arrival and level of service, factors that influence the waiting time of the ship.

1.1. Literature Review

1.1.1. Role of the Port . The port as a sub-system of shipping, and considering that the sail itself is a flag bearer following the trade pattern (ship follows the trade), then the port becomes one of the determining elements of trading activities. An efficiently managed business will encourage trade progress; even industries in the back regions will advance with itself[4]. If observed in the historical development of several metropolitan cities, moreover, in an archipelagic country such as Indonesia, the port will help to enlarge the city. The port triggers the growth of the road network, rail network railroad, and warehousing for distribution or consolidation of commodity goods[5]. The land transportation facility and infrastructure network become a port as an intramode node of land transportation and land-sea intermodal. Service costs at the port managed efficiently and professionally will be become low, so business in other sectors is overgrowing. Port acts as a focal point for the economy and trade and becomes a collection of business entities such as shipping and agency, warehousing, freight forwarding, and transportation of goods.Port has several functions, among others[6]:

1. Gateway. Derived from the word port or port that comes from the Latin word porta has meaning as a gateway or gateway[3, 12]. The port functions as the door that people and goods pass into and out of the harborconcerned. Referred to as the door because the port is an official road or area for trade in goods traffic. Entry and exit of goods must meet customs procedures and quarantine, outside the official way is not justified.
2. Link. From the limits of understanding that have been explained before, the existence of a port essentially facilitates the movement of content goods between land transportation mode and sea transportation mode channeling incoming goods and exit customs areas as quickly and efficiently as possible[2, 15]. In this link function there are at least three important elements namely: (a) channel or movecargo on board to the truck[2, 6, 11]; (b) the transfer operation is rapidit means minimum delay [5, 10]and (c) efficient in the sense of cost[4, 16].
3. Interface. Freight transported via maritime transport is at least acrossthe port area twice, once at the loading port, and once at the portunloading[6, 10]. At the loading port and likewise at the loading port, it is movedfromor to transportation facilities by using various facilities and mechanical equipmentand non-mechanical. Equipment for moving cargo bridging shipsby truck to train or by ship. At the activity, the port function is the interface[9, 11]. At every item moving operation, which consists of ship operations, dock transfer operations, a warehouse used field operations,and handover of goods for conveyance & loading (lifting & transfer) goodsequipment) necessary. In the service of bulk commodities, the interface functionsvery real physical. Loader and unloader equipment connects ships withtrainsor trucks on land. The reliability (reliability) of tools and work methodssystemic is the determining factor of the level of speed, smoothness, and efficiencyport activity.
4. Industrial Entity. A well-organized port will grow and willfertilize other business fields so that the port area becomes a related industrial zonewith port[11, 12, 16].

1.1.2. Port Performance. Port performance can be used to determine the level of serviceport to port users (ships and goods) depending onship service time while in port. High port performance shows that the port can provide excellent service [7]. Based on the Decree of the Director-General of Sea Transportation NumberUM.002 / 38/18 / DJPL-11 dated 15 December 2011 concerning Performance StandardsPort Operational Services, operational service performance is the result of work measurable achieved at the port in carrying out the service of ships, goods,utility facilities and equipment within a specified period and unit.

Performance indicator Services related to port services consist of[1, 5, 11]:

1. Dwelling Time is the amount of time since the submission of the mooring application after the ship arrives at the landing location until the ship is driven towards the mooring[3, 10].
2. Approach Time (AT) is the amount the time used for ships moving from the anchored location to the rope tied inmooring or vice versa[4, 15].
3. Effective Time (Effective Time - ET) is the number of hours for a ship is used for loading and unloading while the ship is onmoorings[7, 12].
4. Berth Time (BT) is the amount of time ready for mooring operationsserving ships[4].
5. Berth Working Time (BWT) is the time for loading and unloading forthe ship is at the dock[3, 14].
6. Receiving and Delivery of containers is the speed of servicesubmissionacceptanceat the container terminal, which is calculated from the time the conveyor arrivesexits recorded at the entrance or exit.
7. Level of Berth Occupancy Ratio (BOR) is a comparison between the time of use of the dock with the time of dayavailable (docks ready for operation) within a specified periodexpressed as a percentage[1, 7].
8. Shed Occupancy Ratio (SOR) is a comparison between the number of users of stacking space and space; the available stacking is calculated in units of tons or daysM^3 day unit[8, 10].
9. Level of Use of Stacking Field (Yard Occupancy Ratio - YOR) is a comparison between the amount of space usageStacking with available stacking space (ready for operation)calculated in units of tons of days or M^3 days[5].
10. Equipment operation readiness is a comparison between quantities of equipment that is ready to operate with the amount of equipment that isavailable in a certain period of time[9].

1.2. Loading and Unloading

Measures of tonnage, speed, and time of loading and unloading operations (time andmotion measurements) is an important indicator for measuring levelsefficiency of operating activities[8]. For international ports usedfour common types of loading and unloading operation indicators, namely[3, 7, 12]: (1) Flow of Goods (Output); (2) Boat service time (Service Time); (3) Ratio of use of berth facilities (Berth Occupancy); (4) Cost of loading and unloading goods (Cost per Ton Handled). The four indicators are closely related to each other, mutuallyinfluence, and form a system in which one indicator is notstand alone in the following picture.

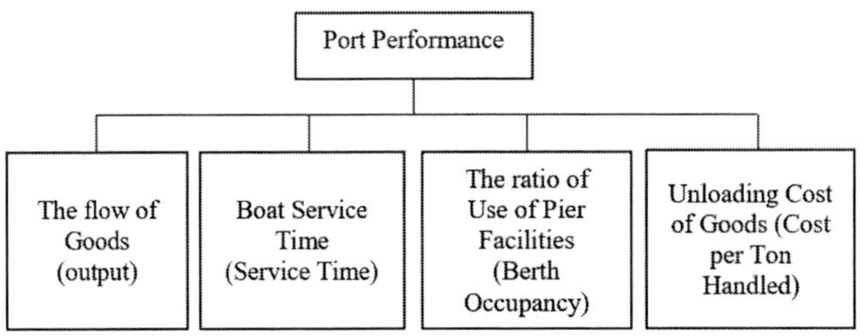

Figure 1. Port Service Indicator [2, 6]

1.2.1. Definition of Loading and Unloading .Dismantling is translated as: "Dismantlingmeans lifting, bringing out all the contents of something, removing all. Meanwhile, according to the PBMAssociation Terminal Operator Communication Forum Jakarta (2002) Unloading is the activity of unloading cargo from a ship,

Demolition is a transfer of goods from one place to another and can also be said unloading of products from the ship to the dock, from the dock to the warehouse orotherwise from the warehouse to warehouse or from the warehouse to new dock transported to ship[9].

Definition of Load contains, fits, fits, enters therein, can contain, load, fill, into, place. The load is the activity of loading cargo on board[10]. Loading and Unloading Activities is the activity of moving goods from the land conveyance, and to carry out the transfer of such activities required the availability of adequate facilities or equipment in a manner or service procedures[11]. Unloading is removal cargo from and above the ship to be piled in or directly transported to the place of the owner of the goods through the harbor dock by using loading and unloading apparatus, both those located at the dock and those located at the ship itself[12]. Loading and unloading is a series of terminal company activities to carry out loading or unloading from and onboard. Work to unload goods from the deck or hold and place it on the dock (kade) or into a barge or the opposite, loading from the dock or in a barge and placing it to the deck or into the hold using a boat crane[13].

2. Method

According to KM No. 25 of 2002 Article 1 concerning Basic Guidelines for CalculationUnloading and Shipping Rates for Shipping to and from ships at the port:

a. Stevedoring. The work of unloading goods from ships todockand barge or truck or load goods from dock and barge or truck toin ships until arranged in the hatch by usingship crane or land crane.
b. Cargodoring. The work of releasing goods from the rope nets (ex-tackle)on the dock and transport from the dock to the warehouse and stockyardthen arrange in the warehouse field or vice versa.
c. I was receiving and delivery. The work of moving goods from the heap and place Stacking in the warehouse field stacking and handing uparranged on top of a vehicle at the door of the warehouse to a yard or stackingotherwise.

The definition of loading and unloading is[4]: (a) Load the Wharf Namely the work of loading goods from above or from the warehouse by using a derrickconveyor to be placed in the hold ship; (b) Unloading the hold Namely the work of unloading on decks or hatches with ships use a derrick/conveyor and put it to the dock or in a warehouse; (c) Load loading Namely service activities load or unload a load from the dock using a derrick or crane or with other loading and unloading tools.While the load data is required for the implementation of loading and unloading are (1) Type and amount of charge; (2) Form of delivery transportation; (3) List of goods or heavy loads, special loads or cargo dangerous.

2.1. Compaction principles

These are a. Protect the ship It was an effort to keep the ship safe during the activities loading and unloading or shipping, for example maintaining stability ships, do not load more than the capacity of the deck, pay attention to SWL (Safe Loading Ability) of an equipment loading and unloading, etc.

(1) Distribution of upright loads is an effort that must be done on a charge when loading so as not to damagethe shipbuilding construction or can sink the ship.
(2) Load distribution horizontally Is an effort made on the current load loading each hatch from the front to the back must be balanced, so that the ship always has a little trim to the back, neither hogging nor sagging.
   a. Trim Is the difference between the first draft and the backdraft on the ship. The selection is the distance between the ship's keel with water level limits.
   b. Hogging Is a condition caused by placement cargo that is concentrated at the ends of the ship; as a result, the ship will break easily if it gets big waves. c) Sagging Is a condition caused by

placement; the cargo is concentrated in the middle of the ship; as a result, the ship will break easily if it gets the waves big.

c. Protect ship crews and laborers from cargo hazards
Ensure that loading is carried out regularly and systematicallyto avoid long hatches, over stowage, over the carriage, and broken stowage [12, 17]
(1) Long hatch is the distribution of load on each hold for each unloading ports are uneven or centered on one hold, resulting in the ship taking too long at the loading port.
(2) Over stowage is a condition where the cargo should be unloaded first at a port closed by another cargo that will be unloadedat the next port.
(3) Over carriage is a condition where the cargo should be unloaded first in a port because of something brought to the portnext.
(4) Broken stowage is the percentage of hatch space that cannot be filled by loadcaused by the charge itself.A bulk carrier, bulk carrier, or bulker is a shipspecifically designed to transport bulk cargo, such as grains,coal, iron ore, and cement in a cargo hold or hold.

2.2. Characteristics of Ship Queues

The existing condition of the port is generally divided into several terminals. The queue analysis flow at the port consists of the pre-analysis stage, the probability distribution test suitability stage, and the analysis stage[5, 17].

1. Estimating Number of Servers (M)
A number of servers are calculated assuming utilizing the length of the existing dock at the port and the average length of the ship with the equation:

$$M = rounddown\left(\frac{docklength}{averagelengthofship + guard}\right) \quad (1)$$

2. Arrival Rate (λ) Arrival rate is the average of many ships coming (k) per day.

$$\lambda = \frac{firstday + \cdots + lastday}{manydays} \quad (2)$$

3. Service Rate (μ)
Service Rate (μ) is the average ship service per day. The length of service for each ship is obtained by the following formula:

$$\Omega = \frac{\sum_{k=1}^{n} Timetoleave - LeaningTime_k}{n} \quad (3)$$

Where:
k: ship serial number; n: many ships get services during an observation

After obtaining the Average Ship Service Time (Ω), the Service Rate (μ) can be found using the following formula:

$$\mu = \frac{1}{\Omega} \quad (4)$$

4. Stages of Test Suitability of Poisson Distribution of Ship Arrival Probability
To calculate the probability of ship arrival per day, the following equation is used:

$$P(x) = \frac{e^{-\lambda}\lambda^{x}}{x!}, \text{ untuk } x = 0, 1, 2, 3, 4, \quad (5)$$

where: $P(x)$ = chance of arrival x ship/day; x = number of ship arrival/day;
λ = average ship arrival/day $e = 2.7183$

5. Ship Service Exponential Distribution Test Exponential test for service time is carried out using the following equation:

$$P(t) = \lambda e^{-\lambda t}, for x \geq 0 \tag{6}$$

where: $P(t)$ = opportunity service time more than t t = service time
λ = average service time e = 2.7183

6. Server Utilization (ρ) is the percentage of server usage using the equation:

$$\rho = \frac{\lambda}{M\mu} \tag{7}$$

Where: M = number of service channels; channel = average service level in each (channel)
λ = average arrival rate

7. The chances are that not one ship in the system uses the equation:

$$P_0 = \frac{1}{\left[\sum_{n=0}^{M-1}\frac{1}{n!}\left(\frac{\lambda}{\mu}\right)^n\right] + \frac{1}{M!}\left(\frac{\lambda}{\mu}\right)^n \frac{M\mu}{M\mu - \lambda}}, \text{ for } M\mu > \lambda \tag{8}$$

8. The average number of ships in the system uses the equation:

$$L_s = \frac{\lambda\mu\left(\lambda/\mu\right)^M}{(M-1)!(M\mu - \lambda)^2} P_0 + \frac{\lambda}{\mu} \tag{9}$$

9. Average waiting time and service using the equation:

$$W_s = \frac{\lambda\mu\left(\lambda/\mu\right)^M}{(M-1)!(M\mu - \lambda)^2} P_0 + \frac{1}{\mu} \tag{10}$$

10. The average ship (unit) that waits for the equation used

$$L_q = L_s - \frac{\lambda}{\mu} \tag{11}$$

11. The average waiting time for the equation used:

$$W_q = W_s - \frac{1}{\mu} \tag{12}$$

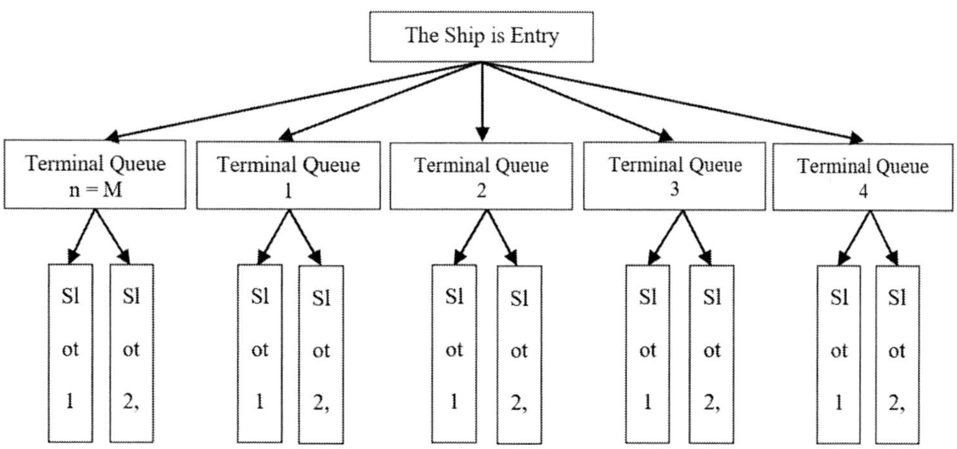

Figure 2. Terminal Queue Concept [3, 7, 15]

2.3. Berth Working Time Parameter

The ship's visit to a port is aimed at carrying out unloading activities load as fast and safely as possible. If you don't need a ship, it won't take long at the port, including times waiting for the availability of facilities, cargo, completion of documents, and local port work schedule[14]. Ship time at port (Ship Turnaround Time or Ship's Time in Port) counted since the ship made use of the anchorage waters where the ship anchored. It was recorded that it had arrived at the local port or when the Pand was on board until leaving the harbor when the guide releases the ship's departure concerned. Ship Turnaround Time or abbreviated as TRT decomposes to Ship time at the dock (Ship Berthing Time or Ship's Time at Berth) abbreviated as BT after deducting the waiting time for the opportunity to lean ona dock (Waiting Time) abbreviated as WT[2, 5, 17].

Ship Berthing Time breaks down into Timeworking boat at the berth (Berth Working Time) abbreviated as BWT after deduction Non-operational timetable abbreviated NOT[1, 6, 12]. Berth Working Time decomposes to Effective ship time (Effective Time) abbreviatedET after deducting the operation stop time (Idle Time) abbreviated as IT.

As for the activities of loading and unloading of cargo ships in the following order[3, 8, 16]: (a) **Ship operation.** The operation to lower cargo directly to the truck or to railroad cars and or barges, and through warehouses or fields buildup; (b) **Quay transfer operation.** The operation of moving goods from the dock, especially to the warehouse or field; (c) **Storage or sheld& yard operation.** Operation of arranging goods in a manner regularly in the warehouse field; (d) **Receiving & delivery operation.**Handover of goods that can be receivedtakes place at the dock location (to the truck or to the barge), and on the landside warehouse or stacking yard.

3. Result and Discussion

Services in calculating terminal operational performance, there are several indicators, especially those relating to ship service at the dock, namely service time. This service time consists of[9, 12]: (1) Berthing time, i.e., the total time spent by the ship during moorings. Berthing time consists of berth working time and not operation time; and (2) Berth working time is the planned time to carry out activities loading and unloading, which consists of effective time and idle time.

Berthing time (BT):

$$BT = BWT + NOT \tag{13}$$

Where: BT = number of hours one boat while in additional

Berth Working Time (BWT):

$$BWT = ET + IT \tag{14}$$

$$BWT = BT - NOT \tag{15}$$

Where:

- Berth Working Time, which is the number of hours a ship is planned for conduct container loading unloading activities while in mooring, not operation time, which is the planned time for not working (nocarrying out loading and unloading activities), such as rest periods, namely 30 minutes eachShift[2, 9, 11]
- Effective time, which is the time used to carry out activities loading and unloading effectively[5, 17]
- Idle time, i.e., time that is not used to carry out activitiesunloading or idle time, such as time wasted whenloading and unloading equipment is broken[3, 7, 8]

Service at the port or general cargo terminal is said to be quality if port managers or equivalent, carry out control functions by seeking waiting time, non-operational time, idle time, and berth. The lowest possible working time is close to zero[4, 15]. At the same time, function control is directed at the loading and unloading speed supported by the right choice for mechanical and non-mechanical devices and other resources. Here the author will focus more and discuss a lot in terms of working berth the time of the many assessments in the waiting time of the ship. According to the author, berth working time has a significant influence inside the port[9]. The working time of ships at the dock, especially during the loading and unloading process, must take place effectively and efficiently to improve the performance of the port itself.

3.1. Port Revenue

Revenue is the amount of money received by a company from an activity that is done, and most of these activities are activities product sales and / or service sales to consumers[15, 17]. The word income in the business world is no stranger. For investors, income is not very important when compared to profits, which is the amount of money that will be received after deducting expenses. Revenues are inflows gross of economic benefits arising from the normal activities of an entity during a period, if the inflows result in an increase in equity that does not originate from investment contributions. Revenues are inflows or completion (or combination of both) of the delivery or production of goods, provide services, or perform other activities, which are the main activities or centra's ongoing activities[16].

3.2. Port Revenue Sources

Facilities services are divided into two, namely necessary facilities and facilitiesancillary services. The basic facilities are intended primarily for servicesships, goods, and passengers; Supporting facilities include public servicesincludes leasing land and buildings/offices, water for ships, electricity, parking lot, and port fitting.Production of basic facilities for ships, including[10, 12]: (1) safety navigation services; (2) anchoring services; (3) guiding services; (4) delay services; (5) mooring or unmooring services; (6) mooring services.

Production of basic facilities for cargo, including[2, 13]: (1) dock services; (2) stevedoring services; (3) transfer services to the warehouse; (4) extra movement services; (5) receiving-delivery services; (6) warehouse services; (7) tool rental services.

While revenue from ship production can be obtained from[7, 11]: (1) Navigation safety services (port dues) are levied on the use of sailing facilities in the form of grooves, vessel traffic control, breakwaters, firefighters, sea pollution advocates, and maritime security; (2) Landing service (anchorage) is imposed on the utilization of anchor area when arrive, wait, or depart; (3) Pilotage services are levied on the use of the functions of sea-guided personnel and guide the city; (4) Towage services are imposed on the use of tugs when the ship mooring, shifting and unmooring; (5) Mooring / unmooring service is charged for the use of gang power mooring gang; and (6) Berthage services are levied on the use of berthing, berthing facilities, jetty.

4. Conclusions

Ship loading and unloading queues can be reduce though maximizing the performance of loading and unloading workers and diverting ships to terminals that arestill not busy and improving services and procedures for handling documents at the port to cope with the increase in the flow of goods, development of shipping lanes to revitalize shipping lanes and port ponds, apply regulations on port properly and maximize existing guide forces in anticipation of increased ship visits, increasing the length of the pier or by diverting fist to lean on supporting ports and Encouraging and maximizing cruise line projects to deepen Shipping Lines and port ponds, and Increase the number of guide and tugboats to anticipate increased ship traffic, placing and using loading and unloading equipment in accordance with its purpose and adding information technology facilities at the port to anticipate an increase in the flow of goods at the port, maximizing the performance of Unloading Workers and placing ships at the dock as well as simplifying document service procedures at the port to reduce the level of dock usage. move existing installation facilities in the shipping lane and maximize pilotage to facilitate ship traffic.

5. References

[1] A. Bacchioni and J. Ramus. 2008. Best Practices in Port Management: An Assessment of Two Ports, no. 4 May.
[2] O. Baccelli, M. Percoco, and A. Tedeschi. 2008. Port Authorities as cluster managers: the case of the Ligurian ports, vol. 39, pp. 44–58.
[3] H. Meersman, E. Van de Voorde, and T. Vanelslander. 2012. Port Congestion and Implications to Maritime Logistics," *Marit. Logist.*, pp. 49–68.
[4] P. Gaur. 2005. Port Planning as a Strategic Tool : A Typology. No. April p. 89.
[5] E. William K. Ruto. 2015. Logistical Factors Influencing Port Performance a Case of Kenya Ports Authority (Kpa). *Int. J. Curr. Res. Rev.*, vol. 7, no. 12, pp. 52–59.
[6] K. McGinley. 2014. Preparing port container terminals for the future: Making the most of intelligent transport systems (ITS). *WIT Trans. Built Environ.*, vol. 138, pp. 419–427.
[7] W. C. Huang, T. C. Kuo, C. T. Wu, and C. C. Kuo. 2012. Port management improvement strategies based on the simulation model. *J. Mar. Sci. Technol.*, vol. 20, no. 5, pp. 492–507.
[8] H. Apriliaanty and H. Evander. 2018. Information Technology in Port Container Terminal: Automation Tally System Implemented in Port of Tanjung Priok,vol. 147, no. Grost, pp. 758–766.
[9] M. Dooms, C. Macharis, and A. Verbeke. 2004. Proactive stakeholder management in the port planning process: empirical evidence from the port of Brussels. *44th Eur. Congr. Eur. Reg. Sci. Assoc.*, p. 34.
[10] E. T. Lawer, J. Herbeck, and M. Flitner. 2019. Selective adoption: How to port authorities in Europe and West Africa engage with the globalizing 'green port' idea. *Sustain.*, vol. 11, no. 18.
[11] E. Van De Voorde and H. Meersman. 2014. Discussion paper Port Pricing: Principles, Structure, and Models. *Norges Handel.*, no. April 2014.
[12] D. Ozer Caylan and Ç. Emiroglu. 2014. the Importance of Strategic Leadership for Port Management: a Delphi Research on Top Managers of Turkish Private Ports. *J. Glob. Strategy. Manag.*, vol. 2, no. 8, pp. 5–5.
[13] B. B. Khiem. 2017. Port Authority Models and Application Possible in Vietnam. *J. Econ. Bus. Manag.*, vol. 5, no. 3, pp. 154–159.
[14] R. Prabowo, M. L. Singgih, P. D. Karningsih, and E. Widodo. 2020. New product development from inactive problem perspective in indonesian SMEs to open innovation.*J. Open Innov. Technol. Mark. Complex.*, vol. 6, no. 1.
[15] B. A. Priyambodho. 2016. Container Port Development Projects on The Java Sea, Indonesia, vol. 5, no. 1.
[16] J. Hollebrands. 2011. *Performance-Based Contracting: A New Tool 1 Introduction.*
[17] Basuki M, Lukmandono L, ZauBeu MM. 2020. Ballast Water Management at Inaport 4th Makasar Based Environmental Risk Assessment. Available at SSRN 3512750.

Acknowledgment

The author would like to thank the financial support given from The Ministry of Research and Technology / National Agency for Research and Innovation of Indonesia under the Magister Thesis Research scheme 2020 (Contract No. 046/SP2H/LT/MULTI/L7/2019 and 046/SP2H/AMD/LT/MULTI/ L7/2020)

IOP Conf. Series: Materials Science and Engineering 1010 (2021) 012020 doi:10.1088/1757-899X/1010/1/012020

Study of catamaran fishing vessel based on wind energy

P Santoso[1], **I Utama**[2]

[1)] Institut Teknologi AdhiTama Surabaya-Indonesia.
[2)] Professor Hydrodinamics of Naval Architect Dept., FTK, ITS, Surabaya-Indonesia.

pramudya05@itats.ac.id

Abstract. Amid the Covid19 Pandemic, the world is currently experiencing a difficult time. The impact of the chain due to soaring crude oil prices on world markets has been felt by all levels of society. Responding to this, austerity movements in various sectors are needed in order to create alternative solutions to the dependence of FO to get through difficult times. In the sea transportation sector, the use of fuel for motorized vessels is not only uneconomical but also environmentally unfriendly. The dependence on fossil fuels energy has led to the development of the concept of energy efficient vessels that use alternative energy sources in moving ships. This is anticipated by the re-use of sails in the form of Sailboats (SB) which are applied as fishing vessels. This paper presents a study of the efficient use of wind energy in SB as a boat mover that does not utilize fuel for the development of environmentally friendly fishing vessels. Technical analysis is done through testing the ship model in the Towing Tank then the results are explored by Naval Architect theories. In conclusion, there is a potential savings in fuel consumption by 100% when compared to ships that have only engine-driven.

1. Introduction

Amid the Covid19 Pandemic, the world is currently experiencing a difficult time[1]. The impact of the chain due to soaring crude oil prices on world markets has been felt by all levels of society. Subsidies by the government to ensure the supply of fuel oil (FO) in the country actually becomes an increasingly heavy burden[2]. Responding to this, austerity movements in various sectors are needed in order to create alternative solutions to the dependence of FO to get through difficult times[3].

At present fishing vessels are still urgently needed to support and sustain food security for billions of people in this world. We know that the mission of a fishing boat is to catch fish from the sea to get fish that meet the quality in appropriate ways and take the fish ashore or to other ships for further processing. In operation a fishing boat must be absolutely safe because even in bad weather the ship must work, so that the problem of total resistance, driving force, fuel consumption, stability and seakeeping are very important concerns.

In general, the operation of a fishing boat is always associated with economic and environmental issues[4]. The economic factor is the cost of fuel because the consumption of fossil fuels (fossil) is quite large, especially for ships primemover, while environmental factors are related to the level of pollution that occurs due to ship operations[5]. The high price of fuel oil is not at all profitable for ship operators. The use of fuel for motorized ships is not only uneconomical anymore, but it is also not environmentally friendly. Economic problems and strong environmental pressure forced ship designers and owners to create more efficient ships so as to minimize the use of boat propulsion[5]. Reduction in the amount of propulsion (and fuel requirements) can be met since the design phase of the ship by creating a more

IOP Conf. Series: Materials Science and Engineering 1010 (2021) 012020 doi:10.1088/1757-899X/1010/1/012020

efficient hull design and propulsion system and ship operational activities including ship operations, such as: Sailing Boats, Solar Electric Ships, Engine Sails Ships[4].

The impact given from the operational activities is very important to be studied and, in the future, it will be used as a basis for developing ship drivers using renewable energy that is efficient and effective[6]. The Ccatamaran hull is very potential to be used as a fishing vessel because it has a large deck area and also has good resistance performance, stability and ship motions (seakeeping) as well as the potential to reduce exhaust emissions and subsequently will be used as a basis for non-fossil energy applications on ships[7].

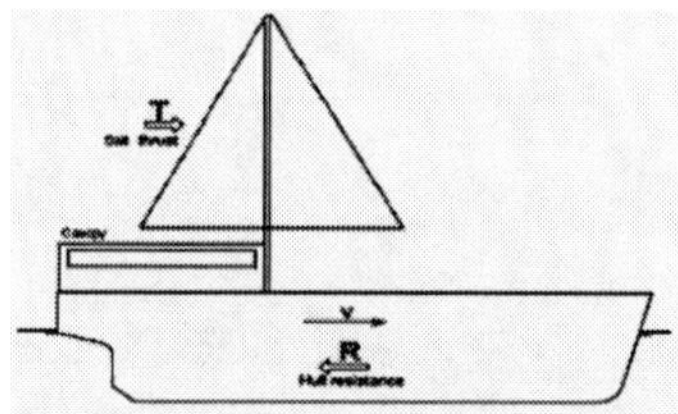

Figure 1. Sailboats Configuration [9].

Figure 1 shows the Sailboats configuration, which the concept of energy conversion in the configuration is to convert wind energy into the required thrust force of the ship through the sail.

2. Methodology

Experimental model test on a symmetrical catamaran has been carried out and reported in[8]. The experimental model test was carried out at ITS towing tank as shown in Figure 2 and the principle particular of the model and full ship was given in Table 1.

Table 1. Principle particular

Parameter	*Catamaran*	*Demihull*
LWL (m)	14.5	14.5
B (m)	7.118	1.318
H (m)	1.44	1.44
d (m)	0.694	0.694
CB	0.434	0.434
Displ. (ton)	11.8	5.9

Figure 2. Experimental model test

Table 2 showed the results of resistance test which indicated that the wider the separation between the hulls (S/L), the lower the total resistance.

Table 2. Result of resistance test

Run No.	V (Knots)	Fr	Catamaran Resistance (kN) S/L = 0.2	S/L=0.3	S/L = 0.4
1	5.788	0.250	1.821	1.659	1.659
2	6.218	0.268	2.141	1.851	2.061

3	6.677	0.288	2.443	2.239	2.348
4	7.051	0.304	2.852	2.678	2.947
5	7.560	0.326	3.460	3.568	3.547
6	8.032	0.347	4.467	3.954	3.766
7	8.384	0.362	4.844	4.345	4.341
8	8.818	0.380	5.149	4.790	4.662
9	9.233	0.398	5.807	5.592	5.515
10	9.813	0.423	7.101	6.448	6.138

Further, the Data's in Table 2 used to explore the Sail catamaran fishing vessel.

3. Result

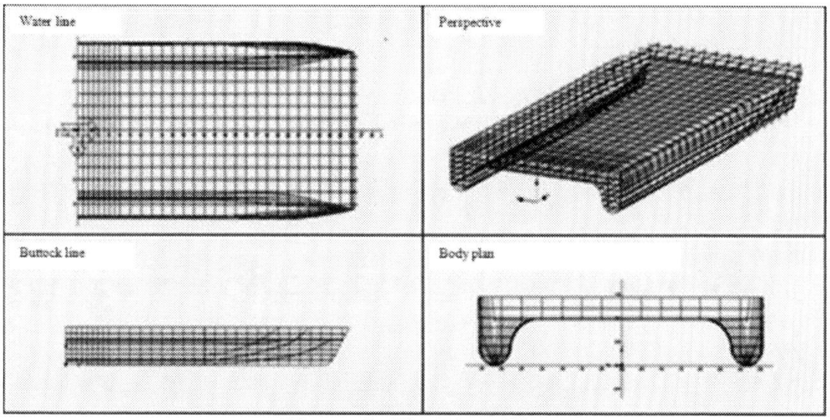

Figure 3. The Lines Plan

Figure 3 shown The lines plan of vessel, is a vertical cross section of the ship (body plan), vertical elongated (buttock line) and horizontal elongated (water line) equipped with absent ordinate information of the ship. Information on this line plan data as shown in Figure 3 is an input for depicting and calculating the Hydrostatic Curve. The making of Lines Plan is carried out using the Design re-use method by entering the main size data's available.

Table 3 The Hydrostatic Properties

Draft Amidships m	0.000	0.050	0.100	0.150	0.20	0.25	0.30	0.40	0.50	0.60	0.694
Displacement t	0.0000	0.288	0.801	1.458	2.22	3.08	4.00	5.99	8.10	10.3	11.8
Heel deg	0.0	0.0	0.0	0.0	0.0	0.0	0.0	0.0	0.0	0.0	0
Draft at FP m	0.000	0.050	0.100	0.150	0.20	0.25	0.30	0.40	0.50	0.60	0.694
Draft at AP m	0.000	0.050	0.100	0.150	0.20	0.25	0.30	0.40	0.50	0.60	0.694
Draft at LCF m	0.000	0.050	0.100	0.150	0.20	0.25	0.30	0.40	0.50	0.60	0.694
Trim (+ve by stern) m	0.000	0.000	0.000	0.000	0.00	0.00	0.00	0.00	0.00	0.00	0
WL Length m	0.332	14.80	14.91	14.99	15.0	15.1	15.1	15.2	15.3	15.3	15.44
Beam max on WL m	3.194	6.743	6.876	6.971	7.04	7.09	7.12	7.13	7.13	7.13	7.138
Wetted Area m^2	0.006	9.234	13.71	17.68	21.3	24.7	28.1	34.6	41.0	47.3	53.63
Waterpl. Area m^2	0.006	8.298	11.529	13.99	15.9	17.450	18.57	20.019	21.099	22.4	24.19
Prismatic coeff. (Cp)	1.181	0.666	0.688	0.703	0.71	0.724	0.734	0.752	0.769	0.78	0.797
Block coeff. (Cb)	0.004	0.056	0.076	0.091	0.10	0.112	0.121	0.134	0.145	0.15	0.16
Max Sect. area coeff. (Cm)	0.003	0.085	0.111	0.129	0.14	0.155	0.165	0.179	0.188	0.19	0.201
Waterpl. area coeff. (Cwp)	0.005	0.083	0.112	0.134	0.15	0.163	0.172	0.184	0.193	0.20	0.219
LCB from 0 pt. (+fwd) m	0.268	5.062	5.282	5.428	5.54	5.643	5.734	5.901	6.054	6.18	6.303
LCF from zero pt.(+fwd) m	0.268	5.255	5.514	5.687	5.83	5.967	6.102	6.368	6.596	6.76	6.885
KB m	-0.201	0.030	0.060	0.089	0.11	0.14	0.17	0.235	0.291	0.34	0.401
KG m	0.525	0.525	0.525	0.525	0.52	0.52	0.52	0.525	0.525	0.52	0.525
BMt m	-467778.6	296.75	148.43	99.11	73.9	58.4	47.8	34.46	26.788	22.1	19.36
BML m	646.687	302.58	166.74	118.1	92.2	75.7	64.2	49.38	40.511	34.9	31.7

Draft Amidships m	0.000	0.050	0.100	0.150	0.20	0.25	0.30	0.40	0.50	0.60	0.694
GMt m	-467779.3	296.25	147.96	98.67	73.5	58.076	47.52	34.175	26.553	21.9	19.24
GML m	645.961	302.09	166.27	117.6	91.8	75.383	63.95	49.097	40.276	34.7	31.57
KMt m	-467778.8	296.78	148.49	99.20	74.1	58.601	48.04	34.701	27.079	22.5	19.76
KML m	646.486	302.61	166.79	118.2	92.3	75.9	64.4	49.62	40.802	35.2	32.10
Immersion (TPc) tonne/cm	0.000	0.085	0.118	0.143	0.16	0.179	0.19	0.205	0.216	0.23	0.248
MTc tonne.m	0.000	0.057	0.087	0.112	0.13	0.152	0.16	0.192	0.213	0.23	0.259
RM at 1°=GMt.Disp.sin(1)	-0.001	1.490	2.069	2.510	2.86	3.127	3.32	3.576	3.755	3.96	4.221

Table 3 shown Hydrostatic properties, is the ordinate data of the ship line plan used to determine the characteristics of the ship to be designed. The results of the hydrostatic calculations are used for the purpose of: 1) Crosscheck main size data provided from previous studies, 2) Besides that it is also used to complete the data needed for development the Coniguration of Sail catamaran fishing vessel.

3.1. Ship Moving Theory

The vessel may move forward due to a sufficient thrust to resist ship resistance at a certain service speed [6].

$$T > RT \quad or \quad T - RT > 0 \tag{1}$$

where T is Thrust (kN) and RT is Total ship resistance (kN).

$$R_T = \frac{1}{2} \rho \, C_T \, (WSA) \, V^2 \tag{2}$$

$$\text{Thrust } (T) = R_T/(1-t) \tag{3}$$

$$\text{where: } t = k_R \, . \, wt \tag{4}$$

$$wt = -0.0458 + 0.3745 C_B^2 + 0.1590 D_w - 0.8635 Fr + 1.4773 Fr^2 \tag{5}$$

$$D_w = \frac{B}{\nabla^{1/3}} \sqrt{\frac{\nabla^{1/3}}{D}} \tag{6}$$

where: t is *thrust deduction factor* for *single screw*, k_R is 0.5 for thin rudder.

3.2. The sail

Sail is one of the propeller props without a propeller on the ship that can work due to the wind force (catch the wind) on the surface of the sail, resulting in a drifting force on the ship at a certain speed[9]. The Sail is one of the propulsion devices on the ship. As with other propulsion devices such as propellers, the sail is attempted to produce an optimal thrust force, in order to produce maximum velocity of the ship[10]. The forces on the Sailing Ship, as the ship moves due to the thrust of the propeller or sail there will be a lift that will lift the hull from the water. In addition, obstacles caused by the aerodynamic resistance of the sail are the forces acting on the sailboat[9].

The sail force (Fs) defined as[11],

$$Fs = q \, . \, As \tag{7}$$

where q is the dynamic wind pressure. The q could be defined as

$$q = \tfrac{1}{2} \, \rho \, £ \, V_{w2} \text{ (ton/m}^2\text{)} \tag{8}$$

where ρ is the air mass density which is equal to γ/g. γ is the weight per unit volume which is equal to 1.2265 t/m^3, g is equal to 9.81 ms^{-2}, £ is the wind pressure coefficient, V_w is the wind speed, and As is the sail area.

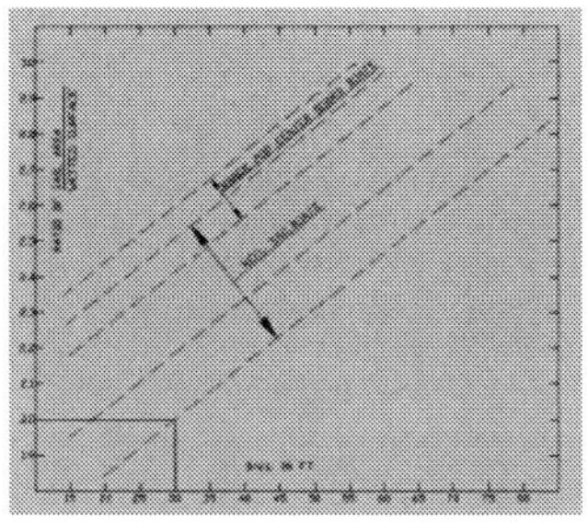

Figure 4. Graph of *SR-LWL*[11]

Figure 4 shown *SR-LWL* relationship which can be used to determine Sail Area with 15 - 80 feet or 5 - 25 m LWL limitation[11]. Determination of Sail Area almost as a comparison of sail area (*As*) with wetted surface area (*WSA*) is between 2.0 and 2.5[10]. Comparison of sail area (As) with wetted surface area (*WSA*) known as sail ratio (*SR*). There is another way according to[11], where the determination of *SR* depends on the LWL of the ship by using the graph.

3.3. Stability

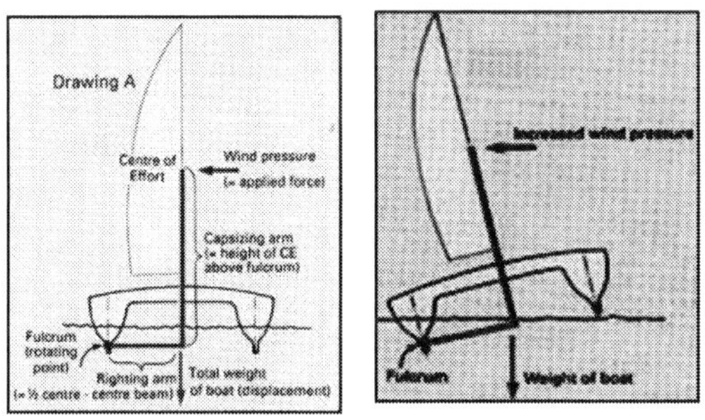

Figure 5. The stability balance of Catamaran Sailboat, [18].

Figure 5. shows the balance of forces and moments acting on a catamaran sailboat. This balance occurs because the values of the moment of return when the ship is shaken (displacement multiplied by the righting arm) is the same as the moment of wind (the force of the wind multiplied by the heeling/capsize arm). Sailboats have two types of stability arms, the static arm is the moment of return when the ship is shaken (righting arm) and the wind force moment arm is working at the centre of the sail (heeling/capsize arm)[12].

$$\text{W Boat x Right arm} = \text{Wind press x Capsize arm} \quad (9)$$
$$\text{(Righting moment) (Capsizing moment)}$$

3.4. Ship motion

Ship motion is the motion response of ship when it receives a disturbance from the outside (external) which in this case is caused by wave blows[13] or in other words is one aspect of hydrodynamics that studies the behaviour of ships on the waves.

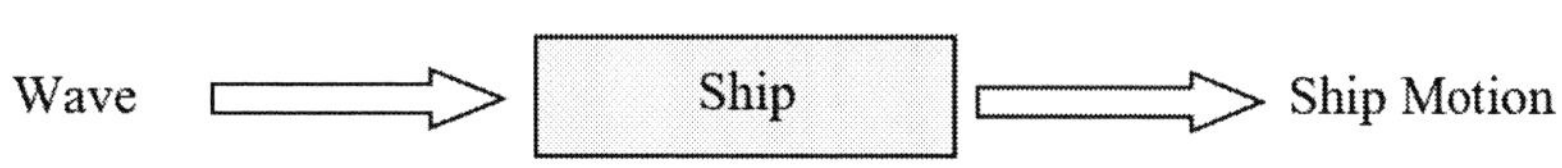

Figure 6. Ilustration of Ship Motion response

Figure 6 shows the illustration of the response of the ship, where waves as the input of the ship hit the ship causing dynamic movement of the ship as its output[13].

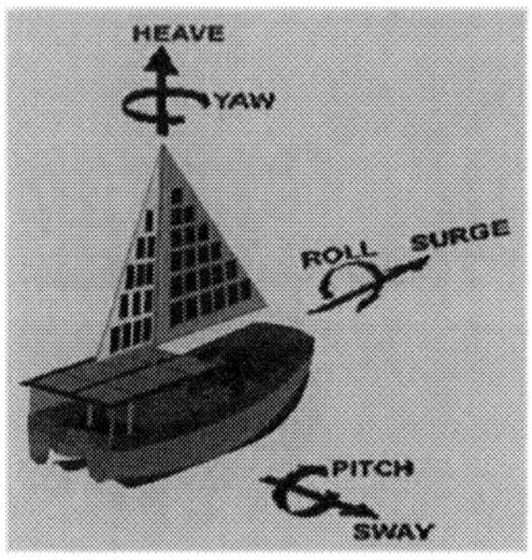

Figure 7. Dynamic Ship Motion with 6 Degrees of Freedom,

Analysis of the dynamic motion of this ship is very dependent on the characteristic wave information where the ship will be operated. The dynamic motion of ship tends to cause a return force that functions to return the ship to its original position. The dynamic motion of the ship is shown in six degrees of freedom, namely: three are linear / translational motion and three are rotational, all movements are based on three axes (x, y, z) as in Figure 7[13]

In ship operations at sea, there are three of the most dominant movements that can cause changes in displacement, namely:

Heave $$(m+a)\ddot{z}+b\dot{z}+cz=F(t) \quad (10)$$

Pitch $$(mk^2+a)\ddot{\theta}+b\dot{\theta}+c\theta=M(t) \quad (11)$$

Roll $$(mk^2+a)\ddot{\phi}+b\dot{\phi}+c\phi=0 \quad (12)$$

In the dynamic motion of the ship parameters measured include: the response of heaving, pitching and rolling movements and the possibility of slamming and deck wetness.

Limits regarding seakeeping performance criteria relate to specific aspects of the ship's response to sea conditions, such as the amplitude of the roll / pitch / heave motion at a certain point (area) on the ship. Every aspect of the movement, if it has a high value will be able to allow the element of the ship to degrade to an unnatural level.

Response Amplitude Operators (RAO) is calculated by following Equation:

$$RAO_{(\omega)}=\frac{\chi}{\zeta_a} \quad (13)$$

4. Results and Discussion

In general, the normal operational of fishing vessels according to[14], are: (1) the ship departs and operates in the port (*departure from port*), (2) the ship goes to the location of the fishing ground (*outward bound*), (3) the ship arrives at the location fishing ground and fishing (*on fishing ground*), (4) the ship leaves the location of the fishing ground to the port (*homeward bound*), and (5) the ship arrives at the port and docked at the port (*arrival at port*).

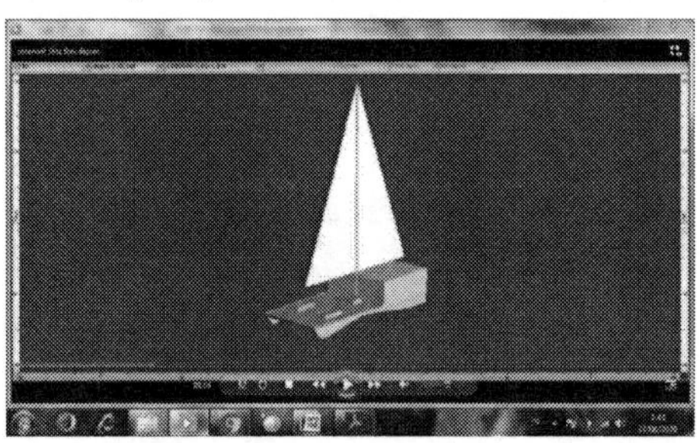

Figure 8. Operational simulation of Sailboat Catamaran Fishing Vessel

Figure 8. Operational simulation of Sailboat Catamaran Fishing Vessel. In its operation, a fishing vessel must be completely safe (*very seaworthy indeed*). Even in bad weather, the ship must work. All work on the fishing boat must be done quickly, starting from the process of catching until the processing of the catch is a function of time. The slow catching process causes the fish to run all (migration), while the sluggish processing of the catch causes the fish to be damaged

Table 4. Results of data development

No.	Parameter	Simbols	Results	Units
1	Displacement	Δ	11.8	ton
2	Lightweight	*LWT*	7.070	ton
3	Payload	*DWT*	4.730	ton
4	Resistances	R_T	6.138	kN
5	Thrust	*T*	6.685	kN
6	Sail Areas	*SA*	125	m^2
7	Cubic number	*CUNO*	55.04	m^3
8	Cargo hold Capacity	*Cap.*	8.5	m^3
9	Fish cargo	*Wc*	1851	kg
10	Gross tonnage	*GT*	15	tonnage
11	Gross tonnage	*GT*	14.45	tonnage
12	Nett tonnage	*NT*	3.03	tonnage
13	Weight of Sail Eqpt.	*Wm*	243	kg
14	Weight of Ship Eqpt.	W_{SE}	727	kg
15	Weight of Fish Proc.Eqpt.	W_{FPE}	339	kg
16	Weight of Fishing gear	W_{FE}	554	Kg

Table 4 shows Results of data development as a engineering analysis is done through testing the ship model in the Towing Tank then the results are explored by Naval Architect theories.

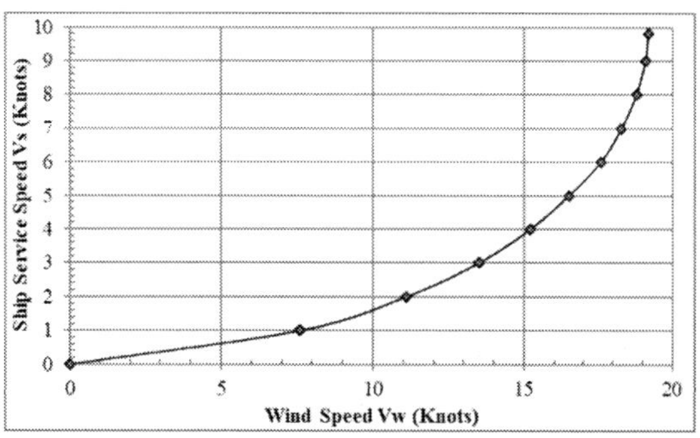

Figure 9. Relationship between *Vw* – *Vs*

Figure 9 shows the graph of the relationship between the wind speed (*Vw*) and the Ship service speed (*Vs*). When the wind speed is captured by the sail to move ship, it will produce the ship service speed. Figure 10 shows the graph of the relationship between the thrust of the ship T with the speed of the ship Vs and the wind speed Va. This graph can be used to calculate the speed of a sailboats. For example, 8 knots (see red line) then the boat's thrust value T will be around 4.5 kN and then use the graph Fig.2 enter the T value of 4.5 kN, then the wind velocity (*Vw*) value is obtained around 15 more knots and 8 knots Ship service speed (*Vs*), or vice versa.

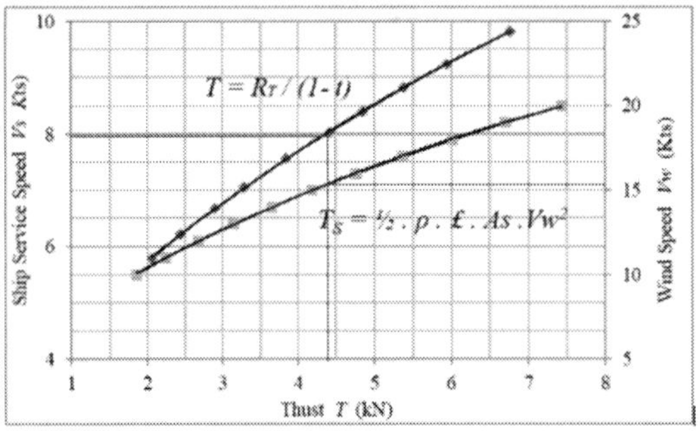

Figure 10. Relationship between T - Vs - Vw

The results of the calculation of Response Amplitude Operators (RAO) for heave and pitch motions as presented in Figures 11 and 12

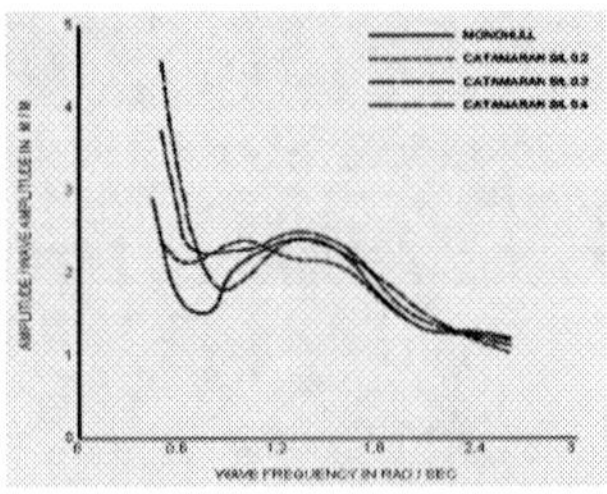

Figure 11. *Heave Motion*

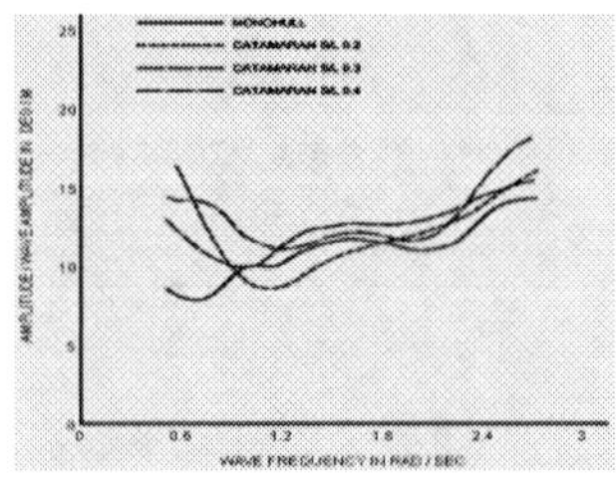

Figure 12. *Pitch Motion* [16]

Motion of catamarans is very dependent on the choice of the distance between the hull (S/L) and in S/L = 0.2 is 18% smaller than the catamaran S / L = 0.4 in the direction of wave 120°, as obtained from the results of experimental studies by [16].

The results of the calculation of the static stability of catamaran fishing vessels with sail propulsor under various operational conditions as shown in Table 5.

Tabel 5. Stability Datas

No	Ship Conditions	LWT (kg)	DWT (kg)	Displ. Δ (kg)	Fishcatch (kg)	SA/Δ	Centre Gravity (m)		Draught, d (m)
							LCG	KG	
0	Design	7070	4730	11800	2880	10.59	7.142	1.308	0.694
1	Departure from port	7070	1850	8920	0	14.01	6.523	1.360	0.525
2*	Outward bound	7070	1540	8610	0	14.52	6.554	1.370	0.506
3	On fishing ground	7070	2670	9740	1440	12.83	6.953	1.346	0.573
4	Homeward bound	7070	3675	10745	2880	11.63	7.288	1.331	0.632
5	Arrival at port	7070	3335	10405	2880	12.01	7.341	1.339	0.612

Note: Condition 2* is the conditions that are considered the most critical stability. However, after checking the stability, it still meets fish catch requirements

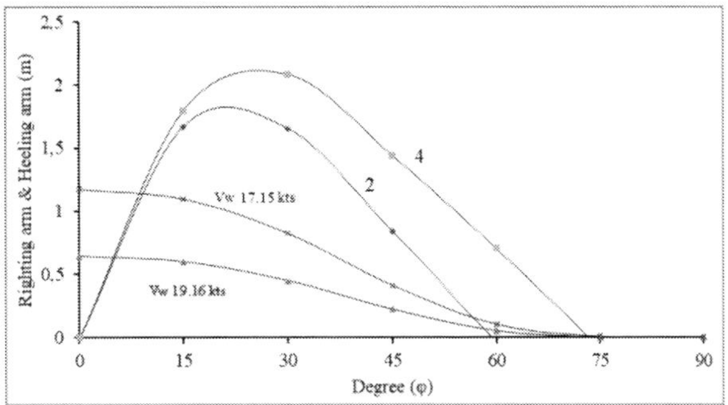

Figure 13. Sailboat stability

Figure 13 shows Sailboat stability when the ship is operational. In condition 2 (*outward bond*), to achieve 9.8 knots service speed (*Vs*) required 17.15 knots wind speed (*Vw*). There is a balance at an angle of about 9° and the value of Sail force (*Fs*) is 5.467 kN.but when the ship is in condition 4 (*homeward bond*), to achieve 9.8 knots service speed (*Vs*) required 19.16 knots wind speed (*Vw*). There is a balance at an angle of about 4° and the value of Sail force (*Fs*) is 6.823 kN.

Table 6. Fish vessel stability requirements according to The Fishing Vessels (Safety Provisions) Rules 1975 and HSC annex 7. IMO 2016.

Table 6 shows the results of calculating the stability requirements of sailboat catamaran fishing vessels under various conditions. The results all meet requirements according to *The Fishing Vessels*

No	Description	Result					Satuan	Requirement	Remark
		deport	p-fg	fg	Fg-p	aport			
1	Area under curve up to 30^0	25.356	25.34	25.35	25.27	25.27	m.deg	$\geq$ 7.4274	pass
2	Area under curve up to 40^0	0.928	0.998	1.085	1.156	1.196	m.rad	$\geq$ 0.090	pass
3	Area between 30^0-40^0	0.245	0.273	0.318	0.355	0.371	m.rad	$\geq$ 0.030	pass
4	Minimum GZ $\geq 30^0$	1.785	1.920	2.131	2.305	2.383	m	$\geq$ 0.200	pass
5	Maximum GZ	18	20	22	26	28	degree	$\geq 10^0$	pass
6	Initial GM	8.018	8.813	8.074	7.489	7.865	m	$\geq$ 0.150	pass

(Safety Provisions) Rules 1975 and HSC annex 7, IMO 2016.

The economic value of FO use on Engine fishing vessels is shown in Figure 14 below, [14].

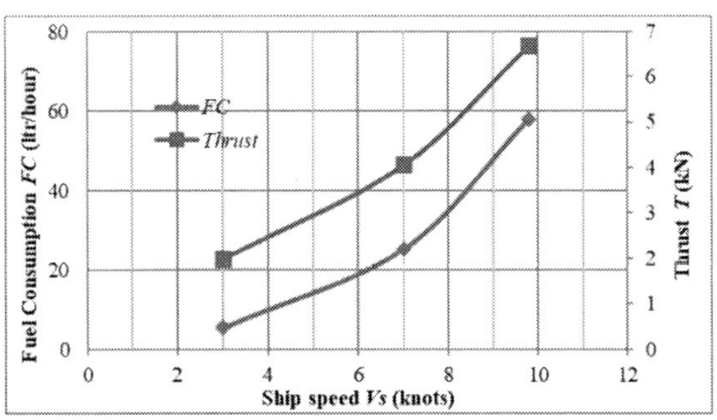

Figure 14. Relationship between Vs-FC-T at 60 kW power.

Figure 14. shows the graph of the relationship between ship speed (*Vs*) - fuel consumption (*FC*) - thrust (*T*). To reach the official speed Vs 9.8 knots with 60 kW engine power (*P*) , 6.685 kN thrust (*T*) requires 57.9 liters/hour fuel consumption (*FC*). So it can be said that the economic value of the ship speed (*Vs*) of 9.8 knots, the thrust (*T*) of 6,685 kN is equivalent to the cost of fuel consumption of 57.9 liters/hour. Furthermore, if the economic value of the use of wind energy is equated with the economic value of the use of fossil energy, the results obtained are a sailboat speed of (*Vs*) 1 knot equivalent to wind speed (*Vw*) of 1.96 knots which also has an equivalent value to the force thrust (*T*) of 0.68 kN which is equivalent to FO of 5.9 liters / hour. So it can be simply concluded that wind speed (*Vw*) of 1 knot is equivalent to FO of 2.95 liters / hour.

5. Conclusions

Application of sail on catamaran fishing vessels is very useful has potency of cost saving about 100% fuels consumption and can reduce the emission of greenhouses gases. The present work apparently portraits of study into the development of more energy efficient and less polluted fishing vessel.

6. References

[1] Keputusan Presiden Republik Indonesia Nomor 12 Tahun 2020 tentang Penetapan sebagai Bencana Nasional - Regulasi | Satgas Penanganan COVID-19, https://covid19.go.id/p/regulasi/keputusan-presiden-republik-indonesia-nomor-12-tahun-2020 (accessed 28 October 2020).

[2] PERPRES No. 47 Tahun 2017 tentang Penyediaan Lampu Tenaga Surya Hemat Energi bagi Masyarakat yang Belum Mendapatkan Akses Listrik [JDIH BPK RI], https://peraturan.bpk.go.id/Home/Details/72904/perpres-no-47-tahun-2017 (accessed 28 October 2020).

[3] Imawan Santosa P, Utama IKAP, D.A W. A Study into the Development of More Energy Efficient and Less Polluted Fishing Vessel. *IJOER* 2017; 3: 75–79.

[4] Sanntosa PI. The Configuration of Solar Sail Catamaran Fishing Vessel. *Global Journal of Research In Engineering*, https://engineeringresearch.org/index.php/GJRE/article/view/1961 (2019, accessed 28 October 2020).

[5] PI S. Techno-Economic Review: The Use of Solar Sail On Catamaran Fishing Vessel. *International Journal of Mechanical Engineering* 2019; 6: 32–37.

[6] Santosa PI. The Configuration of Engine-Sail Catamaran Fishing Vessel. *IOP Conf Ser: Mater Sci Eng* 2019; 462: 012008.

[7] Setyawan D, Utama IK a. P, Murdijanto M, et al. Development of Catamaran Fishing Vessel. *IPTEK The Journal for Technology and Science*; 21. Epub ahead of print 1 November 2010. DOI: 10.12962/j20882033.v21i4.90.

[8] Samuel S, Iqbal M, Utama IKAP. An Investigation into the Resistance Components of Converting a Traditional Monohull Fishing Vessel into Catamaran Form. DOI: https://doi.org/10.14716/ijtech.v6i3.940.

[9] Marchaj CA. *Sail Performance : Techniques to Maximize Sail Power*. 2nd Edition. Camden, Me: International Marine/Ragged Mountain Press, 2002.

[10] Larsson L, Eliasson R, Orych M. *Principles of Yacht Design*. 4th Edition. Camden, Maine: International Marine/Ragged Mountain Press, 2014.

[11] Kinney FS. *Skene's Elements of Yacht Design, Eighth Edition*. Dodd, Mead & Co., 1983.

[12] Catamaran Stability | James Wharram Designs, https://www.wharram.com/articles/how-we-design/catamaran-stability (accessed 28 October 2020).

[13] Bhattacharyya R. *Dynamics of marine vehicles*. New York: Wiley, 1978.

[14] Hind JA. Stability and Trim of Fishing Vessels and Other Small Ships, 2nd Edition | Wiley. *Wiley.com*, https://www.wiley.com/en-us/Stability+and+Trim+of+Fishing+Vessels+and+Other+Small+Ships%2C+2nd+Edition-p-9780852381212 (accessed 28 October 2020).

Gradient based ant spread modification on ant colony optimization method for retinal blood vessel edge detection

F Liantoni[1], N F Rozi[2], T Indriyani[2], W M Rahmawati[2], and R K Hapsari[2]

[1]Department of Computer and Information Education, Universitas Sebelas Maret, Indonesia
[2]Department of Informatics Engineering, Institut Teknologi Adhi Tama Surabaya, Indonesia

febri.liantoni@gmail.com

Abstract. Retinal blood vessels are a very important object to diagnosing illness. Fast, automatic, adaptive, and accurate systems to segment the retinal vessels are particularly useful. The retinal blood vessels can be detected by edge detection. This research will compare the edge detection techniques with adaptive ant colony optimization (Adaptive ACO). Generally, the early ants on the conventional ant colony optimization (ACO) are randomly distributed. This condition can cause the ant distribution imbalances. Based on this problem, the ant distribution modification on ACO is proposed to optimize the ant placement based on the gradient. The gradient values are used to determine the ants' placement. The ants are not randomly distributed but placed in the highest gradient. The so-called Adaptive ACO is expected to be used for better and faster path discovery optimization. The average PSNR of Prewitt edges, Sobel edges, conventional ACO, and the proposed method are 11.605, 11.913, 9.783, and 15.874 respectively. The PSNR of the proposed method has the greatest value than others. It shows that the placement of ants based on gradient can improve edge detection accuracy. The application of the Adaptive ACO method has successfully optimized the result of retinal vessel edge detection.

1. Introduction

Eyes are an important organ for humans that acts as a very sensitive vision. The occurrence of abnormalities in the eyes will be very disturbing. Almost all eye diseases can still be prevented. Eyes disease with a precise initial diagnosis will ease the healing process. The identification of some parts of retinal anatomy is a prerequisite diagnosis of some diseases [1]. Some eye fundus images are used to detect some retinal diseases such as retinal nerve tissue, optical disc, and fovea early. One cause of eye disorders is the abnormal growth of blood vessels in the eyes called neovascularization. This abnormal growth causes the retina to move backward from its original position. If not treated promptly, it will lead to vision loss, a kind of blindness. Neovascularization in the retinal fundus image is associated with changes in bifurcation point (BP), which is the branched point in the blood vessels and crossovers point (CP) from the cross point in the blood vessels [2], [3]. The changes that occur in the retinal blood vessels can be a signal of abnormalities in the eyes. Some abnormalities are characterized by blood vessel disorders in the eyes caused by certain diseases, such as diabetic retinopathy. Abnormalities in the vessels of the eyes can be known quickly and accurately through early detection. Preliminary detection can be done by looking at the abnormal branching of blood vessels, the changing number of retinal blood vessel cut points, etc. Periodic cut point detection that consists of a branch point and cross point of a retinal vessel can be used to detect retinal blood vessel anomalies.

Research that aims to develop and test the early detection system of diabetic retinopathy has been done by researchers with different media and methods. An algorithm with math morphology to detect microaneurysm is proposed in research [4]. Mathematical morphology is chosen because microaneurysm has a distinctive shape. The proposed algorithm consists of 3 parts: preprocessing, detects, and post-processing microaneurysm candidates.

Retinopathic detection using image processing is performed in [5]. This paper shows an audit of the most recent work on the utilization of digital image processing techniques for DR highlight identification. In [6], the detection of retinal vessels using structural predictors of bright lesions is performed. In the study, Gabor filters are applied to the grayscale image to lesion enhancement.

Usually, the image used in detection is not in ideal condition. This condition is a shadow disturbance, photos including blurred images [7]. Various problems with analysis and planning can affect the results of interpellation. Therefore, image processing techniques are required to obtain the ideal image. Many approaches have been used to perform edge detection on an image. Some commonly used methods are Sobel, Prewitt, and Canny [8], [9]. Research using ant colony optimization (ACO) to perform edge detection in the image has been done in [10], [11]. Singh et al [12] approaching Fuzzy-ACO. The number of ants is calculated and placed at the endpoint of the image using a double detector. The Fuzzy derivative technique implements a fuzzy probability factor to determine which next pixel is most likely to be the edge. Wong et al [13] performed improvements to edge detection by combining the Canny method with ACO.

Previous research about ACO by dividing of picture and count the number of ants based on gradient each division area is done [14]. However, the initial ant spreading process is still done randomly as the traditional ACO, thus allowing ants to be trapped under optimal local conditions. In this study, gradient values are used to determine the initial ant placements. Ants are not randomly distributed but placed in a high gradient. At the highest gradient, the ants travel to determine the path first. This method is expected to be used for pathfinding optimization.

2. Methods

This research is done by several steps include the selection of image data, preprocessing, and edge detection. Figure 1 below describes the steps to be performed.

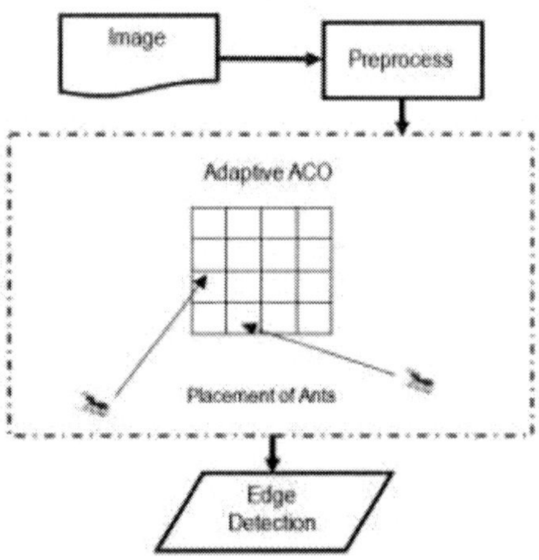

Figure 1. Research model.

2.1. Preprocessing

The used data input in the system is the RGB image. From the RGB image then converted into a gray image [15], [16]. The process of changing the color image into a gray image using lightness and average counting method. In addition to that method, there is also a luminosity method which is the development of the lightness and average method. The luminosity method calculates the value of each color element of the image, red, green, and blue and adding the weight according to the perception of human vision. The calculation is shown in Equation 1.

$$Luminosity = \frac{R*299+G*587+B*114}{1000} \tag{1}$$

Where R represents the red, G represents the green, and B represents the blue channel. The luminosity method is chosen because of the nature of the luminosity method. It is closer to the perception of human vision. After the grayscale image is obtained, the next process is edge detection using the Adaptive ACO method.

2.2. Gradient

The initial step of digital image processing for edge detection is a gradient [8]. The gradient values of the area is indicated by the presence of edges in the area. Gradients are made from transitions or color changes.

$$\nabla f = \begin{bmatrix} G_x \\ G_y \end{bmatrix} = \begin{bmatrix} \frac{\partial f}{\partial x} \\ \frac{\partial f}{\partial y} \end{bmatrix} \tag{2}$$

For the magnitude of the vector shown by Equation 3.

$$\nabla f = \left[\left(\frac{\partial f}{\partial x}\right)^2 + \left(\frac{\partial f}{\partial y}\right)^2\right]^{\frac{1}{2}} \tag{3}$$

The component of the vector gradient itself is a linear operator, but the magnitude of the vector is not a linear operator because it is quadratic and square root operations. In image processing, the derived based operator becomes the basis of calculation. Some examples are using the operator Sobel, Prewitt, Robert [8]. Figure 2 shows a pseudo-convolution kernel that is used to calculate the gradient magnitude (G) quickly.

a_1	a_2	a_3
a_4	a_5	a_6
a_7	a_8	a_9

Figure 2. Pseudo-convolution kernels with the size of 3 × 3

To calculate the kernel on the summing process of the magnitude shown in Equation 4.

$$|G| = |(a_1 + 2 * a_2 + a_3) - (a_7 + 2 * a_8 + a_9)| + |(a_3 + 2 * a_6 + a_9) - (a_1 + 2 * a_4 + a_7)| \tag{4}$$

One method to determine image gradient is the Sobel operator. The first step of a Sobel operator is to estimate the value of horizontal gradient (*G*x) and vertical gradient (*G*y). This is done by kernel operation of the image matrix as shown in Figure 2. The kernel matrix for *G*x is defined as the *K*x operator, while the kernel matrix for *G*y is defined as *K*y, as in Equation 5.

$$Kx = \begin{bmatrix} -1 & 0 & 1 \\ -2 & 0 & 2 \\ -1 & 0 & 1 \end{bmatrix} \quad Ky = \begin{bmatrix} -1 & -2 & -1 \\ 0 & 0 & 0 \\ 1 & 2 & 1 \end{bmatrix} \tag{5}$$

After obtaining the gradient values in the vertical and horizontal direction, then search matrix calculation values that produce image gradient values. The gradient values are also used to determine the angular direction of the edge *(θ)*. The value of θ will affect the treatment of a pixel against its neighboring pixels. The counting edge value is calculated by Equation 6.

$$\theta = \tan^{-1}\left(\frac{Gy}{Gx}\right) \tag{6}$$

After calculating pixel's gradient value, choose gradients that has value equal or more than the threshold. The number of adequate gradients then become the number of ants *(K)*.

2.3. Adaptive Ant Colony Optimization (Adaptive ACO)

Ant Colony Optimization (ACO) is a heuristic method to imitate the behavior of ants to solve discrete optimization problems [17]. In general, pseudocode is shown in Figure 3.

```
Initialize
SCHEDULE_ACTIVITIES
    ConstructAntSolutions
    DoDaemonActions (optional)
    UpdatePheromones
END SCHEDULE_ACTIVITIES
```

Figure 3. ACO pseudocode.

Initialize is a step done at the beginning of the process. The parameters used are the number of ants (K), the number of step of construction (L), iteration (N), pheromone evaporation rate (ρ), decay pheromone (φ), the weight factor of pheromone (α), and heuristic information weight factor (β). The number of ants (K) used as input parameters in this study is based on the gradient. Counting the number of ants is shown in Equation 7.

$$K = Number\ of\ gradients >= threshold \tag{7}$$

The number of ants obtained on the gradient will be the input parameter. Ants are not randomly distributed but are placed on pixels that have a high gradient in the image. Ants on the gradient based pixel position will move to the neighbor pixel to determine the edge. *ConstructAntSolutions* is an ant trip activity. The construction process contains a number of construction steps. Ants will move in an image until the target number of construction steps are formed. In the construction process to (n^{th}) the number of ants (k^{th}) will move from node i to a node j following the probability displacement $P_{i,j}^{(n)}$. This pseudorandom proportional rule is shown in Equation 8.

$$P_{i,j}^{(n)} = \frac{\left(\tau_{i,j}^{n-1}\right)^{\alpha}\left(\eta_{i,j})^{\beta}\right)}{\sum_{j\in\Omega_i}\left(\tau_{i,j}^{n-1}\right)^{\alpha}\left(\eta_{i,j})^{\beta}\right)}\ if\ j\ \in\ \Omega_i \tag{8}$$

Where Ω_i is the neighboring node of the ant given to node i, pheromone update is denoted by τ, and η is the heuristic information.

DoDaemonActions is a construction solution for additional steps before pheromone value updates. This activity is an additional measure of updating pheromone values. This process can not be done by only single ant. *UpdatePheromones* is a pheromone update activity after the construction and daemon actions are performed. Counting local pheromone updates are shown in Equation 9.

$$\tau_{i,j} = (1-\varphi)\tau_{i,j} + \varphi\tau_0 \tag{9}$$

Where φ is a pheromone damage and τ is a pheromone update. The global update of pheromones is shown in Equation 10.

$$\tau_{i,j} = (1-\rho)\tau_{i,j} + \rho\Delta\tau_{i,j} \tag{10}$$

In ACO there is a displacement rule with a probability factor at eight neighboring pixels. The pixels with the maximum probability factor in detecting neighbors have edge pixels. To reduce the repetitive repetition of ants the stopping criteria are made, ie the displacement of the ant will stop if it passes through the path that the other an ant has passed and when all pixels of neighbor (8 pixels) have been passed by all the ants then the displacement will stop [12].

In the next step, a decision is taken on each pixel to determine the edge or not. This is done by implementing a threshold (*T*) on the final pheromone matrix ($\tau^{(N)}$) [11]. In this proposed study, the threshold value (*T*) was adapted based on the Otsu method [18]. Otsu thresholding is implemented to determine the best solution based on the number of pheromones stored in each pixel. The Otsu approach can define a useful variable that distinguish two or more naturally occurring groups. Edge determination is based on the final pheromone matrix value. The Otsu threshold technique will reduce the gray image results to binary images with two possible values for each pixel. From the final pheromone matrix value will show a pixel is an edge or not edge.

The initialization threshold $T^{(0)}$ is chosen based on the mean pheromone value matrix. Furthermore, the pheromone matrix is classified into two categories with a value criterion less than $T^{(0)}$ or more than $T^{(0)}$. The initialization value of $T^{(0)}$ is calculated based on Equation 11.

$$T^{(0)} = \frac{\sum_{i=1:M_1}\sum_{j=1:M_2}\tau_{i,j}^{(N)}}{M_1 M_2} \quad (11)$$

Where M_1 vaporizes the width of the image, while M_2 is the height of the image. The threshold value (T) will be compared with the final pheromone results used as the fitness value in determining the edge or not. The edge determination with a fitness value that determines a pixel as an edge ($E_{i,j}$ = 1) or not ($E_{i,j}$ = 1) is based on the criteria as shown in Equation 12.

$$E_{i,j} = \left\{\begin{matrix} 1, & if\ \tau_{i,j}^{(N)} \geq T^{(l)} \\ 0 & other \end{matrix}\right\} \quad (12)$$

In this research, the number of small construction and iteration parameters aim to speed up the edge detection process. The used values are shown in Table 1.

Table 1. Parameter initialization.

	Parameter	ACO	Adaptive ACO
K	Total number of ants	$\sqrt{width * height}$	Number of gradients ≥ threshold
L	An ant movement steps within each construction step	10	10
N	Total number of construction step	4	4
α	Weighting factor of pheromone information	0.5	0.5
β	Weighting factor of heuristic information	0.1	0.1
ρ	Evaporation rate	0.1	0.1
φ	Pheromone decay coefficient	0.05	0.05
τ_{init}	Initial value of each element of pheromone matrix	0.0001	0.0001
Ω	The tolerable ant's movement range	8	8

There are several steps performed on adaptive ACO edge detection. In this research, adaptive ACO is proposed in the process of initial ants placement based on gradient values. The edge detection process that will be performed on adaptive ACO can be seen in Figure 4.

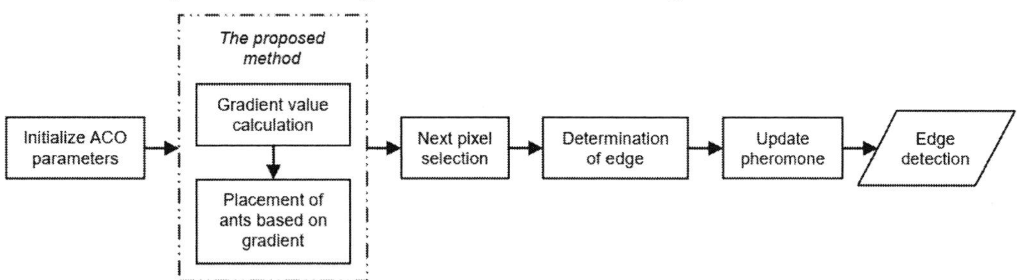

Figure 4. The proposed method.

A numerical assessment can be used to determine the accuracy of the edge detection test results. The testing of image edge detection result is based on Peak Signal-to-Noise Ratio (PSNR) and Mean Square Error (MSE) method. PSNR is a relationship between the ratio of the maximum value and the value of damage (noise) that shows the accuracy of the process. Its value will be high on the good quality image. The calculation of MSE and PSNR values is shown in Equation 13 and 14.

$$MSE = \frac{1}{m\,n}\sum_{i=0}^{m-1}\sum_{j=0}^{n-1}[I(i,j) - K(i,j)]^2 \quad (13)$$

$$\text{PSNR} = 10\,log_{10}\left(\frac{MAX_I^{\,2}}{MSE}\right) \quad (14)$$

3. Results and Discussion

The tests are conducted to determine the success level of the system in producing image edge detection. The pre-processing stage will get the gray image. The results of the pre-processing images are used as input at the image edge detection stage. Testing is done by comparing the image edge detection results between ant algorithm optimization with the conventional edge detection algorithm. An example of image capture is shown in Figure 5.

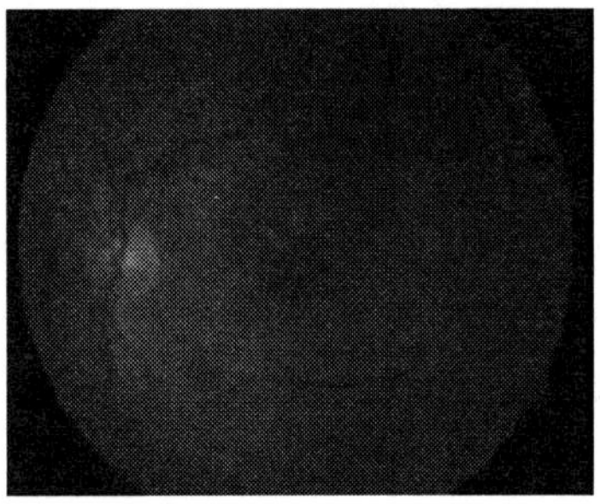

Figure 5. A retinal image in RGB.

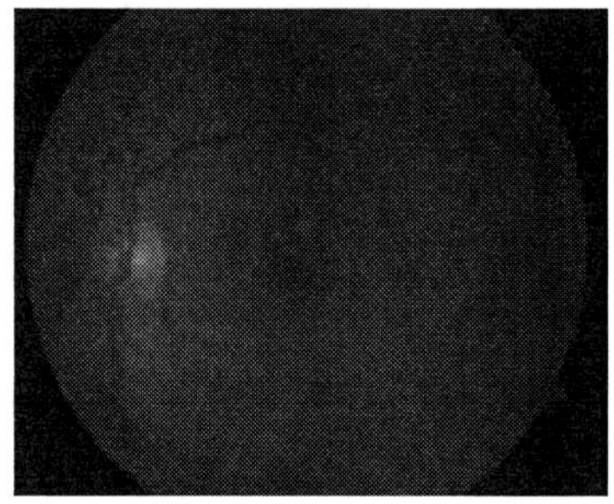

Figure 6. A retinal image in grayscale.

Figure 5 will be used as an input image on the pre-processing phase. The input image in the form of an RGB image is converted to a gray image. The gray image conversion results are shown in Figure 6. After the gray image data obtained then the edge detection performed. The result of the edge detection image will be compared to other selected edge detection method result. The edge detection results sample are shown in Figures 7, and 8.

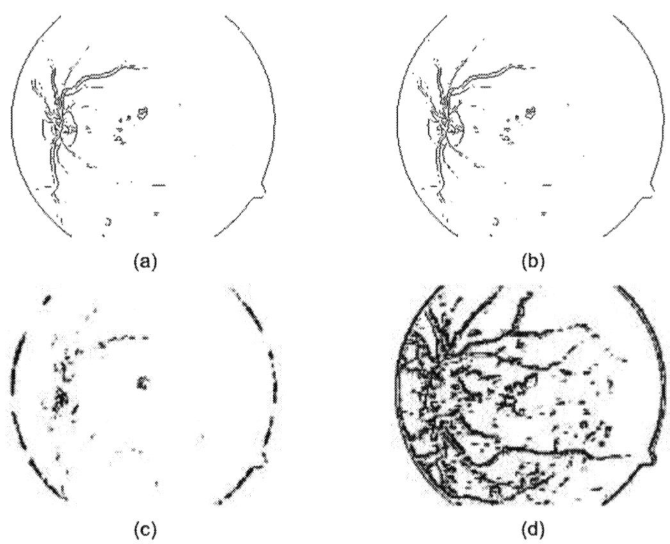

Figure 7. Image 1: (a) Prewitt edges, (b) Sobel edges, (c) ACO, and (d) proposed method.

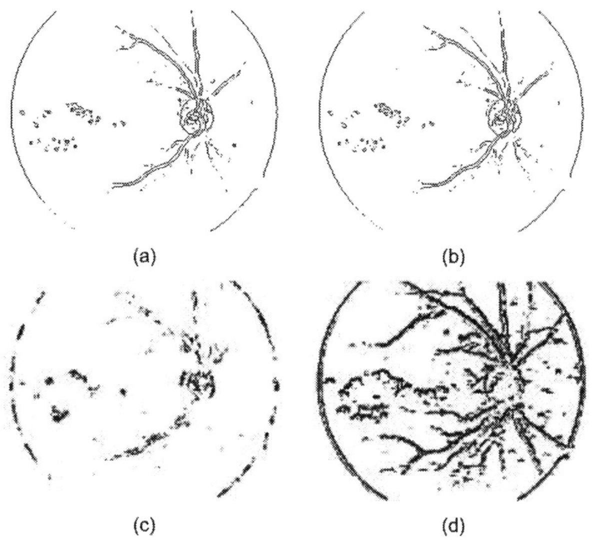

Figure 8. Image 2: (a) Prewitt edges, (b) Sobel edges, (c) ACO, and (d) proposed method.

Based on the testing of the retinal blood vessels, the proposed Adaptive ACO edge detection results provide a more detailed outline than the other edge detection techniques. It proves that the process of placement of ants based on the value of gradient is influential in determining the edge of an image with the Adaptive ACO method. Table 2 and 3 is the value of the test results.

Table 2. MSE Values

Result	MSE			
	Prewitt	Sobel	ACO	Proposed Method
Mean	0.090	0.087	0.103	0.072
Stdev	0.062	0.060	0.081	0.051
Min	0.019	0.018	0.057	0.009
Max	0.217	0.212	0.327	0.167

Table 3. PSNR Values

Result	PSNR			
	Prewitt	Sobel	ACO	Proposed Method
Mean	11.605	11.913	9.783	15.874
Stdev	3.497	3.617	2.638	5.992
Min	6.625	6.994	5.071	9.761
Max	19.111	19.712	16.512	27.322

Tables 2 and 3 show the average MSE and PSNR. The proposed method has the smallest average MSE value compared to conventional edge detection techniques and ACO methods. The average PSNR of the proposed method also has the greatest value compared to conventional edge detection techniques and ACO methods. Conventional PSNR ACO values are smaller than the edge detection techniques (Prewitt and Sobel) method because the used parameter values are minimal.

4. Conclusions

Based on the research, it can be concluded that the optimization of Adaptive ACO algorithm performance for the retinal blood vessel detection process could produce a good feature compared to other edge detection methods. This condition is shown from the results which produce an image with a more detailed edge (has a thicker border). The greatest value of PSNR and the smallest value of MSE of the proposed method compared to the other shows that the process of ants placement by gradient could improve the accuracy of image edge detection results.

5. References

[1] M. Badar, M. Haris, and A. Fatima, "Application of deep learning for retinal image analysis: A review," *Comput. Sci. Rev.*, vol. 35, p. 100203, Feb. 2020.

[2] J. Malek, A. T. Azar, B. Nasralli, M. Tekari, H. Kamoun, and R. Tourki, "Computational analysis of blood flow in the retinal arteries and veins using fundus image," *Comput. Math. with Appl.*, vol. 69, no. 2, pp. 101–116, Jan. 2015.

[3] R. GeethaRamani and L. Balasubramanian, "Retinal blood vessel segmentation employing image processing and data mining techniques for computerized retinal image analysis," *Biocybern. Biomed. Eng.*, vol. 36, no. 1, pp. 102–118, 2016.

[4] A. Purwita and K. Adityowibowo, "Automated Microaneurysm Detection using Mathematical Morphology," *Int. Conf. Instrumentation, Commun. Inf. Technol. Biomed. Eng*, pp. 1–4, 2011.

[5] A. Pattanashetty and S. Nandyal, "Diabetic Retinopathy Detection using Image Processing: A Survey," *Int. J. Comput. Sci. Netw.*, vol. 5, no. 4, 2016.

[6] J. Amin, M. Sharif, M. Yasmin, H. Ali, and S. L. Fernandes, "A method for the detection and classification of diabetic retinopathy using structural predictors of bright lesions," *J. Comput. Sci.*, pp. 153–164, 2017.

[7] M. Foracchia, E. Grisan, and A. Ruggeri, "Luminosity and contrast normalization in retinal images," *Med. Image Anal.*, vol. 9, no. 3, pp. 179–190, 2005.

[8] R. Gonzales and R. Wood, *Digital Image Processing*. Prentice-Hall, Inc., United State, America, 2007.

[9] F. Al-Hafiz, S. Al-Megren, and H. Kurdi, "Red blood cell segmentation by thresholding and Canny detector," *Procedia Comput. Sci.*, vol. 141, pp. 327–334, 2018.

[10] B. Anna and C. Oppus, "Image Edge Detection Using Ant Colony Optimization," *Int. J. Circuits, Syst. Signal Process.*, vol. 4, no. 2, pp. 24–33, 2010.

[11] J. Tian, Y. Weiyu, and X. Shengli, "An Ant Colony Optimization Algorithm For Image Edge Detection," *IEEE Congr. Evol. Comput.*, pp. 751–756, 2008.

[12] G. Singh, N. Kumar, and A. Kumar Verma, "Ant colony algorithms in MANETs: A review," *J. Netw. Comput. Appl.*, vol. 35, no. 6, pp. 1964–1972, Nov. 2012.

[13] Y.-P. Wong, V. Soh, K.-W. Ban, and Y.-T. Bau, "Improved Canny Edges Using Ant Colony Optimization," *IEEE Comput. Graph. Imaging Vis.*, pp. 197–202, 2008.

[14] F. Liantoni, C. K. Kartika, H. M. Tri, K. C. Kirana, and T. H. Muliawati, "Adaptive Ant Colony Optimization based Gradient for Edge Detection," *J. Ilmu Komput. dan Inf.*, vol. 7, no. 2, pp. 76–82, Aug. 2014.

[15] M. Foracchia, E. Grisan, and A. Ruggeri, "Luminosity and contrast normalization in retinal images," *Med. Image Anal.*, vol. 9, no. 3, pp. 179–190, Jun. 2005.

[16] F. Liantoni, R. Indra Perwira, S. Muharom, R. Agung Firmansyah, and A. Fahruzi, "Leaf classification with improved image feature based on the seven moment invariant," *J. Phys. Conf. Ser.*, vol. 1175, p. 012034, Mar. 2019.

[17] M. Dorigo, M. Birattari, and T. Stutzle, "Ant Colony Optimization: Artificial Ants as a Computational Intelligence Technique," *IEEE Computational Intelligence Magazine*, 2006.

[18] N. Otsu, "A threshold selection method from gray level histograms," *IEEE Trans. Syst. Man. Cybern.*, vol. 9, no. 1, pp. 62–66, 1979.

IOP Conf. Series: Materials Science and Engineering 1010 (2021) 012022 doi:10.1088/1757-899X/1010/1/012022

Application of lean manufacturing to improve procurement lead time in the case of the steel industry

M Z P Nugroho[1], S Hasibuan[1], H Adiyatna[1]

[1]Universitas Mercu Buana, West Jakarta, Indonesia

Email: 55318110005@student.mercubuana.ac.id

Abstract. The steel industry is classified as a strategic industry that supports many industries, due to continuous national development. The steel industry must improve the quality of its production processes in order to increase its competiveness. One factor that supports the stability of the production process is the procurement process. Delays in the procurement process can disrupt the smooth production process. The purpose of this research is to minimize waste in the steel procurement process by using the value stream mapping method. This research begins by making current state mapping of the actual procurement process, then analyzed the activity categories including value added activity and non-value added activity. In the beginning, the total time value added activity is 7488 minutes and total time non value added activity is 12240 minutes. An improvement was made in terms of the negotiation process, document review, and waiting for an offer, the result of the total value added activity is 7488 minutes and non-value added activity is 2160 minutes, so the procurement process initially 41 days became 21 days.

1. Introduction

The steel industry is a strategic industry that is used as raw material and supporting material for many industries, for infrastructure (road construction, building construction, and bridge construction) transportation equipment, automotive, or weaponry [1]. Indonesia is one of the major steel users and suppliers. Based on data from the Ministry of Industry, Indonesia's steel consumption reached 12.7 million tons in 2017 and its projected to be 21.4 million tons in 2025 [2]. The Ministry of Industry ensures that the full potential of the steel industry can have a maximum role, because it much needed for national development. The development of the national steel industry is needed to support increased infrastructure development in Indonesia, both by the Government and the private sector which requires large amounts of steel [3].

PT. Krakatau Steel is the one of steel company in Indonesia. To carry out business activities, the company is supported by many functions, the functions are marketing, production and technology, logistics, finance, human resources, and business development. Procurement or logistics is an important function in a company's [4]. Procurement department has an important role for supply the material in the company and responsible to get the quantity and quality of material needed. Timeliness in the procurement process is one of the key factors preventing the disruption of the production process [5]. The current condition of the procurement process in the company is still not optimal. The lead time target for the material procurement process set by the company has not yet been reached.

The purpose of this research is to improve the performance of the procurement process in the case of steel companies using lean manufacturing approach. The value stream mapping method is one of the well-known and widely adopted lean manufacturing tools to reduce waste in the production process in various industries [6][7][8][9].

2. Literature Review

2.1. Procurement

Procurement is the activity of finding, obtaining, buying materials through a tender process or competitive bidding process. This process is done in order to satisfy the buyer in receiving materials at the best price by considering many aspects, these aspects are quality, quantity, time and distance. Procurement is a comprehensive activity of the purchasing process [10]. Purchasing is an activity to obtain material to meet the needs of the company. The purpose of purchasing is different and narrower than procurement, because of purchasing is part of procurement. Purchasing functions include obtaining goods by fulfilling the following provisions: Right product that is right and in accordance with standards and specifications, Right amount with the amount according to needs or orders, Right Time with delivery time according to user needs, Right Price with competitive prices [10].

2.2. Lean Manufacturing for Procurement

Lean manufacturing for procurement is a production practice that considers all the expenditure of available resources to obtain economic value to the procurement process without any waste, and this waste is the target to be reduced [11]. Implementation of lean manufacturing is one of the characteristics that indicates that the company focuses on customer satisfaction, and product results [12]. Lean always sees the value of a product from the customer's point of view, where the value of a product is defined as something that the customer wants to pay. If the company does not want to bear excessive costs, it must be able to find a number of things to be eliminated to avoid waste. Cases that often occur in daily life and operational activities of a company are a waste of time and energy. Other waste may occur because the use of equipment that is considered not so important for the company or during the production process.

2.3. Value Stream Mapping for Procurement

Value stream mapping is a collection of all activities in which there are activities that provide added value activity and non-added value activity is needed to complete the Purchase Request process to become a Purchase Order. This activity is part of the procurement process which includes the flow of the procurement process. Value Stream Mapping is one of the ideal tools as a first step in the process of improving the company that is used to help visualize the procurement process as a whole [13].

Value Stream Mapping consists of a sketch that maps the current state and the Future State. Current state sketches describe the process of material and information flow currently in the process. This visualizes the process of being able to identify waste in the system and encouraging users to take a systematic approach to eliminate waste. A sketch of a future state is a chart showing how to make a lean flow.

The purpose of this mapping is to identify all types of waste along the value stream and to take steps in an effort to eliminate the waste. Taking steps in terms of the value stream means working in a large scope, and improving the entire flow rather than just optimizing the flow in piecemeal. This gives rise to a language that is commonly used in the process of procurement of goods, thus will be able to facilitate more mature decisions in improving the value stream. Value Stream Mapping can provide an optimal turning point for the company. The two main steps in VSM mapping are:

1. Making Current State Map to map the current procurement conditions, so that it can identify any waste that occurs. Current State Map Our starting point is to look at the current process flow before improvements are made.

2. The making of the Future State Map as a proposed design for the improvement of the current state map is a description of the process and information after the information after the improvement is made.

3. Research Method

The steps used in this research intended to solve problems in the procurement process, the steps consist of identification problem, data collection and processing the data, value stream mapping (VSM) analysis, and make the fish bone diagram.

For the data collection stage, it is done by direct observation, taking the procurement history data from SAP, study literature, and interview or discussion with expert. The type of data used is qualitative data that is the lead time of procurement from and the data sources used is primary data obtained directly from the SAP.

For the data processing done with two method, there are value stream mapping methods and fishbone methods. Value stream mapping methods is used for identification waste time in procurement process. Current state value stream mapping shows the current conditions of the procurement process so can identify existing waste time, while the future state value stream mapping show the condition of procurement after improvement [14]. Fishbone methods done by making fishbone diagram for identification the causative factor has not been reach the target time for procurement. Diagrams are made based on the results of discussions and interviews with employees who are experts in the procurement field.

4. Result and Discussion

4.1. Current State Value Stream Mapping

According to [15] making current state value stream mapping intended for recognize and know the existing process. Besides that, current state value state mapping also used to determine the flow of information during the process. Value Stream Mapping (VSM) method is one method that applies efficiently visualization picture to describe the current state of, and can identification the long term vision and can develop the planning of the company for reach the target. Current state mapping made base on the identification of value added activity and non-value added activity in the process of procurement in one of the steel company in Indonesia

Table 1. Identification of value added time and non-value added time in procurement

No	Flow process of procurement in PT. Krakatau Steel	Time (min)	Category
1	The user prepares an approved Purchase Requisition complete with supporting document (materials catalog)	3360	VA
2	Waiting for review	1488	NVA
3	Purchaser receives the Purchase Requisition (PR) file from the User and evaluates the completeness of the PR documents	1200	NNVA
4	Purchaser consult the document to user if it is not complete	2880	NNVA
5	Purchaser chooses a list of bidders who are competent and have good performance in the supply of goods needed by the user	480	VA
6	Purchaser make Request for Quotation (RFQ) based on PR that has been made by the User	1440	VA
7	Purchaser sends RFQ to the List of Tender Participants (DPT) via e-mail	240	NNVA
8	Purchaser is waiting for an offer from the vendor	3360	NNVA
9	Purchaser makes a comparisons of Quotation based on offers sent by vendors according to deadlines	240	NNVA
10	Purchaser make, send, and negotiates to vendor who was appointed as a potential winner until the Minutes of Negotiation are signed and endorsed by the Material Procurement Manager	2832	NNVA

No	Flow process of procurement in PT. Krakatau Steel	Time (min)	Category
11	Making a Purchase Order (PO) until the validation of the PO is in accordance with the negotiation results and submits the completed PO to the vendor so that the vendor sends the goods according to the time specified	2208	VA
	Total	**19728**	

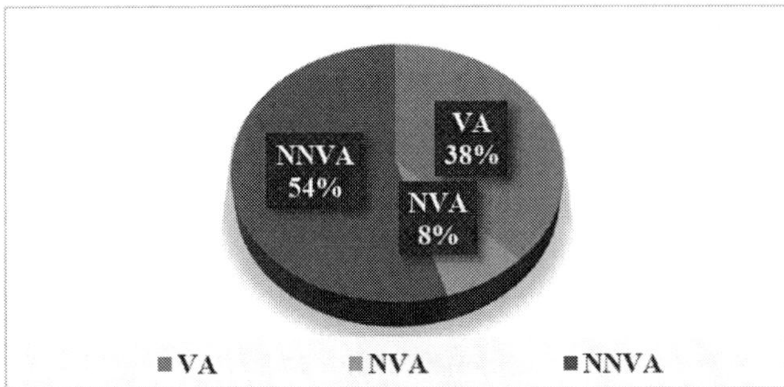

Figure 1. The percentage comparison of VA and NVA value of procurement processes.

The average time needed in the procurement process at PT. Krakatau Steel is 19.728 minutes, if converted in days is 41.1 days. The process of procurement has been delayed for 20 days from the standard procurement process targeted by the company and listed in the Work instruction (WI). At this stage of the process shows that percentage of value added time is 38% and non-value added time is 62% with detail necessary but non-value added time is 54% and non-value added is 8%. Total value added time in procurement process is 7.488 minutes, while the total non-value added time is 12.240 minutes. Value added time is an activity that provides added value to the company, non-value added time is an activity that does not provide added value to the company while necessary but non-value added time activities are needed but do not provide added value to the company.

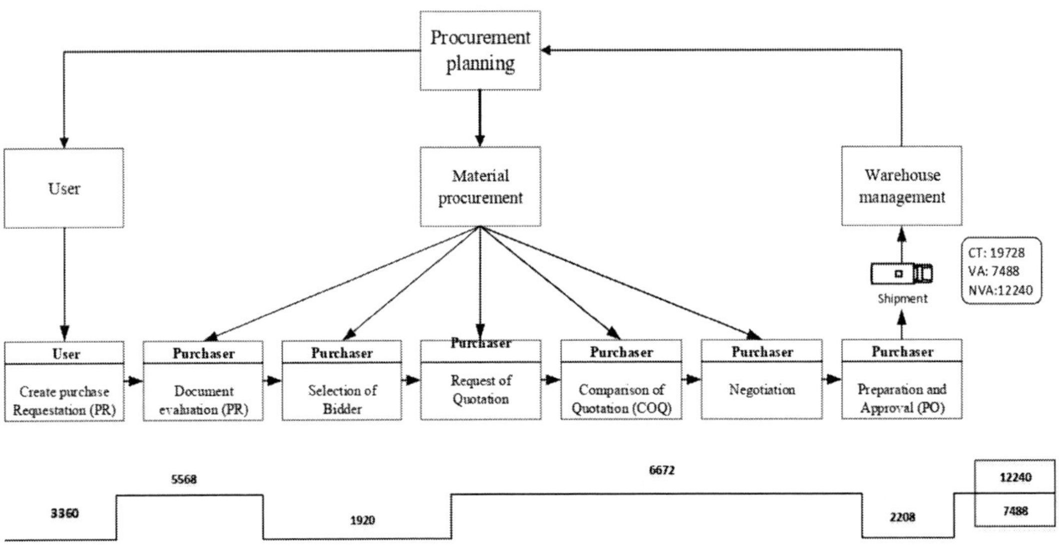

Figure 2. Current State Value Stream Mapping in procurement process.

Figure 2 shows the entire flow both in terms of the process and information contained in the process of procurement in PT. Krakatau Steel starting from the user process by creating a PR based on the notification of the need for goods in the field which is equipped with supporting files in the form of materials catalogue, bill of materials, as well as drawings, PR made after procurement planning check the stock of goods in the warehouse. Then the PR file is continued to the Material Procurement division, and the PR file will be processed by the purchaser. Next the Purchaser evaluates the completeness of the PR file, if the file is complete and complies with the procurement requirements, then the purchaser chooses a list of bidders who are competent and have good performance in the supply of goods needed by the user according to the commodity. After that, purchaser make a RFQ, RFQ is requests for price quotes to vendors or suppliers of goods. The next process the vendor sent by RFQ makes a Quotation containing the price offered at the time of procurement of goods at a specified deadline. After the vendor sends a price quote letter, the purchaser opens the offer with the bid opening team and make Minutes of opening of bids which are signed by the team and approved by the manager of the Procurement Materials Division. After the minutes of opening the offer are signed by the manager, a Comparison of Quotation (COQ) is made by Purchaser so the prospective tender winner was chosen. After that negotiations are carried out by the purchaser to prospective tender winners until the Minutes of Negotiation are signed and endorsed by the Material Procurement Manager. Next, the Purchaser creates a PO and validates the PO based on PR data and negotiation results, and submit it to the vendor so that the vendor sends the goods within the specified time.

After the allotted time, the vendor sends the goods according to the criteria and requirements on the PO. When the goods are received, a conformity inspection is carried out by the Inspector in the Warehouse Management Department. If the items match, then proceed to the Warehouse Management storage area, however, if the item has a discrepancy, the item is rejected and returned to the vendor. Furthermore, goods that have been stored in the warehouse can be taken by the user who ordered the goods by making a reservation for the collection of goods.

4.2. Waste identification in procurement

According to [11] waste is all activities that cannot add value in the process of transforming inputs into outputs along the value stream. Waste identification is obtained through interviews with employees in the vendor management services and internal experts in the procurement department.

Through the results of measuring the time of the observation process carried out, there is some waste that results in a long time in the process of making a Purchase Order (PO). The impact of making a Purchase Order (PO) that is too long is resulting in requests for user goods to arrive late from the specified time. Based on observations and lead time calculations, there are various kinds of waste found in the existing processes. Waste that exists are:

1) *Over Processing*. That is because there is a process that requires procurement in replying to requests from Users via file. The response is given when confirming the continuation of the PR process, whether it is a request to complete supporting PR documents or unclear PR specifications.
2) *The second Waste* is negotiation process by Purchaser, where the negotiation process is included in the necessary non value added activity, because the negotiation process is still done manually that is face to face, so that the determination of the negotiation schedule can be delayed because of the busyness of both the supplier and purchaser.
3) *The third Waste* is waiting for offers from vendors, where the process is waiting for offers from vendors in the category of necessary non value added activity, where the standard is waiting for the offer is 7 days and shipping the offer, and some vendors still send them manually via hard copy of the offer delivered to the procurement department.

From the causes of waste procurement can be analyse with a fishbone diagram, where the fishbone diagram is a tool to identify various kinds of potential causes that occur in a problem and analyse the problem [16]. The fishbone diagram was developed from the results of discussions and interviews with procurement and vendor management employees. Here is a fishbone to describe the flow of the causes of the waste that occurred in the procurement process.

1. Evaluation process of request documents conducted by Purchaser

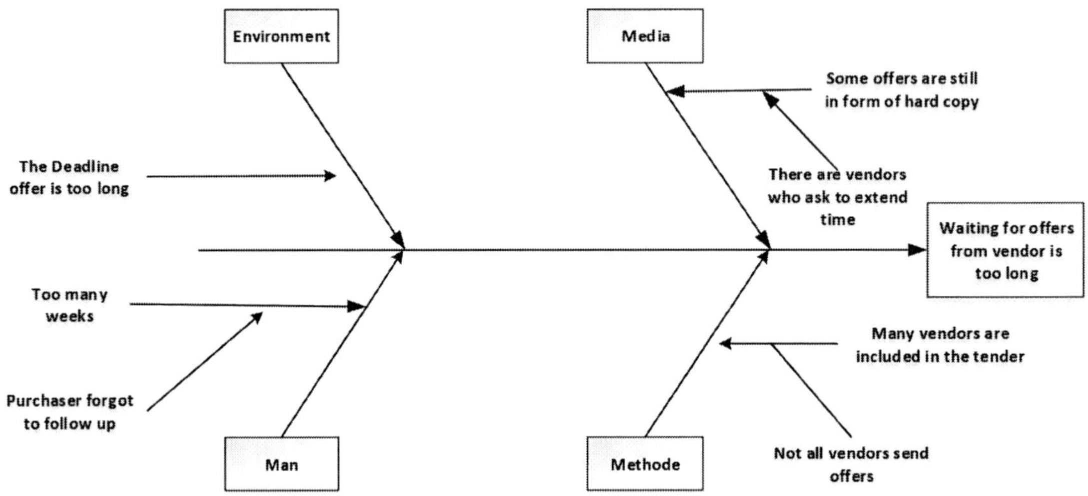

Figure 3 Fishbone causes the length of document review.

2. Negotiation Process By Purchaser

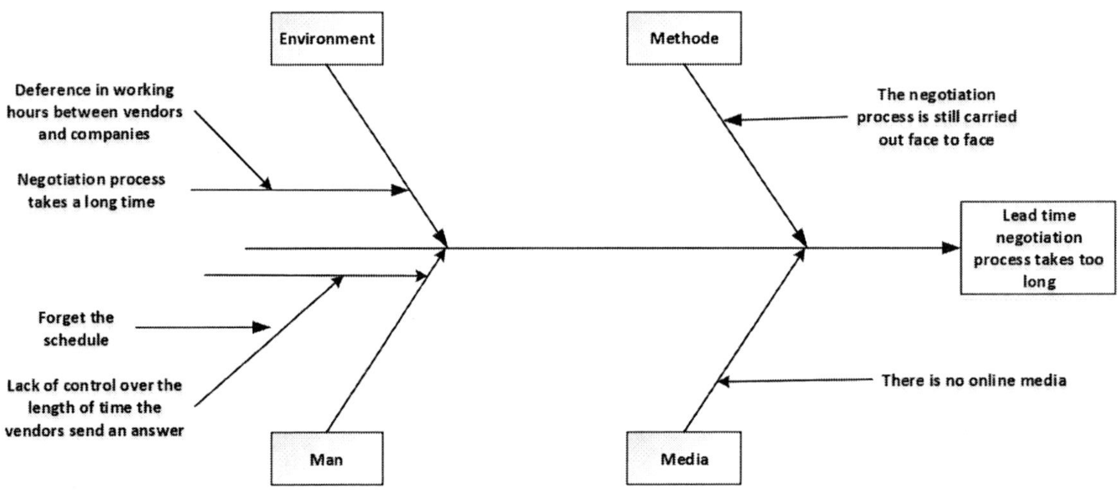

Figure 4. Fishbone causes the length of the negotiation process by the Purchaser.

3. Waiting for Offer from Vendor

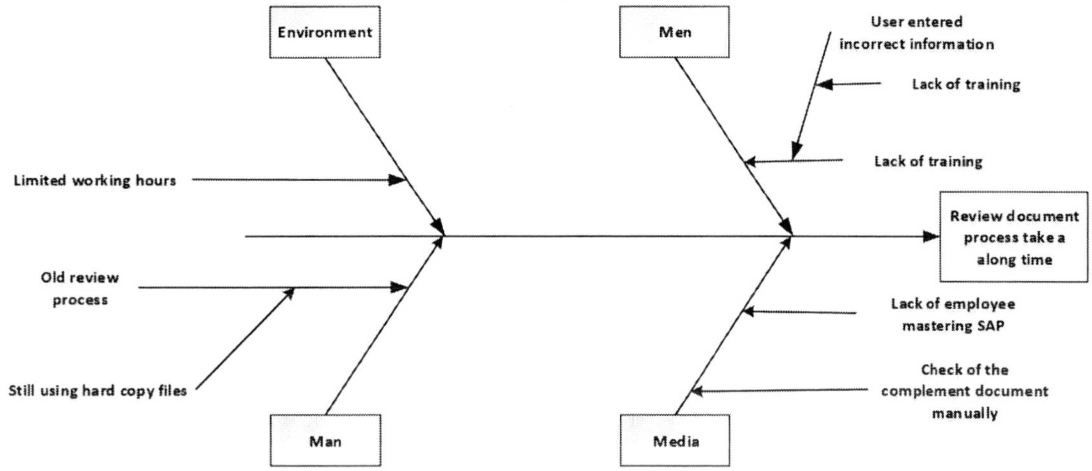

Figure 5. Fishbone causes the duration of the offer from the Vendor.

Table 2. Identify Value Added Time and Non-Value-Added Time procurement

No	Waste	Proposed Improvement
1	The Purchase Request Document Review Process by Purchaser	Provide training to users so that users are not wrong in completing documents, and checklist forms can be made so that no documents are missed
2	Negotiation Process	Creating an e-negotiation system using either WEB or other online applications to facilitate time control of the negotiation process, as well as giving notifications to vendors when negotiation time is almost reached the limit
3	Awaiting Offers from Vendors	The follow-up process uses the tender deadline reminder notification via e-mail

4.3. Proposed Improvement

This section explains the proposed improvements to be able to reduce or even eliminate the waste that is used as the basis for making future state value stream mapping procurement. This proposed improvement was obtained from the results of identification with fishbone.

4.4. Future State Value Stream Mapping

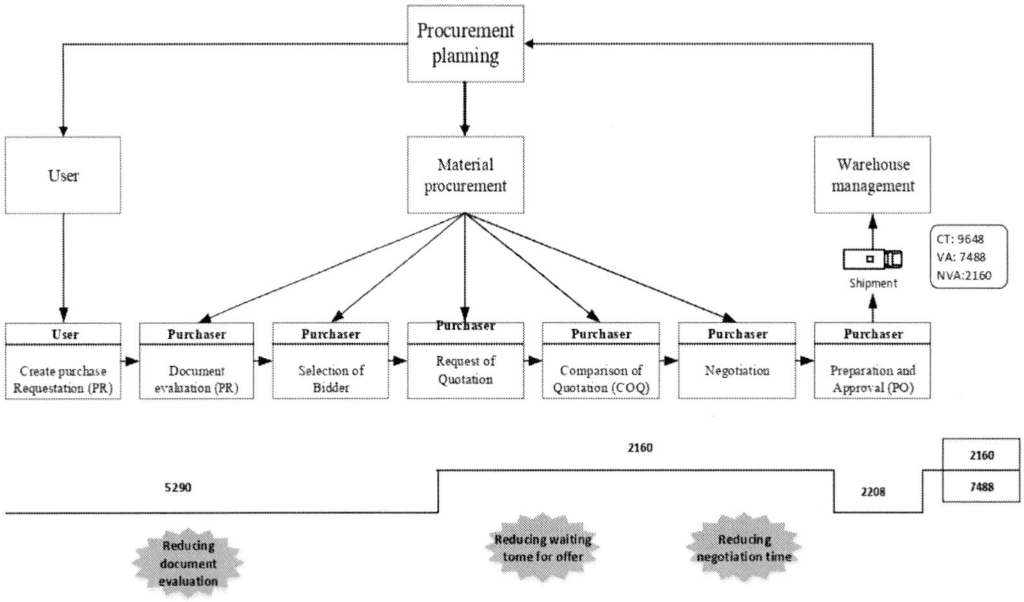

Figure 6. Future state value stream mapping of Procurement.

Future State Value Stream Mapping s a proposal and suggestion for improvement of current state value stream mapping by minimizing or eliminating waste. Future State VSM made based on the proposed improvements that have been made. The process of procurement starts from the user making a PR until the issuance of a PO. Waste elimination in the Future State VSM is done at the Evaluation stage of the document from the user, where by eliminating the time Evaluation of documents from the user can eliminate waste as much as 5568 minutes both in terms of reviewing documents, evaluating documents, and returning documents, this can be done by providing information to the user regarding the completeness of supporting files that must be sent to the procurement department. The next elimination of waste is in the process of waiting for offers from vendors, which originally took 3360 minutes, can be reduced to 1440 minutes, This can be done by providing reminders and following-up of vendors who take part in tenders and offer prices. The next eliminating waste is reducing negotiation time, which originally took 2832 minutes, can be reduced to 240 minutes with notes that negotiations are conducted online without having to face to face. The total time shown in Figure 6 is 9648 minutes or 20.1 days, half a day faster than the company's target of 21 days. And faster than the total time at the VSM Current State which is 41.1 days.

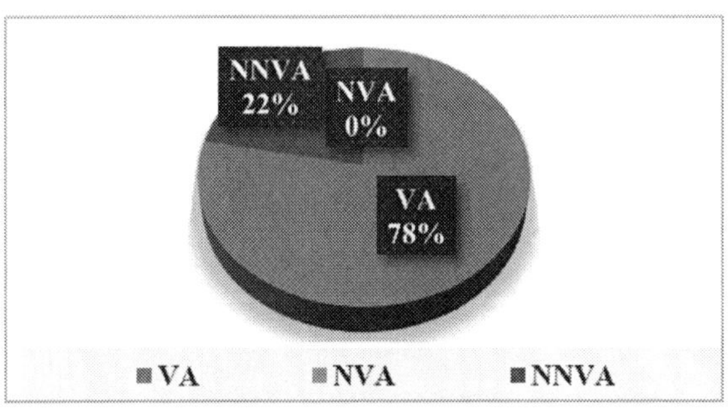

Figure 7. Percentage of VA and NVA values in future state value stream mapping.

In the Future State VSM conditions show that the total percentage of value added time in the procurement process is 78%, while the percentage of total non-value added time is 22%, with details of non-value added time of 0% and necessary non value added time of 22%, with a total value added time of 7488 minutes, while the total non-value added time is 2160 minutes. *Value added time can be interpreted as activities that can provide added value to the company, non-value added time is an activity that does not provide added value to the company*, while necessary but non-value added time is an activity that does not provide added value to the company.

4.5. Comparison of Current Sate VSM with Future State VSM in the procurement process

This section explains the differences between current state value stream mapping and future state value stream mapping in the procurement process which includes the percentage of value added time *the percentage of non-value added time,* total value added time and total non-value added time

Table 3. Comparisons of *Current State Value Stream Mapping* with *Future State Value Stream Mapping* in steel *procurement process*

	Current State VSM	***Future State VSM***
Percentage value added time	38 %	78 %
Percentage non-value added time	62 %	22 %
Total value added time	7488 minutes	7488 minutes
Total non-value added time	12240 minutes	2160 minutes

The total percentage of value added time in the Future State Value Stream Mapping has increased from the Current State Value Stream Mapping, from 38% to 78%, an increase of 40% with a fixed time of 7488 minutes, time at value added time remains the same because value added activity is an activity that can provide added value to the activity, so there is no need to make changes.

The total percentage of non-value added time when the condition of the Future State Value Stream Mapping has decreased from the condition of the Current State Value Stream Mapping i.e. from 62% to 22% decreased by 40% with a total time on the condition of Current State Value Stream Mapping of 12240 minutes to Future State Value Stream Mapping of 2160 minutes so that it decreased by 10080 minutes.

5. Conclusions

The implementation of the value stream mapping method in the procurement process in steel industry can reduce procurement lead time from 41 days to 21 days or decrease in procurement lead time by 50 percent. In the initial state, the percentage of activities that were classified as non-value added time was 62 percent and value added time was 38 percent. The waste activities or non-value added activities identified in the steel procurement process are the process of evaluating PR documents, the process of waiting for offers from vendors, and the negotiation process. The solution provided to anticipate the waste are provide training to users and make checklist forms, create an e-negotiation system using a web system to reduce face to face, and provide notification to the vendor when the offer deadline is near. After the improvement process, the percentage of non-value added activities decreased to 22 percent and value added activities increased to 78 percent.

6. References

[1] Abdulmalek FA, Rajgopal J. Analyzing the benefits of lean manufacturing and value stream mapping via simulation: A process sector case study. *Int J Prod Econ* 2007; 107: 223–236.
[2] Association SI. Indonesia Steel Industry: Development & Opportunities.
[3] Hartati TS. Bab i pendahuluan a. latar belakang. *Anal Strategi PTKrakatau Steel Dalam Menghadapi Persaingan Prod Baja* 2014; 1–10.

[4] Nugroho MZP, Nugroho ME, Susanti M. Study and Analysis of Delays in the Material Procurement Process: A case study of Steel Manufacturing Companies at Indonesia. *Int J Eng Res Adv Technol* 2020; 06: 01–08.
[5] Teichgräber UK, Bucourt M De. Applying value stream mapping techniques to eliminate non-value-added waste for the procurement of endovascular stents. *Eur J Radiol* 2012; 81: e47–e52.
[6] GÜZEL D, KABAKUŞ AK, ŞİRİN MS. A Value Stream Mapping Implementations: A Case of Textile Industry. *Atatürk Üniversitesi İktisadi Ve İdari Bilim Derg* 2018; 763–772.
[7] Garza-Reyes JA, Torres Romero J, Govindan K, et al. A PDCA-based approach to Environmental Value Stream Mapping (E-VSM). *J Clean Prod* 2018; 180: 335–348.
[8] Ikatrinasari ZF, Hasibuan S, Kosasih K. The Implementation Lean and Green Manufacturing through Sustainable Value Stream Mapping. *IOP Conf Ser Mater Sci Eng*; 453. Epub ahead of print 2018. DOI: 10.1088/1757-899X/453/1/012004.
[9] Lacerda AP, Xambre AR, Alvelos HM. Applying Value Stream Mapping to eliminate waste: A case study of an original equipment manufacturer for the automotive industry. *Int J Prod Res* 2016; 54: 1708–1720.
[10] Hasibuan S. Modul 1 Procurement Logistic 2018. In: *Procurement Logistic*. 2018, pp. 2–3.
[11] Luciana L, Lestari YD. Aplication Value Stream Mapping to Minimize Waste in Aircraft Industry. 2015; 4: 1119–1131.
[12] Nugroho MZP. Penerapan lean manufacturing untuk meningkatkan kinerja procurement dengan analisa performance supplier pada kasus industri baja. *Oper Excell* 2020; 12: 169–176.
[13] Stadnicka D, Litwin P. Value stream mapping and system dynamics integration for manufacturing line modelling and analysis. *Int J Prod Econ* 2019; 208: 400–411.
[14] Tyagi S, Choudhary A, Cai X, et al. Value stream mapping to reduce the lead-time of a product development process. *Int J Prod Econ* 2015; 160: 202–212.
[15] Yansen O, Bendatu LY. Perancangan Value Stream Mapping dan Upaya Penurunan Lead time pada Bagian Procurement-Purchasing di PT X. *Titra* 2013; 1: 9–16.
[16] Simbolon JP, Hasibuan S. Perbaikan Manajemen Pergudangan di Perusahaan Penerbangan. 2017; 9: 189–203.

The experiment of ambient wind speed and argon flow rate on tig welding process

Suheni[1], A A Rosidah[1], Z Lillahulhaq[1], I A Ridhlo[1], and I P Wardani[1]

[1]Mechanical Department, Faculty of Technology Industrial, Institut Teknologi Adhi Tama Surabaya, Surabaya, Indonesia.

zain@itats.ac.id

Abstract. Tungsten inert gas (TIG) welding is one of the most popular welding which has high quality welds. TIG welding can be used for all kinds of welding positions and high efficient. During the welding process, the welding specimen and the heat affected zone (HAZ) affected by a series of thermal cycles where the area warms up to a maximum temperature then followed by cooling process. The thermal cycle affects the microstructure of the welding specimen and HAZ. By controlling these parameters to reach the right heat cycle, the mechanical property of the weld metal can be improved and maintain the microstructure of the weld metal. This study is conducted an experiment to determine the effect of different wind velocity around the TIG welding material. Moreover, this study also discuss the effect of shielding gas flow rate on TIG type welding on JIS G 3131 SPHC steel material. The welding area and HAZ the higher the wind speed the hardness value tends to decrease. Because there is a difference in cooling speed and heat input on weld metal and HAZ.

1. Introduction

1.1. Tungsten Inert Gas Welding

Tungsten inert gas (TIG) welding which is commonly known as GTAW (Gas Tungsten Arc Welding) is a process of welding without consuming tungsten electrodes. This welding process use inert shielding gas as a protector to avoid environmental air influence. In the TIG welding process metal smelting occurs due to heat generated by an electric arc between the electrode and electrode filler with the base metal. The TIG welding is one of the most popular welding technologies in manufacturing industries which has high quality welds. TIG welding can be used for all kinds of welding positions and high efficient. Low weld penetration is the main problem in TIG welding [1]. TIG welding has limited thickness of material that can be weld in single pass about 2-3 mm thick plate. In thick material this welding type causes poor productivity resulted by many factors such as low energy input, welding speed and high number of passes required to fill the weld join [2]. There are several factors that affected on welding process result such as filler material, inert gas, welding speed, local temperature welding area, welding penetration, and so on.

A low carbon steel material tests to find the effect of the welding current on the tensile strength, hardness and microstructure of the TIG welding with E6013 electrodes. The welding treatment use current variation of 80 Amperes, 100 Amperes and 120 Amperes in AISI 1045 steel. The strength testing of the welding material showed that higher current resulted in decrement of material strength in tensile strength. Moreover micro-photo testing shows that the greater the current used in welding the lower the porosity will occur[3]. Previous research shows for low carbon steel material using V temper and ER 70S-6 filler that using 100 A current gives 522.2 HVN and 120 A current gives 623.2 HVN. It means that higher current will give higher hardness number [4]. Also, other research states a 28% hardness change of the weld metal, the HAZ showing a 30% hardness difference increase, and no significant change in the base metal when the current [5].

The current and speed of welding process combine as improvement step to maintenance the microstructure of welding material. In an experiment study, the higher current and welding speed resulted better improvement on material microstructure in mild steel (low carbon steel). It results high yield strength and ultimate strength on material compared with base material especially on 75 A with 1.75 rpm welding speed[6]. The optimum value of current and welding speed parameter is adjusted with the welding material used. In Incoloy 800HT plates which tested on TIG welding process in various current and welding speed. The welding process is using various electric current in 90, 110 and 130 A. The welding speed which used in the process is various among 1.2, 1.5 and 1.8 mm/s. The optimum welding parameter which suitable for Incoloy 800HT is using 110 A current and 1.5 mm/s[7]. The welding current has significant influence around 58% and followed by the influence of welding speed 30% and voltage 12%.

Welding process causes temperature raising in local area which causes thermal expansion and shrinkage during cooling time. The thermal stress in local welding area is able to change the when it has short duration of cooling time, the welding material become harder and brittle. This condition results weld cracks and can be endanger the welded construction[8]. Arc welding process is applied to produce hard-faced weld metals on high strength surfaces such as low alloy, grey cast iron, carbon steel, or alloy steel by control some parameter such as welding process, welding parameter, and electrode. By controlling these parameters to reach the right heat cycle, the mechanical property of the weld metal can be improved and maintain the microstructure of the weld metal. The suitable heat cycle of the welding process is required to produce high quality welding joint. Furthermore, the control of heat cycle during the welding process such as input heat, pre-heat, inter-pass, and post-heat temperature is able produce a reinforce particle which resulted increment of the hardness and strength of the weld metal[9], [10], [11].

1.2. Thermal Affected Welding Area

In an experiment hard-faced weld metal was produced on JIS-S50C carbon steel using the welding process parameter was 100 A welding current, 50-450°C of pre-heat and inter pass temperature. The increase of the inter pass temperature resulted the increment of the grain size, but decrease the hardness and the wear resistance of the hard-face weld metal. Low inter pass temperature increase residual stress and lead cracking on the interface between the first weld metal layer and the base metal[12]. Annealing process also could be a heat treatment to increase material hardness after welding process. Experimental research shows annealing process on material ST 42 low carbon steel has significant effect in material hardness increment. The material which is ready welded compared with basic material specimens in impact test. The material is welding in TIG using ASTM E6011 specifically low carbon steel and current used 78-125 A. The annealing heat treatment is various at 600°C, 650°C and 700° C at 30 mins. The welding material with 700°C annealing process, shows slight difference result compare with basic material. The higher the temperature used in annealing process generate higher toughness value which can take after the toughness of the basic material[13].

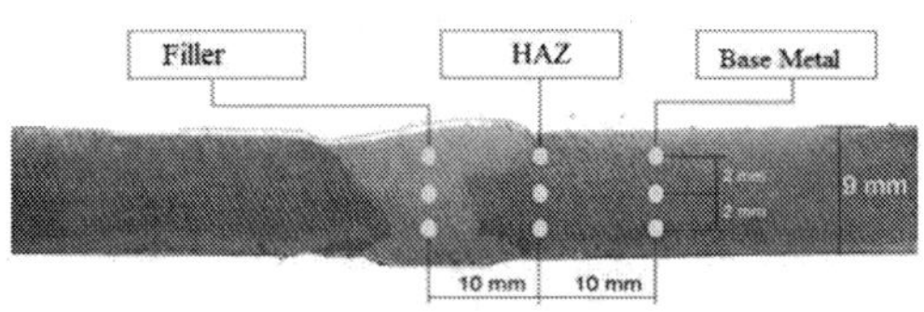

Figure 1. The classification area of temperature cycle effect on TIG welding process.

During the welding process, the welding specimen and the heat affected zone (HAZ) affected by a series of thermal cycles where the area warms up to a maximum temperature then followed by cooling process. The thermal cycle affects the microstructure of the welding specimen and HAZ which shown in figure 1. In the HAZ area, grounding without annealing changes in the micro structure. This change is marked by the growth of crystal grains. In this area the structure is martensitic because the structure is small and elongated. The increase in the value of violence in the HAZ area is the highest. This high increase in hardness value occurs because in this area there is a rapid thermal heating and cooling cycle[14]. The welding specimen undergoes a series of phase transformations during the cooling process where the molten welding metal turns into Ferrite-δ then γ (Austenite) and eventually becomes α (Ferrite). In general, cooling time between 800°C – 500° C. is used as a reference for carbon steel welding, because at this temperature interval a phase transformation from Austenite (γ) to Ferrite or Bainite depends on the cooling speed. The duration of cooling in a certain temperature region of a thermal cycle of a weld greatly influences the welding quality. Some researches have been done to extend the duration of cooling time. The microstructure and mechanical properties of the HAZ region largely depend on the cooling time from temperatures of 800°C to 500°C, while cold cracks in which hydrogen plays an important role are highly dependent on the cooling duration from temperatures of 800°C to 300°C or 100°C.

Beside the parameters mentioned earlier, other parameters derived from the environment can also have an influence on welding results such as temperature and pressure on the environment. Ambient temperature around the weld material has a material influence on the cooling process. The cooling process affects the microstructure of the material and establishes the material strength. The cooling temperature is determine the hardness level of the material, whether the material becomes hard or brittle. The cooling temperature can be determined by ambient temperature, wind velocity or even the weather. This study is conducted an experiment to determine the effect of different wind velocity around the TIG welding material. Moreover, this study also discuss the effect of shielding gas flow rate on TIG type welding on JIS G 3131 SPHC steel material.

2. Experimental set up

This experiment study conducted in JIS G 3131 SPHC steel with 9 mm thickness in single V TIG welding. The material welds on horizontal position or down hand welding. The metal fillers used are ER 70 S-6 with a diameter of 2.4 mm. The argon gas which used in this experiment flows at 9 L/min, 11 L/min and 13 L/min. This welding processes conducted in various in wind speed used are 3 mph, 5 mph, 7 mph. After welding process the material tests on hardness, macro structure and microstructure test.

2.1. Welding Step

- The material cut into 60 mm x 30 mm and clean the surface from the dirt. The less neat remaining pieces polish on oxide with brush wheel hand grinding. The dimension of the material is shown in figure 2. The brush wheel is used to clean the surfaces from oxide contamination and corrosion.
- The V tempers welding made by a hand grinding with 60°. Then grinding the material in the same direction so that the surface of the tempers is flat. After the steel is tempered, the steel is connected with a 2 mm root pass with an angle of 60°. A tack weld is implemented at the ends of the steel to avoid the deformation and clean the dirt with a steel brush so it does not affect the welding results during the root pass, filler pass and capping process.

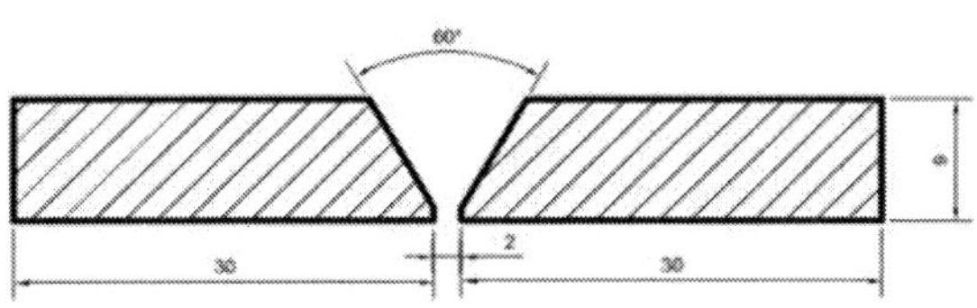

Figure 2. The dimension of TIG Specimen test.

- Set up the welding material in the TIG welding machine. Set the welding on 100 A current and open the Argon gas flow rate of 9 l / m, 11 l / m, and 13 l / m. The filler used is ER 70 S-6 2.4 mm in diameter.
- Set up the fan to reach the wind speed parameters in 3 mph, 5 mph, and 7 mph and measure the speed with anemometer.
- Weld the root pass and does the welding process stable welding speed. Cleaning impurities with a steel brush is an important part and make sure to fill the tempers neatly. This process is completed by capping step.

2.2. Metallography Test

- The hardness test on the specimen is carried out up to three points on one specimen. This test is carried out on the specimen before the welding process and on the specimen after the welding process. The method used in this hardness test is the Rockwell method. Rockwell used in this test is Rockwell C with HRC units and carried out using a 150 kg (981 N) load using a 120° diamond cone indenter for 30 seconds. The material is cut into cross section direction of the welding line after the welding process complete.
- Macro test is carried out after the specimen is cut transversely at both ends so that the base metal, HAZ area and weld metal appear to have no welding defects. After that sandpaper with rubbing paper grade 100, 500, 1000, 1200 and 1500. Sanding is done in the same direction. After sanding process, polishing is done using wool and auto sol paper. Then the etching process is done using NHO_3 liquid. The macro pictures using the camera.
- Microstructure tests conducted in few steps. First the mounting process is carried out by giving a mixture of resin and catalyst in a ratio of 10: 1 which is poured into the cast. When mounting has hardened then the grinding process is done by first adjusting the grinding rotation and water flow rate. The polishing process is done by installing a polishing cloth smeared with alumina paste and adjusting the rotation speed and water flow until the specimen becomes smooth and shiny. The surface of the material is smeared with nitric acid: alcohol = 5%: 95%. Then the specimen is washed, dried and observed using an optical microscope.

2.3. Microstructure point count (ASTM E562)

Point count ASTM E562 is a method to determine the percentage of phases in a metal. By using a formula:

$$\%\, point\ phase = \frac{total\ phase}{total\ point} \times 100\% \quad (1)$$

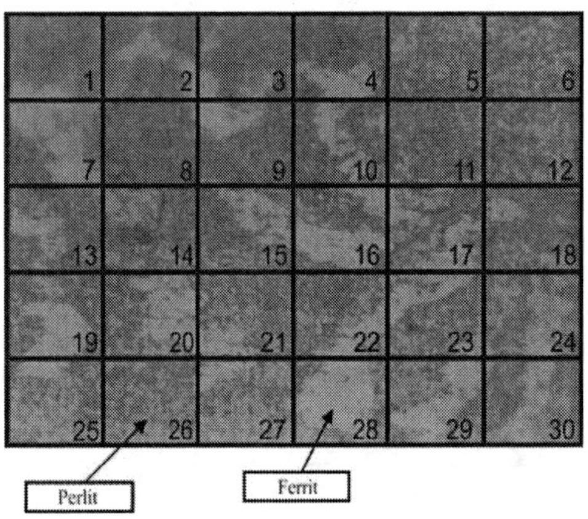

Figure 3. The calculation of ferrite and pearlite at 3 mph wind speed with a discharge rate of 9 L/min on HAZ

Using this method, observation is focused on weld metal and HAZ area. Then, the percentage of ferrite and percentage of pearlite will be obtained. With the data obtained and then made in the form of graphs and then analysed. The microstructure of the weld metal with a discharge rate of 9 L/min is shown in Figure 3. A grid contains a number of points with a certain number that is placed on the microstructure images. The number of points associated with the phase compared to the total points on the grid result the volume fraction of the phase. In order to get more accurate, more points on the grid can be used. The exact point on a particular phase is counted 1 (one), while those who stick to the edge of the phase are counted ½ (half).

3. Result and Discussion

3.1. Rockwell test

Rockwell experiment data were taken at one point in the weld metal area, HAZ and base metal. This comparison aims to determine the strength of the material from each point affected by the thermal cycle effect on the welding process. Based on table 1 the highest hardness value (HRC) is found in the weld metal followed by HAZ and base metal. From the data in table, it can be seen that the highest hardness value (HRC) is found in the weld metal area with a value of 33.25 HRC. Meanwhile the lowest hardness value is found in the base metal area with a value of 25.875 HRC.

The average value of hardness in the weld metal area is relatively decreased with increasing wind speed. The lowest hardness is at 7 mph wind speed with a shielding gas flow rate of 11 L/min with a hardness value of 29 HRC. The highest hardness in the weld metal area is at 3 mph wind speed with a p shielding gas discharge rate of 9 L/min with a hardness of 33.25 HRC. However, at 3 mph wind speeds with a flow rate of 13 L/min the hardness value rises according to the various in flow rate gas.

Table 1. The result of hardness test using Rockwell method

Argon Flow rate (L/min)	Ambient wind Velocity (mph)	Rockwell test (HRC)		
		Weld metal	HAZ	Base Material
9	3	33,25	29	27
	5	30,875	28,5	27,75
	7	30,25	28	27,25
11	3	30,875	30,25	27,375
	5	29,875	28,375	27,7
	7	29	28,25	27,12
13	3	31,875	29,5	27,125
	5	31,5	28,625	27
	7	30,125	28,25	27,12

In the HAZ area the greatest hardness value is 30.25 HRC at 3 mph wind speed with a shielding gas flow rate of 11 L/min and the lowest hardness value is 28 HRC at 7 mph wind speed with a shielding gas flow rate of 9 L/min. From the graph it can be concluded that the welding metal and HAZ the higher the wind speed the hardness value tends to decrease. Because there is a difference in cooling speed and heat input on weld metal and HAZ. The presence of wind speed that directly leads to the weld metal affects the mechanical properties of the steel. Conversely, the greater the flow rate used the average value of violence tends to rise.

Whereas in the main metal area the greatest hardness value is 27.375 HRC at 3 mph wind speed with shielding gas flow rate of 11 L/min and the lowest hardness value is 27 HRC at 3 mph and 5 mph wind speed with shielding gas flow rates of 9 and 13 L/min. In general, the average value of the base metal is relatively the same. On the base metal the average value of hardness tends to be the same and there is a slight difference according to the time of testing the suppressor indenter is not always right on the pearlite or ferrite phase. Sometimes the indenter presses on the pearlite phase so that the hardness rises, and sometimes the indenter presses on the ferrite phase so that the average value of hardness decreases. However, the values of the hardness of the base metal are relatively the same.

3.2. Microstructure test

In this study macro test shows the HAZ area, weld metal, and base metal, in JIS G3131 SPHC material using single V tempers with an angle of 60° currents 100 A. Table 2 shows the point phase of welding material which process on 9, 11 and 13 L/min of argon gas in variation of wind speed 3, 5 and 7 MPH. From this table an increase in the argon gas flow rate in the welding process can cause a decrease in the percentage of phase ferrites in the welding area and the HAZ. The decrease in the percentage of ferrite in the weld specimen is followed by an increase in the percentage of the pearlite structure. Moreover the decrement of the ferrite structure is also influenced by an increase in wind speed around the weld specimen.

Table 2. The result of phase percentage on the microstructure

Wind Speed	Argon Gas Flow rate (L/min)	Welding Area		HAZ	
		% Ferrite	% Perlite	% Ferrite	% Perlite
3 mph	9	56,6	43,4	50	50
	11	43,4	56,6	40	60
	13	43,34	56,66	33,34	66,66
5 mph	9	40	60	46,67	53,33
	11	53,33	46,67	60	40
	13	53,33	46,67	56,67	43,33
7 mph	9	36,66	63,34	36,66	63,34
	11	33,34	66,66	63,34	36,66
	13	40	60	56,66	43,34

The microstructure in the weld and HAZ area is dominated by ferrite and pearlite. The grain size in the HAZ area is bigger than the weld area as a result of the influence of heat on welding. Figure 4, 5 and 6 shows the argon gas discharge of 9 L/min, the structure of pearlite tends to decrease along with the increase in the amount of ferrite. But the pearlite grain boundaries still has big structure size, this is due to the greater wind speed that is 5 mph. At a wind speed of 5 mph the amount of ferrite is finer compared to the speed of 3 mph. The average value of ferrite is 40% and pearlite is 60%. There was a decrease from the previous speed of 3 mph. At a speed of 7 mph according to the calculation, the amount of ferrite was 36.66% and the number of pearlite was 63.34%. The decrease in the amount of ferrite and pearlite is caused by the presence of wind speeds of 7 mph which directly enter the welding area. Thus causing reduced heat input to the metal. The greater the heat input applied to the steel, the greater the amount of pearlite produced. So, the average value of hardness has also increased. At 7 mph the number of pearlite and ferrite decreased significantly.

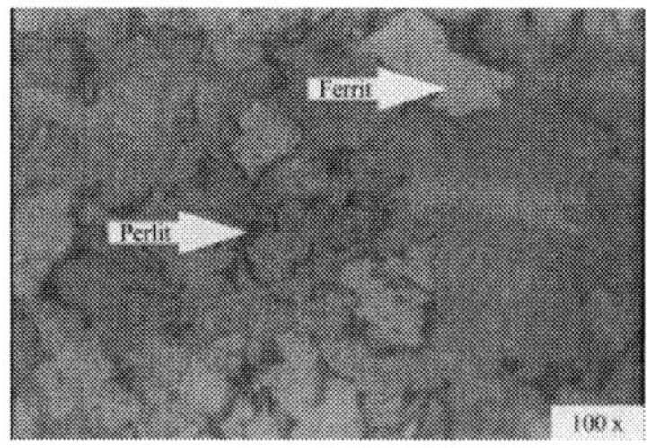

Figure 4. Microstructure of weld metal at gas discharge 9 L/min with wind speed of 3 mph.

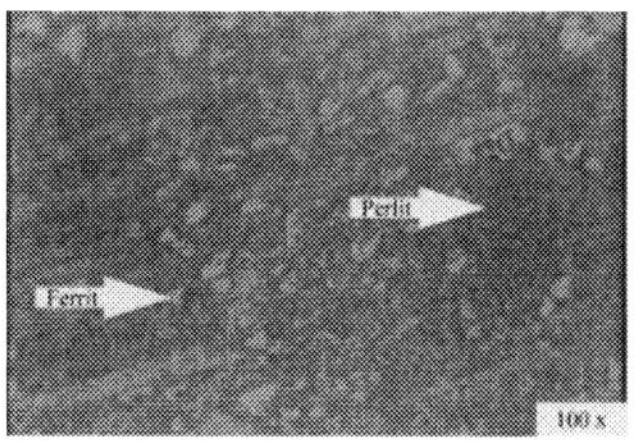

Figure 5. Microstructure of weld metal at gas discharge 9 L/min with wind speed of 5 mph.

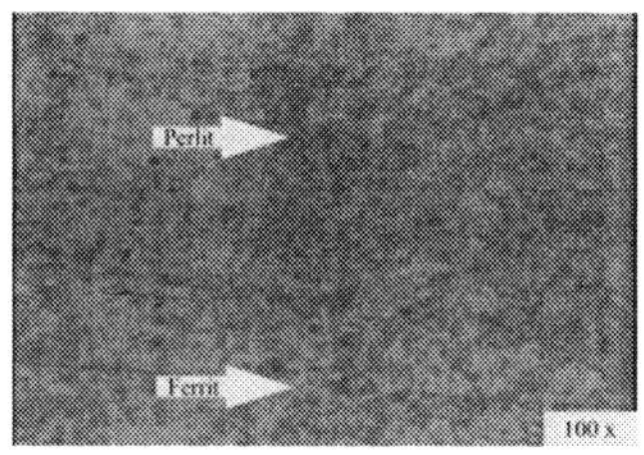

Figure 6. Microstructure of weld metal at gas discharge 9 L/min with wind speed of 7 mph.

3.3. Macro structure test

In this study macro test shows the HAZ area, weld metal, and base metal, in JIS G3131 SPHC material using single V tempers with an angle of 60° currents 100 A. Figure 7, 8 and 9 shows welding material which process on 9 L/min of argon gas in variation of wind speed 3, 5 and 7 mph. It has a wide weld pool but the penetration is not deep enough so that more welding inter pass layers are needed. Root pass in 3 mph variation is difficult to do so at the discharge of 9 L/min using back weld to prevent welding defects. In macro photos of HAZ areas can be seen in darker areas around the weld area. The thermal cycle change microstructure of the welding material. From the figure 7, 8 and 9 it can be seen that the length of the HAZ change along with increasing wind speed. Table 2 shows the change in the length of the HAZ area in the variation of wind speed around the welding material. Based on table 3 the material which is cooling with 5 mph air speed has the longest HAZ length in welding with argon gas discharge 9 L/min. The shortest HAZ length is found in welding material cooled by 7 mph wind speed. This high increase in hardness value occurs because in this area there is a rapid thermal heating and cooling cycle. In the HAZ area of grounding without annealing there has been a change in the microstructure characterized by the growth of pearlite crystal grains. In this area the structure is martensitic because the structure is small and elongated. In the HAZ region the welding with annealed critical temperature has begun to change. In this area the ferrite structure is larger than the pearlite.

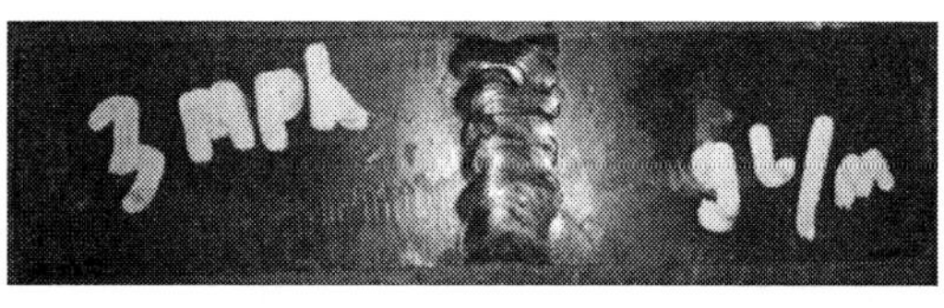

Figure 7. Macro structure of weld metal at gas discharge 9 L/min with wind speed of 3 mph.

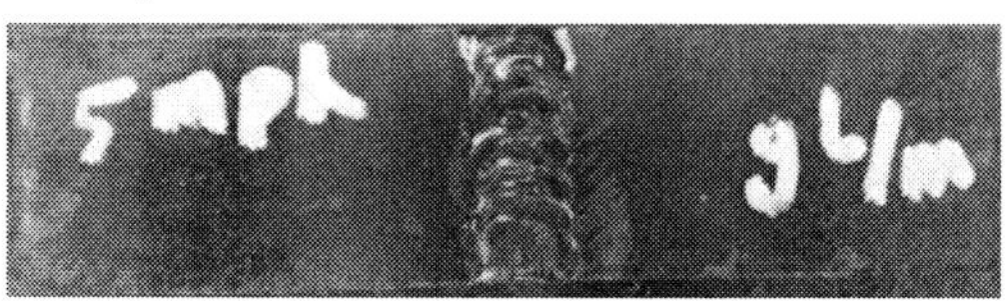

Figure 8. Macro structure of weld metal at gas discharge 9 L/min with wind speed of 5 mph.

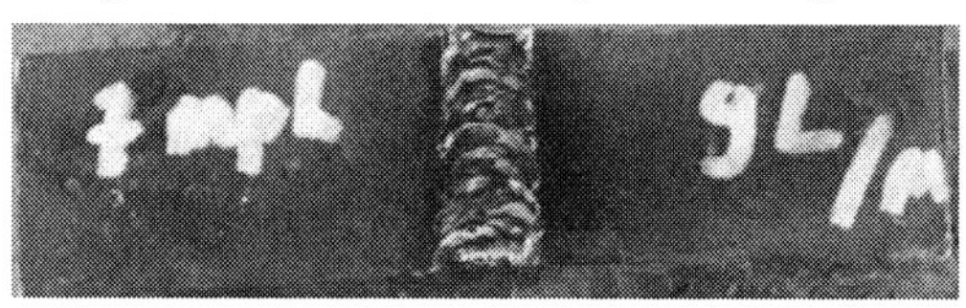

Figure 9. Macro structure of weld metal at gas discharge 9 L/min with wind speed of 7 mph.

Table 3. The result of HAZ length at gas discharge 9L/min with various wind speed.

Wind Speed (mph)	HAZ length (mm)
3	3,05
5	3,42
7	2,82

4. Conclusions

In the TIG welding process the HAZ area gives a slight enhancement of hardness with increase in argon flow rate. This is because during the welding process the HAZ area is affected by the welding heat cycle. Then the HAZ area will experience a rapid decrease in temperature due to the cooling process. During the heat cycle of the weld, the HAZ region changes the microstructure phase by decreasing the amount of ferrite and increasing the structure of the pearlite with increase in wind speed. The macrostructure results reveal the HAZ length getting smaller with the escalation of wind speed.

5. References

[1] Singh A K Dey V and Rai R N, 2017 Techniques to improveweld penetration in TIG welding (A review) in *Materials Today: Proceedings* **4**, 2 p. 1252–1259.

[2] M K AI Y and E D, 2019 Activated flux TIG welding of non- ferrous metals *Indian J. Chem. Technol.* **26**, 2 p. 164–169.

[3] Fernando E, 2019, Pengaruh Variasi Kuat Arus Pengelasan SMAW (Shielded Metal Arc Welding) Terhadap Kekuatan Tarik dan Struktur Mikro Sambungan Las Baja AISI 1045 dengan Menggunakan Elektroda E6013.

[4] Santoso W Widodo E W R Kuswanto A and Setyowati V A, 2018 Analisis Kuat Arus, Jenis Logam Pengisi, dan Jenis Kampuh Las Terhadap Kekuatan Tarik dan Kekerasan Dissimilar Welding dengan Gas Tungsten Arc Welding (GTAW) *J. Tek. Mesin Inst. Teknol. padang* **8**, 2 p. 2–6.

[5] Irawan H Sukendro B S and Anzaldi M, 2017 Studi Eksperimental Efek Kuat Arus Terhadap Distribusi Kekerasan Di Daerah Las Pada Pengelasan GMAW Secara Manual in *Seminar Nasional Sains dan Teknologi Terapan V Institut Teknologi Adhi Tama Surabaya* p. 229–234.

[6] Alhafiz M, 2018 Effects of Electrode Velocity Variations and Selection of Electric Current Against Quality Welding Results Mild Steel on SMAW Welding **57**, September p. 12–16.

[7] Srirangan A K and Paulraj S, 2016 Multi-response optimization of process parameters for TIG welding of Incoloy 800HT by Taguchi grey relational analysis *Eng. Sci. Technol. an Int. J.* **19**, 2 p. 811–817.

[8] Munawar Abbas H and Aminy A Y, Oct. 2017 The Effects of Shielded Metal Arc Welding (Smaw) Welding On The Mechanical Characteristics With Heating Treatment inn S45c Steel *Int. Conf. Nucl. Technol. Sci. (ICoNETS 2017)* **962**, 1 p. 89–93.

[9] Gualco A Svoboda H G Surian E S and Vedia L A d., Oct. 2010 Effect of welding procedure on wear behaviour of a modified martensitic tool steel hardfacing deposit *Mater. Des.* **31**, 9 p. 4165–4173.

[10] Yang K Yu S Li Y and Li C, Jun. 2008 Effect of carbonitride precipitates on the abrasive wear behaviour of hardfacing alloy *Appl. Surf. Sci.* **254**, 16 p. 5023–5027.

[11] Liu D Liu R Wei Y Ma Y and Zhu K, Apr. 2013 Microstructure and wear properties of Fe-15Cr-2.5Ti-2C-xB wt.% hardfacing alloys *Appl. Surf. Sci.* **271** p. 253–259.

[12] Triwanapong S Angthong S and Kimapong K, 2018 Interpass Temperature Affecting Abrasive Wear Resistance of SMAW Hard-Faced Weld Metal on JIS-S50C Carbon Steel *Mater. Sci.*

Forum **950** p. 60–64.
[13] Fazli T Erizon N and Adri J, 2018 Experimental Test of Annealing Process on SMAW at Low Carbon Steel Toughness **1**, 1 p. 32–35.
[14] Miftakhudin Rusiyanto and Masugino, 2012 PENGARUH PROSES ANNEALING PADA SAMBUNGAN LAS *J. Mech. Eng. Learn.* **1**, 1 p. 1–10.

Application of vikor (vise kriterijumska optimizacija i kompromisno resenje) method for identifying amplifier damage

S.Nurmuslimah [1], N Saidatin[1]

[1]Institut Teknologi Adhi Tama Surabaya, East java, Indonesia

puty71naura@gmail.com

Abstract In this era, many electronics have been marketed in Indonesia. One of which is amplifier that is used to amplify the sound of MP3 player, radio, VCD or television. An amplifier is an electronic appliance that obviously has a certain lifespan or lifetime and afterwards damage will occur. Most people go to an electronics handyman to repair the damaged electronic appliance but not all handymen can analyze the damage easily. To overcome this problem, a system was setup to identify the damage of an amplifier. A damage identification on this system applied Vikor method. Vikor is a ranking method using multiple criteria ranking index based on certain measurement closed to the ideal solution. This application cloud be used as supporting consideration in talking the required steps for repairing an amplifier. The test result and calculation analysis using Vikor method obtained the system accuracy of 81.25%.

1. Introduction

In the era now there are many electronic goods are spread out in indonesia.One of which is an amplifier OCL.An amplifier is a the electronic devices that serves as a loudspeaker to the input in the form of a device mp3 player, cd players, radio and others.A an amplifier sure have a period of time because of age or wear and in time this instrument. To overcome the problems most of a person chooses ran to the electronic service to fix an amplifier than buy a new issue. Maybe that is due to economic reasons.An artisan electronic will serve for analysis before repair goods which will be renovated.It is not a bit an artisan serve electronic experienced difficulty at the time of analyzing or conducts a one-by one. Component checking to overcome the problems so the government has built a system that could help know damage to a an amplifier.

2. Methodology

Retrieval system of decision is a in the manner of someone think about something before take a action.To design systems decision-making obtained from a benefits and therefore measures to be taken will be optimal. The support system of decision is a system that has the ability to solve the problems semi-structured by providing information or advice on a decree certain.Each of the components is regulated by a software which has been available or designed specifically.There are several definition about the support system a decision, between others.[1]: defines third party contributions as a population a procedure based on the models to process the data and to help manager in doing decision making.Third party contributions as system based on computer consisting of three components that interacted, namely: a language system (a mechanism that providing a means of communication between user components of the third party contributions). In the decision-making must go through the three phases that is [2]

a. Intelligence :This stage is a process of the search for and detection of scope problems it was and the process of the introduction of a problem.A data input obtained, processed, and tested in order to identify problem.
b. Design: This stage is a process of finding, develop and analyze the act of alternative could be done.This stage covering the process of understand the problem, sent down a solution and test the feasibility solution.
c. Choice : At this stage done the process of selection of a variety of alternative the act of one who may not be executed.The outcome of an election is then implemented in the process of decision making.

Amplifier is a series of electronic components that are used to strengthen power (or power in general). In the audio field, amplifiers will amplify the sound signal that is to strengthen the current signal (I) and voltage (V) electricity from the input into an electric current and a greater voltage (more power) at the output. Power Amplifier OTL model is one of the power amplifier models that are used for small to medium power no more than 100 Watts. Maybe in the past there were still those who used this OTL model power amplifier for sound system devices, but for now it is rarely used. However, today it is still widely used in some electronic devices for producing sound with small power such as televisions, radios, laptops, even cellphones that we use every day also use the type of OTL power amplifier for audio amplifier. One of the characteristics of the OTL type power amplifier model is that the power supply or power supply used is non-symmetrical so it is sufficient to use a battery power supply (at the poles + and -) or an adapter with V (+) and ground (0). [3]

Vikor (*Vise Kriterijumska Optimizacija I Kompromisno Resenje*) is a method that developed to optimize multicriteria of a system of the complex. This method tend to focus on ranking and chose him by using a number of alternatives based on criteria exist. A method of vikor provides multiindex rank with the measurement of the level of proximity ideal for a solution [4].It assumed that every alternative has evaluated based on each function criteria so could be codified rank by comparing between the measurement of the level of proximity to an alternative that ideal. The measurement of multicriteria developed from L_p Metric that has been used as a function of aggregate programming on method[5]. Variety of alternative ($k = 1, \ldots, n$) Symbolized by$a1, a2, \ldots, an$. On the alternative ak, the ratio of the jth symbolized by f_{kj}, Where$f_{kj}$is the value of the function of criteria of the jth for the alternative ak; m is the number of criteria ($j = 1,2, \ldots, m$). Development of the VIKOR method is derived from the shape of $L_p - metric$:

$$L_{p,k} = \left\{\sum_{j=1}^{n} \left[w_j(f_j^* - f_{kj})/(f_j^* - f_j^-)\right]\right\}^{1/p}, 1 \leq p \leq \infty; k = 1,2, \ldots, n \quad (1)$$

In the VIKOR method, $L_{1,k}$and$L_{\infty,k}$used for setting measurement ratings. The solution obtained bymin$_k S_k$is the advantage of the majority group ("the majority" of the described as the average distance, where$p = 1$) and the solution obtained by min$_k S_k$is loss individual minimum. The solution F is the solution that is closest to F* ideal where $\Delta f_1 = f_1^* - f_1^c$ dan $\Delta f_1 = f_2^* - f_2^c$.Refer formula 1.1 Provide a ranking on the alternatives that exist with how to sort or sort the results of the calculation of the index the vikor, starting from the largest value to the smallest value [4]. Then the calculation steps of the VIKOR method will be, the authors describe in detail as follows :

a. Calculate the value of the weight (w_k)on each alternative with how to share between [the total amount of keriteria of one of the alternatives] by [the total number of keriteria of all the alternatives].

$$w_k = \frac{\sum_{j=1}^{n} f_{kj}}{\sum_{k=1}^{n}\sum_{j=1}^{n} f_{kj}} \quad (2)$$

Here we did not use the symbol $w_j$in accordance with the formula reference, or formula (2), because the system that created this is the, a symbol of$j$deifying a line and the symbol k is like a column, in accordance on the tables arrival. So the symbol used isw_k .

b. Set the value $f_j{}^*$and$f_j{}^-$

$f_j{}^*$ = The greatest value of all keriteria on each alternative.

$f_j{}^-$ = The smallest value of all keriteria on each alternative.

c. Calculate the value of the normalization matrix$R_{kj}$by using the following formula :

$$R_{kj} = \frac{(f_j{}^* - f_{kj})}{(f_j{}^* - f_j{}^-)} \quad (3)$$

f_{kj} = The value of the data sample k criteria j

Multiply the result of normalization with the weights criteria

$$Rw_{kj} = R_{kj} \times w_k \quad (4)$$

d. Calculate the value of $S_j$andR_j
$S_j$is the value of the sum of all alternatives in each keriteria, contained in the variableRw_{kj}.

$$S_j = \sum_{k=1}^{n} Rw_{kj} \tag{5}$$

$R_j$is the greatest value of all alternatives in each keriteria, contained in the variableRw_{kj}

$$R_j = \frac{max}{k}\{Rw_{kj}\} \tag{6}$$

e. Set value$S_j{}^*$, $S_j{}^-$, $R_j{}^*$and$R_j{}^-$
$S_j{}^*$ = The greatest value of the variableS_j
$S_j{}^-$ = The smallest value of the variableS_j
$R_j{}^*$ = The greatest value of the variableR_j
$R_j{}^-$ = The smallest value of the variableR_j

f. Calculate the Vikor Index value (Q_j)
Please note that the value of$v = 0{,}5$

$$Q_j = \frac{(S_j - S_j{}^*)}{(S_j{}^- - S_j{}^*)} \times v + \frac{(R_j - R_j{}^*)}{(R_j{}^- - R_j{}^*)} \times (1 - v) \tag{7}$$

Flowchart of the app Decision-Making System to Identify Damage Amplifier OCL was in Figure 1. The explanation of the flowchart of figure 1 is initiated from the data entry expert and stored in the database. Then inputting the symptoms of the damage so that the calculation can be done in the process of Calculation of Vikor[6] where the data to be calculated is data that refers to the Damage Data and select*Amplifier*. Then the flowchart of the process of the Calculation of Vikor was shown in Figure 2. At Figure 2 we can see the process after the value of the data experts and the choice of the symptoms the damage has already been entered, then the next process or the process is looking for the value of the weights of each expert[7], then the process of seeking the best value and the worst in each expert. Then go the process of calculating the Normal value Damage. After the value of S and R obtained then carried out the process of searching for the best value and worst each value of S and R. The next step is to calculate the value of the Index the Vikor way as the formula shown in the picture flow. Having obtained the values of the Vikor, then done process sorting index values of the Vikor from the largest value to the smallest value so that the results obtained from the Steps of the Repair Amplifier[8] based on the value of the Vikor, which has sorted the.

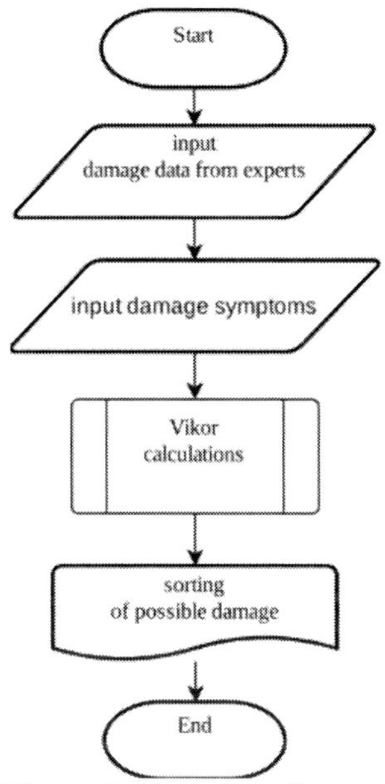

Figure 1. Flowchart System

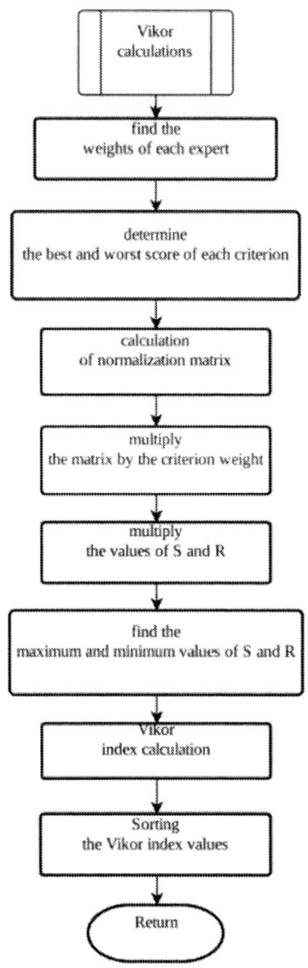

Figure 2. Flowchart of the Calculation of VIKOR

3. Results and discussion

3.1 System Analysis

The system is built can make it easy for a handyman service electronic in recognizing a damage to repair an Amplifier. In other words this system only as a tool to start the job or the first step that should take precedence, i.e. observe, analyze, check the in sequence in accordance with the highest percentage of estimates which approach the logical symptoms of damage to an Amplifier.

Based on this, we need an application that can help a service electronic to get to know the damage and perform the repair device Amplifier based on the criteria that has been determined by using Vikor method. The Data Criteria that is used is the result of the assessment of some Experts the Electronics against damage data that we have given alternative data in table 2.

To start the construction of the system required some of the data component damage that often occurs on a device Amplifier OCL.

1. Alternative Data (Damage)
   Alternative Data (damage) is the data of some component damage that often occurs on an Amplifier OCL.
2. Data Criteria
   Data criteria is data that contains the appraisal-appraiser Experts on the damage data which have been determined.

3.2 Calculation of vikor

Before performing the calculation of VIKOR, firstly find out the data that have been obtained from some of the Expert Electronic. In this case I used the symptoms of the damage to No Noise and 5 Experts to fill in the data of such damage.

a. Find the value of the weights of each Expert
b. Find the value of the maximum and minimum of each Expert.
c. The Calculation Of The Normalization Matrix

Table 1. Weighting of Assessment for Each Expert

	Pakar1	Pakar 2	Pakar 3	Pakar 4	Pakar 5
Bobot	0,2007	0,2007	0,2043	0,1971	0,1971

Table 2. The value of the Maximum and Minimum of Each Expert

	$f_1(.)$	$f_2(.)$	$f_3(.)$	$f_4(.)$	$f_5(.)$
Max	4	4	4	4	4
Min	1	1	1	1	1

Table 3. The Value Of The Results Of The Normalization Matrix

	$f_1(.)$	$f_2(.)$	$f_3(.)$	$f_4(.)$	$f_5(.)$
$A_1(.)$	0,6667	0,6667	0,6667	1	1
$A_2(.)$	0,3333	0,3333	0,3333	0,3333	0
$A_3(.)$	0	0	0	0	0
$A_{17}(.)$	0,6667	0,6667	0,6667	0,6667	0,6667
$A_{18}(.)$	0	0	0	0	0
$A_{19}(.)$	0	0	0	0	0

The multiplication of Normalized Matrix with the Weights of the Criteria ($\boldsymbol{Rw_{kj}}$),example of Multiplication of Matrix Normalization with Weights for **A₁(.)** :

$R_{11} = 0{,}6667 \times 0{,}2007 \quad = 0{,}1338$
$R_{12} = 0{,}6667 \times 0{,}2007 \quad = 0{,}1338$
$R_{13} = 0{,}6667 \times 0{,}2043 \quad = 0{,}1362$
$R_{14} = 1 \times 0{,}1971 \quad = 0{,}1971$
$R_{15} = 1 \times 0{,}1971 \quad = 0{,}1971$

Table 4. Results of the Normalization Matrix Multiplication with Weight

	f₁(.)	f₂(.)	f₃(.)	f₄(.)	f₅(.)
A₁(.)	0,1338	0,1338	0,1362	0,1971	0,1971
A₂(.)	0,0669	0,0669	0,0681	0,0657	0
A₃(.)	0	0	0	0	0
A₁₇(.)	0,1338	0,1338	0,1362	0,1314	0,1314
A₁₈(.)	0	0	0	0	0
A₁₉(.)	0	0	0	0	0

d. Calculate S and R
Calculate S on **A₁(.)** :
$S_{A1(.)} = 0{,}1338 + 0{,}1338 + 0{,}1362 + 0{,}1971 + 0{,}1971 = \mathbf{0,7981}$
Calculate R :
R value is the largest value of the weight of the criteria with normalization data. R value for sample **A₁(.)**is $R_{A1(.)}$= $\mathbf{0,1971}$

e. Calculate the Vikor Index (Q)
Calculate the Vikor Index for$\boldsymbol{Q_1}$or**A₁(.)** :

$$Q_1 = \frac{(0{,}7981 - 1)}{(0 - 1)} \times 0{,}5 + \frac{(0{,}1971 - 0{,}2043)}{(0 - 0{,}2043)} \times (1 - 0{,}5) = \mathbf{0,1184}$$

d. Rank alternatives available by sorting or sorting resultsvikor index calculation, starting from the largest value to the smallest value.

Table 5. S and R values

	S	R
A₁(.)	0,7981	0,1971
A₂(.)	0,2676	0,0681
A₃(.)	0	0
A₁₈(.)	0	0
A₁₉(.)	0	0

Table 6. Maximal and Minimum S and R values

	S	R
Max	1	0,2043
Min	0	0

Table 7. Ranking S, R, and Q

	S	R	Q
$A_1(.)$	0,7981	0,1971	0,1185
$A_2(.)$	0,2676	0,0681	0,6995
$A_3(.)$	0	0	1
$A_{17}(.)$	0,6667	0,1362	0,3333
$A_{18}(.)$	0	0	1
$A_{19}(.)$	0	0	1

Table 8. Ranking of S, R, and Q with Big to Small Order based on Q

	S	R	Q
$A_3(.)$	0	0	1
$A_8(.)$	0,9319	0,2007	0,0428
A7(.)	0,9331	0,2043	0,0335
A15(.)	1	0,2043	0

From table 8. we get the data that the possible sequence of damage to Soundless Amplifiers is starting from the possibility of having the largest percentage to the possibility of having the smallest percentage in the following order:

A3(.) :Power Disconnect
A5(.) : Speaker cable is broken
A10(.) : Ceramic resistor (Resistor Fuse) on the amplifier output is damaged
A11(.) : The fuse broke
A12(.) : Speaker out of order
A14(.) : Power button is broken
A16(.) : Transformer is broken
A18(.) : The final amplifier transistor is broken
A19(.) : The voltage stabilizing transistor on the Power Supply is damaged
A2(.) : Diode Bridge on the Power Supply is damaged
A4(.) : The RCA cable in the input channel has broken
A6(.) : The capacitor in the input section is damaged
A17(.) : The transistor in the Driver section is damaged
A9(.) : Potential volume damaged
A13(.) : Lead on the component legs is loose
A1(.) : Diode in the Driver section is damaged
A8(.) : Ground to ground cable broken
A7(.) : The capacitor in the Power Supply section is damaged
A15(.) : Hot Power Supply Transformer

3.3 Expert Data Input Form

The Expert Data Input Form contains a list of the types of damage that are often experienced by Amplifiers and data from electronic experts, namely predetermined assessment range data. Each expert has different assessment criteria, so then the data that has different criteria will be carried out the calculation process using the Vikor method so as to produce an output index Vikor as a decision maker in repairing a damaged amplifier

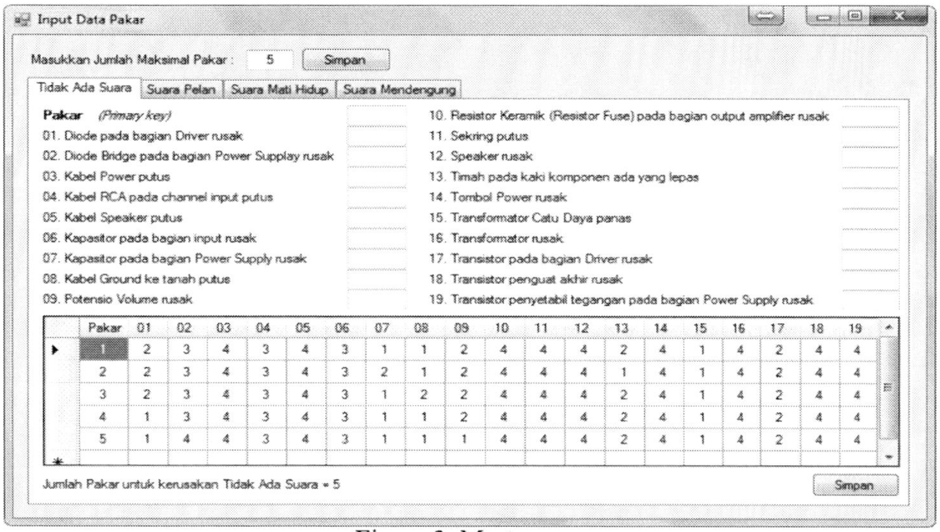

Figure 3. Master page

3.4 Process Form

This Process Form contains a menu of symptoms that have to be chosen one of them and there is a Process button to do a calculation of error and produce an Output or Report in accordance with the input of damage symptoms that have been selected.

3.5 Calculation Result Form

Calculation Result Form is the final result data from the system calculation that will be made as a reference to repair the Amplifier damage.

3.6 Form Report

Data that has been inputted on the Expert Data Input Form will be displayed again as a preview of the damage data that has been inputted by the expert.

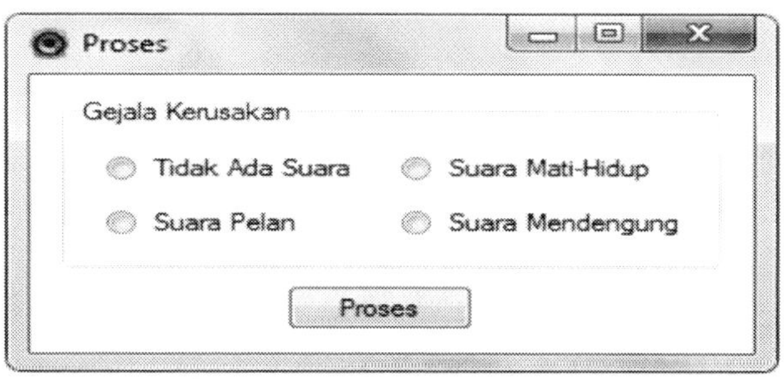

Figure 4. Process page

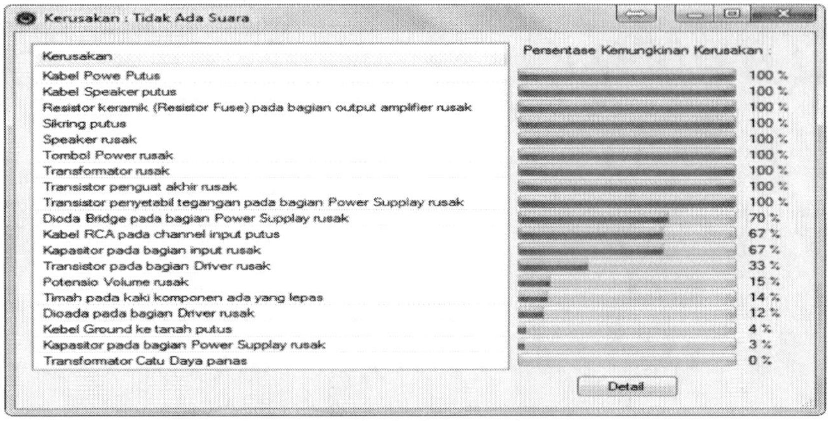

Figure 5. Calculation Results Page

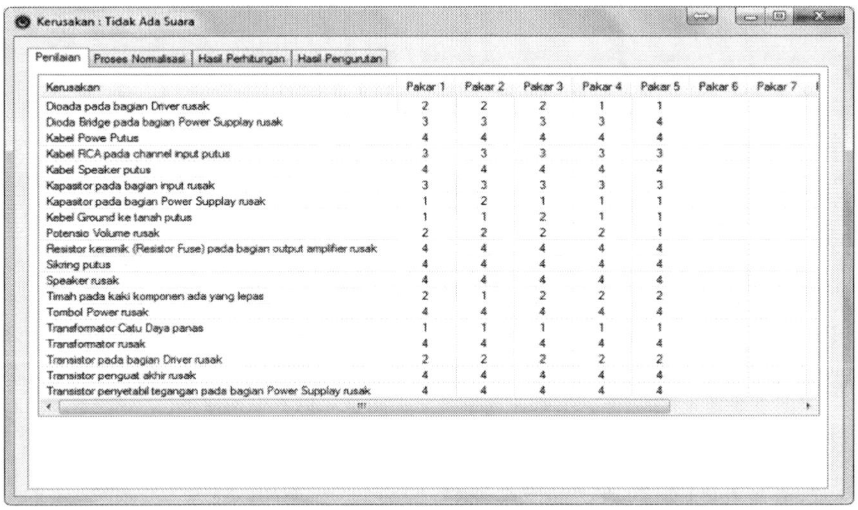

Figure 6. Report Page (Tab: Rating)

4. Conclusion

From the description in the previous chapters, it can be concluded that the Application of the Vikor Method (Vise Kriterijumska Optimizacija I Kompromisno Resenje) To Identify Amplifier Damage has the following conclusions:

1. Application of the Vikor Method (Vise Kriterijumska Optimizacija I Kompromisno Resenje) to Identify Damage Amplifiers can be a supporting consideration for making decisions on steps to improve the Amplifier.
2. The results of testing for the Vikor Method Application (Vise Kriterijumska Optimizacija I Kompromisno Resenje) To Identify the Damage Amplifier has an accuracy value of 81.25%.

5. References

[1] Turban E dan JEA. *Instructor's Manual for Decision Suppor System and Intelligent System.* 6 Edition. Upper Saddle River. NJ: Prentice Hall, 2001.

[2] Mintzberg H, Simon HA. The New Science of Management Decision, Revised Edition. *Adm Sci Q* 1977; 22: 342.

[3] Frank,D Petruzella. *Elektronik Industri*. Yogyakarta: Penerbit Andi, 2001.

[4] Opricovic S. *Multicriteria Optimization of Civil Engineering Systems*. 1998.

IOP Conf. Series: Materials Science and Engineering 1010 (2021) 012024 doi:10.1088/1757-899X/1010/1/012024

[5] Zhang N dan GWei. Extension of VIKOR method for decision making problem based on hesitant fuzzy set. *ELSEVIER* 2012; 37: 4938 – 4947.
[6] Chatterjee P, Chakraborty S. A comparative analysis of VIKOR method and its variants. *Decis Sci Lett* 2016; 5: 469–486.
[7] 1st International Conference on Green and Sustainable Computing (ICoGeS) 2017. *J Phys Conf Ser* 2018; 1019: 011001.
[8] Hidayat R. Penerapan Audio Amplifier Stereo Untuk Beban Bersama Dan Bergantian Dengan Menggunakan Saklar Ganda Sebagai Pengatur Beban. *J Tek Elektro Unnes* 2013; 5: 141163.

IOP Conf. Series: Materials Science and Engineering 1010 (2021) 012025 doi:10.1088/1757-899X/1010/1/012025

The improvement of sea highway route by using parallel insertion and exhaustive search

I Subiantoro[1], Lukmandono[2], P I Santoso[3] and H C Suroso[4]

[1]Postgraduate Student, Master of Industrial Engineering Department, Institut Teknologi Adhi Tama Surabaya, Indonesia
[2,3,4]Lecturer, Master of Industrial Engineering Department, Institut Teknologi Adhi Tama Surabaya, Indonesia

irfan.subiantoro18@gmail.com

Abstract. Indonesia contains of island and ocean which perform some problems in logistics delivery such as create a gap between west side and east side of Indonesia. A sea highway program were made by president to send logistics to be more effective and efficient. This program has been implemented for four years recently but still has many adjusment such as creating the best route of sea highway. This research try to create a better route of sea highway by using tour constraction to create an initial route in two steps, clark and wright saving matriks and parallel insertion. The second method is tour improvement aims to generate a new route after initial route by using exhaustive search. Five route of sea highway are generated in tour constraction and two route has been adjusted in the tour improvement because of a better score came up. The improvement of fourth route from Depot-Malahayati-Batam-Telukbayur-Depot becomes Depot-Malahayati-Telukbayur-Batam-Depot and second route still being the same but there also another option of route which has 18 score different from the best route score , those are Depot-Sampit-Samarinda-Balikpapan-Pontianak-Depot.

1. Introduction

As technology becomes more sophisticated and human mobility needs keep growing, the logistics demand also increasing. The consumer want to get goods as soon as possible so that the lead time requirement should be shorter. By this condition, Indonesia has a main problem to increase the shipment of logistic from one city to another because Indonesia contains of thoushands island.

The region of Indonesia mostly contains of ocean as much as 63% with the coastline 80.791 km^2. Meanwhile, the 37% contains of mainland with 17.499 islands. This condition of Indonesia that own lots of island which is surrounded by the ocean, the sea highway is one of the solution to support logistics transportation to be spreaded all over Indonesia and also able to reduce the logistics cost in the society [1]. In 2019 there are 18 route of sea highway with the frecuency target and different day service on each route. However, the more sea highway route the more fuel that will be consumed, the more cost on loading unloading material, and also the amount of labour that will be taken. All of those variabel will increase the logistics cost so that a method such as Vehicle Routing Problem (VRP) should be implemented in this case to create a new route to be more efficient. Vehicle Routing Problem is a problem in the distribution to define vehicle route with certain capacity from one or two port to fullfill the consumen needs [2][3].

The improvement of sea highway route have not been implemented because this program has just running for four years and everything still on adjustment. However some research about sea highway have been created such as using AHP method to analyse the effectiveness and efficiency factor in the third year of sea highway implementation [4], analysing the consistency of route and capacity of the shipment [5] and giving a recommendation about improving the efficiency about route, frequency, and capacity.

Tools that will be used in this research is implementation of Vehicle Routing Problem which has been proven the effectiveness of creating a better route. Parallel insertion and Exhaustive Search will be implemented in this case to find out a better route of sea highway. The parallel insertion has been proven to get a better route compare with sequential version and also exhaustive search has better route to improve the previous option as long as there are enough time to calculate [6].

1.1. Theory

1.1.1. Vehicle Routing Problem. Vehicle Routing Problem (VRP) is a problem in distribution which has the objective to create an optimal route especially for vehicle that has been known the capacity to fulfil the consumer expectation in the certain location and certain demand [7]. VRP also can be defined as a problem to design optimal delivery route from several depots to several supplier by fulfilling certain constraints [8]. Both definition empezhised by [6] that VRP is a problem which focus on distribution of logistics from depot to the consumer. The solution of VRP is some routes to deliver the logistics by certain vehicle. Each route travelled by one vehicle and the vehicle should be back to the early port. There are some general objective VRP such as minimize the transportation cost, minimize the vehicle, balance the route and minimize the penalty due to unpleasant service of consumer.

The characteristic of VRP can be seen as follows [9][8]:

1. The travel route of vehicle should be started and finished in the early depot
2. There are some place that the demand should fullfill once
3. If the capacity of vehicle has been used and unable to serve the next consumen, the vehicle should turn back to the main depot and fullfill the capacity to serve the next consumen
4. The main objective of this problem is to minimize the total distance of vehicle by arrange the sequence of consumen and when the vehicle turn back to the main depot to fulfill the capacity

1.1.2. Tour Construction. In this step the early solution of sea highway route will be created. The method which used to create early route is construction method especially Clarke and Wright Saving Algorithm. This algorithm categorize as construction method because it works by gradually insert the consumen into a route. The saving means some distance can be reduced by combine two consumen into one [3]. The two route that having the greatest saving which has chance to create as the early route[6][10][11].

The algoritm of saving matriks categoriezed as heuristik algoritm which is not given the optimal solution in the certain problem. The illustration concept of saving matrix can be draw in the diagram below [7] :

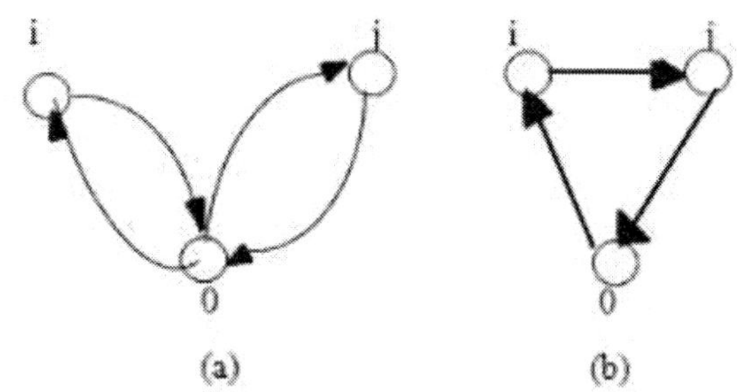

Figure 1. Saving Matriks Illustration

I, j = consumer i, consumer j
O = depot

$$Da = Coi + Cio + Coj + Cja \quad (1)$$
$$Db = Coi + Cij + Cjo \quad (2)$$
$$Sij = Da - Db = (Coi + Cio + Coj + Cja) - Coi + Cij = Cio + Coj - Cij \quad (3)$$

Da = cost or total distance of two route (consumer route i plus consumer route j)
Db = cost or total one route (consumer route i and j)
Coi = cost or distance from depot to consumer i
Cio = cost or distance from i to depot
Coj = cost ot distance from depot to consumer j
Cjo = cost or distance from consumer j to depot
Cij = cost or distance from consumer i to consumer j
Sij = the saving cost or distance between consumer i and j

There are two version to create the route after doing saving matriks. Sequential version which create route gradually by looking at every consumen saving list to try to put in to the route. The priority of this method is creating the first route by the greatest saving list. Meanwhile parallel version can be created more than one every choice on the saving list. Priority of parallel version are focus on the greatest saving list[12][6].

1.1.3. Tour Improvement. After getting the earlier route from tour construction, the next step is to optimize the route by tour improvement. The improvement were done by switch the node which stay on the solution (single route improvement). The location of sequential node moving by certain rules. During the movement if there is the best solution so that the earlier solution will be replaced by the new solution.

Exhaustive search is an approaches to problem solving with brute-force enumeration [13]. This method is simple based on the mathematical argument. An exhaustive search examines every search point inside the search region and gives the possible match. The basic principal of this methode is divide the search process into sequentioal steps and choosing the next step search direction based on the current step result[14]. However, this method considered inelegant, boring, error prone, and exhausting. When the option is too much then this method is unrecomended to be applied [6].

2. Materials and Method

Method that has been used in this research are divided into two step, tour construction and tour improvement. Tour construction used for defining the early route of sea highway which choosen Clark and Wright Saving Matriks as a tools to create the route. The Clark and Wright Saving Matriks done by parallel insertion due to it has better solution rather than sequential version. The next step is tour improvement which aims to optimize the earlier solution. Exhaustive search were choosen to solve this problem because it still feasible to get the best solution by looking at each sequential step to create the optimal solution.

2.1. Study Design and Overview

The data were got from ministry of maritime database that has been share into documents. There are 22 port which became consume and there is a port that became the main depot. The main depot located in Belawan the westernmost Indonesia. Distance Matrix that has been collected can be seen as follows:

Table 1. Destination Matriks location 1-10

		DESTINATION (Nm – Nautical Mile)										
			1	2	3	4	5	6	7	8	9	10
ORIGIN		Belawan	Tj. Priok	Tj. Perak	Makassar	Sorong	Batam	Malahayati	Jambi	Telukbayur	Palembang	Panjang
Belawan	0	0	1021	1119	1427	2243	372	251	630	784	673	1102
Tj. Priok	1	1021	0	358	750	1492	468	977	334	495	225	100
Tj. Perak	2	1119	358	0	431	1226	725	1325	640	1005	540	455
Makassar	3	1427	750	431	0	754	999	1569	967	1167	887	842
Sorong	4	2243	1492	1226	754	0	1644	2176	1656	849	1595	1578
Batam	5	372	468	725	999	1644	0	573	155	245	246	398
Malahayati	6	251	977	1325	1569	2176	573	0	649	490	756	886
Jambi	7	630	334	640	967	1656	155	649	0	200	108	255
Telukbayur	8	784	495	1005	1167	849	245	490	200	0	289	400
Palembang	9	673	225	540	887	1595	246	756	108	289	0	153
Panjang	10	1102	100	455	842	1578	398	886	255	400	153	0
Tj. Emas	11	1033	217	195	547	1298	617	1168	518	700	32	317
Pontianak	12	696	393	503	670	1316	330	893	352	541	326	406
Sampit	13	1104	422	281	416	1101	583	1154	561	761	491	490
Banjarmasin	14	1151	488	257	309	1011	689	1260	662	862	586	567
Balikpapan	15	1421	660	461	279	867	779	1340	789	986	728	732
Samarinda	16	1481	700	510	308	846	792	1347	812	1008	757	769
Pantoloan	17	1555	841	578	266	683	955	1507	973	1169	915	916
Kendari	18	1793	949	636	202	552	1152	1720	1143	1343	1068	1036
Tenau Kupang	19	1806	1018	665	390	724	1347	1924	1293	1488	1198	1119
Ternate	20	2048	1295	1107	595	252	1400	1931	1431	1625	1376	1379
Ambon	21	1994	1281	960	531	250	1476	2035	1475	1675	1403	1371
Jayapura	22	2819	2036	1696	1282	555	2190	2732	2203	2422	2134	2106

Table 2. Destination Matriks location 11 – 22

		DESTINATION (Nm – Nautical Mile)										
		11	12	13	14	15	16	17	18	19	20	21	22
ORIGIN		Tj. Emas	Pontianak	Sampit	Banjarmasin	Balikpapan	Samarinda	Pantoloan	Kendari	Tenau Kupang	Ternate	Ambon	Jayapura
Belawan	0	1033	696	1104	1151	1421	1481	1555	1793	1806	2048	1994	2819
Tj. Priok	1	217	393	422	488	660	700	841	949	1018	1295	1281	2036
Tj. Perak	2	195	503	281	257	461	510	578	636	665	1107	960	1696
Makassar	3	547	670	416	309	279	308	266	202	390	595	531	1282
Sorong	4	1298	1316	1101	1011	867	846	683	552	724	252	250	555
Batam	5	617	330	583	689	779	792	955	1152	1347	1400	1476	2190
Malahayati	6	1168	893	1154	1260	1340	1347	1507	1720	1924	1931	2035	2732
Jambi	7	518	352	561	662	789	812	973	1143	1293	1431	1475	2203

		DESTINATION (Nm – Nautical Mile)											
		11	12	13	14	15	16	17	18	19	20	21	22
ORIGIN		Tj. Emas	Pontianak	Sampit	Banjarmasin	Balikpapan	Samarinda	Pantoloan	Kendari	Tenau Kupang	Ternate	Ambon	Jayapura
Telukbayur	8	700	541	761	862	986	1008	1169	1343	1488	1625	1675	2422
Palembang	9	32	326	491	586	728	757	915	1068	1198	1376	1403	2134
Panjang	10	317	406	490	567	732	769	916	1036	1119	1376	1371	2106
Tj. Emas	11	0	421	305	329	511	558	677	748	801	1118	1078	1811
Pontianak	12	421	0	265	371	454	470	633	829	1044	1085	1151	1887
Sampit	13	305	265	0	106	242	279	427	582	779	888	914	1643
Banjarmasin	14	329	371	106	0	182	230	354	482	674	809	816	1548
Balikpapan	15	511	454	242	182	0	50	186	382	669	647	697	1415
Samarinda	16	558	470	279	230	50	0	162	386	695	619	688	1398
Pantoloan	17	677	633	427	354	186	162	0	255	610	461	530	1235
Kendari	18	748	829	582	482	382	386	255	0	377	407	335	1070
Tenau Kupang	19	801	1044	779	674	669	695	610	377	0	699	478	1106
Ternate	20	1118	1085	888	809	647	619	461	407	699	0	274	803
Ambon	21	1078	1151	914	816	697	688	530	335	478	274	0	734
Jayapura	22	1811	1887	1643	1548	1415	1398	1235	1070	1106	803	734	0

Demand of each location or consumer also need to be outlined as one of constrain in this research. The demand will be illustrated in TEUS (Twenty-foot Equivalent UnitS) which usually used for container or logistics metrices. The demand data has been collected and can be seen bellow :

Table 3. *Demand of each location/consumer*

Location/ Consumer	Demand (TEUS)	Location/ Consumer	Demand (TEUS)
Tj. Priok	132510	Pontianak	159303
Tj. Perak	225514	Sampit	159303
Makassar	67547	Banjarmasin	229438
Sorong	47850	Balikpapan	159303
Batam	253369	Samarinda	159303
Malahayati	159303	Pantoloan	159303
Jambi	159303	Kendari	159303
Telukbayur	159303	Tenau Kupang	159303
Palembang	159303	Ternate	159303
Panjang	159303	Ambon	159303
Tj. Emas	159303	Jayapura	159303

IOP Conf. Series: Materials Science and Engineering 1010 (2021) 012025 doi:10.1088/1757-899X/1010/1/012025

2.2. Software

The main software that used in this research is Ms. Excel 2013 especially on Excel Solver to calculate the saving matriks and create an optimal route. Netpas Distance also been used to define the distance between depot to consumer and consumer to consumer. Distance metrices illustrated in Nautical mile (Nm).

3. Result and Discussion

3.1. Saving List

Saving list can be found by calculating the destination and formula below :

$$Sij = Cio + Coj - Cij$$

$$S\ sorong, Jayapura\ (S\ 4{,}22) = C4, o + Co, 22 - C4{,}22$$

$$= 2243 + 2819 - 555$$

$$= 4507\ Nm$$

Table 4. Sample of Saving List

Consumer i	Consumer j	Saving (Nm)
Sorong	Jayapura	4507
Ambon	Jayapura	4079
Ternate	Jayapura	4064
Sorong	Ternate	4039
Sorong	Ambon	3987
Ternate	Ambon	3768
Kendari	Jayapura	3542
Tenau Kupang	Jayapura	3519

3.2. Creating earlier route

After getting saving list result, a new route can be created by using parallel insertion method which choosing the highest saving list to be executed firstly. A new route has been solved by Ms.Excel and can be outlined below :

Table 5. New Sea Highway Route

Number of Route	Sequence Route
First Route	Depot – Jambi – Panjang – Tj.Priok – Tj. Emas – Palembang - Depot
Second Route	Depot – Pontianak – Samarinda – Balikpapan – Sampit – Depot
Third Route	Depot – Banjarmasin – Pentoloan – Kendari – Tenau Kupang – Tj. Perak - Depot
Fourth route	Depot – Malahayati – Batam - Telukbayur – Depot
Fifth route	Depot – Makassar – Ambon – Jayapura – Sorong – Ternate - Depot

3.3. Improvement route by using exhaustive search

There are five route that will be processed by exhaustive search algorithm. Fisrtly the enumeration list should be noted sistematically, evaluated every possibility of solution one by one, when all the enumeration list has been proceed then announce the best score.
The enumeration list can be created by using formula below :

$$Number\ of\ enumeration = (n-1)!$$

So that the numbers enumeration of each route are below :

Table 6. Enumeration

Number of Route	Calculation	Result
First Route	(6-1) ! = 5!	120
Second Route	(5-1) ! = 4!	24
Third Route	(6-1) ! = 5!	120
Fourth route	(4-1) ! = 3!	6
Fifth route	(6-1) ! = 5!	120

Table 7. Enumeration fourth route

Tour route	Score
Depot – Malahayati – Telukbayur – Batam – Depot	251 + 490 + 245 + 372 = 1358
Depot – Malahayati – Batam- Telukbayur – Depot	251 + 573 + 245 + 784 = 1853
Depot –Telukbayur - Malahayati – Batam – Depot	784 + 490 + 573 + 372 = 2219
Depot – Telukbayur – Batam – Malahayati –Depot	784 + 245 + 573 + 251 = 1853
Depot – Batam – Malahayati – Telukbayur – Depot	372 + 573 + 490 + 784 = 2219
Depot – Batam – Telukbayur – Malahayati - Depot	372 + 245 + 490 + 251 = 1358

Table 8. Enumeration second route

Tour Route	Score
Depot – Pontianak – Samarinda - Balikpapan - Sampit – Depot	2562
Depot – Pontianak – Samarinda – Sampit – Balikpapan – Depot	3108
Depot – Pontianak – Sampit – Samarinda – Balikpapan – Depot	2711
Depot – Pontianak – Sampit – Balikpapan –Samarinda – Depot	2734
Depot – Pontianak – Balikpapan – Sampit – Samarinda – Depot	3152
Depot – Pontianak – Balikpapan – Samarinda – Sampit – Depot	2583
Depot – Samarinda – Pontianak – Balikpapan – Sampit – Depot	3751
Depot – Samarinda – Balikpapan – Pontianak – Sampit – Depot	3354
Depot – Samarinda – Balikpapan – Sampit – Pontianak – Depot	2734
Depot – Samarinda – Pontianak – Sampit – Balikpapan – Depot	3879

Tour Route	Score
Depot – Samarinda – Sampit – Pontianak – Balikpapan – Depot	3900
Depot – Samarinda – Sampit – Balikpapan –Pontianak – Depot	3152
Depot – Balikpapan – Pontianak – Samarinda – Sampit – Depot	3728
Depot – Balikpapan – Samrinda – Pontianak – Sampit – Depot	3310
Depot – Balikpapan – Sampit – Samarinda – Pontianak – Depot	3108
Depot – Balikpapan – Sampit – Pontianak – Samarinda – Depot	3879
Depot – Balikpapan – Samarinda – Sampit – Pontianak – Depot	2711
Depot – Balikpapan – Pontianak – Sampit – Samarinda – Depot	3900
Depot – Sampit – Pontianak – Balikpapan – Samarinda – Depot	3354
Depot – Sampit – Balikpapan – Samarinda – Pontianak – Depot	2562
Depot – Sampit – Balikpapan – Pontianak – Samarinda – Depot	2966
Depot – Sampit – Balikpapan – Pontianak – Samarinda – Depot	3751
Depot – Sampit – Samarinda – Pontianak – Balikpapan – Depot	3728
Depot – Sampit –Samarinda – Balikpapan – Pontianak - Depot	2583

4. Discussion

From table 7 can be concluded the best solution of sea highway route with the lowest score 1358 are Depot- Malahayati – Telukbayur – Batam – Depot and Depot – Batam – Telukbayur – Malahayati – Depot. The first solution before implementing tour construction obtained score as much as 1853 which is the second lowest route in tabel 6. This might not significant difference before and after implementing exhaustive search but still worthy to try.

Table 8 shown the lowest score is 2562 with the best route Depot – Pontianak – Samarinda – Balikpapan – Sampit – Depot and Depot – Sampit – Balikpapan – Samarinda – Pontianak – Depot. By using exhaustive search also can be seen the other possibility that might be choosen by the researcher. For example, there are slightly different score as much as 2583 which has 18 different score so it can be choosen as consideration. Knowing many possibility has some advantages such as the election of route can be added by some other consideration. The consideration might be the easiest route due to weather in the sea, the schedule of each port or destination and so on. However for some highly possibility of permutation such as first, second, and fifth route, It spent so much time to calculate each enumeration. This prove that exhaustive search is not recommended for higher level of research.

5. Conclusions

Tour construction generate five sea highway route option by using clark and wright saving matriks and parallel instruction. The five sea highway route still able to be improved by using exhaustive search method so that some initial route are changed. The improvement occured on the fourth route which is the route from tour construction was *Depot – Malahayati – Batam - Telukbayur – Depot* becomes *Depot- Malahayati – Telukbayur – Batam – Depot.* The second route also spot an interesting highlight, there is no improvement in the second route but there is another option route that can be offered those are *Depot – Sampit –Samarinda – Balikpapan – Pontianak - Depot* which only 18 score different from earlier route (*Depot – Pontianak – Samarinda – Balikpapan – Sampit – Depot*). This prove that exhaustive search able to improve the initial route by creating the enumeration. Nonetheless, exhaustive search are not recommend for big permutaion possibility due to time consuming.

6. References

[1] A. Lukman and L. Adam, "Maritime Connectivity Policy in Indonesia," *Politica*, vol. 6, no. 1, 2015.

[2] M. A. Clark-wright, P. Studi, T. Industri, F. Teknik, and U. Diponegoro, "Optimasi Rute Angkutan Publik dengan Menggunakan," pp. 8–9, 2017.

[3] Lukmandono, M. Basuki, M. J. Hidayat, and F. B. Aji, "Application of Saving Matrix Methods and Cross Entropy for Capacitated Vehicle Routing Problem (CVRP) Resolving," *IOP Conf. Ser. Mater. Sci. Eng.*, vol. 462, no. 1, 2019.

[4] F. Kristini, A. N. Bambang, W. Handoko, and A. A. Priadi, "Efektivitas dan Efisiensi Program Tol Laut berbasis AHP (Studi Kasus: Pelabuhan Tahuna)," *War. Penelit. Perhub.*, vol. 31, no. 2, pp. 75–82, 2019.

[5] D. D. Andilas and L. A. Yanggana, "Pelaksanaan Program Tol Laut Pt Pelayaran Nasional Indonesia," *J. Manaj. Transp. Dan Logistik*, vol. 4, no. 1, p. 1, 2017.

[6] I. Maulana and R. Arifati, "PERENCANAAN RUTE PENGIRIMAN MENGGUNAKAN METODE PARALLEL INSERTION DAN EXHAUSTIVE SEARCH PADA PT . STARMASS LOGISTICS UPN " VETERAN " JAKARTA UPN " VETERAN " JAKARTA."

[7] P. Nusmesse, A. Rahawarin, and D. B. Paillin, "Usulan Penentuan Rute Dalam Pendistribusian Bbm Bersubsidi (Premium) Pada Pt . Pertamina Tbbm Wayame Ambon Ke Spbu Di," *Arika*, vol. 10, no. 1, pp. 1–14, 2016.

[8] I. A. Soenandi, J. Joice, and B. Marpaung, "Optimasi Capacitated Vehicle Routing Problem with Time Windows dengan Menggunakan Ant Colony Optimization," *J. Sist. dan Manaj. Ind.*, vol. 3, no. 1, p. 59, 2019.

[9] C. Roch and S. Langer, "The Capacitated Vehicle Routing Problem," *Digit. Welt*, vol. 3, no. 2, pp. 30–33, 2019.

[10] L. Octora, A. Imran, and S. Susanty, "Pembentukan Rute Distribusi Menggunakan Algoritma Clarke & Wright Savings dan Algoritma Sequential Insertion," *Reka Integr.*, vol. 2, no. 2, pp. 1–11, 2014.

[11] N. Ikfan and M. Ilyas, "Penentuan rute Transportasi Terpendek untuk Meminimalkan Biaya Menggunakan Metode Saving Matriks," *J. Ilm. Tek. Ind.*, vol. 12, no. 2, pp. 165–178, 2013.

[12] F. Ahmad and H. F. Muharram, "Penentuan Jalur Distribusi Dengan Metode Saving Matriks," *Competitive*, vol. 13, no. 1, p. 45, 2018.

[13] J. Nievergelt, "Exhaustive search, combinatorial optimization and enumeration: Exploring the potential of raw computing power," *Lect. Notes Comput. Sci. (including Subser. Lect. Notes Artif. Intell. Lect. Notes Bioinformatics)*, vol. 1963, no. January 2000, pp. 18–35, 2000.

[14] H. M. Hang and Y. M. Chou, *Motion Estimation for Image Sequence Compression*. Woodhead Publishing Limited, 2012.

Acknowledgments

The author would like to thank the financial support given from The Ministry of Research and Technology / National Agency for Research and Innovation of Indonesia under the Magister Thesis Research scheme 2020 (Contract No. 046/SP2H/LT/MULTI/L7/2019 and 046/SP2H/AMD/LT/MULTI/ L7/2020)

IOP Conf. Series: Materials Science and Engineering 1010 (2021) 012026 doi:10.1088/1757-899X/1010/1/012026

The design concept of bamboo in micro housing as a sustainable self-building material

Nareswarananindya[1], S H Laksono[1], A N Ramadhani[1], A Budianto[2], I Komara[3], A I D Syafiarti[3]

[1]Department of Architecture, Institut Teknologi Adhi Tama Surabaya, Indonesia
[2]Department of Chemical Engineering, Institut Teknologi Adhi Tama Surabaya, Indonesia
[3]Department of Civil Engineering, Institut Teknologi Adhi Tama Surabaya, Indonesia

nareswarananindya@itats.ac.id

Abstract. Bamboo is building material that is well known in Southeast Asia, which is renewable, environment-friendly, widely available and fast-growing, as other materials resource such as timber is diminishing and restricted due to deforestation. While timber demand is rapidly increasing, unfortunately, the supply is depleting. Therefore, bamboo can be suitably replacing timber and other materials in construction and other works—moreover, bamboo grown abundantly in Indonesia and easily found in each district as local material. Using local materials is essential in this fast-growing world. Local materials can reduce carbon footprint and support sustainable development efforts. Also, bamboo construction also technically eco-friendly and cost-effective with durable and have high sustainability impact. Besides, the land available for residential housing is currently running low. This problem requires innovation in terms of space effectiveness in design. As a two- or three-story compact bamboo house concerning sustainable local material. This research analyses the user configuration to proceed bamboo as local material to construct a compact house model that is green and durable. These works initiate nature sustainability buildings to fit into natural surroundings.

1. Introduction

Land extinction is one of the main problems in the urban world, including Surabaya city, which is known as a metropolis area. Land for housing is minimal, resulting in high land prices that are no longer affordable for low to middle-class society. Innovations are necessary to accommodate urban housing needs which continues to increase. One of them is a compact house concept that has an affordable price. The compact house concept has been applied for a long time in residential design in Indonesia, starting with the emerging phenomenon of the Tiny House Movement in America around 1978. There are four aspects of compact house concept, namely space diversity, furniture, color, and insulation [1].

Besides, the use of local materials and resources in construction is essential to perform a sustainable development [2]. Bamboo is local material that can use as an alternative material for construction which is considered as one of the most sustainable building materials. The characteristic of bamboo material can be explored as an initial approach to design a building, whether to be applied as a building structure, shell, or interior [3, 4]. Bamboo is widely used due to possesses a vast advantageous usage scheme which is environmentally friendly, in respect to timber which is gathered within 10–30 years, bamboo can be collected within 3–5 years, which are the fastest-growing plants in the world has grown up to 60 cm or more in a day [5]. The use of natural materials to be environmentally friendly in addition to the use of bamboo is also carried out by other researchers by seeking fuel from plants [6–10] , making activated carbon from various plants [11–14] and making Fashion products from used paper [11]. In this term, bamboo can be a local material that is suitable for developing compact houses with affordable prices. Though, the market seems to be sceptical about its use for constructive materials. This might be explained by the fact that some structural evaluations still need to be proven. On the other hand some local products such as cabinetmakers and handicrafts makers have already adopted bamboo (laminated) as a valuable material to make their products and its proven durable. These issues mainly apply in Indonesia whereas needs to take a fundamental different approach to Bamboo which is normally know as a temporary structures. As a consequence, this research needs to clarify technical data and comply with structural standars as well as to illustrate the bamboo as a main structural building material.

There are various study about bamboo-based construction materials that were cultivated, such as laminated bamboo, bamboo-reinforced concrete, and many others [12, 13]. As the history of bamboo in laminated uses traces back only 10 to 15 years and development are still going on [14]. Laminated is known as a process to make a use of different layers of wood, glued and pressed together. This principlas is applied to bamboo to turn out to be a more labour intensive process regarding the assemblage of multiplicity of strips [2]. Bamboo is more durable, with 11 types of curing processes. The processes are Harvesting, Soaking, Washing Lime, Painting, Soaking with a salt solution, Vertical Soak and Diffusion (VSD) with boric acid, Soaking with camphor detergent, filling or immersing in kerosene and used oil solutions, and injection with Sarpeco chemical solutions [15].

Due to early studies, additional studies are required to better understand the socio-economic aspects of bamboo-based construction materials as well as the development study on bamboo material application in a built environment [16]. Hence, this research conducts the exploration of bamboo as local material to design a compact house that suits the needs of the urban community at affordable prices to adjust the socio-economic aspects of developing country. Bamboo building needs to be developed with a compact house concept to overcome the scarcity of land in urban areas where land is not widely available and has extremely overpriced. Bamboo is sustainable local material that can easily found in the Surabaya area, environmentally friendly, and has a low price. The development of this bamboo house model also can solve the housing backlog problem Indonesia as well as the affordability issues. Further, starting to grow bamboo locally could both improve air and soil quality striving for a sustainable future which conclude, growing, harvesting, processing and constructiong bamboo has to be intensified locally.

2. Theoretical Review

2.1. Bamboo in the application of construction

2.1.1. Bamboo materials and products

Bamboo is one of the local materials which is categorized as a sustainable material due to the rapid spread and fast-growing time. Compared with timber, bamboo has stronger strength and affordable prices. Some types of bamboo have rapid growth on average 90-120 cm per day, while bamboo's nature that often used as construction usually takes 3-6 years to grow [5]. It has to be taken into account that bamboo grows on its existing roots and therefre has no need to be replatend. Most other wooden species will die after being harvested which means need time to grow (See Figure 1). The time for bamboo to fully grow is 3 to 6 years. That is different from pine and oak trees respectively need 30 and 60 years. Within the same bamboo can be harvested 12 times more than oak [17]. Some literatures inform variety of bamboo's type which commonly used as structural materlas in the local area of East Java region, listed in Table 1.

Figure 2 illustrates an overview of a modernized bamboo by biobased composite materials that usually constructed nowdays. The use of all biodegradable product provide a great potential for further development especially in local construction.

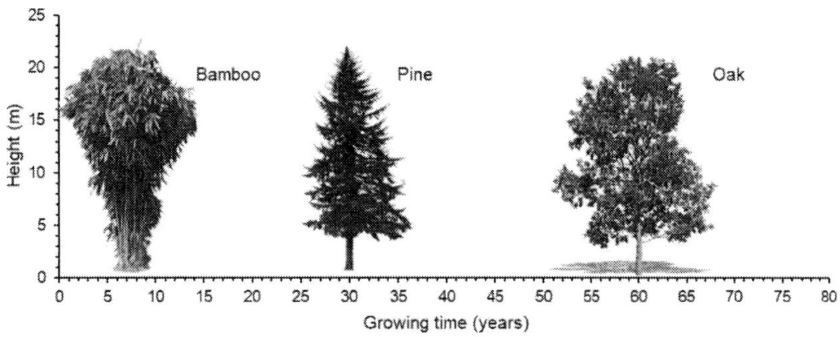

Figure 1. Growth graph of varieties of threes

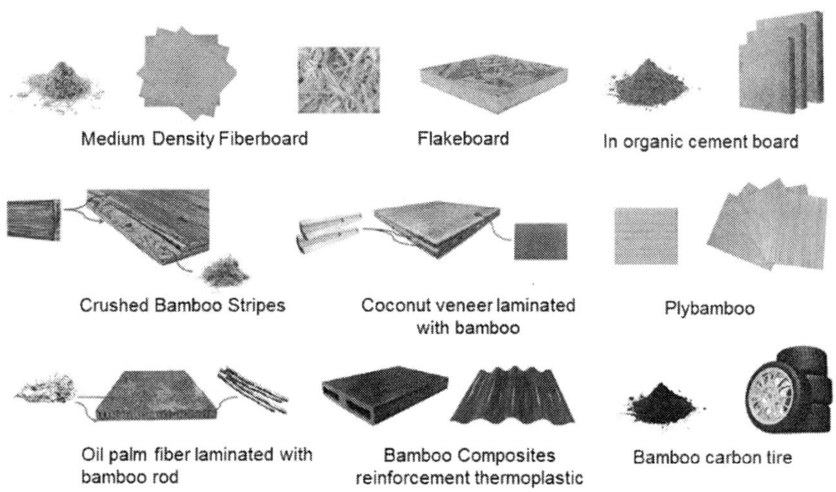

Figure 2. Modernized bamboo product as composites

Table 1. Bamboo types [18]

Bamboo Type	Height (m)	Diameter (cm)	Thickness (cm)	Internode Length (cm)	Utility
Petung Bamboo	20-25	>20	< 20	40-50	Column, beam, batt, rinse
Wulung/ Black Bamboo	15-20	6-14	< 8	40-50	Wall
Apus Bamboo	22	4-15	< 15	20-60	Columns, beams, floors, walls, roofs, furniture
Ori Bamboo	25-30	5-15	10-20	20-40	Stucco construction, floor
Ampel Bamboo	10-20	4-10	7-15	20-45	Ornament
Ater Bamboo	22	5-10	< 8	40-50	Wall, furniture

Table 2. Structural properties of bamboo compared to other materials [19]

kN/cm^2	Spruce	Bamboo	Steel
Elastic modulus	1100	2000	2100
Compressive	4.3	6.2 – 9.3	14
Tension	8.9	14.8	16
Bending	6.8	7.6 – 27.6	14
Shear	0.7	2.0	9.2

2.2.2. Structural Properties and processing techniques

To support the capacity as a building materials, the performance of bamboo is enhanced e.g., compression 1/3 of tension strength, tension which needs 3 times as strong as in compression, splitting easily along the length, easy to work and construct, ability to laminate and lastly can be modified into different appearances (see Table 2) [19–21]. However, the weakness has to be concerned such as low bending capacity if no laminated, weather sensitive (the special treatment needed) and hard to make nail joints since the splitting easily along the length. In addition, to support the better natural durability, a certaint chemical treatment needs to be given, i.e., treated in submerged watter or adding chemical compound in the process of curing, resurfacing, soaking or boiling. The process of treatment more tricky than steel where the simply conceptual process is created, e.g., strengthen the connection [22, 23].

For applications in buildings, bamboo with intact conditions will last longer than cleaved bamboo. This is because, in cleaved bamboo, fungi and pests can quickly enter into bamboo pores. As to avoid the entry of fungus and parasites, the preservation process is necessary before the bamboo usage. There are two kinds of bamboo preservation process, which are natural processes and chemicals. According to the paper theme about sustainability, natural preservation is chosen. The natural way of preservation [24] includes:

- Curing: this method is done by cutting bamboo stems at the bottom but remain the leaves and branches. Then the bamboo clump is stored in a particular room within a specific time. Because of the assimilation process, the leaves still survive, the starch content of bamboo segments is reduced.
- Resurfacing: this method is done by sprinkling bamboo sticks with lime. The aim is to slow down the absorption of water so that the resistance of bamboo to mold is higher.
- Bamboo Soaking: this method is done using bamboo stems soaked for one month in freshwater, brackish water, calm seawater, or flowing water so that the starch in bamboo is reduced. Bamboo soaking should be done after the bamboo is dried in a standing condition, then soaked entirely. Bamboo that has been immersed in the water must be pale (not yellow-green or black) and sour.
- Boiling: boiling bamboo at a temperature of 55-60 degrees celsius for 10 minutes will cause the starch to undergo complete gelatinization that is amylose insoluble in water. Boiling at 100 degrees Celsius for 1 hour is effective enough to reduce attacks from pests.

From the literature, we can conclude that with the correct preservation process, bamboo buildings can last up to 40 years. As for the joints in bamboo construction, it is better to avoid using nails. And for sustainability reason, it is better to use natural material and methods such as rope, nut, and bolt techniques, so it does not cause damage to the bamboo material itself. It is also important to minimize the danger of fire. To slow the spread of the light that burns vertically, the bamboo arrangement can be made by arranging bamboo horizontally or diagonally.

2.3. Micro housing

The micro housing is a dwelling that is sized to meet its context, and its occupants need with little excess space. This concept is also popular as a small house, tyiny house or micro apartment. The ideal size for a small home seems to be 18 square meters for 1-2 people (the size of a one-car) or 36 square meters for a 4-person family. Four key design elements in a small house that influence liveability [25], which are:

- Trade space for a place, maximize the quality of space by paying attention to the layout, function, and design of space.
- Durability, maximize the capacity of space & furniture to be multifunctional.
- Air and light, maximize lighting & natural air.
- Privacy and noise reduction, minimize noise and interference from outside.

This concept was derived from research conducted in North America, which used a mixed-method between literature reviews, interviews, and direct survey observation in several samples of small houses. From this literature review, it can be concluded that the small house has some prime functional areas which are kitchen and dining area, bathing area, sleeping area, living area, and storage area. Some spaces can be designed as public or shared spaces, such as parks, laundry, fitness rooms, etc.

3. Result and discussion

3.1. Design concept

Figure 1 shows the diagram design for the bamboo house. This bamboo house is designed according to the standard size of a small house which is 36 square meters (figure 3.a) [26]. Ground floor modules adapted to the size of a carport, which is 4.5 meters x 4.5 meters (figure 3.b). As a response to the scarcity of land in urban areas, design development is made vertical (figure 3.c). The base of the building is designed elevated to maximize open space and rainwater absorption. The shape concept of this building is dominated by triangular geometry, where triangles are known as the sturdiest in forming a space frame (figure 3.d). The last figure (figure 3.e) explains the position of vertical circulation in the bamboo house. The complementary elements such as window, wall, and door are designed to adjust to triangle shape to fit in.

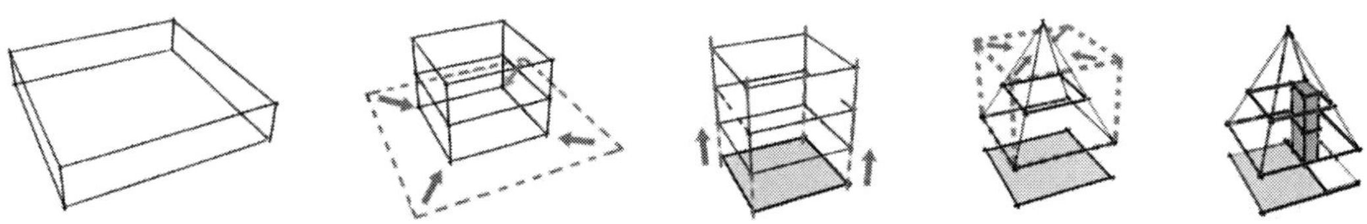
(a) (b) (c) (d) (e)

Figure 3. Design diagram for the bamboo house, (a) small house standard size 36 square meters, (b) ground floor modules, (c) vertical development, (d) triangular geometry concept, (e) vertical circulation position of bamboo house

This small house has to accommodate behavior setting of the user, which are include some domestic activity such as eating, sleep, gathering, and the need of storage based on user's action [27]. The small house must have an efficient layout to accommodate all activities. Some multifunctional rooms are also needed to be flexible spaces that can adjust to the user's needs. Due to the small house theory, the floor plan concept used in this building is open plan; thus, the spaces in this house can be used multi functionally [28].

3.2. Plan design: a small house

This bamboo house plan applied the small house theory, where the open-plan plan concept is used to make some multifunctional room and space. Figures 4 and 5 depict the ground floor, which consists of a carport, bathroom, and foyer room that can be used as a terrace, living room, or an intermediate room. Carport design is designed with minimal pavement by selecting paving grass material to maximize the open space and rainwater absorbing.

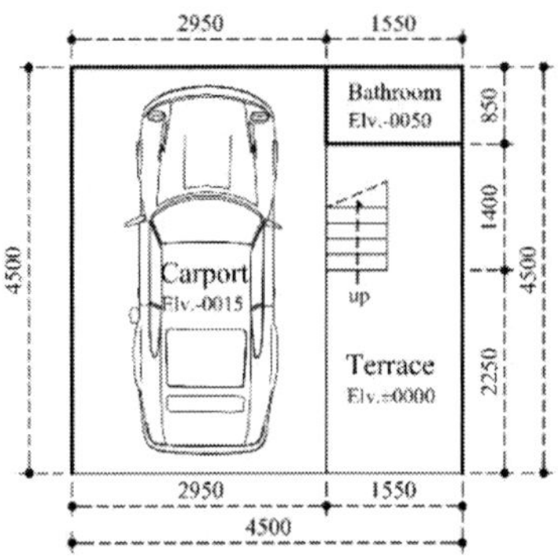

Figure 4. First-floor plan

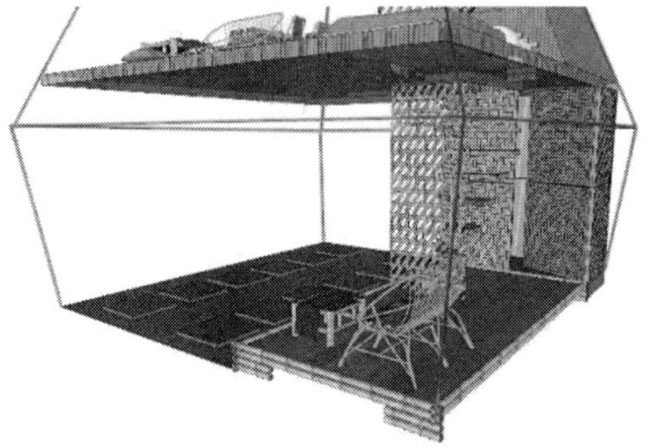

Figure 5. First-floor axonometric

Moreover, figures 6 and 7 show the second-floor design, which consists of a service area and multifunctional area. Kitchen area is made vertically parallel to the bathroom in ground floor to facilitate the shaft, and plumb the multifunctional space is quite broad; thus, it can be used as a living room during the daytime and change its function as an additional bedroom in the night time.

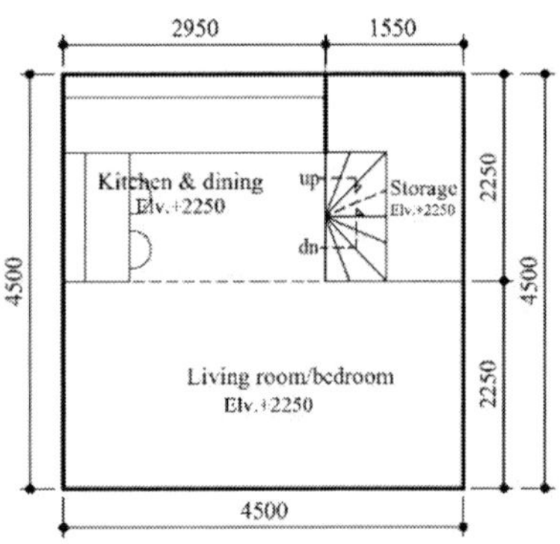

Figure 6. Second-floor plan

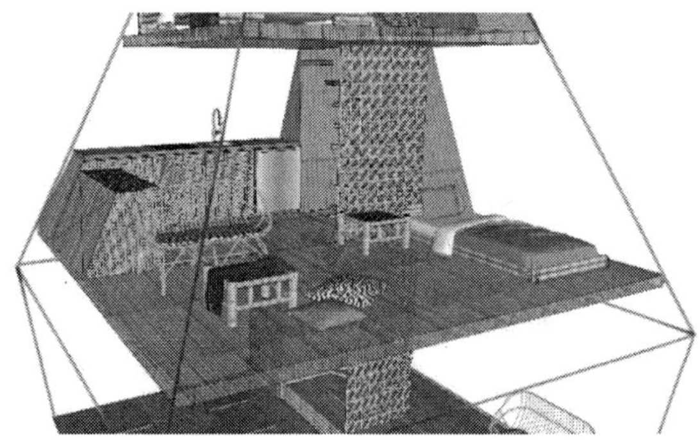

Figure 7. Second-floor axonometric

Figures 8 and 9 show the third floor is a mezzanine that functioned as the main bedroom area. According to small house theory, custom furniture that fits in with house design is needed. The furniture in this room is a bed and wardrobe, which is made custom to adjust to a compact size. All this three-level house is 36 m^2 in total, thus comply with the minimum size standard of a small house (figure 10).

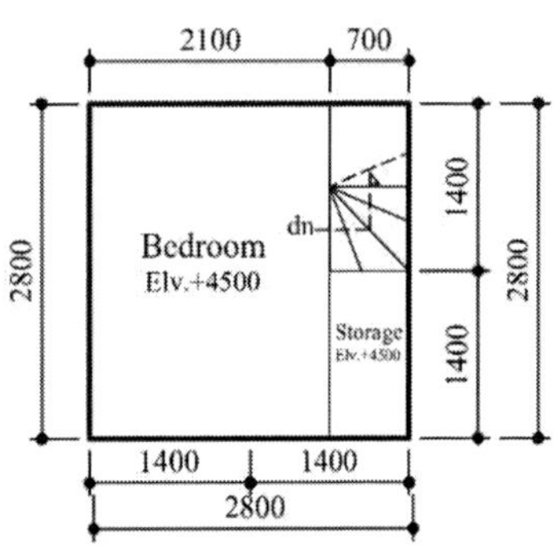

Figure 8. Third-floor plan

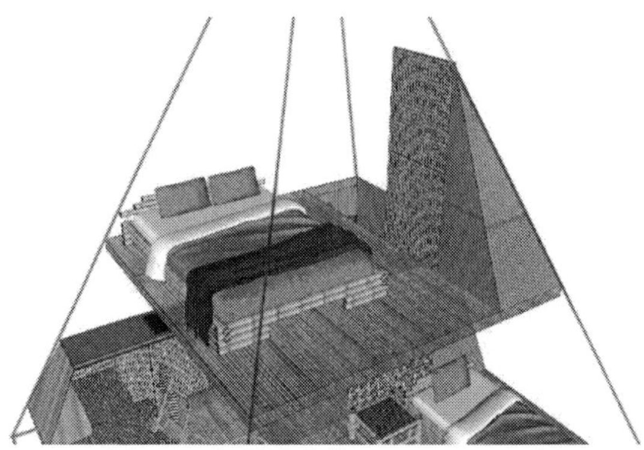

Figure 9. Third-floor axonometric

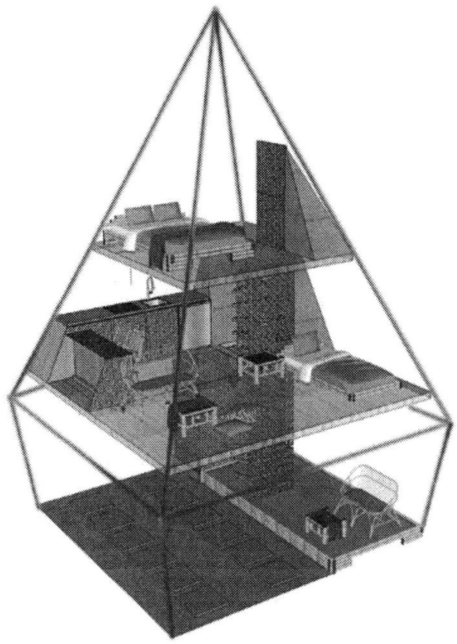

Figure 10. Interior axonometric – the concept of bamboo compact house

3.3. Structure design

Figures 11 and 12 show the structure concept of the house design. The first level is designed elevated to minimize the footprint and maximize rainwater absorption. It uses bamboo as the main construction, including primary beams and column, while the joints are using palm fiber lashing—the foundations using a concrete base which joined with bamboo's column. Meanwhile, the shape concept applied triangle geometry because the triangle is the most sturdy and stable in forming a space frame structure.

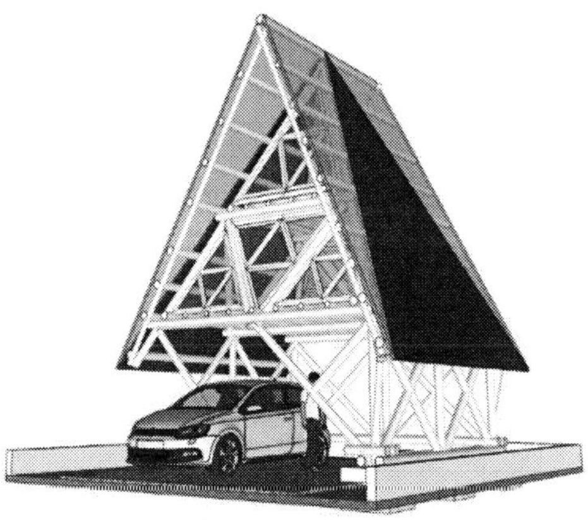

Figure 11. The structure concept of the bamboo house – normal view

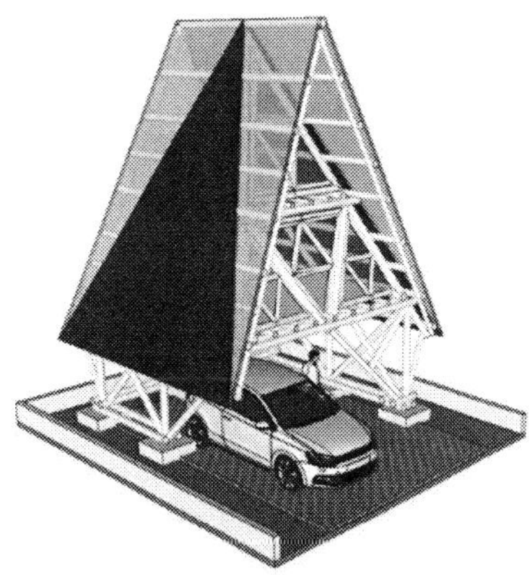

Figure 12. The structure concept of the bamboo house – bird eye view

4. Conclusions

A micro housing with bamboo material has been designed in this study. The design concept proposes a sustainable construction using local content and compact design to solve the scarcity of land in an urban area. The design applied small house theory which is the efficiency of layout design, the use of a multifunctional room and furniture, good air conditioning and lighting, and protection from outside disturbances such as noise. However, this bamboo house consists of three-level which comply with the minimum size standard of a small house. This house uses bamboo as the main construction is emerging with triangle geometry shape, though not widely accepted yet.

5. References

[1] Putri PU, Prianto E. Kajian Prinsip Compact House Design pada Rumah ber-Arsitektur Konvensional di Semarang. *J Penelit Dan Pengabdi Kpd Masy UNSIQ* 2016; 3: 148–158.

[2] Ramadhani AN, Faqih M, Septanti D. Development Concept Of Urban Housing Renewal Based On Sustainable Tourism: A Case Study Of Kampung Tambak Bayan, Surabaya. 2017; 6: 9.

[3] Nareswarananindya N. Eksplorasi Material Glulam pada Perancangan Shelter menggunakan Saluran Kreativitas Focus on Material. *BORDER* 2019; 1: 83–96.

[4] Nareswarananindya, Noerwasito T, Purnomo H. Eksplorasi Material Laminated Veneer Lumber pada Selubung Bangunan Struktural. *ResearchGate*, https://www.researchgate.net/publication/336845407_Eksplorasi_Material_Laminated_Veneer_Lumber_pada_Selubung_Bangunan_Struktural (2014, accessed 19 August 2020).

[5] Bhonde D, Nagarnaik DPB, Parbat DDK, et al. Physical and Mechanical Properties of Bamboo (Dendrocalmus Strictus). 2014; 5: 5.

[6] Budianto A, Prajito DH, Budhikarjono K, et al. Rekayasa Katalis Komposit Berbasis Zeolit untuk Proses Cracking Palm Oil menjadi Biofuel : Pembuatan katalis Pd-HZSM-5.

[7] Budianto A, Sumari S, Udyani K. Biofuel production from nyamplung oil using catalytic cracking process with Zn-HZSM-5/γ alumina catalyst. 2015; 10: 10317–10323.

[8] Budianto A, Sumari S, Pambudi WS, et al. Uji Coba Produksi Biofuel dari RBD Stearin dalam Reaktor Fixed Bed dengan Metode Cracking. 2019; 6.

[9] Budianto A, Sumari S, Wahyudi WSP and. Production of Various Chemicals from Nyamplung Oil with Catalytic Cracking Process. *Indian Journal of Science and Technology*, https://indjst.org/ (2018, accessed 16 October 2020).

[10] Budianto A, Prajitno DH, Budhikarjono K. *Biofuel Production from Candlenut Oil Using Catalytic Cracking Process with Zn/Hzsm-5 Catalyst*.

[11] Budianto A, Kimia T. Peningkatan Export Produk Berbahan Kertas Bekas Semen di Surabaya Jawa Timur. *Semin Nas Sains Dan Teknol Terap V* 2017; 2017: E7–E12.

[12] Krötsch S. Bamboo as a Building Material in Kenya. Nairobi, Kenya: School of Architecture and Building Sciences, Jomo Kenyatta University of Agriculture and Technology, 2013, pp. 1–16.

[13] Piazza M, Lobovikov M, Paudel S, et al. World Bamboo Resources - A thematic Study Prepared in The Framework of The Global Forest Resources Assessment 2005. Epub ahead of print 2007. DOI: 10.13140/RG.2.1.1042.3764.

[14] Nurdiah EA. The Potential of Bamboo as Building Material in Organic Shaped Buildings. 2016; 9.

[15] Zuraida S, Larasati D. The Assessment of Marginal Prevention Costs on Bamboo Preservation Method. *Procedia Environ Sci* 2015; 28: 789–798.

[16] Manandhar R, Kim J-H, Kim J-T. Environmental, Social and Economic Sustainability Of Bamboo and Bamboo-Based Construction Materials in Buildings. *J Asian Archit Build Eng* 2019; 18: 49–59.

[17] Bakar ES, Ashaari Z, Jusoh MZ, et al. A New Conversion Method of Bamboo Culm into Flat Sheet for Laminated Bamboo Board Manufacturing. *INTROPica*.

[18] Muhsin A, Febriany LM, Hidayati HN, et al. Material Bambu sebagai Konstruksi pada Great Hall Eco Campus Outward Bound Indonesia. *REKA KARSA*; 3. Epub ahead of print 1 March 2015. DOI: 10.26760/rekakarsa.v3i3.719.

[19] Gutu T. A Study on the Mechanical Strength Properties of Bamboo to Enhance Its Diversification on Its Utilization. 2013; 2: 6.

[20] Janssen JJA. *Designing and Building with Bamboo*. China: The International Network for Bamboo and Rattan (INBAR), 2000.

[21] Flander KD, Rovers R. One laminated bamboo-frame house per hectare per year. *Constr Build Mater* 2009; 23: 210–218.
[22] Komara I, Wahyuni E, Suprobo P, et al. Assessing the Tensile Capacity of Cold-Formed Steel Connections using Self-Drilling Screws and Adhesive Materials. *Int J Adv Sci Eng Inf Technol* 2018; 8: 397.
[23] Komara I, Wahyuni E, Suprobo P, et al. Micro-Structural Characterization of the bond strength capacity of adhesive material in the alternative of cold-formed steel frame system. *IOP Conf Ser Mater Sci Eng* 2019; 462: 012004.
[24] Hidajat A, Fatharani F, Martika D, et al. Kajian Penggunaan Material Kayu dan Bambu Ditinjau Dari Pengawetan dan Perawatan. *REKA KARSA*; 2. Epub ahead of print 2014. DOI: 10.26760/rekakarsa.v2i4.615.
[25] Evans H. *Research Study Exploring Best Practices and Lessons Learned with Small Market Units*. Project Report, Real Estate Institute of British Columbia, 2015.
[26] Suryo MS. Analisa Kebutuhan Luas Minimal Pola Rumah Sederhana Tapak Di Indonesia. *J Permukim* 2017; 12: 116–123.
[27] Ramadhani AN, Faqih M, Hayati A. Behaviour Setting and Spatial Usage Analysis on Sombo Low Cost Flat's Corridor. *J Archit* 2017; 16: 61-74–74.
[28] Alfirevic D, Simonovic-Alfirevic S. Open-plan in housing architecture: Origin, development and design approaches for spatial integration. *Arhit Urban* 2016; 45–60.

The design of the ship's fuel estimation simulator uses a case study of the bung tomo trainer ship

D Wiratno[1], A Mirianto[1], T Cahyadi[1], Z Zuhri[2] and N V Harini[3]

[1]Department of Nautics, Politeknik Pelayaran Surabaya, Surabaya 60294, East Java, Indonesia
[2]Department of Mathematics, Universitas Islam Negeri Sunan Ampel, Surabaya 60237, East Java, Indonesia
[3]Department of Mathematics Education, Universitas Islam Negeri Sunan Ampel, Surabaya 60237, East Java, Indonesia

novitavindri19@gmail.com

Abstract. The calculation of ship operating costs is a major element in determining tariffs. The determination of high or low tariff is very important in the shipping business. In other words, determining the proper ship operating costs is the key to the success of the shipping business. There are several components in determining the operational costs of a ship, one of which is the calculation of fuel needed to cover a certain distance. Usually for new routes, the calculation uses the formula specified by the Minister of Transportation No. KM. 58 of 2003. However, this manual calculation requires a lot of time and human error factors will result in inaccurate calculation results. So, we need a more effective and efficient way to calculate the amount of fuel needed to cover a certain distance. In this study, a simulator was developed that uses the concept of a spherical triangle to determine the distance and direction of a ship and then integrates it with the formula of the Minister of Transportation to get the results of the fuel calculation. From the results of 2 trials using the Bung Tomo Training Ship, the calculation results were faster and more accurate. Even more, the simulator users can simply click two points as the beginning and end of the ship's journey. Thus, the simulator can simplify the process of calculating ship operating costs, especially the determination of new routes in the shipping business.

1. Introduction

The determination of vessel operating costs is a major element in determining tariffs. The determination of high and low tariffs is very important in the shipping business. Engine fuel is one component of ship operating costs. Bung Tomo Training Boat is a 1200GT Special Purpose training ship which requires engine fuel on the cruise route. All costs for the use of engine fuel must be calculated in detail. To determine the right objectives for passengers so that the company gets maximum revenue.

However, the problem that must be faced is the calculation of the amount of engine fuel needed is still using the manual method, namely how to calculate manually with the formula. This method takes a long time and can only be done by some people. Researchers are interested in optimizing the calculation time so that it is faster and the determination of the amount of fuel oil can be done by anyone.

The method to be carried out in this research is to create an application that can quickly calculate the amount of engine fuel needed from one place to another simply by clicking on those two places. Henceforth, this application is called as Simulator. A simulator is a tool for the process of implementing the model into a computer program (software) or electronic circuit and executing the software in such a way that its behavior mimics or resembles a real system [1]. The simulator is designed in the form of software to calculate the estimated operational costs of the ship when the ship will sail [1] [9] [10].

This study aims to describe the process and the results of a good simulator design [2]. A good simulator is the fuel output calculation is the same as the manual method using the formula from the transportation ministry. However, the calculation time with the help of the simulator is faster than the manual method using the formula. Besides, the simulator is said to be good if it is easy to use by anyone.

2. Method

There are three stages in making this simulator. The first stage is the search for ship data (ship type data, ship engine, and ship engine fuel consumption data). In this study, the Bung Tomo training boat was used as a case study. The second step is storing data that has been inputted as a database in my SQL. The third stage is programming the simulator in PHP my Admin. At this stage, the researcher enters the formula from the transportation minister about how to calculate the engine fuel needed to reach a certain distance. The amount of fuel oil consumption can be determined using the following equation:

$$W_{FL} = \frac{(Pbme.bme + Pae.bae)\,S}{V.10^{-6}.Add} \quad (1)$$

$$W_{FE} = (Pae.bme).wp.10^{-6} \quad (2)$$

With:
W_{FL} = Fuel consumption at sea (Kw)
W_{FE} = Fuel consumption at the port (Kw)
$Pbme$= Main Engine Power
Pae = Auxiliary Engine Power
bme = Weight of lubricant in main engine (1.2-1.6gr/Kwh)
bae = Weight of lubricant in auxiliary engine (1.2-1.6gr/Kwh)
S = Distance (Mile)
V = Speed (Knot)
Add = Reserve factor (1.3 – 1.5)
wp = Time at the port (hour)

The map display used in this simulator is google map. The user can utilize the map display to determine the shipping route by clicking on the map. Besides, it can be inputted by the location coordinates for the shipping route.

3. Results and Discussion

Our discussion begins by identifying the diesel engines of ships that consume fuel. Here are the details (see Table 1):

Table 1. Data of Bung Tomo Training Boat

Type of engine	Status	RPM	Fuel consumption (Liter)
Main Engine I (SS)	Full Speed	1200	25/hour
	Idle Speed	700	54/ hour
	Manoeuvre	900	25/ hour
Main Engine II (PS)	Full Speed	1200	25/ hour
	Idle Speed	700	54/ hour
	Manoeuvre	900	25/ hour
Generator SS (STBD)	Idle Speed	720	30/ hour

Type of engine	Status	RPM	Fuel consumption (Liter)
Generator MS (Middle)	Idle Speed	720	30/ hour
Generator PS (port)	Idle Speed	720	30/ hour
Generator Harbour	Idle Speed	600	25/ hour
Main Engine AX	Full Speed	50	15/ hour
Main Engine AX	Full Speed	100	15/ hour

This simulator development uses PHP My Admin as a prototype/display, My SQL as a database, and Google map as a digital map [3][4][5]. The simulator design is made/designed to facilitate the user in operating it. This simulator is a software to calculate ship engine fuel that can be used easily by users [6][7][8]. Figure 1 follows the simulator display design.

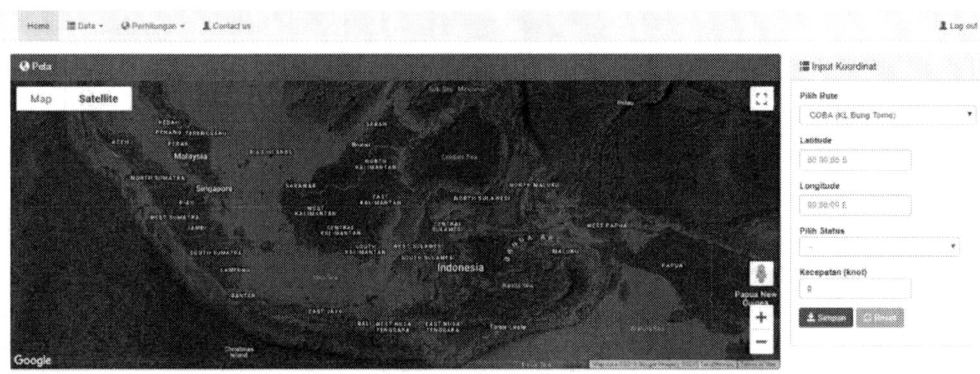

Figure 1. Main Page Map

To be able to determine Bung Tomo Training Boat's engine fuel consumption, firstly we input the ship's data as follows (see Figure 2 and Figure 3):

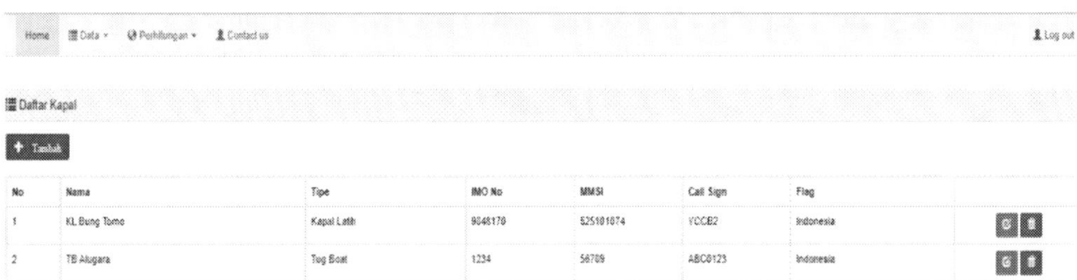

Figure 2. Input Data of Boat

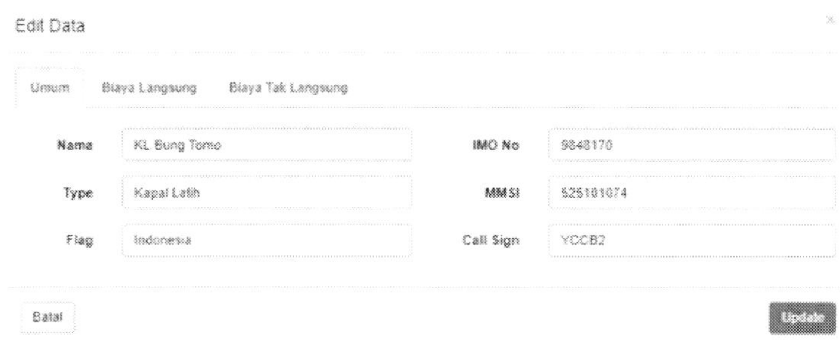

Figure 3. Input Data of Bung Tomo Training Boat's

Then we input Machine Rate Consumption Data (see Figure 4) and Data for Ships Route Plans (see Figure 5) as follows:

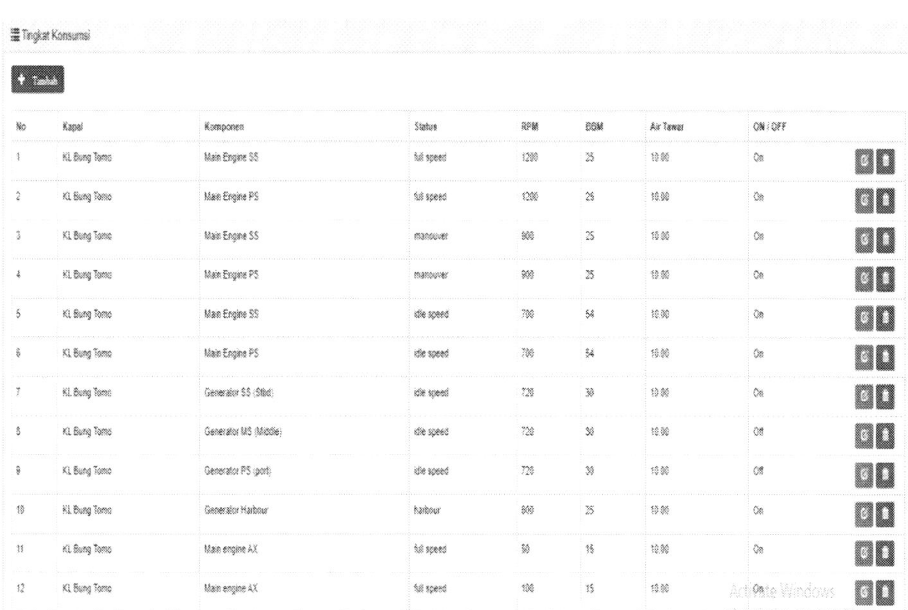

Tingkat Konsumsi

No	Kapal	Komponen	Status	RPM	BBM	Air Tawar	ON / OFF
1	KL Bung Tomo	Main Engine SS	full speed	1200	25	10.00	On
2	KL Bung Tomo	Main Engine PS	full speed	1200	25	10.00	On
3	KL Bung Tomo	Main Engine SS	manouver	900	25	10.00	On
4	KL Bung Tomo	Main Engine PS	manouver	900	25	10.00	On
5	KL Bung Tomo	Main Engine SS	idle speed	700	54	10.00	On
6	KL Bung Tomo	Main Engine PS	idle speed	700	54	10.00	On
7	KL Bung Tomo	Generator SS (Stbd)	idle speed	720	30	10.00	On
8	KL Bung Tomo	Generator MS (Middle)	idle speed	720	30	10.00	Off
9	KL Bung Tomo	Generator PS (port)	idle speed	720	30	10.00	Off
10	KL Bung Tomo	Generator Harbour	harbour	800	25	10.00	On
11	KL Bung Tomo	Main engine AX	full speed	50	15	10.00	On
12	KL Bung Tomo	Main engine AX	full speed	100	15	10.00	On

Figure 4. Input Data of Machine Rate Consumption

No	Nama Kapal	Nama Rute	Keterangan
1	KL Bung Tomo	COBA	SEMINAR HASIL
2	KL Bung Tomo	SUB to BJM	Tes 1
3	KL Bung Tomo	SUB - JKT	SURABAYA TO JAKARTA
4	KL Bung Tomo	SMG to BJM	Uji Coba 2

Figure 5. Input Data of Ships Route Plans

After the data input process is done, then we test the simulator that we have made.

3.1. *First Trial*

The first trial was conducted by shipping routes from the port of Surabaya to the port of Makassar. We can see in Figure 6.

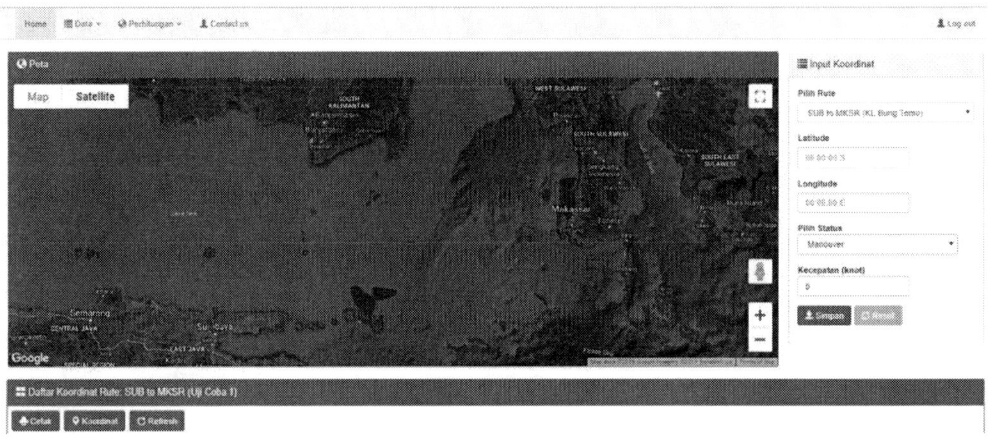

Figure 6. Route of the First Trial (Surabaya-Makassar)

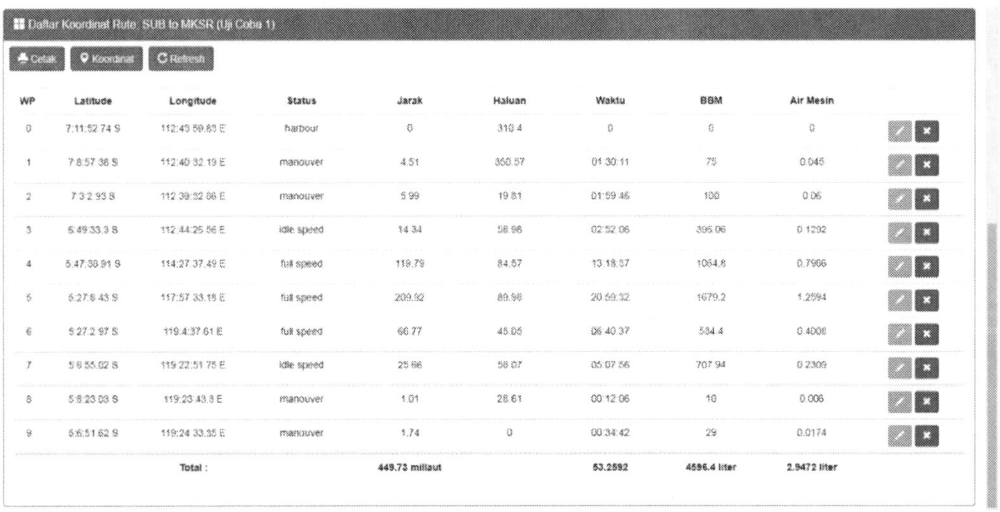

Daftar Koordinat Rute: SUB to MKSR (Uji Coba 1)

WP	Latitude	Longitude	Status	Jarak	Haluan	Waktu	BBM	Air Mesin
0	7:11:52.74 S	112:43:59.83 E	harbour	0	310.4	0	0	0
1	7:8:57.36 S	112:40:32.19 E	manouver	4.51	350.57	01:30:11	75	0.045
2	7:3:2.93 S	112:38:32.86 E	manouver	5.99	19.81	01:59:46	100	0.06
3	6:49:33.3 S	112:44:25.56 E	idle speed	14.34	58.96	02:52:06	395.06	0.1292
4	5:47:38.91 S	114:27:37.49 E	full speed	119.79	84.57	13:18:37	1064.8	0.7986
5	5:27:6.43 S	117:57:33.18 E	full speed	209.92	89.98	20:59:32	1679.2	1.2594
6	5:27:2.97 S	119:4:37.61 E	full speed	66.77	45.05	06:40:37	534.4	0.4008
7	5:6:55.02 S	119:22:51.75 E	idle speed	25.66	58.07	05:07:56	707.94	0.2309
8	5:8:23.03 S	119:23:43.3 E	manouver	1.01	28.61	00:12:06	10	0.006
9	5:6:51.62 S	119:24:33.35 E	manouver	1.74	0	00:34:42	29	0.0174
		Total :		449.73 millaut		53.2592	4596.4 liter	2.9472 liter

Figure 7. Consumption List of Fuel

Based on the first trial data (see Figure 7) the trip from Surabaya to Makassar obtained a distance of 449.73 miles with a travel time of 53 hours 25 minutes, thus requiring 4596.4 liters of fuel on the engine.

3.2. *Second Trial*

The second trial was conducted by shipping routes from the port of Semarang to the port of Banjarmasin. We can see in Figure 8.

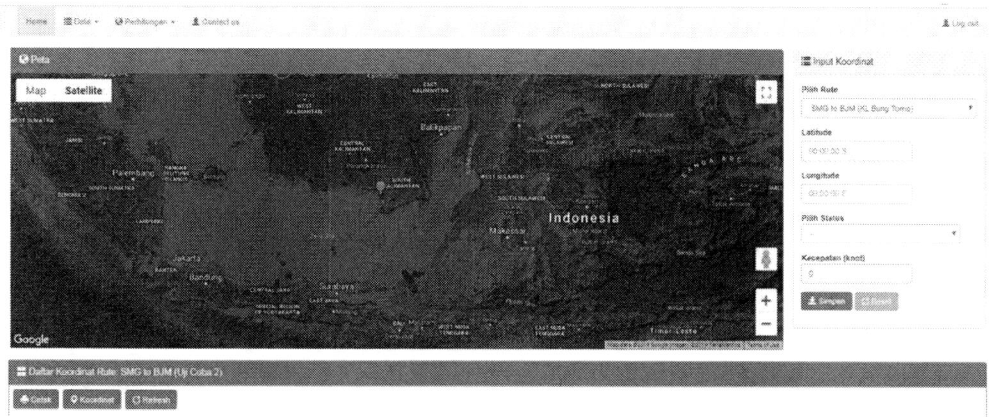

Figure 8. Route of the Second Trial (Semarang-Banjarmasin)

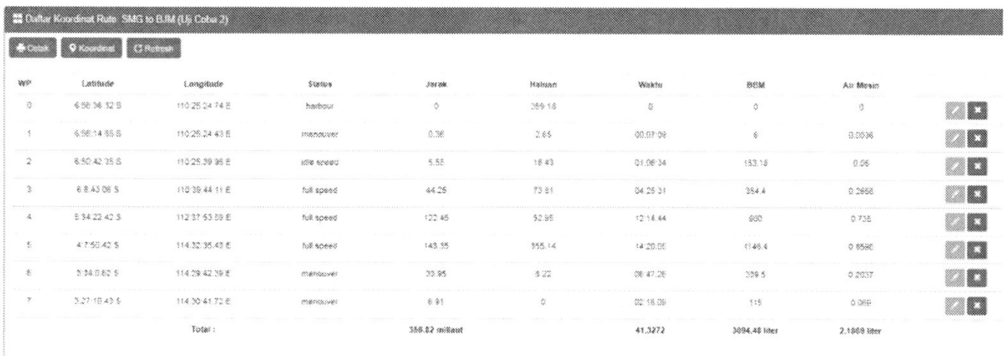

Daftar Koordinat Rute: SMG to BJM (Uji Coba 2)

WP	Latitude	Longitude	Status	Jarak	Haluan	Waktu	BBM	Air Mesin
0	6:58:36.32 S	110:25:24.74 E	harbour	0	359.18	0	0	0
1	6:58:14.85 S	110:25:24.43 E	manouver	0.38	2.65	00:07:09	6	0.0036
2	6:50:42.35 S	110:25:39.96 E	idle speed	5.55	18.43	01:08:34	153.18	0.06
3	6:8:43.06 S	110:39:44.11 E	full speed	44.25	73.81	04:25:31	354.4	0.2658
4	5:34:22.42 S	112:37:53.59 E	full speed	122.45	52.95	12:14:44	980	0.735
5	4:7:50.42 S	114:32:35.43 E	full speed	143.35	355.14	14:20:05	1146.4	0.8598
6	3:34:0.82 S	114:29:42.39 E	manouver	33.95	8.22	08:47:26	339.5	0.2037
7	3:27:18.43 S	114:30:41.72 E	manouver	6.91	0	02:18:09	115	0.069
		Total :		356.82 mililaut		41.3272	3094.48 liter	2.1869 liter

Figure 9. Consumption List of Fuel

Based on the second trial data (see figure 9) the trip from Semarang to Banjarmasin obtained a distance of 356.82 miles with a travel time of 41 hours 32 minutes, thus requiring 3094.48 liters of fuel on the engine.

4. Conclusion

Based on the description and analysis of the results and discussion, it appears that the Simulator can display the ship's distance and direction of the ship so that the Simulator can display fuel consumption correctly. The time needed for the calculation is not as long as the manual calculation using the formula from the transportation ministry, because the results can be directly obtained after the data input process is carried out. This simulator can be used by anyone with a guarantee that the calculation of the amount of fuel used will be the same as a manual calculation using a formula.

5. References

[1] Bambang Sridadi. *Pemodelan dan Simulasi Sistem*. 1st Edition. Bandung: Informatika Bandung, 2009.

[2] S. M. Zinchenko, P. S. Nosov, V. M. Mateichuk, et al. USE OF NAVIGATION SIMULATORS FOR DEVELOPMENT AND TESTING SHIP CONTROL SYSTEMS. *Kluwer Academic Publisher* 2019; 350–355.

[3] Xiaoxia W, Chaohua G. Electronic chart display and information system. *Geo-spat Inf Sci* 2002; 5: 7–11.

[4] Hecht H. The Requirements of Precise Navigation for the Electronic Chart Display and Information System (ECDIS). 1997, pp. 728–735.
[5] Harris J. Handbook of mathematics and computational science, https://www.academia.edu/4578430/Handbook_of_mathematics_and_computational_science (accessed 27 October 2020).
[6] Heinich R, Molenda M, Russell JD, et al. *Instructional Media and Technologies for Learning*. 7th Edition. Upper Saddle River, N.J: Prentice Hall, 2001.
[7] Anderson RH. *Pemilihan dan pengembangan media untuk pembelajaran*. Rajawali, Jakarta, 1987.
[8] Schwier R, Misanchuk ER. *Interactive multimedia instruction*. Englewood Cliffs, N.J.: Educational Technology Publications, 1993.
[9] Solas. (2010). *The International Convention on* Safety *Of Life at Sea, The Fundamental IMO-convention on maritime safety*: London.
[10] International Maritime Organization. 2010. *Standart* Of *Training Certification And Watchkeeping for Seafarers (STCW)* 1978 Amandemen Manila. London.

IOP Conf. Series: Materials Science and Engineering 1010 (2021) 012028 doi:10.1088/1757-899X/1010/1/012028

Identification of automatic guided vehicle (agv) based on magnetic guided sensor for industrial material transfer

Y A Prabowo[1], R I Imaduddin[2], W S Pambudi[1], R A Firmansyah[1], A Fahruzi[1]

[1]Institut Teknologi Adhi Tama Surabaya, East Java, Indonesia.
[2]Nurul Jadid University, East Java, Indonesia.

Email: agungp@itats.ac.id

Abstract. Automatic Guided Vehicle (AGV) is a material transfer tool that is now widely used in the industry. One of the advantages is that it can transfer material automatically without the need for an operator who has to drive it. AGV using a magnetic guided sensor is one of the most widely used types. So that the tool can run according to the planned trajectory, it is necessary to design a controller that matches its characteristics. The design of the controller is needed identification of the model of the plant to know the characteristics of the plant. Identification with the Extended Least Square method is a form of parameter estimation recursively through a quadratic error gradient formulation with a criterion function involving all errors. The structural approach model used linearly uses Auto-Regressive eXogenous (ARX) and Auto-Regressive Moving Average eXogenous (ARMAX). The validation process of the model uses the Means Square Error (MSE) criteria and the FIT percentage. Based on the results obtained by the best plant with ARMAX model structure with MSE values 82.14 and percentage of FIT value 89.98%.

1. Introduction

Automatic Guided Vehicle (AGV) is a device that has experienced rapid development in recent years. AGV is widely used for material transfer in industries such as automotive, medicine, cigarettes, etc. In the process of transferring materials usually use vehicles operated by humans such as the Forklip. The use of AGV allows it to transfer material automatically without using human operators. So that it can improve efficiency in production and can reduce errors by humans in carrying out work that has a high degree of saturation [1–3] The latest technology allows AGV movements to be monitored online, as well as to calculate the amount of material that has been moved.

The mechanism of guidance on AGV to work automatically varies including RFID, Magnetic Sensor, and Laser [4, 5]. The mechanism most often used is using a magnetic sensor. One of the advantages of the track installation process using magnetic tape is that it is easier to apply and has a low cost [6, 7]. There are several types of AGV models among the most frequently used for material transfer applications, i.e. the tow or tugger type [4]. The drive and steering control mechanism for movement of the AGV there are several types including the skid drive, which consists of two driving motors as well as steering control.

For AGV to run well following the planned track, reliable control strategies are needed. Before designing the controller, the identification of the AGV plant is needed. The identification process aims to find out the characteristics of the plant. In case know the characteristics of the plant, it will easily determine the strategy and control method that will be used [8, 9]. In AGV with the skid drive mechanism is a type of Multiple Input Single Output (MISO) system. Therefore, a structural model approach to MISO is needed to obtain a system model that is close to the nature of the actual characteristics.

The structural approach models of ARX (Auto-Regressive eXogenous) and ARMAX (Auto-Regressive Moving Average eXogenous) can be used for modelling MISO systems [10]. The parameter estimation method can be used using the Extended Least Square. This method is also often used for Single Input Single Output (SISO) systems [11]. As a validation of the identification process, the Mean Square Error (MSE) and the best FIT criteria can be used. The MSE criterion allows the value of the estimated error obtained to be as small as possible. While the best FIT criteria allow a model that can resemble the real plant [12].

In this study, the plant identification process will be carried out on the AGV that has the characteristics of the MISO system using the extended least square method. The structure of the approaching model that is designed using ARX and ARMAX will be compared. As a validation of the model that has been obtained, MSE and FIT criteria are used to determine the best model obtained.

2. Automatic Guided Vehicle (AGV)

AGV has been very widely used in the industrial world, especially to carry out the task of transferring materials in warehouses, assemblies, etc. This use cannot be separated from the development of mobile robot technology which is the initial idea of making this tool [1] .Guiding methods to be able to walk according to the track are also diverse. In this study, the method used to guide using magnetic sensors. This selection is because it is easier for the track installation process simply by using magnetic tape which tends not to disturb the density of the industrial environment [6]. Figure 1 shows the AGV design used.

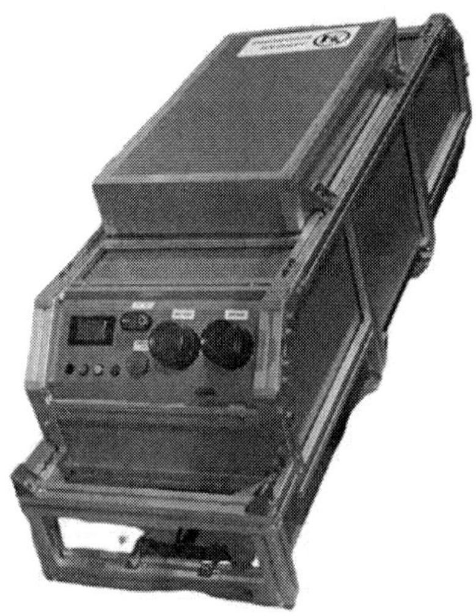

Figure 1. Automatic Guided Vehicle

The AGV of the drive and steering control mechanism to get through the planned trajectory is the most important thing. The control depends on the type and mechanism used. In the design that is made using a type of tow or tugger relative to the medium load class maximum of 200 kg. For the drive and steering mechanism that is used to use the skid drive. This mechanism allows two motors are used as a drive as well as for steering. So that this mechanism is included in the Multiple Input Single Output (MISO) system. Where there are two inputs are control signals from the Brushless motor driver and one output is the position of the AGV deviation.

3. Identification

3.1. Extended Least Square

The plant parameter identification method is performed by measuring the input and output of data online and offline. The input signal measurement data u (k) is obtained by dynamically giving the signal to the system. Then the results of the response from the system in the form of output data y(k) will be recorded along with the input signal. In the identification process carried out in this study using data offline. The approach taken considers the plant is a black box with no known parameters. Plant modeling is used in the form of a stochastic system model with a linear structured model approach. The identification method used is using the Extended Least Square. This method is a recursive method through the quadratic error gradient formulation with the criteria function involving all errors. The basic model of a system is represented in equation (1) below

$$\boldsymbol{y(k) = \varphi^T(k-1)\theta} \tag{1}$$

where,

$$\boldsymbol{\varphi^T(k-1) = \begin{bmatrix} -y(k-1) \\ -y(k-1) \\ u(k-1) \\ u(k-2) \end{bmatrix} \quad \theta = \begin{bmatrix} a_1 \\ a_2 \\ b_0 \\ b_1 \end{bmatrix}} \tag{2}$$

where, $\boldsymbol{\varphi^T(k-1)}$ is measurement of input-output data and θ is parameters plant

The plant parameter values will continue to be calculated until the error meets the minimum criteria as shown in equation (3). The estimation is then performed to obtain the parameter value that satisfies equation (4).

$$\boldsymbol{J_{min} = \frac{1}{2}e^2} \tag{3}$$

$$\boldsymbol{\widehat{\theta}(k) = \theta(k-1) + F(k)\{y(k) - \varphi^T\theta(k-1)\}\varphi(k-1)} \tag{4}$$

3.2. Structure Model of MISO System

MISO system is a system that has more than one input and only one output. In this AGV system, the input consists of two pieces, and the input consists of one input. The system structure model in the discrete domain can be represented in equation (5) and can be shown in Figure 2.

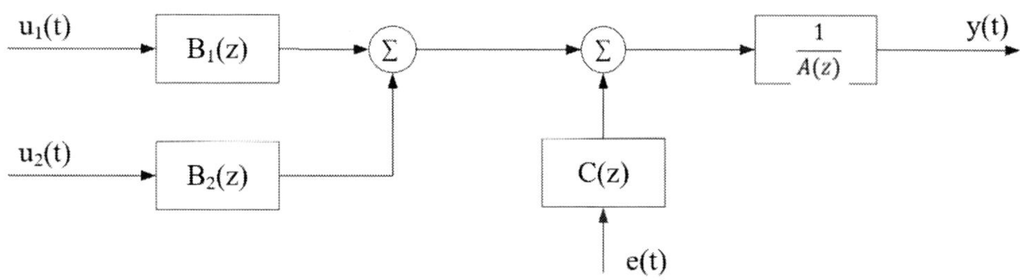

Figure 2. Structure MISO System

$$A(z)y(t) = B_1(z)u_1(t) + B_2(z)u_2(t) + C(z)e(t) \tag{5}$$

The structural model that can be used in this system can be selected by the ARX structure model approach. The structure of this model is a form of stochastic system which considers external factors such as noise or interference in the input and output bookkeeping process [8]. The equation of the structure of this model is represented in equation (6) below

$$A(z)y(t) = B_1(z)u_1(t) + B_2(z)u_2(t) + C(z)e(t) \tag{6}$$

In addition to the structure of this model, there is also a model structure. The most fundamental difference with the ARX structure is that there is a dynamic disturbance found in the system [10]. The equations of the structure of this model are represented in equation (7) below

$$A(z)y(t) = B_1(z)u_1(t) + B_2(z)u_2(t) + C(z)e(t) \tag{7}$$

3.3. *Validation*

In order to know the equation of the model that is close to the real condition with the real plant, each model that has been obtained needs to be carried out a validation process. Validations that are often used in the modeling process include the MSE (Mean Square Error) criteria and the percentage of fit [12]. MSE is a criterion that shows how much the error deviation is from zero. Mathematically shown in equation (8).

$$\mathbf{MSE} = \sqrt{\frac{\sum_{i}^{n}(y_i - \hat{y}_i)^2}{n}} \tag{8}$$

where, $\mathbf{y_i}$ is output measurement data, $\hat{\mathbf{y}}_\mathbf{i}$ is output estimation data and n is number of data.

The percentage of fit criteria is a criterion that shows the compatibility of the system of identification results with the actual plant. Mathematically shown in equation (9).

$$\mathbf{Best\ Fit} = 100\left[1 - \frac{\mathrm{norm}(\hat{y}_i - y_i)}{\mathrm{norm}(y_i - \bar{y})}\right]\% \tag{9}$$

where, $\mathbf{y_i}$ is output measurement data, $\hat{\mathbf{y}}_\mathbf{i}$ is output estimation data and $\bar{\mathbf{y}}$ is mean value data.

4. Result and Discussion

The AGV identification process is carried out using three input and output measurement data with varying amounts of data. Input or reference signals are given in the form of analog brushless motor driver analog signals totaling 2 channels. The form of the input and output signals as shown in Figure 3.The output signal is in the form of a deviation from the AGV movement to the magnetic tape where the deviation is measured by a magnetic sensor. For the sampling time used in the process of taking both data every 0.01 second.

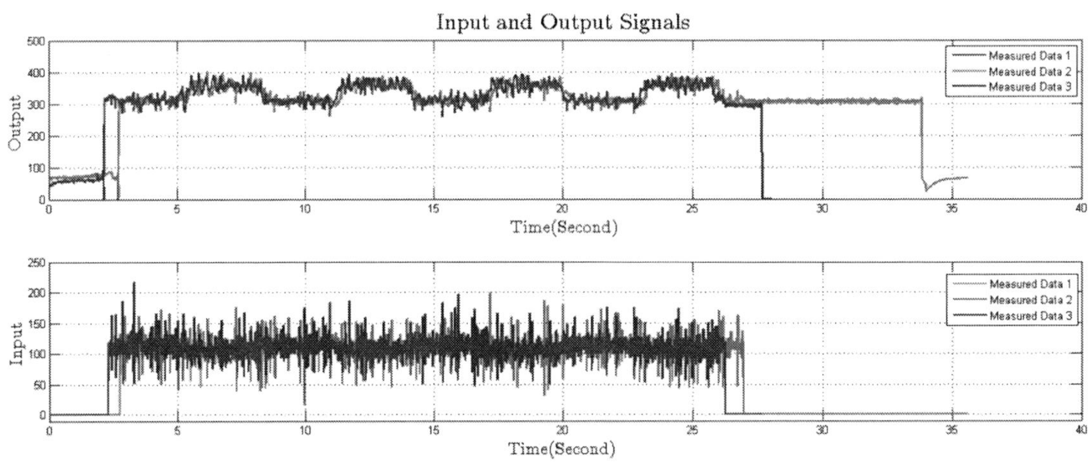

IOP Conf. Series: Materials Science and Engineering 1010 (2021) 012028 doi:10.1088/1757-899X/1010/1/012028

Figure 3. Input-Output Measurement Data

The desired model structure in this identification will be selected with the ARX and ARMAX models. In the ARX structure, the desired system model has the number of pole 4, the number of zero 4, and the dead time 1. Whereas for the ARMAX structure model the desired system model has the number of pole 4, the number of zero 3, and dead time 1. The process of identifying the plant model is done offline using measurement data input and output through the process of estimating the parameters of the Extended Least Square method. After the plant model is obtained, a validation process will be carried out using the Means Square Error (MSE) criteria and fit ratio presentation (FIT). The plant model to be chosen must have the smallest MSE value. The FIT percentage criteria must have the largest value close to 100 %.

In the identification process using the first variation of data with the number of data 2768. Based on the identification process, the ARX structure the MSE value is 81.39 and the percentage value of FIT is 88.41 %. For the ARMAX structure, the MSE value is 81.42 and the percentage value of FIT is 88.41 %. Then, using the second data variation with the number of data 2560. Based on the identification process, the ARX structure the MSE value is 83.19 and the percentage value of FIT is 89.92 %. For the ARMAX structure, the MSE value is 82.14 and the FIT percentage value is 89.98 %. Then, identification process using the third variation of data with the number of data 2805. Based on the identification process, the ARX structure the MSE value is 83.19 and the percentage value of FIT is 89.92 %. For the ARMAX structure, the MSE value is 82.14 and the FIT percentage value is 89.98 %. Figure 4 shows the results of the model estimation for all measurement data. The overall results of the identification process using variations in data changes and the structural model approach can be presented more fully in Table 1.

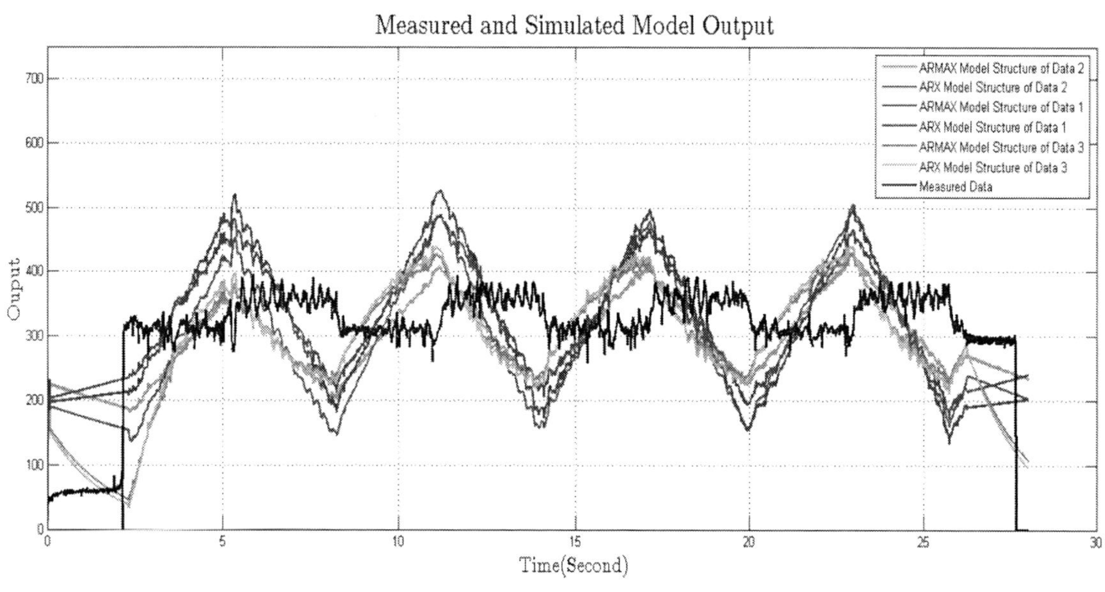

Figure 4. Result of Model Estimation

Table 1. Result of Identification

Data	Structure	Model (Discrit)	MSE	FIT
1	ARX	$A(z) = 1 - 1.063\,z^{-1} - 0.874\,z^{-2} + 0.7864\,z^{-3} + 0.5684\,z^{-4}$ $B_1(z) = 0.2828\,z^{-1} + 0.6702\,z^{-2} + 0.2171\,z^{-3} + 0.04413\,z^{-4}$ $B_2(z) = -0.09782\,z^{-1} - 0.09849\,z^{-2} + 0.00356\,z^{-3} + 0.04413\,z^{-4}$	101.2	73.97 %
	ARMAX	$A(z) = 1 - 1.031\,z^{-1} - 0.7249\,z^{-2} + 0.7626\,z^{-3} + 0.4436\,z^{-4}$ $B_1(z) = 0.6926\,z^{-1} + 0.8358\,z^{-2} + 0.3849\,z^{-3}$ $B_2(z) = -0.2217\,z^{-1} - 0.3289\,z^{-2} - 0.1013\,z^{-3}$ $C(z) = 1 + 1.152\,z^{-1} + 0.5499\,z^{-2} + 0.2711\,z^{-3}$	73.53	77.81 %
2	ARX	$A(z) = 1 - 0.9727\,z^{-1} - 0.1838\,z^{-2} + 0.1225\,z^{-3} + 0.03377\,z^{-4}$ $B_1(z) = 0.09846\,z^{-1} + 0.09513\,z^{-2} - 0.07827\,z^{-3} + 0.0007531\,z^{-4}$ $B_2(z) = -0.1018\,z^{-1} - 0.008699z^{-2} + 0.03157\,z^{-3} + 0.03328\,z^{-4}$	81.39	88.41 %
	ARMAX	$A(z) = 1 - 1.613\,z^{-1} + 0.8417\,z^{-2} - 0.1637\,z^{-3} - 0.06511\,z^{-4}$ $B_1(z) = 0.1131\,z^{-1} + 0.004832\,z^{-2} - 0.04377\,z^{-3}$ $B_2(z) = -0.09184\,z^{-1} + 0.007001\,z^{-2} + 0.01459\,z^{-3}$ $C(z) = 1 - 0.6346\,z^{-1} + 0.4406\,z^{-2} - 0.07688\,z^{-3}$	81.42	88.41 %
3	ARX	$A(z) = 1 - 0.7164\,z^{-1} - 0.407\,z^{-2} - 0.05785\,z^{-3} + 0.1822\,z^{-4}$ $B_1(z) = 0.002861\,z^{-1} + 0.0876\,z^{-2} + 0.04482\,z^{-3} + 0.0001309\,z^{-4}$ $B_2(z) = -0.07994\,z^{-1} - 0.02877\,z^{-2} - 0.03347\,z^{-3} + 0.01933\,z^{-4}$	83.19	89.92 %
	ARMAX	$A(z) = 1 + 0.4929\,z^{-1} - 0.5375\,z^{-2} + 0.977\,z^{-3} + 0.02437\,z^{-4}$ $B_1(z) = 0.06146\,z^{-1} + 0.1127\,z^{-2} + 0.09943\,z^{-3}$ $B_2(z) = -0.07767\,z^{-1} - 0.1098\,z^{-2} - 0.0583\,z^{-3}$ $C(z) = 1 + 1.314\,z^{-1} + 0.7272\,z^{-2} - 0.13761\,z^{-3}$	82.14	89.98 %

5. Conclusions

Based on the identification process that has been carried out through the parameter estimation and validation stages, the system that approaches the plant has the smallest MSE and the largest percentage of fit value. So that the structural model derived from the third data with the ARMAX structural approach with the number of pole 4 and zero 3 is the best plant model. Where has a validation value of MSE 82.14 and FIT 89.98 %.

6. References

[1] Zhou X, Chen T, Zhang Y. Research on Intelligent AGV Control System. In: *2018 Chinese Automation Congress (CAC)*. 2018, pp. 58–61.

[2] Firmansyah RA, Pambudi WS, Suheta T, et al. Implementation of Artificial Neural Networks for Localization System on Rescue Robot. In: *2018 Electrical Power, Electronics, Communications, Controls and Informatics Seminar (EECCIS)*. 2018, pp. 305–309.

[3] Draganjac I, Miklic D, Kovacic Z, et al. Decentralized Control of Multi-AGV Systems in Autonomous Warehousing Applications. *IEEE Trans Automat Sci Eng* 2016; 13: 1433–1447.

[4] Yu R, Zhao H, Zhen S, et al. A novel trajectory tracking control of AGV based on Udwadia-Kalaba approach. *IEEE/CAA J Autom Sinica* 2017; 1–13.

[5] Su S, Dai H, Cheng S, et al. Improved Magnetic Guidance Approach for Automated Guided Vehicles by Error Analysis and Prior Knowledge. *IEEE Transactions on Intelligent Transportation Systems* 2020; 1–10.

[6] Liu F, Li X, Wang Y. Design of Automatic Guided Vehicle Motion Control System Based on Magnetic Navigation. In: *2018 Chinese Control and Decision Conference (CCDC)*. 2018, pp. 4775–4779.

[7] Pambudi WS, Zuliari EA, Firmansyah RA, et al. Identification of Flip Folder Model on Folding Machine. *IOP Conf Ser: Mater Sci Eng* 2019; 462: 012009.

[8] Prabowo YA, Pambudi WS, Imaduddin IR. Identification of the Flip Folder Folding Machine Using Artificial Neural Network with Nonlinear Autoregressive Exogenous Structure. 2020; 5: 6.

[9] Ruslan FA, Zain ZM, Adnan R, et al. Flood water level prediction and tracking using particle filter algorithm. In: *2012 IEEE 8th International Colloquium on Signal Processing and its Applications*. Malacca, Malaysia: IEEE, pp. 431–435.

[10] Bedoui S, Abderrahim K. ARMAX Time Delay Systems Identification Based on Least Square Approach. *IFAC-PapersOnLine* 2015; 48: 1100–1105.

[11] Muroi H, Adachi S. Model Validation Criteria for System Identification in Time Domain. *IFAC-PapersOnLine* 2015; 48: 86–91.

[12] Maarif E, Moyo T. Driving control module for low cost industrial automated guided vehicle. *IOP Conf Ser: Mater Sci Eng* 2019; 535: 012016.

The list order of construction risk contract for small-scale construction service in surabaya, indonesia using analytical hierarchy process (ahp)

Mohamad F. N. Aulady[1], Felicia T. Nuciferani[1,*], D Listyaningsih[1] , Y P Putra[1] , and Feri Harianto[1]

[1] Departement of Civil Engineering, Institut Teknologi Adhi Tama Surabaya,Indonesia

* Felicia@itats.ac.id

Abstract. Small-scale construction service providers dominate the construction industry in Indonesia. However, along with Indonesia's economic growth, small-scale construction service providers owner has a chance to expanse their business. Based on their experience, small-scale construction service providers have limitations. They cannot take a priority, whereas the risky of the contract construction. This study aims to make a list order of the construction risk based on medium-scale construction service providers. With this study, we want to help share the medium-scale experience to small-scale construction service providers in terms of risk contracts in Indonesia to learn and anticipate it. We conduct risk identification by literature and make list order analyzing using Analytical Hierarchy Process (AHP). AHP method can provide a single and easy-to-understand of unstructured problems. This research indicates that mismatch between the volume of work in the Bill of Quantity and field conditions is the most risk in terms of contract construction. However, the lump sum contract also identified the risky contract based on medium-scale construction service providers.

1. Introduction

There are 109,964 construction service providers in Indonesia, and 90.66% are small-scale construction service providers [1]. The Construction industry in Indonesia plays a crucial role in Indonesia's economic growth [2]. Indonesia's construction industry accounts for 10.5% of the value of Indonesia's Gross Domestic Product and grows 6.2% every year [3]. With the support of this profitable economic growth, small-scale construction service providers should be able to take advantage of this to develop into a broader provider of construction services. A study states that small-scale construction service providers want to be bigger, so small-scale construction service providers must improve their knowledge and skills for the bidding, field management, and business management processes [1]. Therefore, construction contracts as a reference to make a construction that is a part of the bidding and business management process that must be quasi-managed well by small-scale construction service providers [4]. Not only that, but the number of construction disputes that must be resolved by using construction contracts as a reference also makes the construction contract more critical to understand [4]. To understand the contract thoroughly, the first thing to do is know the risks of existing construction contracts. Knowing the risk means that we can know, analyze, and control the risk in each of the company's activities to obtain higher effectiveness and efficiency [5]. However, after knowing the contract risk, Small-scale construction service providers still have to decide their priority because there are some contract risk factors. In this case, Small-scale construction service providers lack experience.

However, there a few studies related to contracting risk priority in Indonesia. Adha (2011) researched the building operate transfer (BOT) contract scheme between the government and the private sector [6]. The study results showed that in a contractual relationship, the government as a party to a BOT contract does not have the same position as another party. Whereas in 2013, Lestari examined the comparison of construction contracts in Indonesia with International Construction Contracts. The results showed that the format used both in construction contracts in Indonesia and international construction contracts is comparable. The difference is in the difference between several terms [7]. Therefore, this study aimed to determine the risk order in construction service contracts based on mid-scale service providers' experience in Surabaya by using the Analytical Hierarchy Process (AHP) method. This research intends to cover the gap between Small-scale construction experience to categorized risk contracts in Indonesia. As a result that the Small-scale construction can focus on anticipating the risk of construction contracts in Indonesia. Furthermore, we also analyze which construction contract tends to be a risk in the Indonesian construction industry.

1.1. Literature Review

The Analytic Hierarchy Process (AHP) method is a measurement theory. AHP is used to derive the ratio scale of several paired comparisons that are discrete or continuous. Pair-wise comparisons are obtained through actual measurements as well as relative measurements of the degree of importance; this is a useful method for obtaining a ratio scale of things that were previously difficult to measure, such as opinions, feelings, behaviors, and beliefs that begin with the creation of a hierarchical or network structure of the problem to be studied [8].

The use of the AHP method in this study refers to the advantages of the AHP method, which provides a single, easy-to-understand, flexible model for a wide range of unstructured problems [8]. Also, AHP uses a deductive approach and system in solving complex problems [9]. This method has also been used in various scientific fields such as supplier selection [10], risk assessment on the Toll Road Project [11], resource placement issues, and others [12]. The AHP method used begins by distributing questionnaires to several respondents. Respondents used in this study were medium-scale construction service providers in Surabaya. Surabaya was chosen because Surabaya is the second-largest city in Indonesia, and many construction projects are being carried out here. In 2017 there were 22 shopping centers completed [13].

2. Methods

The research began by distributing questionnaires to medium-scale construction service providers in the city of Surabaya. There were 5 of mid-scale construction service providers interviewed to obtain priority scale. The respondent for each construction service provider is wonted to handle the procurement document. The respondent has a Minimum of two years' experience and thirty years of a full experience. However, all respondents are graduate from undergraduate school. Subsequently, we collect the construction contract's risk factor to generate the questionnaire form—the risk factors used as a basis for some research that focuses on risk identification in Indonesia. There thirteen factors that are related to the current situation, as shown in Table 1.

Table 1. Risk identification of risk contract in Indonesia based on Suanda (2008) [14]

No.	Code	Risk Identification
1	K11	A mismatch between the volume of work in the Bill of Quantity and field conditions
2	K4	There are different contract between both party
3	K7	Obstruction of the process of handing overwork.
4	K10	The owner of the construction project does not pay the penalty of delay payment
5	K12	Termination by service owners
6	K1	Claim proses are complicated or unclear
7	K5	The construction implementation method is not comparable with the contract purpose
9	K13	There is a delay in signing the contract
9	K6	There is unlimited penalty fee on the contract
10	K2	The unbalanced contract between both party
11	K8	Claims of price adjustments because of changes in government regulations are rejected.
12	K3	Time target on the contract are not reliable
13	K9	The owner of the project cannot afford to finance the project.

There are 4 AHP principles, namely Decomposition, Comparative Judgment, Synthesis of Priority, and Logical Consistency. The principle of decomposition is the principle of solving a whole problem into a hierarchical form, a decision-making process where each element or element is interconnected. The decision hierarchy structure can be categorized as complete and incomplete. The form of the decomposition structure, namely, the first level is the decision goal (Goal), the second level is the criteria, and the third level is the alternatives [8]. Furthermore, the factors and types of contracts are arranged into a hierarchical form with three levels consisting of:

1. The first level, which is the objective in the form of a risk level construction contract,
2. The second level is risk indication with code K1 - K13 (Table 1), and
3. The last level is an alternative in the form of lump-sum contract, lump-sum - fixed price (joint contract), and fixed-price contracts.

The questionnaire is interview form based. With the result, the expert can easily understand the survey. Afterward, we were processing the questionnaire based on the AHP calculation [9]. Questionnaire data processing by tabulating the respondents' answers and looking for the mean value for each element being compared. Then pair-wise comparisons were performed for each level element. To see the consistency of the respondents, a consistency analysis using

$$CI = \frac{\lambda max - n}{n - 1} \tag{1}$$

Where:
CI = Consistency ratio
λmax = Eigenvalue
N = Matrix ordo

If the CI value was zero, then the pair-wise comparison matrix was consistent. The limit to the inconsistency that has been set by Saaty [15] is determined by using the Consistency Ratio (CR), which is a comparison of the consistency index with a random index value (RI).

$$CR = \frac{CI}{RI} \tag{2}$$

Where:
CR =Consistency ratio
R I =Random index

According to the Saaty [15], the random value for this case is 13.56. However, If the pair-wise comparison matrix with a CR value is less than 0, 100, then the respondent's inconsistency in making decisions is still acceptable, and if not, then the assessment needs to be repeated.

3. Result and Discussion

3.1. Result

Based on 13 indications of risk factors carried out on small projects in Surabaya, the results are in table 2.

Table 2 The Result Risk Contract List Order Using AHP

No.	Code	Weight	Rank	No.	Code	Weight	Rank
1	K11	0.217	1	8	K13	0.053	8
2	K4	0.159	2	9	K6	0.047	9
3	K7	0.107	3	10	K2	0.045	10
4	K10	0.079	4	11	K8	0.042	11
5	K12	0.064	5	12	K3	0.040	12
6	K1	0.057	6	13	K9	0.034	13
7	K5	0.056	7				

Based on 13 indications of risk factors carried out on small projects in Surabaya, the results are in table 2.

Table 2 shows the result of the list order thirteen risk identification based on the AHP method. The result indicates that the medium-scale construction service providers are concerned about their mismatch between the volume of work in the Bill of Quantity and field conditions, and both parties have a different contract. In the third place is the risk of obstruction of the process of handing overwork. However, the lowest risk for the medium-scale construction service providers is if the project owner cannot afford to finance the project. The weight of risk rank number one and number two reaches 0.058 points, and the gap between number two and three are 0.052 points. At this time, each factor's big gap indicates that each respondent may have a different opinion due to their experience and conditions. However, from risk rank number six until twelve, the result indicates that the weight is mostly similar. This value is also indicated in this point; the respondent has a little different perception. In other words, each respondent has some opinion that the risk is can categorized as medium-low risk level. Moreover, in Table 3, the risky contract in the medium-scale construction service providers view is lump sum type, following by combining lump sum and fixed price. However, the fixed price is believed as a low-risk contract. The gap point between each contract type is typically significant. The gap between lump sum and contract number two reaches 0.39 points or more than a half weight point. Simultaneously, the gap between contract Lump sum and fixed price and the fixed price has the same gap range difference. This difference indicates that each respondent has a similar argument.

Table 3. The List Order of Riskiest Contract Using AHP

No.	Contract Type	Weight	Rank
1	Lump-sum	0.64	1
2	Lump-sum and fixed price	0.25	2
3	Fixed Price	0.11	3

3.2. Discussion

Analytical Hierarchy Process (AHP) is a method of handling complex and unstructured problems arranged in the form of a hierarchy [8], with a rating scale of 1 to 9. AHP is used to determine the priority of several criteria using pair-wise comparative analysis of each criterion up to calculating the Consistency Index (CI) and Consistency Ratio (CR). In order to make decisions, it is essential to know how right the consistency is with the consistency criterion of 10%, then the calculation results are declared correct, and it can be interpreted that the respondent is consistent

A mismatch between the contract volume with the conditions on the ground is a risk that is considered the riskiest by medium-scale construction service providers in Surabaya. This is in line with some of Diekman's studies that state discrepancies in field conditions accounted for 15% of total dispute resolution cases. Also, 6 of the 13 existing risks lead to contract dispute resolution. Therefore, small-scale construction service providers must be prepared with preventive measures if there is a mismatch between the volume of work in the field and the construction contract or other risks that lead to a construction contract dispute settlement. Poerdyatmono concluded that if there is a dispute in the construction industry, then five things can be done. Namely through the channels of consultation, negotiation, mediation, conciliation, a legal opinion by the arbitration agency, or a combination of the five channels by the level of need [16]. In the field implementation, the settlement of construction service disputes is often carried out in consultation, negotiation, mediation, and conciliation, while institutional arbitration and through the court are avoided as far as possible. Therefore, in anticipating disputes, small scale construction service providers must have good mediation skills. This mediation function can be carried out by appointing or seeking a good negotiator. The purpose of a mediator's involvement in the negotiation process is to help the parties to the dispute in the negotiation process so that the dispute can be resolved [17]. Argues that the mediator's role is to help the parties find the problems that arise in their dispute and then understand the difference between their wants and needs and the desires and needs of the other party. That is when deciding to make an offer. In other words, small scale service providers should anticipate the risks of a contract by appointing or providing a Reliable negotiator.

The type of lump-sum contract is also considered to be the riskiest contract. In line with Presidential Regulation No. 16 of 2018 [18] concerning which states that the providers of goods/services entirely bear all risks. In another sense, if a small scale construction service provider takes a contract with the type of lump sum, then all risks must be prepared by them. Indeed, this will be very difficult for small scale construction service providers. Then, the alternative can be done to transfer the risk to another party. In developed countries like China, the transfer of construction risk through insurance is commonly used. In Indonesia, the use of insurance is the awareness of the company [19]. This situation makes small-scale construction service providers have the choice not to use this. However, small-scale construction service providers have to realize the importance of construction insurance to prevent losses. The owner of the project might mention construction insurance in the contract. This set-up might be beneficial for two parties. In small-scale construction service providers' view, they will feel secure if they have some problem with the project. On the other hand, the owner also felt secure if there any problem; construction service providers are still capable of continuing the project due to the insurance's assistance. However, this suggestion might have a limitation. Mention insurance in the contract might increase the cost. It also contradicts the lump sum definition that all risks must accept by the construction service providers [18]. In this situation, as the regulator, the government has initiated the new regulation to accommodate both parties and the benefits of the insurance.

4. Conclusion

This research aims to share the experience of medium-scale construction service providers in Surabaya in terms of construction contract risk for small-scale construction service providers. We make ranking for each risk identification using AHP and nominated the high-risk level for each factor. The result indicates that a mismatch between the volume of work in the Bill of Quantity and field conditions is the highest level, followed by any different contract between both parties. For the type of contract, lump-sum was indicated as the high-risk contract. However, for the anticipate of contract construction, small-scale construction service providers should have good negotiators to help the parties to the dispute in the negotiation process so that the dispute can be resolved in terms of the type, the risk of construction contract type. The Indonesian government has to initiated new regulations to accommodate both parties and the benefits of the insurance. Moreover, we need a review of the beneficial insurances for future research in terms of reducing the risk of contract construction in Indonesia.

5. References

[1] A. Asnudin. KONSEP PENGEMBANGAN KONTRAKTOR SKALA KECIL [Small Scale Contractor Development Concept]. *Jurnal SMARTek* 2005; 3: 209–219.

[2] Agung M. The Contribution Of The Construction Industry To The Economy Of Indonesia: A Systemic Approach.

[3] Tamin RZ, Tamin PF, Shahab F, et al. Improving Indonesian Construction Consulting Services*. *Journal of Engineering and Technological Sciences* 2015; 47: 189–200.

[4] Wulfram I. Ervianto. *Manajemen Proyek Konstruksi (Edisi Revisi)*. III, 1st Published., http://andipublisher.com/produk-0501001336-new-cover-manajemen-proyek-konstruksi-ed.html (accessed 28 October 2020).

[5] Jannadi O, Almishari S. Risk Assessment in Construction. *Journal of Construction Engineering and Management-asce - J CONSTR ENG MANAGE-ASCE*; 129. Epub ahead of print 1 October 2003. DOI: 10.1061/(ASCE)0733-9364(2003)129:5(492).

[6] Lalu Hadi Adha. Kontrak Build Operate Transfer Sebagai Perjanjian Kebijakan Pemerintah Dengan Pihak Swasta [Build Operate Transfer Contract as a Government Policy Agreement with Private Parties]. *Jurnal Dinamika Hukum* 2011; 11: 548–558.

[7] I GUSTI AGUNG AYU ISTRI LESTARI. PERBANDINGAN KONTRAK KONSTRUKSI INDONESIA DENGAN KONTRAK KONSTRUKSI INTERNATIONAL [Comparison of Indonesian Construction Contracts with International Construction Contracts]. *Jurnal GaneÇ Swara* 2013; 7: 64–69.

[8] Saaty T. Decision making with the Analytic Hierarchy Process. *Int J Services Sciences Int J Services Sciences* 2008; 1: 83–98.

[9] Saaty TL. How to make a decision: The analytic hierarchy process. *European Journal of Operational Research* 1990; 48: 9–26.

[10] Deng X, Hu Y, Deng Y, et al. Supplier selection using AHP methodology extended by D numbers. *Expert Systems with Applications* 2014; 41: 156–167.

[11] Zayed T, Amer M, Pan J. Assessing risk and uncertainty inherent in Chinese highway projects using AHP. *International Journal of Project Management* 2008; 26: 408–419.

[12] Ramanathan R, Ganesh LS. Using AHP for resource allocation problems. *European Journal of Operational Research* 1995; 80: 410–417.

[13] Ainul Yaqin, Muhammad, Purwanita, Setijanti. Increasing Customer Intention to Visit Mall in Surabaya- an Architectural Perspective. Seoul, http://www.uia2017seoul.org/ (2017).
[14] Budi Suanda. *Pengelolaan risiko kontrak terhadap kinerja biaya pada proyek konstruksi (studi kasus PT.PP)*. Universitas Indonesia, 2008.
[15] Avenhaus R (ed). *Quantitative Assessment in Arms Control: Mathematical Modeling and Simulation in the Analysis of Arms Control Problems*. Springer US. Epub ahead of print 1984. DOI: 10.1007/978-1-4613-2805-6.
[16] Poerdyatmono B. ALTERNATIF PENYELESAIAN SENGKETA JASA KONSTRUKSI. *Oktober* 2009; 8: 78–90.
[17] Lovenheim P, Guerin L. *Mediate, Don't Litigate*. Nolo, 2004.
[18] President Regulation No. 16. Procurement of Government goods / services. 2018.
[19] Slamet SR. KESEMPURNAAN KONTRAK KERJA KONSTRUKSI MENGHINDARI SENGKETA. *Lex Jurnalica*; 13, https://ejurnal.esaunggul.ac.id/index.php/Lex/article/view/1773 (2016)

Classification of community opinion on the use of the Transjakarta bus based on twitter social network using naïve bayes method

B. D Meilani[1], R K Hapsari[2], I F Novian[1]

[1] Department of Information System, Faculty of Electrical and Information Technology, Institut Teknologi Adhi Tama Surabaya, East Java, Indonesia
[2] Department of Informatics, Faculty of Electrical and Information Technology, Institut Teknologi Adhi Tama Surabaya, East Java, Indonesia

budanis@itats.ac.id

Abstract. At this time, social networking is becoming a very popular communication tool among Indonesians ranging from children to adults. One of them is Twitter, which has quite a lot of users. This social network is used by users to write their opinions on a matter. One of the topics discussed in the Twitter social network is the Trans-Jakarta bus. In this study, the opinions of Trans-Jakarta bus users listed on Twitter are only limited to positive and negative opinions. An application designed to predict positive or negative opinions using the Naïve Bayes algorithm. The Naïve Bayes algorithm was chosen because it only requires a small amount of training data to estimate averages and variable variations. The developed application can make it easier for decision-makers towards Trans-Jakarta to find out public opinion on Trans-Jakarta bus services. The resulting system accuracy is 73% because the amount of training data used is only 62.5% of the total data of 50 data. While the test data that has been used are only 30 data. The proportions of the training and testing data affect the level of accuracy, so if you want a higher level of accuracy you can reproduce the training data used.

1. Introduction

Currently, the Intelligent Transport System (ITS) is aggressively developing urban transportation that is comfortable and easily accessible to the public. With this public transportation, all levels of society can take advantage. So that it can unravel congestion which has been a problem in big cities like Jakarta. "Based on BPTJ data, currently the average speed of vehicles in busy times is 10 km/hour, decreasing every year and causing economic losses. The impact of traffic congestion in Jakarta alone reaches the US $ 1 billion in 2010 and is expected to increase to the US $ 6.5 billion by 2020 if this condition is not corrected" [1].

To overcome traffic congestion in Jakarta, the transfer of the use of private transportation to use public transportation. One of them is the public transportation Trans-Jakarta bus which is the transportation choice of the people of Jakarta. "Our target for 2017 is 185 million people, meaning that it will increase by 50%, the increase in the number of passengers will increase every day" [2]. In 2015 there were 207,000 daily Trans-Jakarta bus passengers and the number increased by 50% in 2016. With the increase in Trans-Jakarta users, many users gave their opinions. Public opinion on Trans-Jakarta transportation is very much, wherein one month there are about ± 100 opinions that enter. The opinion is in the negative and positive classification. With the existence of public opinion towards Trans-Jakarta, the Trans-Jakarta party can find out the level of community satisfaction with trans-Jakarta transportation. To get an opinion on Trans-Jakarta one way is to use social media networks, namely Twitter. At the official Twitter development conference, management delivered statistics on the site and Twitter users, the statistics stated that in April 2010, "Twitter has 106 million accounts and as many as 180 million unique or different users each month. The number of Twitter users has 106 million accounts and an increase of 300,000 users every day" [3].

Research in the field of opinion mining began to bloom in 2002. Turney in 2002 conducted research on opinion mining by using data in the form of consumer review data on a product. By implementing the Semantic Orientation method using Pointwise Mutual Information (SO-PMI). The best results were achieved, with an accuracy of 84% of the motor vehicle review data and an accuracy of 66% for the film review data [4]. While research on the classification of sentiment towards the review of the movie has been done by Dhande and Patnaik using the Naive Bayes algorithm, Neural Network, and Naive Bayes classifier Neural. From the results of the final research that was tested using the three algorithms mentioned that Naive Bayes produces higher accuracy than Neural Network [5].

Naive Bayes is one of the methods in the classification of documents. Naive Bayes is a classification with probability and statistical methods proposed by the British scientist Thomas Bayes. At this time the classification of text using the Naïve Bayes algorithm has experienced many improvements such as reducing the dimensions of word features to improve algorithm performance. In addition to the small training data, rare word features appear randomly. The research found that the precision results from 80% increased to 85%, the recall ratio from 81% increased to 83%, and the f-measure value in the classification from 81% increased to 84% [6].

Many successful studies using the Naive Bayes method include Data Mining Applications to produce Student Graduation Patterns with the Naive Bayes Method [7] and Classification Identification of Factors that Cause Inaccuracy in Students Graduation with the Naive Bayes Classifier Method [8].

2. Method

Data mining is an interdisciplinary field that is part of the process of finding greater knowledge. In data mining includes pre-processing tasks such as extraction, cleaning, mixing, data reduction and construction features as well as post-processing steps such as pattern and model interpretation, hypothesis confirmation and generation [9].

2.1. *Texts Mining*

Text mining is an activity to retrieve data where the source of data is obtained from Word documents, PDFs, quoted text, which is obtained by extracting information, tracking topics, summarizing, categorizing, grouping, and linking concepts *[10]*. In-text mining, the process of taking information from textual data that has high quality and can find out the problems in the text of a particular topic. Text mining in sentiment analysis can identify emotionally about a statement *[11]*. Sentiment analysis often also known as opinion mining is a computational study of opinions, sentiments, attitudes, and emotions that are presented in a text *[12]*.

According to Feldan and Sanger text mining can be considered a new research subject. Text mining can provide solutions to problems such as processing, organizing or grouping, and analyzing large amounts of unstructured text *[10]*.

Text mining can also be assumed as an extension of the data mining or knowledge-discovery database (KDD), which is used to find interesting patterns of large data. But text mining has a more specific field than data mining because data is in the form of unstructured text. Besides text mining also refers to the discipline of computer science that is used to deal with natural language such as exploiting techniques and methodologies from the field of information search. Text mining relies on information retrieval, natural language processing, and information extraction *[13]*.

2.2. Texts Preprocessing

Pre-processing data is the process of cleaning and preparing texts for classification *[14]*. Online text usually contains a lot of noise. So it is necessary to do the text preprocessing stage. This stage is the stage where the data is prepared to be data that is ready to be analyzed, which aims to obtain satisfactory results. This process also aims to eliminate problems that can affect the results of the data mining process.

Preprocessing is an important part of this research system because the results obtained depend on whether or not we do pre-processing data. Preprocessing data will be very important to use on social networking data because most of the sentences used are not formal, unstructured, and have a large noise. To get good results, the raw data will be preprocessed in various stages. The preprocessing stage is shown in Figure 1, which consists of four stages, namely *[15]*:

1. Case Folding
   The process of the folding case that converts the entire text in the document into lowercase.
2. Cleansing
   The cleansing process is removing characters that are considered invalid, invalid characters include words preceded by characters ('@'), hashtags ('#'), links ('https') and symbol characters such as (`~!@#$%^&*()-_=+[{]}\|;:'"<,>.?/).

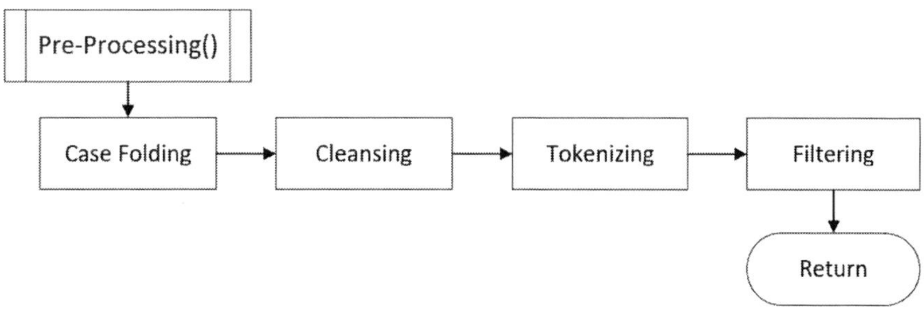

Figure 1. Data preprocessing flow

3. Tokenizing
   The tokenizing process is cutting a document or sentence into parts of words, the sentence to be cut is a sentence separated by a space character, the process aims to facilitate the weighting of each word that appears.
4. Filtering
   The filtering process is taking important useful words and conjunctions that do not count in the number of word occurrences

2.3. Naïve Bayes

Naïve Bayes is a method proposed by a British scientist named Thomas Bayes. This technique is a probability and statistic that can predict future opportunities based on previous events, with a strong assumption of independence from each condition. Naïve Bayes for each class of decisions calculates the probability of condition that the class of decisions is correct. Given the vector of object information, this algorithm assumes that the object's attributes are independent **[16]**.

Naïve Bayes is used in the classification process of documents to be classified into positive and negative classes. In this research, the dataset used is data obtained from Twitter after being processed first. The dataset is divided into two data, namely training data and testing data. Training data is data that the classes have already known, while testing data is data whose classes are not yet known.

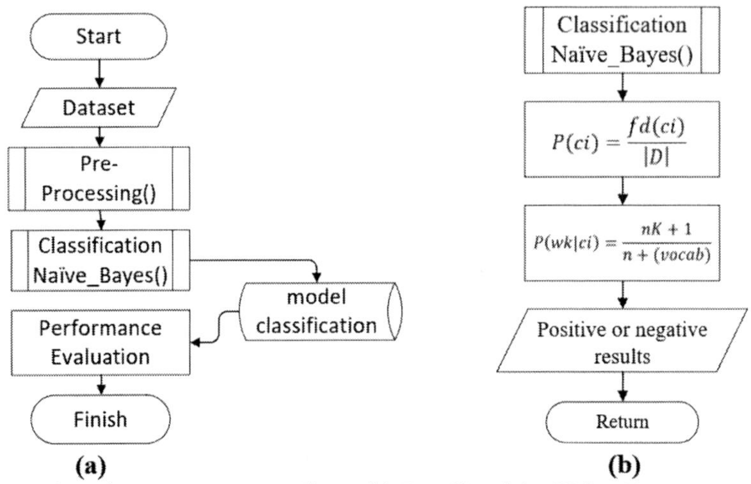

Figure 2. (a) Main system process flow; (b) Details of the Naive Bayes process flow

The process flow in the main system is shown in Figure 2a, while the details of the Naïve Bayes classification sub-process are shown in Figure 2b.

The naïve Bayes algorithm process begins by calculating the prior value which is the probability of class appearance compared to the total number of classes. The calculation of the prior value is written in formula 1.

$$P(c_i) = \frac{fd(c_i)}{|D|} \tag{1}$$

where,

c_i = document category
$fd(c_i)$ = number of documents that have ci category
D = the total number of training documents

The prior calculation is done for as many classes as there are. Then count the number of unique terms in the training data from all classes and count the appearance of each term in the document, by counting the number of occurrences of the word in each category (c_i). The next calculated value is the number of words in each class, the words that are counted are not unique words but all the words in the category. After getting these values, the next step is to calculate the probability of each word against each existing category. The calculation formula is written in formula 2.

$$P(wk|c_i) = \frac{(nK+1)}{(n+(vocab))} \tag{2}$$

where,

w_k = words that come out in the ci category
nK = the number of occurrences of words in the ci category
n = the total number of words in the ci category
$vocab$ = total number of words

3. Result

In testing this research used 30 test data. As well as a comparison between the results of the classification system and the results of the classification which is done manually. The test data is taken randomly from the dataset. The test results of the 30 opinion data are shown in Table 2.

Table 1. Test Result

No	ID People's sentiment	Classification	System prediction	The Result
1	PS-001	Negative	Negative	True
2	PS-002	Negative	Negative	True
3	PS-003	Negative	Positive	False
4	PS-004	Negative	Positive	False
5	PS-005	Negative	Positive	False
6	PS-006	Negative	Negative	True
7	PS-007	Negative	Positive	False
8	PS-008	Negative	Negative	True
9	PS-009	Negative	Negative	True
10	PS-010	Negative	Positive	False
11	PS-011	Negative	Positive	False
12	PS-012	Negative	Negative	True
13	PS-013	Negative	Negative	True
14	PS-014	Negative	Positive	False
15	PS-015	Negative	Negative	True
16	PS-016	Positive	Positive	True
17	PS-017	Positive	Positive	True
18	PS-018	Positive	Positive	True
19	PS-019	Positive	Positive	True
20	PS-020	Positive	Positive	True
21	PS-021	Positive	Positive	True
22	PS-022	Positive	Positive	True
23	PS-023	Positive	Positive	True
24	PS-024	Positive	Positive	True
25	PS-025	Positive	Negative	False
26	PS-026	Positive	Positive	True
27	PS-027	Positive	Positive	True
28	PS-028	Positive	Positive	True
29	PS-029	Positive	Positive	True
30	PS-030	Positive	Positive	True

Based on the results of testing of the testing data, it is used as a reference to calculate the accuracy of the system. Accuracy is calculated by formula 3.

$$accuracy = \frac{\text{the amount of correct data}}{\text{amount of testing data}} x100\% \tag{3}$$

The level of accuracy is obtained from:

$$accuracy = \frac{22}{30} x100\% = 73\%$$

So that the resulting accuracy is 73%.

The resulting system accuracy is 73% because the amount of training data that has been used is only 62.5% of the total data, namely 50 data. While the test data that has been used are only 30 data. The proportions of the training and testing data affect the level of accuracy, so if you want a higher level of accuracy you can reproduce the training data used.

4. Conclusions

Based on the development of the Trans-Jakarta user opinion classification application on the Twitter social network using the Naïve Bayes algorithm, it can be concluded that: The classification application has been able to classify 30 test data with an accuracy of 73%, and this application has made it easier for Trans-Jakarta directors to find out public opinion about Trans-Jakarta buses so as to improve the quality of Trans-Jakarta buses.

5. References

[1] P. Noni, 2017, "ITS Indonesia : penerapan teknologi untuk integrasi transportasi sudah mendesak.,". [Online]. Available: Tribunnews.com.

[2] Joseph, 2017. "Pengguna Transjakarta di Tahun 2017," Kompas.com, Jakarta.

[3] Yarrow, M. Radke, J. A. Clausen and P. R. Robbins, 2010, "The Social Meaning of Mental Illness," *Journal of Social Issues,* vol. 11, no. 4, pp. 33-48.

[4] P. D. Turney, 2002, "Thumbs Up or Thumbs Down? Semantic Orientation Applied to Unsupervised Classification of Reviews," in *the Association for Computational Linguistics 40 Anniversary Meeting*, New Brunswick, N.J.

[5] L. L. &. P. P. G. K. Dhande, 2014 "Analyzing Sentiment of Movie Review Data using Naive Bayes Neural Classifier," *International Journal of Emerging Trends & Technology in Computer Science (IJETTCS),* vol. 3, no. 4, pp. 313-320.

[6] L. Yuan, 2010, "An Improved Naive Bayes Text in Chinese Information Processing," *Henan Polytechnic University, Jiaozuo, China.*

[7] B. D. Meilani and N. Susanti, 2015, "Aplikasi Data Mining untuk Menghasilkan Pola Kelulusan Siswa dengan Metode Naive Bayes," *Jurnal Ilmiah NERO*, vol. 1, no. 3.

[8] B. D. Meilani, S. Wahyudiana, A. Y. P. Putri and A. Pakarbudi, 2019, "Klasifikasi Identifikasi Faktor Penyebab Ketidaktepatan Masa Lulus Mahasiswa dengan Metode Naïve Bayes Classifier," in *Seminar Nasional Sains dan Teknologi Terapan*, Surabaya.

[9] M. J. &. M. J. W. Zaki, 2014, Data Mining and Analysis : Fundamental Concepts And Algorithms, New York: Cambridge University Press.

[10] R. &. S. J. Feldman, 2007, The Text Mining Handbook Advanced Approachies in Analyzing Unstructured Data, New York.: Cambridge University Press.

[11] A. Z. Khan, M. Atique and V. Thakare, 2015, "Combining Lexicon-based and Learning-based Methods for Twitter Sentiment Analysis.," *International Journal of Electronics, Communication and Soft Computing Science & Engineering (IJECSCSE),* pp. 89-91.

[12] L. B, "Sentiment Analysis and Opinion Mining, 2012, " *Synthesis Lectures On Human Language Technologies,* vol. 5, no. 1.

[13] A. Hotho, A. Numberger and G. PaaB, 2005, " A Brief Survey of Text Mining," *Computional Linguistics and Language Technilogy.*

[14] E. Haddi, X. Liu and L. Shi, "The Role of Text Pre-processing in Sentiment Analysis, 2013, " *First International Conference on Information Technology and Quantitative Management,* pp. 26-32.

[15] S. Aji, 2017, "Sistem Klasifikasi Opini Pengguna Maskapai Penerbangan Indonesia Pada Jejaring Sosial Twitter Menggunakan Metode K-Nearest Neighbor," Malang.

[16] O. &. Delen, Advanced Data Mining Techniques, 2008, USA : Springer-Verlag Berlin Heidelberg.

Microwave assisted hydro-distillation (mhd) of citronella oil from lemongrass plants (*cymbopogon nardus*): effect of distiller size on oil yield

Z Ma`sum[1], A Altway[1], M Mahfud[1]*

[1]Chemical Engineering Departement, Institut Teknologi Sepuluh Nopember, Surabaya 60111, Indonesia

mahfud@chem-eng.its.ac.id

Abstract. The essential oil from *Cymbopogon nardus* is widely used in industry, especially medicine and perfume. The microwave assisted method has been successfully applied to a laboratory extraction scale, namely Microwave Hydrodistillation. Therefore, it is necessary to increase the distiller capacity to increase the amount of essential oil obtained. The purpose of this study was to increase the scale of the extraction space through hydrodistillation made from microwaves to produce citronella oil with high yield. It was extracted compared to three distillers 1, 2 and 3 L. The feed to solvent ratio marked as F/S was 0.05 - 0.15 g / and microwave power was 400 – 800 W and extraction time 20 – 60 min. The yield obtained slightly decreased with increasing distiller volume for power parameters and F/S ratio. The highest extraction results using the MHD method occurred at 0.5 cm of material size, 800 W of power and 0.05 g / ml of F/S ratio was 1.81%. GC-MS analysis of citronella oil obtained from the dry leaf extract of Cymbopogon nardus showed that the main bioactive composition of citronella oil was citronell, geraniol, and citronellol at 73.76%. Therefore, it makes sense to extract citronella oil from the dried leaves of Cymbopogon nardus by using a larger distiller size.

1. Introduction

Essential oil is also known as etheric oil or volatile oil produced from some variety plants by some extraction methods. The oil is volatile at room temperature without decomposition of components, has a bitter taste, smells good and distinctive smell in accordance with the type of plants, is generally soluble in organic solvents and insoluble in water [1].

Essential oils are commodities desirable in various industries such as in the perfume, cosmetics, pharmaceutical and food, beverage and beverage industries and can also be used as natural insecticides [2]. As one of the megabiodiversiri centers, Indonesia produces 40 types of 80 types of essential oils traded on the world market. Of these, 13 have entered the world essential market, namely patchouli, lemongrass, cloves, ginger, nutmeg, pepper, cinnamon, sandalwood, jasmine, vetiver, cananga, cajuput, and kemukus. Most of the essential oils produced by farmers are exported with a market share of patchouli 64%, kenanga 67%, vetiver 26%, fragrant lemongrass 12%, nutmeg 72%, clove 63%, ginger 0.4%, and pepper 0.9% from world exports [3]. In the world of trade, this commodity is seen as having a strategic role in producing primary and secondary products and still exists despite price fluctuations, including citronella oil. However, the essential oil industry in Indonesia is still undeveloped so that extraction of essential oils has not been optimal both in terms of quantity and quality [4,5].

Citronella is a plant that has many benefits, including as a producer of essential oils. This plant is easily cultivated and does not require special treatment, so that people who are interested in developing citronella cultivation have great opportunities. It is currently widely cultivated in tropical and subtropical regions in Asia (Indonesia, Malaysia, India, Ceylon), Africa (Madagascar) and almost all countries in Latin America. At present, 300-350 tons of oil have been produced in India over the past 6-8 years, especially in areas that receive good rainfall and are distributed throughout the year such as in the states of Assam, Karnataka, Maharashtra, Tamil Nadu, and West Bengal, due to these area it is suitable for planting Citronella [6–8]. Therefore, the development of fragrant citronella oil has great potential in Indonesia, moreover the fragrant raw material of citronella is abundant. The production of fragrant citronella oil in Indonesia is produced from various regions dominated by the provinces of Nanggroe Aceh Darussalam, East Java, West Java and Lampung with an area of 3,492 hectares [3].

The extraction of essential oils contained in fragrant lemongrass on an industrial scale is currently carried out by steam and water distillation. The selection of this distillation system is because the material used is in the form of leaves and stems so that the essential oil produced is more, the distillation is shorter and the distilled material does not burn [4]. However, in this system there are disadvantages, namely the yield obtained cannot be optimum and the time taken is longer.

Different methods can be used for extracting essential oils from plant materials, for example hydro-distillation (HD), steam distillation, Soxhlet extraction, and distillation extraction simultaneously [9,10] But these molecules are thermally sensitive and susceptible to chemical changes. This extraction method can cause the loss of some volatile compounds, low efficiency extraction, long extraction times, degradation of unsaturated compounds or esters through thermal or hydrolytic effects and toxic solvent residues in extracts. To overcome this problem several new environmentally friendly technologies for extracting essential oils have been developed by researchers [11]. These techniques usually use less solvents and energy, such as supercritical fluids, ultrasound and microwave extraction. The microwave assistance process has been successfully applied in extracting essential oils on a laboratory scale and in various configuration schemes. Microwave-assisted hydrodistilation (MHD) and solvent-free microwave extraction (SFME) are apparently industries important because of their simple configuration, economy and high product yield. [11].

The problem that remains unclear for the new technique is how to increase the capacity of refiners. Several studies have been carried out to increase the capacity of refiners for Microwave-assisted hydrodistillation systems, but the results have not been satisfactory [12]. Therefore, this research is specifically for Microwave-assisted hydrodistillation systems studying the effect of the size of the distillers together with the microwave power and the ratio of feed to solvents for the extraction of fragrant citronella essential oils (*Cymbopogon nardus*)

2. Material and methods

2.1. Materials

The materials used referred to lemongrass leaves (*Cymbopogon nardus*) obtained from Pacet Mojokerto, East Java, Indonesia and distilled water as a solvent in the distillation process. The chemicals such as Ethanol, n Hexane and Na_2SO_4 were obtained from PT. Brataco Chemicals (Surabaya branch).

2.2. Experimental apparatus for Microwave Hydro-Distillation

This study applied microwave hydro-distillation for citronella oil extraction. EMM2308X, Electrolux microwave oven was used. It has a maximal power of 1000 W and a wavelength frequency of up to 2450 MHz. Its dimensions referred to 48.5 cm x 37.0 cm x 29.5 cm. It was modified by drilling a hole at its top. A 1000 mL round bottom flask connected to Clevenger condenser through the hole was placed inside. The hole was closed with a PTFE to prevent the entry of any contaminants.

2.3. Experimental Procedure

The extraction work carried out consisted of the following steps. The plant material was packed in the distiller and then water was added in a sufficient amount. The samples extracted strung together with Clevenger apparatus. The pre-heating was measured from the moment the microwave was turned on till that of the evaporation beginning. The steam which came out from the extraction chamber was condensed by a condenser apparatus and transported to a separator, where the oil separated automatically from the water and collected a vial bottle. The water in the extraction yield was removed using anhydrous sodium sulfate and cooled to 4°C. The maximum temperature in the distillery was boiling water at atmospheric pressure. The yield was followed every 10 min after preheating. The process continued until reaching the maximum extraction time of 150 min. The extraction was repeated three times for each variabel. The extraction yield of Citronella Oil was calculated according to the equation given :

$$Yield\ (\%, w/w) = \frac{mass\ of\ extracted\ citronella\ Oil}{mass\ of\ dry\ citronella\ leaves}\ x\ 100\% \quad (1)$$

2.4. Operating conditions and variables

- Operating Pressure : Atmospheric
- Materials size : 0.5 cm
- Solvent volume : 0.5 Distiller volume
- Feed to solvent volume ratio (F/S) : 0.05, 0.1, 0.15 g / ml.
- Microwave power : 400, 600, 800 Watt.
- Distiller size : 1000, 2000, 3000 mL

2.5. GC-MS Identification

Citronella oils composition was determined by gas chromatography coupled to mass spectrometry (GC–MS) analysis Instrument Agilent 6890N Network GC System, Agilent J & W Capillary GC Column (30 m x 0.320 mm x 0.25μm). The sample was shaken and taken 0.11 μL and injected into the instrument.. The conditions maintained were : carrier gas Alpha Gaz Helium Ultra pure; split 1:100; Injection temperature 280°C; oven temperature programmed from 50 to 230°C at 5°C min^{-1} and from 230 to 280°C at 10°C min^{-1}. Identification of Citronella Oil component.

3. Results and discussion

3.1 Effect of microwave power on citronella oil yield

Power is the amount of energy per unit time (joules / second). Daya has an influence on citronella oil yield in the extraction process. In the extraction process using microwave, microwave power acts as a controller of the amount of energy that will be emitted by a magnetron and then captured by water to be converted into heat energy. This power helps the process of the release of essential oils from plant material. The MHD method is closely related to processing power, where the greater the power produced by the microwave, the increase in temperature of the system will more quickly reach the boiling point of water. High power and low time can cause thermal degradation of the product so that it helps the process of removing essential oils from the material to be extracted. The profile of power to yield used by the MHD method in the extraction process can be seen in Figure 1.

In this study it was seen that the most optimal microwave power to produce citronella oil yield varied. Based on Figure 1, the tendency of the increase in yield is seen as the power increases so that it can be said that the greater the energy received by the material to be converted into heat, causing the yield of essential oils to be obtained more and more. Optimum power occurs at a power of 800 W on the MHD method for a yield of 1.312%.

In the MHD method, fragrant lemongrass is submerged by the solvent so that there is an increase in the heating speed in the citronella by direct absorption of microwave energy which causes the pressure gradient to be higher in the material when the microwave power is increased. This happens because microwave energy significantly influences molecular interactions between compounds and others.

High power affects the temperature of the material to be increased and the yield can decrease if more power is used than the optimum point. The rate of evaporation becomes faster so that the oil glands in citronella overheating and cell walls undergo lysis. This causes the material to experience dry product combustion or can be said to be scorched and can reduce yield gains and damage the composition of the essential oils taken. Material characteristics also have an important role in this regard, especially in the pseudo stem section. Because the pseudo stem part is more easily degraded at high temperatures.

Microwave power needs to be controlled to avoid loss of product quality. Furthermore, the oil that has been obtained will be diffused with water solvent and evaporates with the solvent vapor which is then condensed. This study also showed the occurrence of microwave heating in the rapid mass transfer in the sample due to the formation of heat which results in a large vapor pressure difference between the product center and the product surface [10].

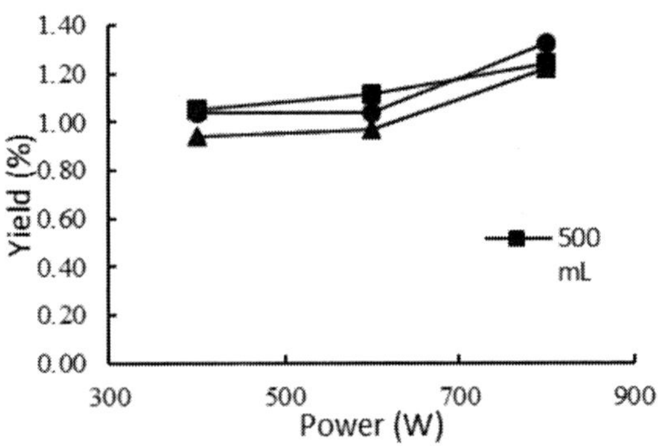

Figure 1. Effect of power on yield
(F / S = 0.1 Material Size 0.5 cm)

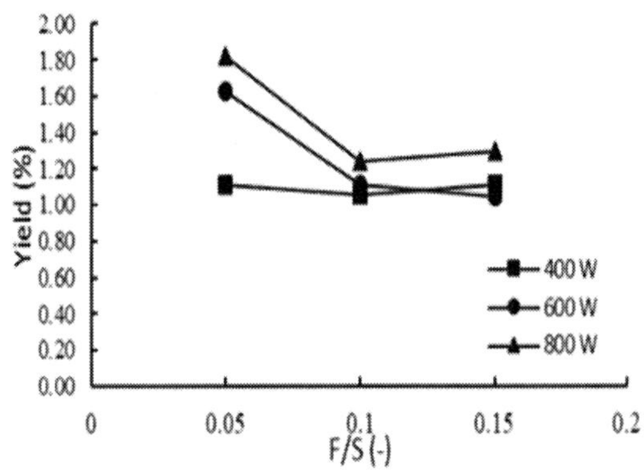

Figure 2. Effect of F/S ratio on yield
(D = 2L, size 0.5 cm)

3.2. Effect of raw materials to distiller ratio on citronella oil yield

In this study, the ratio of raw materials used per volume of solvents for the MHD method is 0.05; 0.1; 0.15; g / ml so that the mass of material used in a 1000 mL flask with a solvent volume of 500 mL is 25, 50 and 75 grams. The material mass ratio affects the density of the material present in the distillation flask. The effect of the ratio of raw materials to yield can be seen in Figure 2. The figure 2 shows that the highest yield was obtained at the lowest F / S ratio (F / S = 0.05), while at the higher F / S ratio the relative yield values were constant. Hence the highest yield value is in the ratio of the weight of the material: the solvent is 0.05 which is equal to 1.81%. Based on the graph, it can be seen that the yield tends to decrease as the mass of raw material increases. This is caused by the increasing mass of material causing the material in the flask to become more dense so that the heat distribution that occurs in the flask is not optimal, as a result part of the material surface is not optimally exposed to heat. The less the mass of the material, the more hollow the mass density of the material, heat can enter through the gaps between the material, heating surface area to increase. This causes the yield to increase as the mass of the extracted material decreases. In the MHD method the amount of solvent added must be in accordance with the optimal ability of the solvent to diffuse to bind the oil that comes out of the lysis oil glands due to microwave heating.

3.3. Effect of distiller size on citronella oil yield

In this study, the size of the material used was the length of fragrant lemongrass 0.5, 1, and 2 cm at each power and ratio. The effect of material size on fragrant lemongrass yield can be seen in Figure 4. Microwave power is also applied to the distillers' different volumes of 1L, 2 L and 3 L. In figure 3 shows that with the increase in the volume of the distillers there is a decrease in yield even though the amount of oil obtained in quantity increases. This shows that with the increase in the size of the distiller, the microvawe power divided per unit of solvent or per unit distiller volume will decrease, but this decrease is not as large as the increase in the size of the distiller. This laboratory experiment provides sufficient justification to start increasing the scale of the microwave energy industry as a tool for large capacity extraction. But what needs to be considered is the heat distribution that arises when the wave source is fixed but the volume of the distiller increases.

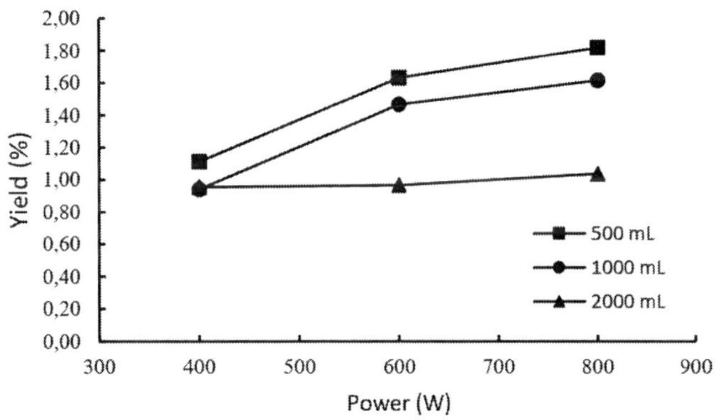

Figure 3. Effect of distiller size on yield (F/S = 0.05 g/ml, material size = 0.5 cm)

3.4 Gas Chromatography / Mass Spectrometer (GC-MS) Analysis Results

To determine the components contained in citronella essential oils, Gas Chromatography - Mass Spectrometer (GC-MS) analysis is used. GC-MS analysis not only shows the components contained in citronella oil, but also the levels of these components. Based on the analysis that has been done, the results shown in table 1. discussion of the results obtained, focused on the main components of citronella oil, namely Geraniol, Citronellal and Citronelol.

In the Microwave Hydro-Distillation method, it was found that the content of the main components of citronella oil was greater than generally obtained from conversional distillation. From table 1 shows that the largest components of citronella oil obtained from the Microwave Hydro-Distillation method are Geraniol, Geraniol Formate, Citronellal and Citronellol. From the% composition obtained, it can be seen that the Geraniol and Citronellal content from the MHD method is above the standard maximum limit of 69.73% and 7.82%. The component that meets the standard of the MHD method is Citronellol which is 3.3%.

Ranitha (2014) reports that the chemical profile of Lemongrass essential oil from Malaysia and other scientists' reports, shows that there is a significant quantitative difference between the chemical profiles of the main components of essential oils extracted by conventional hydrodestillation methods and microwave hydrodestillation methods [10]. In Hydrodestillation and microwave hydrodestillation methods, the number of components obtained is more than the conventional method. From the results of GC-MS analysis of the oil in this study, it shows that the number of citronella oil components is 25 as shows in table 1. This difference shows that the MHD method is more effective in extracting essential oil content both in terms of yied and oil quality.

Table 1. Gas Chromatography - Mass Spectrometer (GC-MS) analysis

No.	Components	Compositions
1	O-Methan- 8-ol	0,32%
2	Linalool	1,09%
3	Rose oxide trans	0,19%
4	alpha.- terpinolene	0,33%
5	Iso pulegol	1,42%
6	(R)-(+)-Citronellal	7,82%
	4-Methyl- 1,4-Heptadiene	0,42%
10	4-Methyl- 7-oxabicyclo [4.3.0] nonan-8-one	0,18%
11	Geraniol formate	12,42%
12	(R)-(+)-. beta.-Citronellol	3,3%
13	Geraniol	69,73%
17	Cis-2,6-Dimethyl-2,6-octadiene	0,44%
18	Trans-Caryophyllene	0,35%
21	Caryophyllene oxide	1,99%
22	Camphene	0%
23	Methyl Isougenol	0%
24	Borneol	0%
25	Limonene	0%
	TOTAL	100%

4. Conclusions

The results of extraction of fragrant lemongrass essential oils by the Microwave Hydro-Distillation (MHD) method show that increasing the power will increase the yield of oil. The optimal power in this study is 800 W. Likewise, the smaller the ratio of the mass of the material to the solvent (F / S) the greater the results obtained. And the maximum F / S for the Microwave Hydro-Distillation method is 0.05 g / ml. Optimal extraction conditions for 1 L distillers are 800 watts of power, F / S ratio of 0.05 g / ml and yield of 1.81%. Whereas in the case of an increase in the size of the distillers, it is shown that an increase in the size of the distillers results in a decrease in oil yield, this occurs because the microvawe power is divided per unit of solvent or per unit of the volume of the distillers will decrease, but this decrease is not as much as an increase in the size of the distillers. In the Microwave Hydro-Distillation method, it was found that the content of the main components of citronella oil was greater than generally obtained from conversional distillation.

5. References

[1] Guenther E. 1990 *Essential Oils Volume IVB (in Bahasa Indonesia), Penerjemah Ketaren., Universitas Indonesia Press. Jakarta*
[2] Ministry of Trade R I 2011 *Indonesian Essential Oils : The Scents of Natural Life* vol 1st
[3] Hernandez-Lambraño R, Pajaro-Castro N, Caballero-Gallardo K, Stashenko E and Olivero-Verbel J 2015 Essential oils from plants of the genus Cymbopogon as natural insecticides to control stored product pests *J. Stored Prod. Res.* **62** 81–3
[4] Clain E, Baranauskien R, Kraujalis P, Kazernavi R and El C 2018 Industrial Crops & Products Biore fi ning of Cymbopogon nardus from Reunion Island into essential oil and antioxidant fractions by conventional and high pressure extraction methods **126** 158–67
[5] Mahfud, Mahfud; Sabara Z 2018 *Industri Kimia Indonesia.pdf* (DEEPUBLISH)
[6] Katiyar R, Gupta S and Yadav K R 2011 Cymbopogon winterianus : An Important Species for Essential Java Citronella Oil and Medicinal Value *Natl. Conf. For. Biodivers. Earth's Living Treasure* 115–8
[7] Singh A and Kumar A 2017 Cultivation of Citronella (Cymbopogon winterianus) and

evaluation of its essential oil, yield and chemical composition in Kannauj region *Int. J. Biotechnol. Biochem.* **13** 139–46
[8] Hamzah M H, Che Man H, Abidin Z Z and Jamaludin H 2014 Comparison of citronella oil extraction methods from Cymbopogon nardus grass by ohmic-heated hydro-distillation, hydro-distillation, and steam distillation *BioResources* **9** 256–72
[9] Golmakani M and Rezaei K 2008 Comparison of microwave-assisted hydrodistillation with the traditional hydrodistillation method in the extraction of essential oils from Thymus vulgaris L . **109** 925–30
[10] M. R, Nour A H, Sulaiman Z A, Nour A H and S. T R 2014 A Comparative Study of Lemongrass (Cymbopogon Citratus) Essential Oil Extracted by Microwave-Assisted Hydrodistillation (MAHD) and Conventional Hydrodistillation (HD) Method *Int. J. Chem. Eng. Appl.* **5** 104–8
[11] Kusuma H, Putri D, Dewi I and Mahfud M 2016 Solvent-free microwave extraction as the useful tool for extraction of edible essential oils *Chem. Chem. Technol.* **10** 213–8
[12] Ma'sum Z, Kusuma H S, Altway A and Mahfud M 2019 On the effect of the ratio of the distiler volume and that of the microwave cavity on the extraction of Cymbopogon nardus dried leaves by microwave hydrodistillation *J. Chem. Technol. Metall.* **54** 778–86

Acknowledgment
This research is supported by research funding from the Directorate General of Research and Technology Strengthening Research and Development of Higher Education (RISTEK-DIKTI) of the Republic of Indonesia through the Basic Research scheme.

Extraction of anthocyanin pigment from *hibiscus sabdariffa* l. by ultrasonic-assisted extraction

Y Yuniati[1], P E Elim[1], R Alfanaar[1], H S Kusuma[2] Mahfud[2]

[1]Department of Chemistry, Universitas Ma Chung, Malang, Indonesia
[2]Department of Chemical Engineering, Institut Teknologi Sepuluh Nopember, Surabaya, Indonesia

mahfud@chem-eng.its.ac.id

Abstract. Roselle flower contains various active compounds including organic acid, hydroxycitric acid, hibiscus acid, anthocyanin, flavonoids and polysaccharides The anthocyanin compounds in rosella flowers are the most important pigment of vascular plants, harmless and that make it interesting is a soluble pigment in water media Anthocyanin extract is one of the most successful natural sensitizers for solar cells, because this organic dye produces a high yield of photons to current conversion and is easy to obtain and cheap. Anthocyanins can be potential as sensitizers, because they range in the light spectrum from red to blue. The purpose of this study was to obtain anthocyanin pigments at the optimum extraction conditions from roselle flowers. The optimization of the method was carried out by analyzing the effect of ultrasonic wave frequency, process temperature, extraction time, and mass ratio of roselle flowers to the volume of distilled water on the anthocyanin concentration. The qualitative and quantitative compositions of the anthocyanin extract of rosella flowers were analyzed by UV-Vis spectrophotometer. The results showed that the optimal operating conditions for the extraction process were known at a temperature of 40 °C, a frequency of 24 kHz, a time of 5 minutes, and a ratio of 1: 25.

1. Introduction

One of the pigments that can be used as a sensitizer is anthocyanin. Anthocyanins can be obtained from various natural sources such as rosella flowers. Rosella flowers contain various active compounds, including organic acids, hydroxycitric acid, hibiscus acid, anthocyanins, flavonoids and polysaccharides [1–4]. The anthocyanin compounds in roselle flowers are the most important pigments of vascular plants, are harmless and what makes them attractive are pigments that are easily soluble in water [5]. Anthocyanin extract is one of the most successful natural sensitizers for solar cells, because this organic dye produces high conversion yields of photons to high currents and is easy to obtain and cheap [6]. Anthocyanins can be potential as sensitizers, because they range in the light spectrum from red to blue. Anthocyanins also have carbonyl and hydroxyl groups so that they can bind to the TiO_2 surface [7].

Anthocyanins are natural secondary metabolite compounds that give color to several types of fruits, vegetables and plants [8]. This pigment is classified as a flavonoid compound which has a function as a natural antioxidant. Antioxidants are substances that can ward off or prevent oxidation reactions from free radicals. This compound is soluble in water and is polar so that it can be dissolved in polar solvents such as ethanol, acetone, chlorophome, and water chloride or formic acid [8].

Anthocyanins are obtained from several sources such as eggplant [9], pomegranate [10], red cabbage [11], and strawberries. It was reported that the anthocyanins obtained from Rosella were delphinidin and cyanidin complexes. These functional groups contribute to the color stability of the roselle flower extract [12]. Rosella has many uses in food, but there are other uses such as use in solar cells. Roselle was chosen as a potential candidate rich in anthocyanins, and can be used as a good source for producing red dye for food [13, 14].

Extraction of natural dyes can be carried out by various methods such as the Soxhlet method, conventional solvent extraction, microwave-assisted [15, 16] and assisted ultrasonic or sonication [17]. Sonication is an extraction method that uses wave energy so that the formation of small bubbles is caused by the transmission of ultrasonic waves to diffuse solvents into the cell walls.The advantages of the sonication process are higher yield and faster extraction time [17].

The purpose of this study was to obtain anthocyanin pigments at the optimum extraction conditions from roselle flowers. The optimization of the method was carried out by analyzing the effect of ultrasonic wave frequency, process temperature, extraction time, and mass ratio of roselle flowers to the volume of distilled water on the anthocyanin concentration. The qualitative and quantitative compositions of the anthocyanin extract of rosella flowers were analyzed by UV-Vis spectrophotometer.

2. Material and methods

2.1. Materials

In this work rosella flower was obtained from Larise Pharmacy in Malang city. The chemicals used HNO_3, 96% ethanol, KI, I_2, and ethylene glycol for this research were obtained from PT. Brataco Chemicals (Surabaya branch).

2.2. Experimental apparatus

The equipment in this research includes: glassware, and sonicator bath (Mosinix USA brand) used for anthocyanin extraction from rosella flowers. Analytical equipment like XRD (X'Pert Pro diffractometer) was used to determine crystallinity of sample. Sample was radiated with Cu Kα (k = 1,5418 Å) for every q = 0,020. FTIR sample spectrum was obtained from Shimadzu FTIR-8400s within a range 400 - 4000 cm^{-1} and prepared as KBr pellet. Diffusion reflectance spectrum UV-Vis (UV- Vis DRS) in range of wavelength 250 - 800 nm by used spectrophotometer (JASCO, V-760 Diffuse Reflectance Model).

2.3. Experimental Procedure

Rosella flower samples or materials are crushed using a blender. After that, 0.5 g of rosella flowers were weighed, and added with 50 mL of distilled water. The extraction process was carried out in a temperature range of 30 to 80 °C for 1-5 minutes with a sonicator (Mosinix, USA) at a frequency of 24-40 Hz bath type. and a variety of solvents with a ratio of rosella and distilled water from 1:25 to 1: 125. The total anthocyanin content (TAC) was determined by the pH-differential method [18]. Hence the anthocyanin extract obtained was measured by a UV-Vis spectrophotometer. The results of the extraction process will then be analysed qualitatively using a UV-Vis spectrophotometer and later quantitative testing will be carried out using the equation:

$$Total\ anthocyanin\ (ppm) = \frac{((A \times MW \times FP \times 1000))}{(\varepsilon \times l)} \quad (1)$$

Explanation:
A : Absorbance
MW : molecular weight
FP : dilution factor
ε : coefficient of extinction
l : cuvette width

3. Results and discussion

3.1 FTIR Absorption spectra of anthocyanin

At the beginning of the study, the rosella used was prepared by crushing it first using a blender to have a larger surface area and causing more contact between the sample and the solvent so that more extraction results were obtained[13]. Then the extraction of anthocyanin pigments from rosella flowers was carried out by sonication method on various variables, namely wave frequency, process temperature, extraction time, and the mass-volume ratio of rosella flowers and aquadest solvent. Aquadest solvent is used because rosella contains glycosides from anthocyanins which are polar so that polar solvents are also used to attract anthocyanins in rosella. Sonication is an extraction method that uses wave energy so that the formation of small bubbles is caused by the transmission of ultrasonic waves to diffuse solvents into the cell walls. [14]

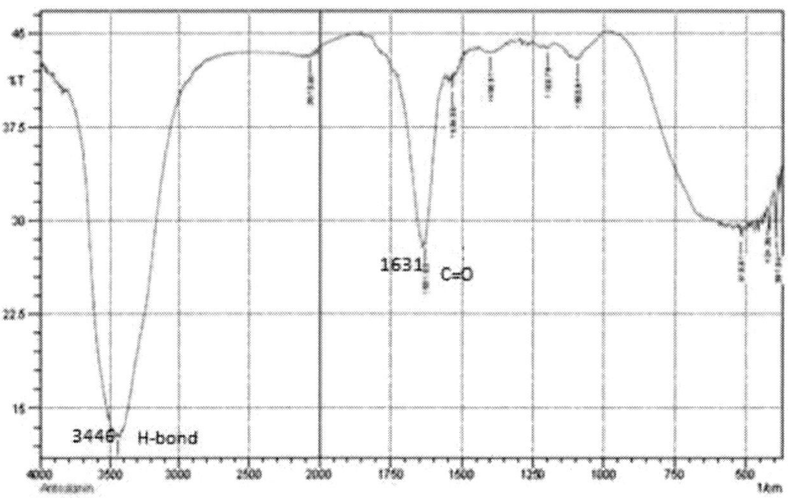

Figure 1. Anthocyanin FTIR Spectra Image

Anthocyanin FTIR Spectrum Image is presented in Figure 1. Figure 1 shows the FTIR spectrum of the spectral range in the wave band 4000 - 400 cm^{-1}, which indicates the presence of absorption at different wave numbers. There is a wide absorption with a strong intensity which indicates the presence of an OH group at wave number 3446 cm^{-1}. Spectrum results also show a sharp absorption with moderate intensity at the wave number 1631 cm^{-1} there is a sharp absorption which is thought to be the absorption of the C = O double bond. Overall, these functional groups correspond to the functional groups contained in the anthocyanin basic framework[18].

3.2 UV-Vis Absorption spectra of anthocyanin

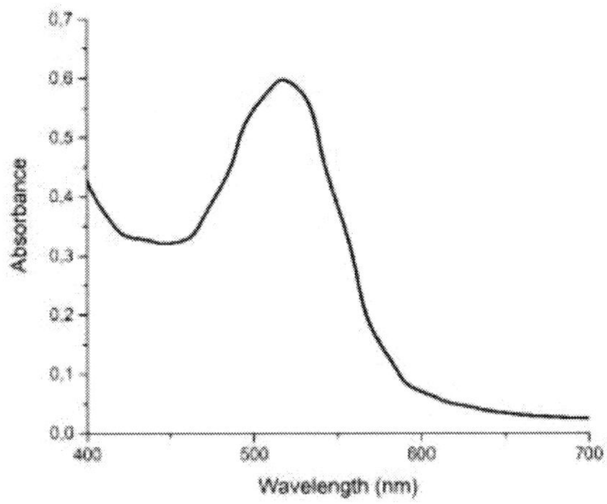

Figure 2. The absorption spectra of anthocyanin extracts

The absorption spectra of anthocyanin extracts obtained from various variables can be seen in Fig. 2 (a-d). Based on the UV-Vis spectrophotometer test, it is known that the wavelength obtained is 516-517 nm. This is appropriate and indicates the presence of anthocyanin compounds in the extraction results, where the maximum absorption limit of anthocyanin pigments is 505-535 nm [5].

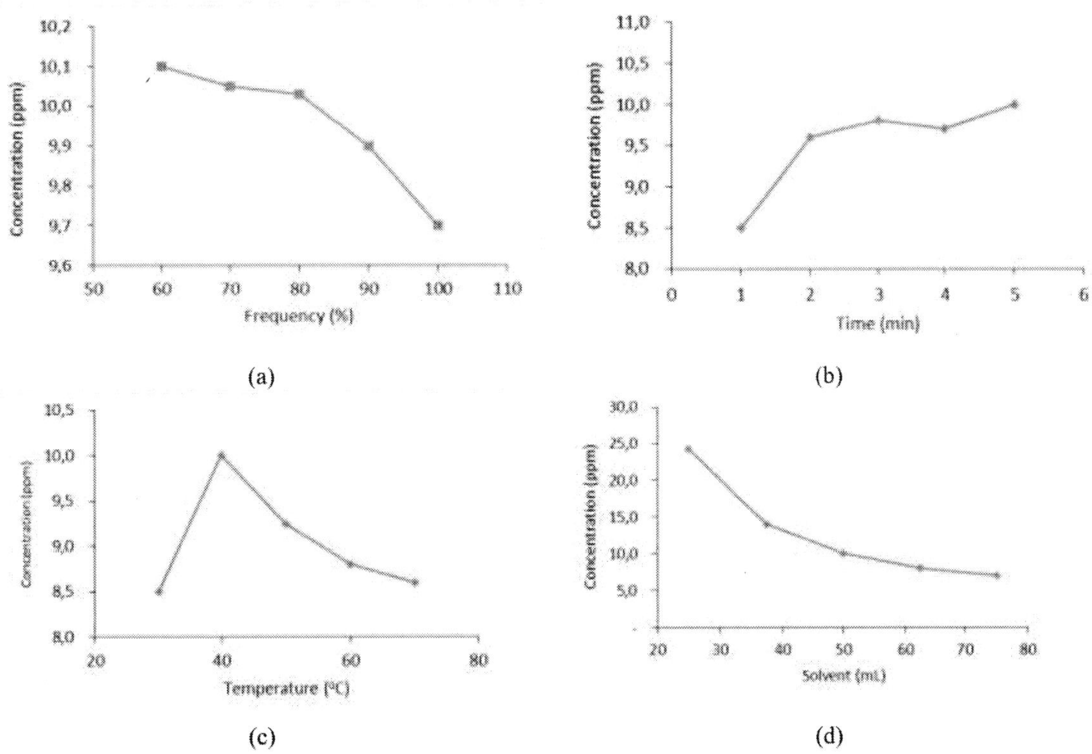

Figure 3. Results of extraction optimization using variables (a) frequency (b) time (c) temperature and (d) solvent volume

In the experiments that have been carried out, it was found that the optimal extraction as shown in Figure 3 (a-d) will use a frequency of 60% or the equivalent of 24 kHz, if using a high frequency such as 100% it can damage the pigments in the cell so that the concentration can decrease. In addition, the optimal extraction uses a temperature of 40 ° C, if the temperature is too high it will damage temperature sensitive pigments while low temperatures do not have sufficient energy. At high temperatures it will damage delfinidin 3-O-sambubiose and cyanidin 3-O-sambubiose which is an anthocyanin group to become gallic and protective acids, while those that are more easily degraded are delfinidin 3-O-sambubiose compounds compared to cyanidine 3-O- sambubiose [19]. For the optimal time done in 5 minutes, so the longer the time the concentration will also increase. Then for the ratio used to get optimal results using a ratio of 1:25 it will get the optimal extraction and have a high anthocyanin concentration.

4. Conclusions

Anthocyanin extraction from rosella flowers can be obtained by the ultrasonic assisted method in a short time. Based on the FTIR spectrum of the spectral range in the wave band 4000 - 400 cm-1, it shows that there is a functional group that corresponds to the functional group contained in the anthocyanin basic framework. Meanwhile, based on uv-vis spectrophotometer analysis, anthocyanin extract absorption spectra were obtained with a peak at a wavelength of 516-517 nm, which indicates the presence of anthocyanin compounds. The use of a frequency of 60% or the equivalent of 24 kHz is very good, if using a high frequency such as 100% it can damage the pigments in the cell so that the concentration can decrease. If extraction using high temperatures will damage pigments that are sensitive to temperature while low temperatures do not have enough energy.The longer the time, the concentration will also increase. Then for a smaller ratio (more solvent) the yield will decrease.The optimal operating conditions for the process of extracting roselle as a source of anthocyanin are at 40 ° C, 24 kHz frequency, 5 minutes time, and 1:25 ratio..

5. References

[1] Da-Costa-Rocha I, Bonnlaender B, Sievers H, et al. Hibiscus sabdariffa L. - A phytochemical and pharmacological review. *Food Chemistry* 2014; 165: 424–443.

[2] Alara OR, Abdurahman NH. Microwave-assisted extraction of phenolics from Hibiscus sabdariffa calyces: Kinetic modelling and process intensification. *Industrial Crops and Products* 2019; 137: 528–535.

[3] Zhou H, Wu L, Gao Y, et al. Dye-sensitized solar cells using 20 natural dyes as sensitizers. *Journal of Photochemistry and Photobiology A: Chemistry* 2011; 219: 188–194.

[4] Setyawati H, Darmokoesoemo H, Ningtyas ATA, et al. Effect of metal ion Fe(III) on the performance of chlorophyll as photosensitizers on dye sensitized solar cell. *Results in Physics* 2017; 7: 2907–2918.

[5] Castañeda-Ovando A, Pacheco-Hernández M de L, Páez-Hernández ME, et al. Chemical studies of anthocyanins: A review. *Food Chemistry* 2009; 113: 859–871.

[6] Suhaimi S, Shahimin MM, Siddick SZM, et al. Effect of Harum Manis mango as natural photosensitizer at different extracting temperature on performance of dye-sensitized solar cells (DSSCs). *RSM 2015 - 2015 IEEE Regional Symposium on Micro and Nano Electronics, Proceedings*. Epub ahead of print 2015. DOI: 10.1109/RSM.2015.7355011.

[7] Hao S, Wu J, Huang Y, et al. Natural dyes as photosensitizers for dye-sensitized solar cell. *Solar Energy* 2006; 80: 209–214.

[8] Pervaiz T, Songtao J, Faghihi F, et al. Naturally Occurring Anthocyanin, Structure, Functions and Biosynthetic Pathway in Fruit Plants. *Journal of Plant Biochemistry & Physiology*; 05. Epub ahead of print 2017. DOI: 10.4172/2329-9029.1000187.

[9] Calogero G, Bartolotta* A, Marco G Di, et al. Vegetable-based Dye-Sensitized Solar Cells Chemical Society Reviews Vegetable-based Dye-Sensitized Solar Cells. *Chemical Sovciety Reviews* 2015; 44: 3244–3294.

[10] Richhariya G, Kumar A, Tekasakul P, et al. Natural dyes for dye sensitized solar cell: A review. *Renewable and Sustainable Energy Reviews* 2017; 69: 705–718.
[11] Chien CY, Hsu BD. Optimization of the dye-sensitized solar cell with anthocyanin as photosensitizer. *Solar Energy* 2013; 98: 203–211.
[12] Shalini S, Balasundaraprabhu R, Satish Kumar T, et al. Enhanced performance of sodium doped TiO2 nanorods based dye sensitized solar cells sensitized with extract from petals of Hibiscus sabdariffa (Roselle). *Materials Letters* 2018; 221: 192–195.
[13] Dayang S, Irwanto M, Gomesh N, et al. Potential of roselle and blue pea in the dye-sensitized solar cell. *AIP Conference Proceedings*; 1885. Epub ahead of print 2017. DOI: 10.1063/1.5002439.
[14] Aryanti N, Nafiunisa A, Wardhani DH. Conventional and ultrasound-assisted extraction of anthocyanin from red and purple roselle (Hibiscus sabdariffa L.) calyces and characterisation of its anthocyanin powder. *International Food Research Journal* 2019; 26: 529–535.
[15] Febriana ID, Gala S, Mahfud M. Ultrasound assisted extraction of natural dye from jackfruit's wood (Artocarpus heterophyllus): The effect of ethanol concentration as a solvent. *AIP Conference Proceedings* 2017; 1840: 1–5.
[16] Qadariyah L, Gala S, Widoretno DR, et al. Jackfruit (Artocarpus heterophyllus lamk) wood waste as a textile natural dye by micowave-assisted extraction method. In: *AIP Conference Proceedings*. 2017. Epub ahead of print 2017. DOI: 10.1063/1.4982324.
[17] Febriana ID, Kusuma HS, Gala S, et al. The effect of temperature on extraction of Swietenia mahagoni by ultrasound–assisted extraction (UAE) method. *ASEAN Journal of Chemical Engineering*; 16.
[18] Mónica Giusti M, Wrolstad RE. Characterization and Measurement of Anthocyanins by UV-visible Spectroscopy. *Handbook of Food Analytical Chemistry* 2005; 2–2: 19–31.
[19] Sinela A, Rawat N, Mertz C, et al. Anthocyanins degradation during storage of Hibiscus sabdariffa extract and evolution of its degradation products. *Food Chemistry* 2017; 214: 234–241.

IOP Conf. Series: Materials Science and Engineering 1010 (2021) 012033 doi:10.1088/1757-899X/1010/1/012033

Wind speed prediction using extreme learning machine and neural network for resolving uncertainty in microgrids

A Seprijanto[1], M Syai'in[2], D F U Putra[1], N H Rohiem[3], N P U Putra[3], M Munir[3]

[1] Department of Electrical Engineering, Sepuluh Nopember Institute of Technology, Surabaya, Indonesia
[2] Department of Marine, Shipbuilding Institute of Polytechnic Surabaya, Surabaya, Indonesia
[3] Department of Electrical Engineering, Institut Teknologi Adhi Tama Surabaya, Indonesia

adisuup@ee.its

Abstract. Wind energy is one of the several types of renewable energy that exist today. however, wind energy has a high degree of uncertainty due to weather effects. Wind speed prediction is needed to determine the energy that wind turbines can produce at each unit. For optimizing wind speed schedulling, the accuracy of wind speed prediction is considered. Extreme learning machine (ELM) and neural network (NN) is implemented to predict hourly wind speed for 24 hour and power generation from wind turbines can produce. Wind speed probability data is taken from sidrap wind farms in indonesia. To determine the performance of wind predictions based on the error value between actual and predicted, mean absolute percentage error (MAPE) is applied.

1. Introduction

Indonesia as the archipelagic country, International Renewable Energy Agency (IRENA) has an estimated total potential energy in Indonesia for onshore wind energy around of 9,3 MW. Java-Bali, Sulawesi and Nusa Tenggara regions has dominated total potential wind energy in Indonesia with an estimated that nearly of 85 percent.

Wind speed has unstable character [1-9] and causes the result of its generation from wind turbines (WTs) to be unstable. Probability data of wind speed in a wind farm [2] is very important information to know the character of wind speed and power can be generated by WTs based on times series[3-7].

Based on the probability data, wind speed prediction can be done by studying the character. Some prediction methods like NN[9-11], IPSO-BP[9], Markov[7] and hybrid[10] have been carried out and shown promising results. Promising results and accuracy of the prediction method are important to help the optimizing process of WTs scheduling.

In this paper, extreme learning machine (ELM) and neural network (NN) are implemented as a prediction function of wind speed hourly for 24 hours. ELM and NN are artificial intelligence methods that consist of 3 layers (input layer, hidden layer, output layer) but both methods have different construction.

2. Typical of Wind Turbines (WTs)

Typical generator of WTs and probability data of wind speed are the part of WTs system in this paper

2.1. *Typical Generator of WTs*

Generally, WTs convert kinetic energy to electricity using the equation as shown in eq.1, but based on the typical data of the WTs, the equation to convert energy is shown in eq.2. In this paper WTs with 140 kW capacity is used. Typical data specification of this WTs as shown in Table 1. 4 WTs installed and can generate 560 kW of the maximum capacity.

$$P = 0.5\rho A C_p V^3 \quad (1)$$

With: P is power generation (Watt), ρ is air density (kg/m^3), C_p is coefficient power, V is wind speed (m/s)

Table 1. Typical data of WTs

Power rate (P_r)	140 kW
Wind speed cut in (V_{ci})	3 m/s
Rating speed (V_r)	15.01 m/s
Wind speed cut off (V_{co})	17 m/s
a	-0,015
b	0,033
c	-0,9
d	-2,1
e	7,1

$$P_{out} = aVw^4 + bVw^3 + cVw^2 + dVw + e \quad (2)$$

With: P_{out} is power generation (Watt), V_w is wind speed (m/s)

2.2. *Probability of Wind Speed Data*

In fig.1 is shown the pattern of the probability of wind speed data at the geographical location in sidrap wind farms during december 2015[12]. Wind probability data is time series data in hourly for 24 hours and in a month. The pattern data as shown in Fig.1 shows the character of the wind speed increase towards morning, decrease toward night and forming the same repetition.

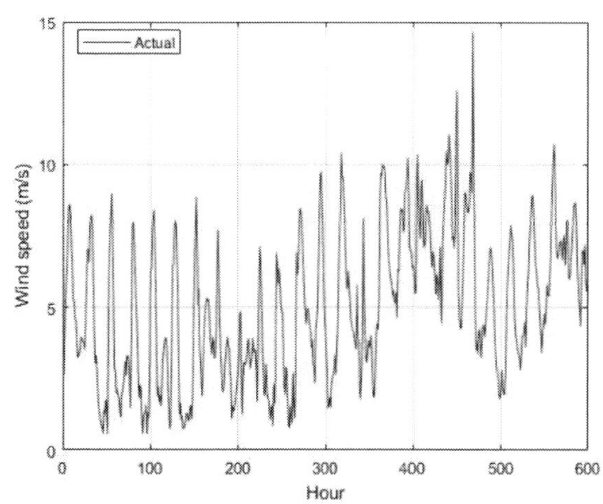

Figure 1. A month probability of wind speed data

3. Methodology

ELM and NN in this paper are constructed by 6 inputs, they are current air temperature (CAT), current air humidity (CAH), current wind speed (CWS), 2 until 4 hours previous wind speed (PWS) and single output as representation of the wind speed prediction as shown in Figure 2.

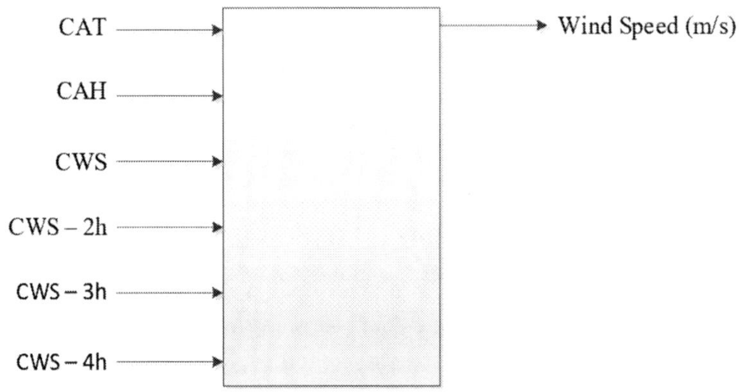

Figure 2. ELM and NN architecture

Input and output training data for the both method using same format. There are 600 columns data in this paper, contains data probability wind speed from first day in december until day 25, so the table display (as show in Table 2 and Table 3) format is taken as an outline.

Tabel 2. Input data train format for ELM and NN

Hour	00.00	01.00	02.00	04.00	>>	22.00	23.00
CAT	24.10	27.80	29.10	29.50	>>	23.50	22.90
CAH	88.78	61.58	53.88	51.20	>>	96.43	92.42
CWS	3.35	2.40	3.33	4.46	>>	3.01	2.65
CWS – 2h	2.95	3.35	2.40	3.33	>>	1.69	3.01
CWS - 3	2.89	2.95	3.35	2.40	>>	1.39	1.69
CWS - 4	2.70	2.89	2.95	3.35	>>	2.10	1.39

Tabel 3. Output data train format for ELM and NN

Hour	00.00	01.00	02.00	04.00	>>	22.00	23.00
Target	2.4	3.33	4.46	5.67	>>	2.65	1.54

3.1. Extreme Learning Machine (ELM)

3.1.1. Design the ELM Structure

The ELM has a different mathematical equation with NN because ELM in this paper consists of 1 hidden layer called matrix H, Beta matrix and single output (as shown in Fig. 3) to predict the hourly wind speed for 24 hours. ELM is constructed by single layer with 600 neurons, because the input data are 600 and the H matrix will be 600 x 600.

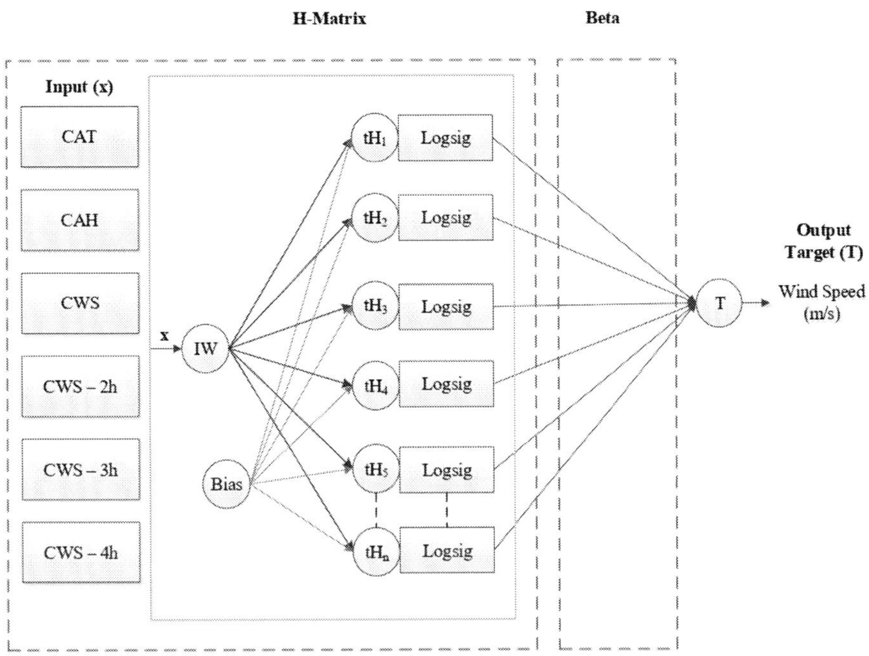

Figure 3. ELM structure

3.1.2. Determining ELM Weight Value

Final goal to finish ELM procedure is to find beta (β) and the input weights (IW) are determined by random value as shown in table 4 and likewise the bias input weight (BIW) as shown in table 5.

Table 4. Weight data of ELM from input layer to hidden layer

Neuron	IW 1	IW 2	IW 3	IW 4	IW 5	IW 6
	to Neuron					
1	0.6232	0.4867	0.9641	0.7356	0.2313	0.6496
2	0.1616	0.0285	0.8676	0.1617	0.7891	0.0848
3	0.2404	0.1791	0.8624	0.3878	0.5765	0.6823
4	0.3351	0.9201	0.7363	0.3633	0.1681	0.5237
5	0.2100	0.4473	0.6638	0.9678	0.0123	0.8361
>>	>>	>>	>>	>>	>>	>>
598	0.7580	0.8613	0.8190	0.3904	0.2196	0.6230
599	0.9003	0.4094	0.6128	0.9880	0.6854	0.1143
600	0.4492	0.8106	0.0533	0.6224	0.9132	0.0429

Table 5. Bias data of ELM from input layer to hidden layer

Neuron	BIW
1	0.211249
2	0.840057
3	0.641954
4	0.008514
>>	>>
598	0.589435
599	0.344251
600	0.445729

3.1.3. Find beta (β) Value

To find beta (β) value, it can be determined by using formula as shown in eq.3[13] below :

$$H\,x\,\beta = T \tag{3}$$

With, H is H Matrix, β is Beta and T is Target

By multiplying value inverse H matrix and T matrix, β can be determined as shown in eq.4. After calculaing β value using eq.4, the value of β as shown in table 6.

$$\beta = H^{-1}\,x\,T \tag{4}$$

Table 6. β value

Neuron	beta (β)
1	1797764484.99
2	-6043934929.45
3	-8147597789.76
4	47078160254.40
5	2901319991.93
>>	>>
588	-110572679878.06
589	-5994258175.21
600	16960226705.63

3.2. Neural Network (NN)

3.2.1. Design The NN Structure

NN structure in this paper construct by 100 neurons to predict the hourly wind speed for 24 hours, with the construction of NN as shown in fig.4.

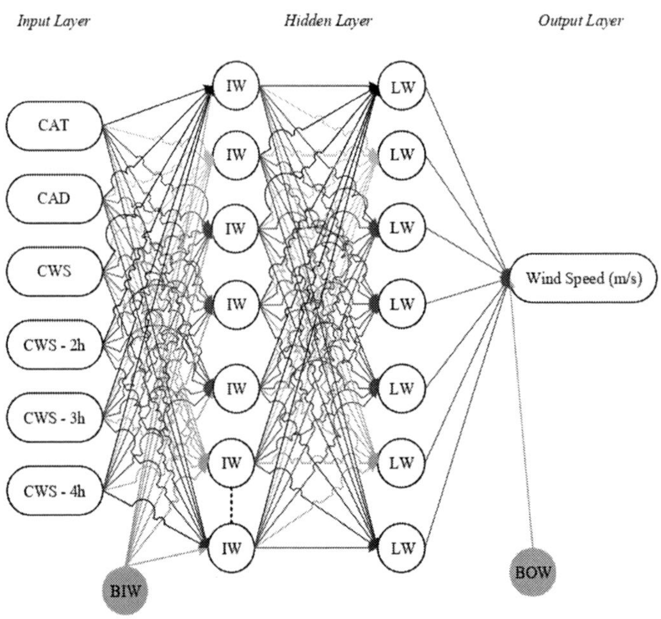

Figure 4. NN structure

3.2.2. Find NN Weight Value

Training process is required to find the weight value of NN. There are 4 types weight value of NN, they are input weight (IW), bias input weight (BIW), bias output weigth (BOW) and layer weight (LW). In this paper training process using matlab. The training process has been completed and successful with an indication of the performance section showing green as shown in Fig.5.

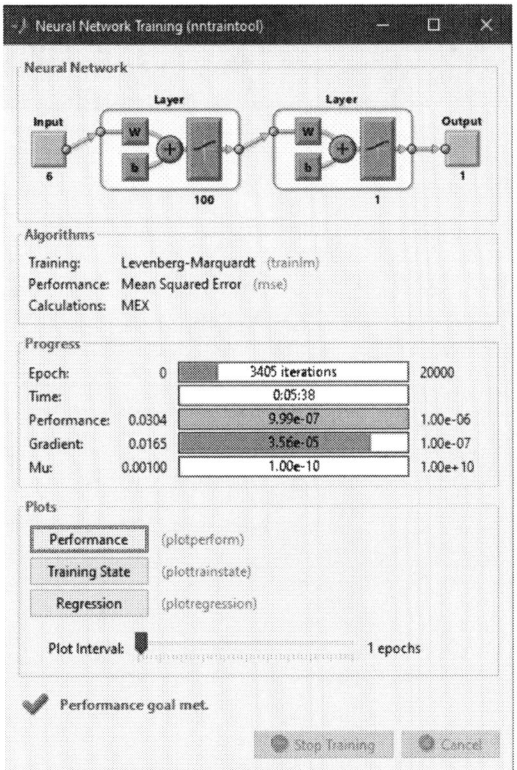

Figure 5. NN training process

To determine the output of NN after all weights is known, it can be detemined by using formula as shown in eq.5[13] below :

$$\textit{Wind Speed} = \text{Logsig (LW x Logsig (IW x Input + BIW) + BOW)} \qquad (5)$$

4. Simulation and Data Analysis

Both methods are tested to predict hourly wind speed for 24 hours and compares with the actuals data. As the first-day test on December 26th, in fig.6 there are 2 sides color line, blue and orange, to illustrate the comparison of actuals data and predictions data from both methods for 24 hours. On December 26 for 24 hours, ELM produces a maximum error for each hour of prediction of 0.0381% while NN produces a maximum error of 0.1923%. The next-day until the last day in december ELM produces maximum errors of 0.0463%, 0.0328%, 0.0512%, 0.0410% and 0.0525% and NN produces maximum errors of 0.1993%, 0.0132%, 0.0322%, 0.0973% and 0.0663%. To know the accuracy of hourly wind speed prediction for 24 hours from both methods mean absolute percentage error (MAPE) is applied.

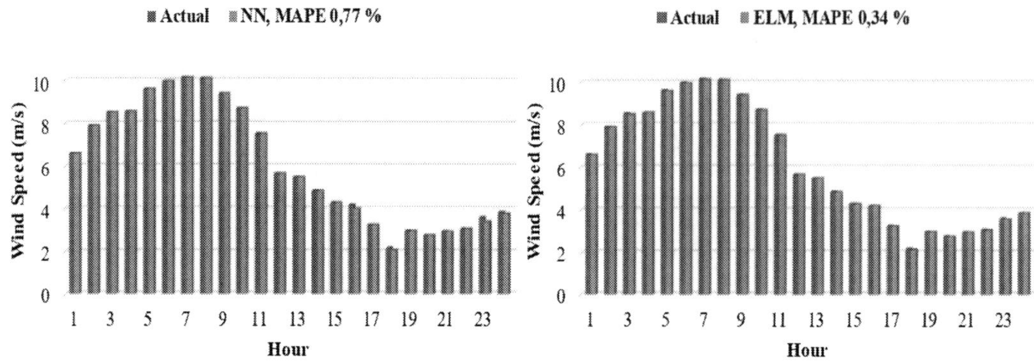

Figure 6. Results prediction from both method, ELM and NN, on December 26th

In table 7, On the first and second day of testing, ELM shown minimum MAPE values of 0.3470% and 0.5392 compared to NN with MAPE values of 0.7707% and 0.9422. But in the last 4 days of testing, NN shown minimum MAPE values of 0.0673, 0.1672, 0.8521 and 0.6489 compared to ELM with MAPE values of 0.2609, 0.2676, 0.9063 and 0.7242. To illustrate the difference of the MAPE results between NN and ELM for 6 times of testing as shown in fig.7. The average of MAPE results in 6 days test from both methods shown under 1%. ELM shown minimum result of 0.5075% in average and NN shown little bit higher of 0.5747% in average.

Table 7. Comparison error forecasting between ELM and NN

Numb.	Date	MAPE (%)	
		ELM	NN
1	December 26, 2015	**0.3470**	0.7707
2	December 27, 2015	**0.5392**	0.9422
3	December 28, 2015	0.2609	**0.0673**
4	December 29, 2015	0.2676	**0.1672**
5	December 30, 2015	0.9063	**0.8521**
6	December 31, 2015	0.7242	**0.6489**
Average MAPE (%)		**0.5075**	**0.5747**

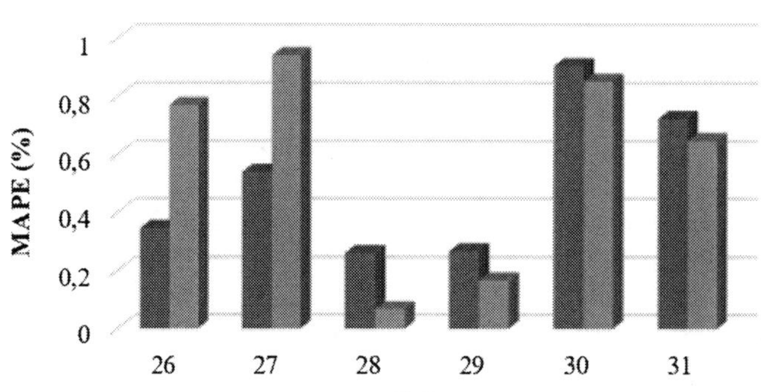

Figure 7. Comparison chart of error prediction results between NN and ELM.

5. Conclusions

Both methods, NN and ELM shown promissing result as prediction function in 6 days test with 6 inputs and single output as the construction's. All MAPE results shown under 1% error from both methods. ELM performs the best results than NN on the first-two-days test with minimum MAPE values of 0.3470% and 0.5392. Otherwise, on the last-four-days test NN performs the best results than ELM with minimum MAPE values of 0.2609, 0.2676, 0.9063 and 0.7242. From all of the testing results data, NN performs better performance on day-ahead wind speeds prediction than ELM with the same architecture.

6. References

[1]. Restrepo, J.F. and F.D. Galiana, *Assessing the Yearly Impact of Wind Power Through a New Hybrid Deterministic/Stochastic Unit Commitment.* IEEE Transactions on Power Systems, 2011. **26**(1): p. 401-410.

[2]. Pinson, P., et al., *From Probabilistic Forecasts to Statistical Scenarios of Short-term Wind Power Production.* Vol. 12. 2009. 51-62.

[3]. Reddy, V., et al. *Hybrid Approach for Short Term Wind Power Forecasting.* in *2018 9th International Conference on Computing, Communication and Networking Technologies (ICCCNT).* 2018.

[4]. Quan, H., D. Srinivasan, and A. Khosravi, *Incorporating Wind Power Forecast Uncertainties Into Stochastic Unit Commitment Using Neural Network-Based Prediction Intervals.* IEEE Transactions on Neural Networks and Learning Systems, 2015. **26**(9): p. 2123-2135.

[5]. Logenthiran, T. and D. Srinivasan. *Short term generation scheduling of a Microgrid.* in *TENCON 2009 - 2009 IEEE Region 10 Conference.* 2009.

[6]. Baby, C.M., K. Verma, and R. Kumar. *Short term wind speed forecasting and wind energy estimation: A case study of Rajasthan.* in *2017 International Conference on Computer, Communications and Electronics (Comptelix).* 2017.

[7]. Hermias, J.P., K. Teknomo, and J.C.N. Monje. *Short-term stochastic load forecasting using autoregressive integrated moving average models and Hidden Markov Model.* in *2017 International Conference on Information and Communication Technologies (ICICT).* 2017.

[8]. Billinton, R., et al., *Unit Commitment Risk Analysis of Wind Integrated Power Systems.* IEEE Transactions on Power Systems, 2009. **24**(2): p. 930-939.

[9]. Zhang, Y., et al., *Wind Speed Prediction of IPSO-BP Neural Network Based on Lorenz Disturbance.* IEEE Access, 2018. **6**: p. 53168-53179.

[10].Nair, K.R., V. Vanitha, and M. Jisma. *Forecasting of wind speed using ANN, ARIMA and Hybrid models.* in *2017 International Conference on Intelligent Computing, Instrumentation and Control Technologies (ICICICT).* 2017.

[11].Bantupalli, M.K. and S.K. Matam. *Wind Speed forecasting using empirical mode decomposition with ANN and ARIMA models.* in *2017 14th IEEE India Council International Conference (INDICON).* 2017.

[12].EMD International A/S, D., financed by the Environmental Support Programme 3 (ESP3) / Danida. *Wind Energy Resources of Indonesia.* 2014-2017 [cited 2019 01-05-2019].

[13].Oktavian Hanggara, P., et al., *Prototype control and monitoring system safety device from leakage ammonia at marine loading arm with comparison of Neural Network (NN) and Extreme Learning Machine (ELM) method.* 2017. 85-89.

Acknowledgements

The authors would like to thank *The Penelitian Dasar Unggulan Perguruan Tinggi (PDUPT) from Kementrian Riset Teknologi dan Pendidikan Tinggi Republik Indonesia* for financial support.

IOP Conf. Series: Materials Science and Engineering 1010 (2021) 012034 doi:10.1088/1757-899X/1010/1/012034

The effect of polyethylene glycol addition on the synthesis of solid acid catalyst and its applications in esterification reaction

S Julaika[1,2], **W Widiyastuti**[1], **H Setyawan**[1*]

[1] Department of Chemical Engineering, Faculty of Industrial Technology and System Engineering, Institut Teknologi Sepuluh Nopember, Surabaya, Indonesia
[2] Department of Chemical Engineering, Faculty of Industrial Technology, Institut Teknologi Adhi Tama Surabaya, Surabaya, Indonesia

sheru@chem-eng.its.ac.id

Abstract. Synthesis of a catalyst by a sol-gel method is commonly conducted using a template to form pores to increase the number of active sites. In this work, mesoporous silica-carbon composites were prepared by sol-gel method using polyethylene glycol (PEG) 6000 as the template. Sodium silicate, commercially known as water glass, a cheap material, was used as the silica source. Here, PEG 6000 served as both the carbon source and the template. The silica-PEG composite was then carbonized at 550°C for 1 h to obtain silica-carbon composite. The sulfonate group was grafted into the composite by sulfonation with sulfuric acid at 100°C for 5 h. The sulfonated silica-carbon catalyst had a surface area of approximately 1490.8 m^2/g. The very high surface area was able to hold a large amount of sulfonate group resulting an ion capacity of approximately 2.09. The composite was successfully be used as heterogeneous catalyst for esterification of acetic acid with ethanol with a conversion of 90.27%, higher than that of commercially Flotrol F-007.

1. Introduction

Many methods have been used to synthesize solid catalysts, including precipitation and coprecipitation, sol-gel, and simple mixing (catalyst alloy) [1]. The sol-gel method seems to attractive from the viewpoint of the possibilities for manufacturers to design the catalysts according to their uses in the reaction. In addition, it also provides many possibilities to improve the internal structures and characteristics of the catalyst. The most common materials for use as the host are based on silica or alumina.

The sol-gel method has been long used to prepare porous materials for catalysts. For the case of silica-based catalyst, the commonly used silica precursor is based on alkoxides such as tetramethyl orthosilicate (TMOS) and tetraethyl orthosilicate (TEOS) [2]–[6]. However, the alkoxides are expensive and toxic that hinder for their commercial application. The initial stage of making this catalyst begins with making mesoporous silica composites, then impregnation is carried out to insert carbon in it, after which the desired functional groups are grafted according to the use of the catalyst in a reaction. Lu, et al, 2010 used hydroxyethylsulfonic acid to graft sulfonate groups on the silica-carbon composite mesopore to produce a solid acid catalyst for acetalization reaction [2]. Valley, et al, 2012, used 98% sulfonic acid to graft sulfonate groups on mesopore-silica-carbon composites (SBA-15, KID-6, and mesocellular silica) to produce solid acid catalysts for esterification reaction [4]. Especially for esterification reaction, this reaction requires acid catalyst, mainly solid acid catalyst which is easily separated from the reactant and product. Ion exchange resins is also used for esterification reaction [7]–[10], in addition to silica-carbon-based solid catalysts [4], [6], [11], and carbon-based solid catalysts [12]–[15].

In this paper, we propose to use an inexpensive sodium silicate as the silica source and PEG 6000 as the carbon source and template to prepare mesoporous silica-carbon composite. First, silica-PEG composite was prepared using a one-pot sol-gel synthesis from sodium silicate and PEG. The obtained composite was then calcined to convert PEG into carbon. The proposed method offers an advantage over the typical two-step route: (i) synthesis of mesoporous silica using template and (ii) impregnation the carbon source into the mesoporous silica. The use of one material with two functions would reduce the step of template removal and carbon source impregnation. This silica-carbon composite serves as a support for the solid catalyst. Furthermore, the performance of the catalyst is evaluated for esterification reaction of acetic acid with ethanol.

2. Materials and Methods

2.1 Materials

The materials used for the synthesis of catalyst were industrial grade sodium silicate solution (water glass; 28% SiO_2, SiO_2:Na_2O = 3.3) provided by PT. PQ Silica Indonesia, polyethylene glycol (PEG 6000; reagent grade, Merck), hydrochloride acid (HCl; 37%, Merck), sodium hydroxide (NaOH; Merck), sulfuric acid (H_2SO_4; 98%, Merck), and demineralized water. They were reagent grade and used as received without further purification.

2.2 Procedure

2.2.1. Preparation of silica-PEG composite. Silica-PEG composites were synthesized by following our developed procedure [16] with some modifications. Briefly, 30 mL of sodium silicate solution containing 8.0% of silica was added to 75 mL of acidic PEG 6000 solution (0,1; 0,25; 0,5, and 1 g of PEG 6000 in 30 ml of demineralized water and 45 ml of chloride acid 1 M) under stirring vigorously. The pH was adjusted to 4 by adding NaOH solution drop by drop under stirring. Then, it was degassed in a vacuum desicator to releasegas bubble present in the solution, aged for 1 h at room temperature to form a gel, and dried in a freeze dryer for 48 h.

2.2.2. Preparation of silica-carbon composite. To produce silica-carbon composite, the silica-PEG composite was carbonized in a tubular furnace under flowing nitrogen gas at 550°C for 1 h.

2.2.3 Sulfonation of silica-carbon composite to produce catalyst. The catalyst was produced by sulfonation of silica-carbon composite using sulfuric acid under nitrogen atmosphere. One gram of the silica-carbon composite was dispersed in 25 mL 98% sulfuric acid. The mixture was heated at 100°C under constant stirring for 5 h. After heating, the mixture was diluted in demineralized water and cooled to room temperature. The solid was filtered from the mixture and washed with demineralized water to remove the residual sulfuric acid. The washing was finished when white precipitate could not be observed by adding $BaCl_2$. The solid sample was then dried with vacuum pump and stored in a desiccator for characterization.

2.2.4 Characterization. To determine of the characteristic of catalyst, the samples were analyzed by nitrogen adsorption-desorption at its boiling point using a gas adsorption-desorption instrument (Nova 1200, Quantachrome). The samples of catalyst were degassed at 300°C under flowing of nitrogen gas for 3 h prior to the measurement. The specific surface area was evaluated using the multi-point Brunauer-Emmet-Teller (BET) method at $P/P_0 < 0.3$. The average pore size was estimated from the desorption profile using the Barrett-Joyner-Halenda (BJH) model. The total pore volume was calculated at P/P_0 approached unity.

The ionic capacity of the catalyst-samples was determined by acid-base titration method. The sample was immersed in 25 ml of 1 N NaOH solution for 48 h. The solid sample was filtered, and the filtrate was diluted 10 times with demineralized water and titrated with standardized 0.1 M HCl solution to determine the remaining NaOH concentration after immersion. The ionic capacity was then calculated by the number of moles of NaOH reacted with the sulfonate group divided by the initial mass of sample immersed.

2.2.5 Performance test of catalyst. To find out the performance, the catalysts was used for esterification reaction of acetic acid with ethanol in a three-neck flask under reflux. 40 mL of ethanol (99.9%, p.a., Merck) was reacted by 20 mL acetic acid (100%, glacial, Merck) with a loading catalyst of 0.2 g. The reaction was carried out for 3 h under constant stirring at 80°C. During reaction, 1 mL of aliquots was sampled from the reaction mixture for analysis. The acetic acid concentration in the aliquot was quantitatively analyzed using a gas chromatography (GC-model, Agilant 6890 series) equipped with a column (DB-WAX 122-7032, Agilent) and a flame ionization detector (FID). The column was operated at 12°C/min ramp from 35°C up to 240°C, kept for 5 min. The carrier gas was nitrogen with a flow rate of 2.5 mL/min and the volume of sample injected was 0.5 μL. The concentration of acetic acid was determined by external calibration and the reaction conversion was calculated by

$$X_A = \frac{C_{A0} - C_A}{C_{A0}} \times 100\% \quad (1)$$

where C_{A0} and C_A are, respectively, the initial and final concentrations of acetic acid. The same procedure was also conducted to Flotrol F-007, a commercial resin, as a comparison.

3. Result and Discussion

The preparation of catalyst started from the synthesis of silica-PEG composite as a basic material for catalyst support. Some amount of PEG 6000 (0.1, 0.25, 0.5, and 1 g) was added to 8% sodium silicate solution which was expected to serve both as a template to form pores in the silica-carbon composite and as a carbon source. One-pot synthesis in this case refers to the synthesis stage of silica-carbon composites and template removal was conducted simultaneously. The carbonization process was carried out under inert flowing nitrogen gas to release the volatile matters and to maintain the formed carbon adhering on silica surface. During the process, removal of the volatile matters left the previously occupied sites empty space to provide pores inside the silica-carbon composite. The formation of pores at this early stage is very important because it would significantly affect the characteristics of the catalyst including surface area and ionic capacity, and in turn, the overall catalyst performance.

The next process was the process of grafting sulfonate groups to porous silica-carbon composites to produce catalysts with the active sites of sulfonate groups that act as acid catalysts in the esterification reaction. This sulfonate group was grafted to silica-carbon composites using sulfuric acid reagents in the presence of atmospheric nitrogen. The presence of nitrogen serves to replace oxygen, so that oxidation was not occurring during the sulfonation process. The sulfonate group attaches to the silica-carbon composite with the van der Waals bond between carbon and sulfonate so that the sulfonate group will be able to exchange ions when used as a solid acid catalyst in the esterification reaction. An important parameter influencing the amount of sulfonate groups grafted on silica-carbon composites are the specific surface area of the material.

Variation of adding of PEG 6000 in the process of synthesis silica-PEG composites produce different pores in silica-carbon composites, as well as in the pores of the catalyst. The main effect of variations by adding of PEG 6000 is on the internal characteristics of the catalyst, such as pore size, pore volume, and specific surface area. Where these characteristics affect the performance of the catalyst in its application to the esterification reaction.

3.1 The effect of PEG 6000 addition to characteristic of catalyst

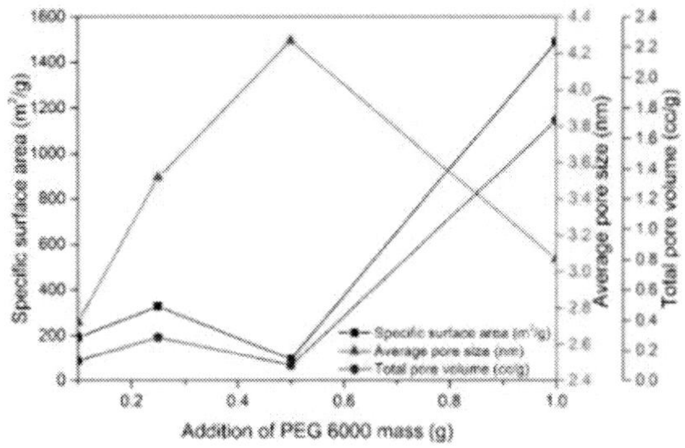

Figure 2. The effect of PEG 6000 addition to surface area, average pore size, and total pore volume of the catalyst

The addition of PEG 6000 at the synthesis of silica-carbon composites affects the characteristics of the catalyst, because PEG 6000 functions as a template and carbon source. The effect of PEG 6000 occurs mainly on the pore characteristics of the catalyst. This can be seen in Figure 2. Fig. 2 shows the effect of adding PEG 6000 to specific surface area, average pore size, and total pore volume on catalyst. Specific surface area, average pore size, and total pore volume ranged from 94.52 to 1490.76 m^2/g, 2.72 to 4.27 nm, and 0.10 to 1.72 cc/g, respectively. Increasing the addition of PEG 6000 tends to increase the specific surface area, average pore size, and total pore volume of catalyst. But, there are the differences among the specific surface area and the total pore volume with the average pore size. The addition 0.5 g of PEG 6000 shows a decrease in the specific surface area and total pore volume of the catalyst, but an increase in the average pore size, which is 94.52 m2/g, 0.10 cc/g and 4.27 nm, respectively. This might be due to the addition of more PEG 6000 which no longer functions as a template, but instead wraps silica [17] which causes the inner pores to not form properly.

3.2 The effect of PEG 6000 addition to the ability of the catalyst to exchange ions and convert the esterification reaction

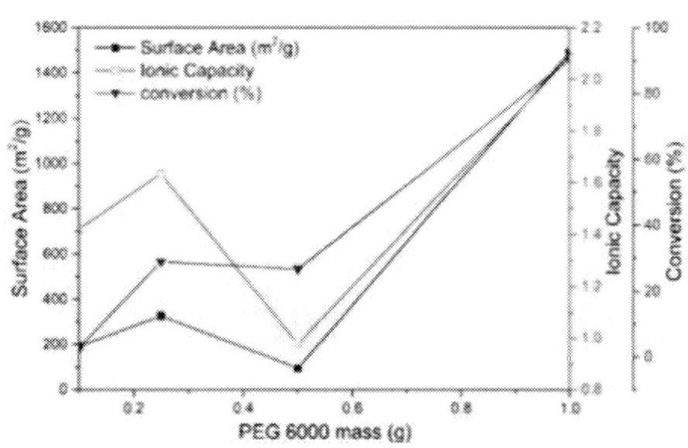

Figure 3. The effect of PEG 6000 addition to surface area of the catalyst, ionic capacity of the catalyst and conversion of the catalyst in esterification reaction

The addition of PEG 6000 tends to increase the specific surface area of the catalyst, where on the catalyst surface there is an active side of the catalyst, which is where the sulfonate groups attach to the catalyst. This sulfonate group plays a role in the reaction, in this case the esterification reaction. The ability of the sulfonate group as an ion exchange in this reaction was measured by ion capacity analysis. Meanwhile, the performance of the catalyst can be tested by involving the catalyst in the esterification reaction between acetic acid and ethanol which is reacted at 80C for 180 minutes. In this reaction, the conversion of acetic acid will be obtained which shows the performance of the catalyst in the reaction. The relationship between surface area, ionic capacity, and conversion of acetic acid in the reaction to the addition of PEG 6000 can be seen in Figure 3.

Figure 3 shows the effect of adding PEG 6000 to the catalyst surface area, the ionic capacity of the catalyst and the catalyst. Like the tendency for a specific catalyst surface area is the ionic capacity of the catalyst and the acetic acid conversion. Ionic capacity and reaction conversion are in the range of 0.98 to 2.09 and 2.75% to 90.27%, respectively. The figure shows that the specific surface area of the catalyst is proportional to the ionic capacity of the catalyst and the ability of the catalyst to convert acetic acid and ethanol in an esterification reaction On addition 0.5 g of PEG 6000, the specific surface area of the catalyst decrease and this is followed by ionic capacity and catalyst conversion ability. This shows that the greater the specific surface area will allow the number of sulfonate groups grafted on the sulfonated catalyst so that the catalyst has a better ability to convert acetic acid and ethanol into ethyl acetate.

3.3 The comparison of catalyst performance with commercial resins

Commercial resins are frequently used by researchers as catalysts. For the esterification reaction, the type of resin used is usually cation exchange resin or also known as acid exchange resin. This is due to the presence of a sulfonate group in the resin which functions as an active group in ion exchange. Several types of resins used by researchers in the esterification reaction include amberlyst 15 [7]–[10], amberlyst 16 [10], amberlyst 131 [10], Indion 130 [9], Indion 190 [9], etc. To determine the performance of this catalyst in the esterification reaction, a comparison was made with the performance of a commercial resin, namely Flotrol F-007, and quantitatively it can be seen from the conversion of acetic acid which is converted to ethyl acetate when reacted with ethanol.

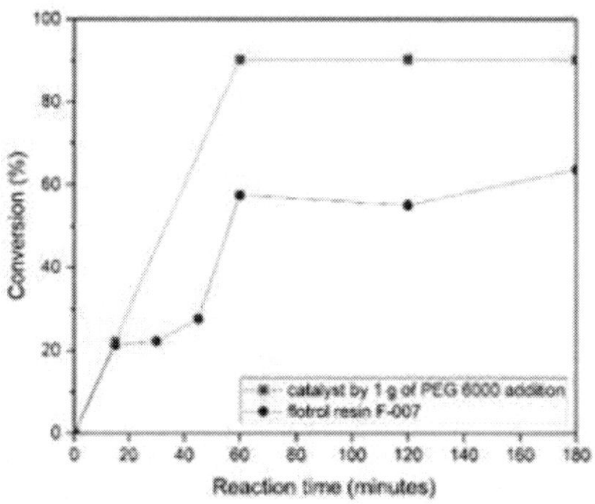

Figure 4. Performance of catalyst and Flotrol F-007 in esterification reaction at 80°C

Figure 4 shows the performance of the catalyst in the esterification reaction between acetic acid and ethanol at a temperature of 80 °C reaction temperature for 180 minutes using 0.2 g of catalyst prepared with the addition 1 g of PEG 6000 and by using a commercial resin, flotrol F-007. This figure shows that the equilibrium conversion occurs 60 minutes after the reaction begins and then tends to be almost fixed. The results of the reaction conversion using Flotrol F-007 were lower than the use of catalyst prepared with the addition 1 g of PEG 6000. At 60 minutes, the conversion achieved by Flotrol F-007 reached 57.46% and at 180 minutes it reached 63.66% while the catalyst with the addition of 1 gram of PEG 6000 reaching 90.27% for up to 180 minutes.

4. Conclusion
The addition of PEG 6000 in the initial step of the synthesis of catalyst, synthesis the silica-PEG composite, tends to increase the specific surface area to 1490.76 m^2/g and increase the catalyst ion capacity to 2.09. The catalyst was successfully converted acetic acid and ethanol with a conversion of 90.27% and higher than Flotrol F-007.

5. References
[1] G. Ertl, H. Knozinger, and J. Weitkamp, *Preparation of Solid Catalysts*. 1999.
[2] Y. Lu, X. Liang, and C. Qi, "Synthesis of novel carbon/silica composites based strong acid catalyst and its catalytic activities for acetalization," *Bull. Mater. Sci.*, vol. 35, no. 3, pp. 419–424, 2012, doi: 10.1007/s12034-012-0292-8.
[3] C. He *et al.*, "Templated Silica with Increased Surface Area and Expanded Microporosity: Synthesis, Characterization, and Catalytic Application," *Chem. Eng. J.*, vol. 162, no. 3, pp. 901–909, 2010, doi: 10.1016/j.cej.2010.06.037.
[4] P. Valle-vigón, M. Sevilla, and A. B. Fuertes, "Sulfonated mesoporous silica – carbon composites and their use as solid acid catalysts," *Appl. Surf. Sci.*, vol. 261, pp. 574–583, 2012, doi: 10.1016/j.apsusc.2012.08.059.
[5] X. Tian *et al.*, "Sulfonic acid-functionalized mesoporous carbon/silica as efficient catalyst for dehydration of fructose into 5-hydroxymethylfurfural," *RSC Adv.*, vol. 6, pp. 1–10, 2016, doi: 10.1039/C6RA20304C.
[6] E. M. Björk *et al.*, "Mesoporous silica and carbon based catalysts for esterification and biodiesel fabrication — The effect of matrix surface composition and porosity," *"Applied Catal. A, Gen.*, vol. 533, pp. 49–58, 2017, doi: 10.1016/j.apcata.2017.01.007.
[7] S. I. Kirbaslar, Z. B. Baykal, and U. Dramur, "Esterification of Acetic Acid with Ethanol Catalysed by an Acidic Ion-Exchange Resin," *Turky J Engin Env. Sci*, vol. 25, no. 837, pp. 569–577, 2001.
[8] V. C. Nguyen, N. Q. Bui, P. Mascunan, T. T. H. Vu, P. Fongarland, and N. Essayem, "Esterification of aqueous lactic acid solutions with ethanol using carbon solid acid catalysts: Amberlyst 15, sulfonated pyrolyzed wood and graphene oxide," *Appl. Catal. A Gen.*, vol. 552, no. December, pp. 184–191, 2018, doi: 10.1016/j.apcata.2017.12.024.
[9] P. E. Jagadeeshbabu, K. Sandesh, and M. B. Saidutta, "Kinetics of Esterification of Acetic Acid with Methanol in the Presence of Ion Exchange Resin Catalysts," *Ind. Eng. Chem. Res.*, vol. 50, pp. 7155–7160, 2011.
[10] R. Tesser, L. Casale, D. Verde, M. Di Serio, and E. Santacesaria, "Kinetics and modeling of fatty acids esterification on acid exchange resins," *Chem. Eng. J.*, vol. 157, no. 2–3, pp. 539–550, 2010, doi: 10.1016/j.cej.2009.12.050.
[11] Shagufta, I. Ahmad, and R. Dhar, "Sulfonic Acid-Functionalized Solid Acid Catalyst in Esterification and Transesterification Reactions," *Catal. Surv. from Asia*, vol. 21, no. 2, pp. 53–69, 2017, doi: 10.1007/s10563-017-9226-1.
[12] J. Janaun and N. Ellis, "Role of silica template in the preparation of sulfonated mesoporous carbon catalysts," *Appl. Catal. A Gen.*, vol. 394, no. 1–2, pp. 25–31, 2011, doi: 10.1016/j.apcata.2010.12.016.
[13] R. Liu, X. Wang, X. Zhao, and P. Feng, "Sulfonated ordered mesoporous carbon for catalytic preparation of biodiesel," *Carbon N. Y.*, vol. 46, no. 13, pp. 1664–1669, 2008, doi: 10.1016/j.carbon.2008.07.016.
[14] P. P. Upare *et al.*, "Chemical conversion of biomass-derived hexose sugars to levulinic acid over sulfonic acid-functionalized graphene oxide catalysts," *Green Chem.*, vol. 15, no. 10, p. 2935, 2013, doi: 10.1039/c3gc40353j.
[15] M. Fauziyah, W. Widiyastuti, and H. Setyawan, "Sulfonated carbon aerogel derived from coir fiber as high performance solid acid catalyst for esterification," *Adv. Powder Technol.*, no. January, 2020, doi: 10.1016/j.apt.2020.01.022.
[16] H. Setyawan, M. Yuwana, and R. Balgis, "PEG-templated mesoporous silicas using silicate precursor and their applications in desiccant dehumidification cooling systems," *Microporous*

Mesoporous Mater., vol. 218, pp. 95–100, 2015, doi: 10.1016/j.micromeso.2015.07.009.
[17] H. Setyawan and R. Balgis, "Mesoporous silicas prepared from sodium silicate using gelatin templating †," *Asia Pacific J. Chem. Eng.*, 2011, doi: 10.1002/apj.

Acknowledgements
The authors are grateful for the financial support provided by the Ministry of Research, Technology, and Education Higher of the Indonesia through Fundamental Research Grant (No. 839/PKS/ITS/2018 and 827/PKS/ITS/2019). One of the authors (S.J.) would like to thank the Ministry of Research, Technology and Higher Education of Indonesia for the doctoral scholarship. The authors also want to thank Ms. Aidatu Rosyida and Ms. Febrita Shela Utami for assistance in the experiments.

IOP Conf. Series: Materials Science and Engineering 1010 (2021) 012035 doi:10.1088/1757-899X/1010/1/012035

Effect of fly ash to water-cement ratio on the characterization of the concrete strength

E Susanti[1], H Istiono[1], I Komara[1], D Pertiwi[1], Y Septiarsilia[1], F K Syahputra[1]

[1] Department of Civil Engineering, Faculty of Civil, environmental, and planning, Institut Teknologi Adhi Tama Surabaya, East Java, Indonesia.

ekasusanti@itats.ac.id

Abstract. Fly ash (FA) is burning coal product which is used as supplementary cementitious materials (SCM) as the conjunction with cement. The use of FA is considered by its pozzolanic properties i.e. SiO_2, Al_2O_3 and CaO to hardened concrete. The potential of FA as SCM has been well-known over the globe but that significant utilization has to be studied based on its compound's characteristics and pozzolanic activity. For this case, the amount of FA is added 30% by mass of total cementitious material, concrete with fly ash (NFA) and another blended without FA as normal concrete (NC) made of ordinary Portland cement. This amount is indicated as dosage in moderate to high volume of FA classification. This replacement pairs with water to cement ratio (W/C) with the value of 0.47 and 0.30 e.g., NC1, NC2 and NFA1, NFA2, respectively. These W/C uses considering effect FA on water demand of concrete proportioned for equal slump. The results showed that NC2 and NFA2 in the age of 28-day inform higher compressive strength, 30% and 28%, respectively. Another indication from the comparison of NC2 and NFA2 were also in good agreement that the use of FA decrease the compressive strength around 3% to 6%.

1. Introduction

Fly ash (FA) is a coal combustion residual which appears as a by-product from processing combustion in steam generator of industrial thermal power plants. The amount of FA is enormously cause serious problems which is related to the environment pollution [1]. Figure 1 inform the increasing cases which is factories pulverized the FA in each year. In fact, the problem is now successfully solved to use a pollutant of FA as a replacement of cement. Many researchers studied the variety of FA in the mixture composition to substitute cement started in low dosage to very high dosage [2]–[4] . Historically, most common used of FA in concrete normally at levels ranging from 10% to 25% by mass of the cementitious materials but it still depends on the pozzolanic activity [5]. Many researchers give the assumption that type of FA is also considered by the location and production type of coal pulverisers which can be deficient between sources and collection methods[6]–[9]. This parameter condition effects the performance of FA in concrete that is followed by the influence of its physical and chemical properties. The most used specification is followed by ASTM C618 [10]. This specification characterises the FA into two clusters based on its chemical compound is illustrated in Figure 2. Due to the differentiation including the match of the raw material, very often, the potential of FA as a replacement does not meet the required standard.

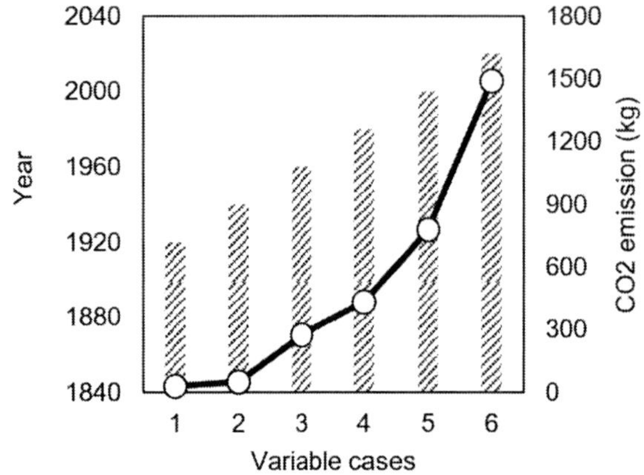

Figure 1. Emission product by factory due to pulverises process [1]

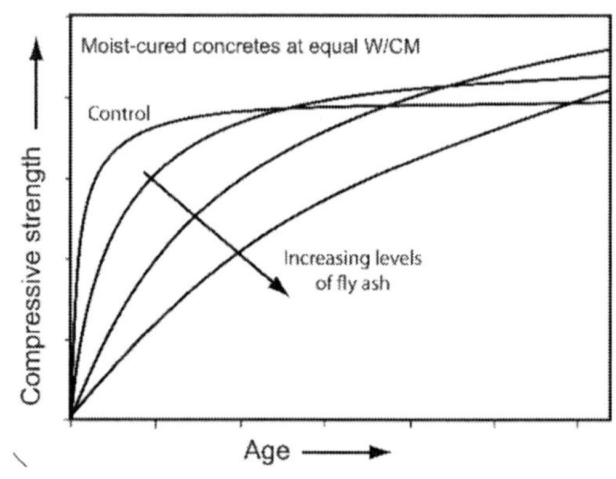

Figure 2. Effect of development FA on compressive strength [11]

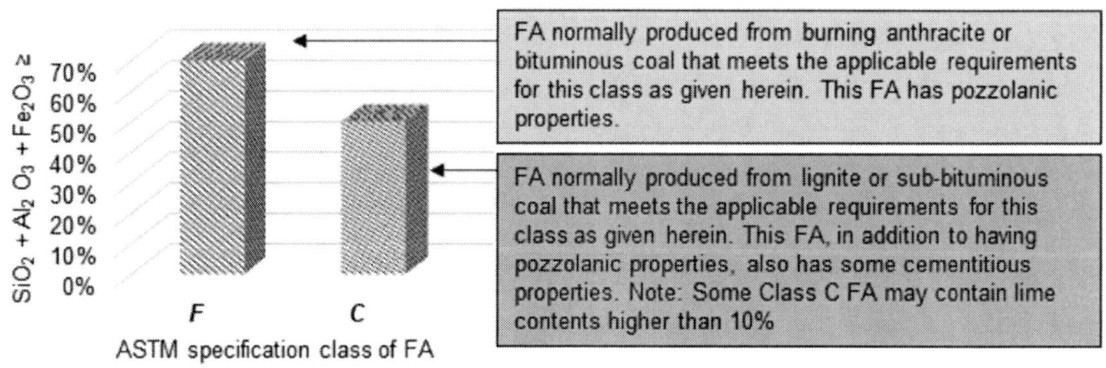

Figure 3. ASTM C618 classification [10]

In this case, the FA from PLTU Paiton Probolinggo is investigated. The mass pollutant product in this PLTU inevitable become a problem due the lack of distribution and allocation. The previous studies have been investigated the same FA but implying a low capacity with high dosage characterization [12]. In this current research, the dosage of FA will remain the same 30% from the total of cementitious materials using two composition of water to cement/binder e.g., 0.47 and 0.3 which is categorized as a high-volume dosage classification. This classification of dosage levels of FA follows the recommendation of Portland cement association (PCA). The chemical compound will first be evaluated and followed to observe the concrete strength behaviour in variety of ages, 14-day, 28-day and 56-day.

1.1. Applications of FA in Concrete

The use of FA in the concrete construction was first designed for a mass concrete application where the early strength was required and where the condition of low-heat generation was implied [13]. The use of this concrete was also considered to support the durability by increasing its mechanical properties [11]. Figure 2 inform the chemical content with the amount of total oxides for FA class F over 70% and FA class C over 50% (SiO_2+ Al_2O_3 + Fe_2O_3) according to ASTM C618 [10].

FA is also favoured to concrete through the use of blended hydraulic cement consisting of Portland cement (PC), FA and other cementitious materials. One type of the concrete that well-paired with the optimization of FA is engineered cementitious composite (ECC) [14]–[17]. Blended cement containing

FA are widely available in Indonesia. Most industrial company implies this composition to support the production and provide an affordable mixing portion. In addition to maximize the use of pollutant of FA which is considered as the biggest pollutant distributed by the process, known as 75% by product [18].

1.1.1. Effect of FA on the fresh concrete properties. In some cases, the use of superplasticizer is necessary due to a very low water content to obtain a workability. Various researchers have successfully produced the optimum composition of FA concrete using locally materials and admixture [19], [20]. It is mainly considered to the proportion of FA on the concrete mixture. A well-portioned mixture using FA will improve the workability when compared to PC concrete with the same slump. This is highlighted, with a given slump, FA concrete consolidates better than conventional PC. In addition, FA also advances the cohesiveness and lessen the segregation [21]. It is emphasizing that all the benefits of FA will only be grasped in well-portioned of concrete mixture. The fresh concrete properties are greatly influenced by the mixture compositions including the type and amount of cementitious material, water to content ratio, aggregate classification, the condition of entrained air and the use of superplasticiser. One of the studies improved rheological properties of FA in high-volume has been conducted by Bouzoubaa (2001) [22], [23].

1.1.2. Effect of FA on the hardened concrete properties. Concrete is normally proportioned to achieve a certain minimum strength at specified age, 14-day, 28-day (typical age of the concrete) and 56-day. Figure 3 illustrates the effect on compressive strength of replacing a certain mass of PC with a low-calcium FA class C and managing a constant content of W/C. The early-age strength is decreased as the level of replacement increased. However, to maintain the development in a long term, the FA has to be controlled in each ages of strength and the quality control of curing has to be provided. The behaviour of FA development shows in the Figure 3 which strength parity with the PC is achieved is greater at higher lever of FA. The FA content can affect the ultimate of compressive strength.

2. Experimental methods

2.1. Test Specimens

A series of cylinder compressive strength specimens with two matrix constituents were made and tested in the concrete laboratory of ITATS in Surabaya. The specimens were used 300mm×150mm dimension and divided by the ages e.g., 14-day, 28-day and 56-day and tested by using electric hydraulic pump compression. Each age was distributed by using 3 specimens to give a proper behaviour. These specimens, hereafter referred to as NC and FA specimen which was concerned with two variety of W/C content, 0.47 and 0.30 namely NC1 and NC2 and also followed by NFA1 and NFA2. The FA was taken from PLTU Paiton Probolinggo which were categorized its chemical compound by using X-ray Fluorescence (XRF) using Rigaku RIX-3000 model (see Figure 5).

2.2. Material Properties and Fabrication

The concrete mixture used to fabricate the specimens are presented in Table 1. The mixtures for both NC and NF used two different water-to-cement ratios of 0.47 and 0.30, respectively. The materials were firstly confirmed their characterization follow the specification standard from various ASTM aspects (see Table 2). The coarse aggregate was taken from Pasuruan (Banyu Biru) with a maximum aggregate size of 20 mm, fine aggregate (Lumajang) with a size < 3 mm and PC (PCC-Gresik), and superplasticizer to help the workability. One mixing was conducted for 5 cylinders using automatic mixer with 0.1 m^3 capacity.

Each specimen was casted into steel cylinder 300mm×150mm and then cured in fully submerged water. The fresh concrete before casting were poured to control the workability using slump cone test in accordance with ASTM C143 [24]. A summary of concrete compressive strength is accordance to ASTM C39 and presented in Figure 5 followed by the detail of each specimen in Figure 6 to Figure 11.

Table 1. Summary of concrete mixture and slump value

Details	W/C 0,47		W/C 0,3	
	NC1	NFA1	NC2	NFA2
Fine Aggregate (kg/m^3)	628	628	628	628
Coarse Aggregate (kg/m^3)	1166	1166	1335	1335
Cement (kg/m^3)	436	305,2	436	305,2
Fly Ash (kg/m^3)	-	130,8	-	130,8
Water (kg/m^3)	205	205	131	131
Superplasticizer (g/m^3)	-	-	5,6	5,6
Slump (cm)	12	14	10	12

Note: coarse aggregate is crushed granite in <20mm with fine aggregate <3mm

Table 2. Characteristic materials of fine and coarse aggregate compared to required standard

Test	Fine Agg.	Req. standard	Coarse Agg.	Req. standard	Guideline code
Moisture (%)	2,1	1-5	1,62	1-5	ASTM C 556-89
Relative Density (gr/cm^3)	2,67	1,6-3,3	2,78	1,6-3,3	ASTM C 127-93
Absorption Capacity	2,14	Max 5	1,23	Max 4	ASTM C 556-89
Bulk Density (Loose) (kg/dm^3)	1,41	0,4-1,9	1,35	0,4-1,9	ASTM C 29/C29M-91[25]
Bulk Density (Compact) (kg/dm^3)	1,54	0,4-1,9	1,5	0,4-1,9	ASTM C 29/C29M-91
Mud content (%)	3,52		2,1	Max 5	ASTM C 117-95
Organic Impurities	No.3	No.6	-	-	ASTM C 40-92
Fineness Modulus	Zone 2 Fm=3,05	2<Fm<3,1	Fm=7,3 5	6,5<F m<8	ASTM C 136-95A ASTM C 33

Table 3. Summary of compressive strength obtained from 300×150 mm cylinder

Specimen		NC1+NFA1 (days)			NC2+NFA2 (days)		
		14	28	56	14	28	56
NC1+NFA1	Mean (MPa)	24.69	27.12	28.37	23.33	26.16	27.26
	Standard of deviation (SD in MPa)	0.31	2.43	3.69	0.17	2.83	3.93
NC2+NFA2	Mean (MPa)	27.94	31.01	32.17	25.95	29.18	30.11
	Standard of deviation (SD in MPa)	0.52	3.14	4.24	0.19	3.23	4.16

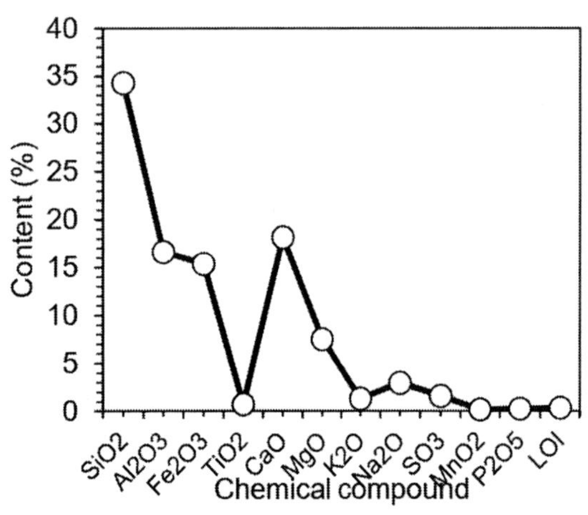

Figure 4. Chemical compound of Paiton FA class C

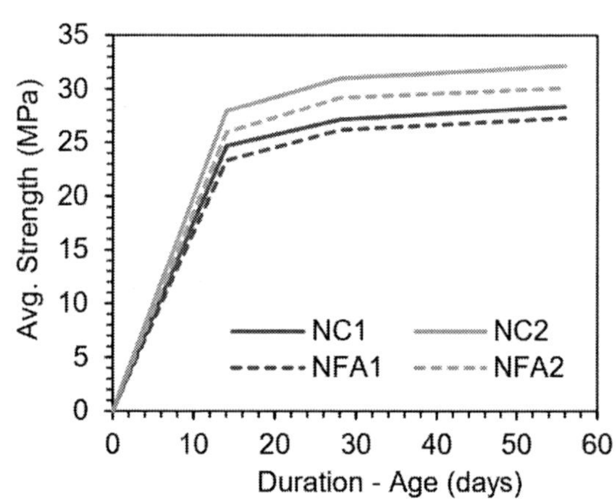

Figure 5. Average of compresive strength due to miture proportioned

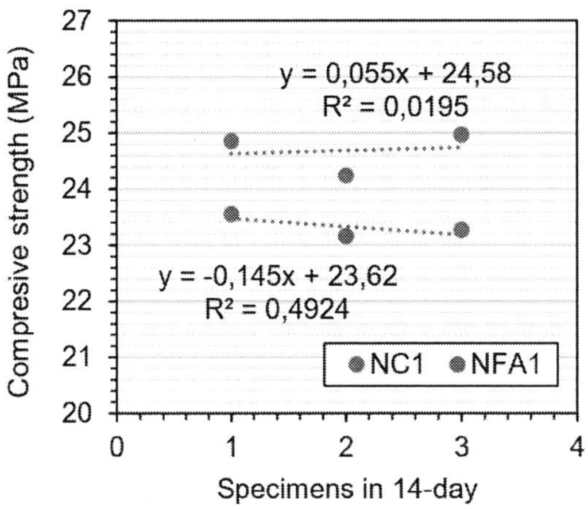

Figure 6. Compressive strength W/C 0.47 – 14 days

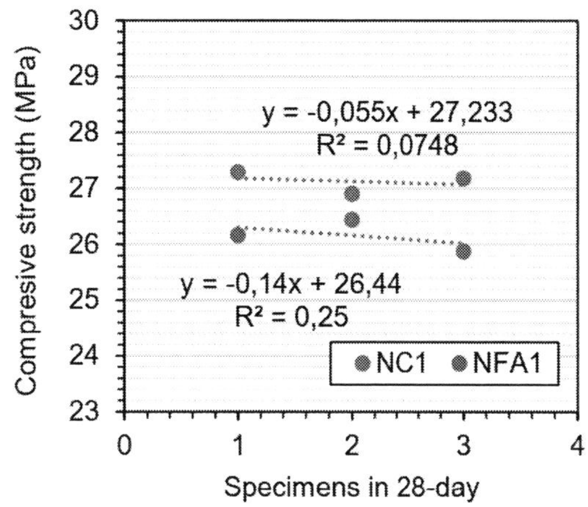

Figure 7. Compressive strength W/C 0.47 – 28 days

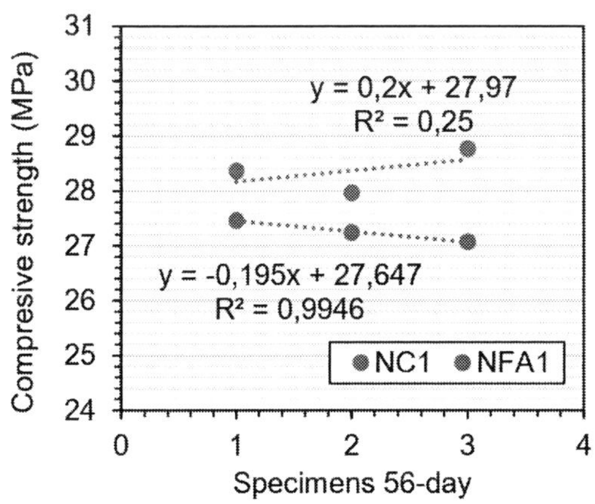

Figure 8. Compressive strength W/C 0.47 – 56 days

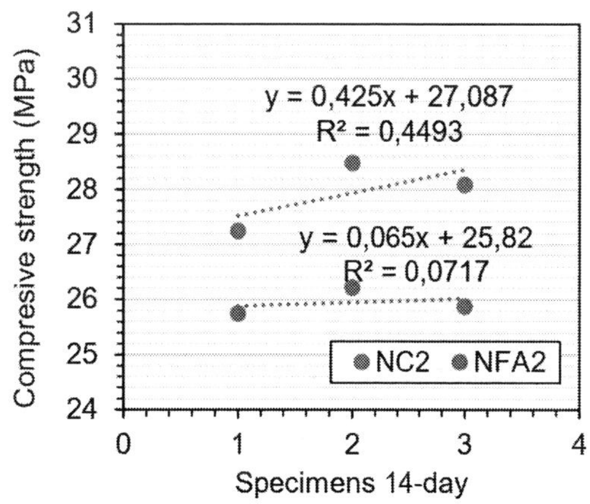

Figure 9. Compressive strength W/C 0.30 – 14 days

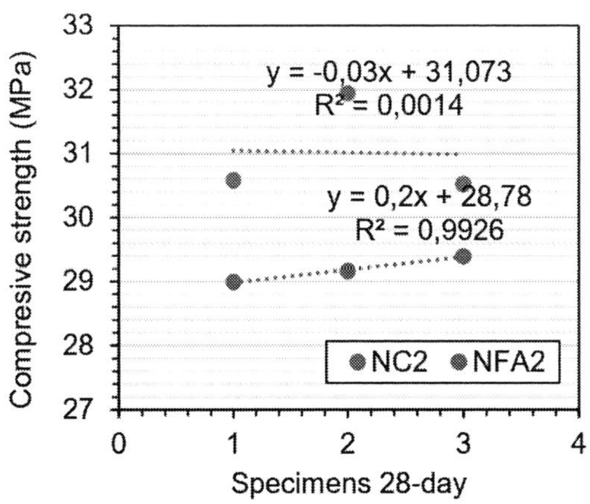

Figure 10. Compressive strength W/C 0.30 – 28 days

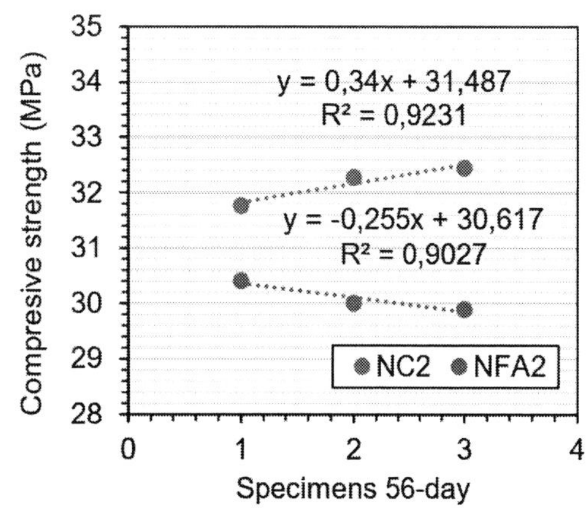

Figure 11. Compressive strength W/C 0.30 – 56 days

3. Results and discussion

A series of compressive strength of W/C 0.3, NC1 and NFA1 were listed in Figure 6 to Figure 8, and W/C 0.47, NC2 and NFA2 were listed in Figure 9 to Figure 11. The value was then divided by the age under curing exposure time, 14-day, 28-day and 56-day. Among all the mixtures, NC1, NC2 and NFA1, NFA2, for 14-day compressive strength achieved no maximum strength, then rises in the 28-day and 56-day. In 14-day NC1 shows lesser 27% compressive strength than NC2 and gradually increasing in 28-day and 56-day with a point 30% and 32%. The NC2 presents a good agreement with the portion of W/C 0.30 which is smaller than NC1 with 0.47. The same behaviour is also illustrated by FA1 and FA2 but informing the lower performance than NC1 and NC2. The FA1 compares to FA2 deponently increase by the 14-age, 28-age and 56-age in about 25%, 28% and 29%, respectively.

PC normally achieves its maximum strength in 28-day and increase insignificantly. However, FA imply a bit different than PC. FA increases gradually when the age of the concrete increasing. The greater portion of FA also gives the pattern of the performance including the class type of FA. In fact, this portion of the mixture using 30% replacement of cement still give a good result according to design specification. It is proved by composition of NFA in both W/C. NFA follow the raising of compressive strength when the age of curing increasing. These NFA considers effect FA on water demand of concrete proportioned for equal slump. In addition, NC2 and NFA2 in the age of 28-day inform higher compressive strength, 30% and 28%, respectively. Another indication from the comparison of NC2 and NFA2 were also in good agreement that the use of FA decrease the compressive strength around 3% to 6%.

Table 4 informs the variety of modulus of rupture and modulus of elasticity of concrete specimens which are classified in two different W/C ratios. Results shows that as far as the compressive strength is concerned in both NC and NFA, regardless of W/C, produced concrete with an acceptable ultimate compressive strength, f'_c . However, f'_c decreases as the W/C increased. Results also indicate that modulus of fracture, f_r and modulus of elasticity, E_e for concrete made by PC is always higher that the corresponding FA. This observation agrees the findings of another research conducted by Alfian [20]. This condition may be attributed to the composition of W/C.

As the information given from Table 4, the compressive strength is customarily used to understand many of the engineering properties of the concrete which have concerned in a good performance such as f_r and E. Measured values of f_r *of* NC and NFA is accordingly to Table 4 which is divided by the effect of the ages. The result concedes with the f'_c to f_r. The evaluation of f_r and E_e is according to ACI formula [21].

Table 4. Modulus of rupture and modulus of elasticity of the specimen

Type of Mixture	W/C	Strength (Mpa)			Modulus of rupture (Mpa)			Elasticity (Gpa)		
		14-day	28-day	56-day	14-day	28-day	56-day	14-day	28-day	56-day
		24.86	27.29	28.37	3.49	3.66	3.73	23434	24553	25034
NC1	0.47	24.24	26.9	27.97	3.45	3.63	3.70	23140	24377	24857
		24.97	27.18	28.77	3.50	3.65	3.75	23486	24503	25210
		23.56	26.16	27.46	3.40	3.58	3.67	22813	24039	24629
NFA1	0.47	23.16	26.44	27.24	3.37	3.60	3.65	22619	24167	24530
		23.27	25.88	27.07	3.38	3.56	3.64	22672	23910	24454
		27.24	30.58	31.77	3.65	3.87	3.95	24530	25991	26491
NC2	0.3	28.48	31.94	32.28	3.74	3.96	3.98	25082	26562	26703
		28.09	30.52	32.45	3.71	3.87	3.99	24910	25965	26774
		25.75	28.99	30.41	3.55	3.77	3.86	23850	25306	25918
NFA2	0.3	26.22	29.16	30.01	3.58	3.78	3.83	24067	25380	25747
		25.88	29.39	29.9	3.56	3.79	3.83	23910	25480	25700

$$f_r = 0.7(f'_c)^{0.5} \tag{1}$$

$$E_c = 4700\ (f'_c)^{0.5} \tag{2}$$

The comparison for both NC and NFA is similarly presented the same and also well-informed in Table 4. The tendency of the result is small but comparable to each mixture proportioned. In general, the empirical relationship of the Eq. 1 and Eq. 2 have the benefits in simplicity, for the exact evaluation, experimental program has to be handed.

4. Conclusions

Based on the result conducted on variety of compressive strength age stage vs the water-to-cement ratios using FA class C as a replacement, the following conclusions can be drawn:

1) The rate gaining in strength of FA with variety of 0.47 and 0.30 W/C are observed to be lower than the corresponding PC concrete.
2) FA provides a satisfactory result concerned the high-volume replacement as 30% which is nearly reached the same capacity as PC concrete.
3) FA with a low calcium content with 30% replacement shows no optimum strength and implies a similar behaviour with NC.
4) Modulus of fracture and modulus elasticity for both NC and NFA is comparable to the compressive strength which illustrates the similar behaviour.
5) The application of FA using high-volume mixture as a replacement of cement reduces CO_2 emission that impact to the safety environment and energy conservation.

5. References

[1] Andrew R M, 2019 Global CO2 emissions from cement production, 1928-2018 *Earth Syst. Sci. Data* **11**, 4 p. 1675–1710.
[2] Halstead W J, 1986 Use of Fly Ash in Concrete. *Natl. Coop. Highw. Res. Program, Synth. Highw. Pract.* **96**, Reapproved p. 1–34.
[3] Atiş C D, 2003 High-volume fly ash concrete with high strength and low drying shrinkage *J. Mater. Civ. Eng.* **15**, 2 p. 153–156.
[4] Bentz D P, 1812 Best Practices Guide for High-Volume Fly Ash Concretes : Assuring Properties and Performance NIST Technical Note 1812 Best Practices Guide for High-Volume Fly Ash Concretes : Assuring Properties and Performance.
[5] Alsayed S H and Amjad M A, 1996 Strength, Water Absorption and Porosity of Concrete Incorporating Natural and Crushed Aggregate *J. King Saud Univ. - Eng. Sci.* **8**, 1 p. 109–119.
[6] Ghosh A Ghosh A and Neogi S, 2018 Reuse of fly ash and bottom ash in mortars with improved thermal conductivity performance for buildings *Heliyon* **4**, 11 p. e00934.
[7] Nuaklong P Wongsa A Sata V Boonserm K Sanjayan J and Chindaprasirt P, 2019 Properties of high-calcium and low-calcium fly ash combination geopolymer mortar containing recycled aggregate *Heliyon* **5**, 9 p. e02513.
[8] Wattimena O K Antoni and Hardjito D, 2017 A review on the effect of fly ash characteristics and their variations on the synthesis of fly ash based geopolymer *AIP Conf. Proc.* **1887**, September.
[9] Islam M M Islam M S Rahman M A and Das A, 1970 Strength Behavior of Mortar Using Slag As Partial Replacement of Cement *MIST J. GALAXY* **3**, September 2010 p. 98–106.
[10] Specification S, 2002 Standard Specification for Coal Fly Ash and Raw or Calcined Natural Pozzolan for Use p. 1–3.
[11] Thomas B M Ph D Eng P Engineering C and Brunswick N, Optimizing the Use of Fly Ash in Concrete.

[12] Cross D and Stephens J, 2008 Evaluation of the Durability of 100 Percent Fly Ash Concrete *Biosci. Bioeng.* **93**, November 2006 p. 101–103.

[13] Lamine Zeggar M Azline N and Azizi Safiee N, 2019 Fly ash as supplementry material in concrete : A review *IOP Conf. Ser. Earth Environ. Sci.* **357**, 1.

[14] Oktaviani W N Tambusay A Komara I Sutrisno W Faimun F and Suprobo P, 2020 Flexural Behaviour of a Reinforced Concrete Beam Blended with Fly ash as Supplementary Material *IOP Conf. Ser. Earth Environ. Sci.* **506** p. 012042.

[15] Komara I Tambusay A Sutrisno W Suprobo P and Iranata D, 2019 The Investigation study of improving Durability Performance of Marine Infrastructure by using the Engineered Cementitious Composite in *The 14th International Student Conference on Advanced Science and Technology (ICAST) 2019* p. 8–12.

[16] Komara I Tambusay A Sutrisno W and Suprobo P, 2019 Engineered Cementitious Composite as an innovative durable material: A review *ARPN J. Eng. Appl. Sci.* **14**, 4 p. 822–833.

[17] Bastian M A Tambusay A Komara I Sutrisno W Irawan D and Suprobo P, 2020 Enhancing the Ductility of a Reinforced Concrete Beam using Engineered Cementitious Composite *IOP Conf. Ser. Earth Environ. Sci.* **506** p. 012044.

[18] M. D.J. Sumajouw and Rangan B V., 2006 LOW-CALCIUM FLY ASH-BASED GEOPOLYMER CONCRETE : REINFORCED BEAMS AND COLUMNS By Faculty of Engineering Curtin University of Technology *GC3, Fac. Eng. Curtin Univ. Technol.* p. 1–120.

[19] Abdun-Nur E A *et al.*, 1987 Use of Fly Ash in Concrete. *ACI Mater. J.* **84**, 5 p. 381–409.

[20] Sivasundaram V Carette G G and Malhotra V M, 1990 Selected properties of high-volume fly ash concretes *Concr. Int.* **12**, 10 p. 47–50.

[21] Wight Chair J K *et al.*, 2005 Consulting Members BUILDING CODE REQUIREMENTS FOR STRUCTURAL CONCRETE (ACI 318-05) AND COMMENTARY (ACI 318R-05) REPORTED BY ACI COMMITTEE 318 ACI Committee 318 Structural Building Code † Deceased **2003**.

[22] Bouzoubaâ N and Lachemi M, 2001 Self-compacting concrete incorporating high volumes of class F fly ash: Preliminary results *Cem. Concr. Res.* **31**, 3 p. 413–420.

[23] Bouzoubaâ N Bilodeau A Tamtsia B and Foo S, 2010 Carbonation of fly ash concrete: Laboratory and field data *Can. J. Civ. Eng.* **37**, 12 p. 1535–1549.

[24] ASTM C143/C143M, 2015 Standard Test Method for Slump of Hydraulic-Cement Concrete *Astm C143* 1 p. 1–4.

[25] Astm:C29/C29M-09, 2009 Standard Test Method for Bulk Density (" Unit Weight ") and Voids in Aggregate *ASTM Int.* **i**, c p. 1–5.

IOP Conf. Series: Materials Science and Engineering 1010 (2021) 012036 doi:10.1088/1757-899X/1010/1/012036

Cold milling machine productivity analysis for determining operator efficiency values in road maintenance projects

S Choiriyah[1], T M C Agusdini[1], G A Prasetyo[1]

[1]Department of civil engineering, Institut Teknologi Adhi Tama Surabaya, East Java, Indonesia

siti.choiriyah@itats.ac.id

Abstract. In construction work uses various types of heavy equipment to make road construction and maintenance, one of which is Cold Milling Machine. The use of cold milling machine is influenced by the operator efficiency factor in its operation and will have an impact on the productivity of the machine. The purpose of this study was to determine the productivity ratio between old and new operators and the number of efficiency factors for cold milling machine operators. The variable reviewed is the result of excavation from the cold milling machine operated by the operator. The research method used is direct observation in the field then the data is processed using t-test analysis. The results of this study are there are differences in skills between operator A and operator B based on the results of the t value (2.846> 1.812) greater than t table. The result of the calculation of the efficiency factor of cold milling machine operators is that by using a comparison of theoretical and actual productivity, it gets a very good category with a value of 0.84 for old operators and for new operators obtains a value of 0.48 which is categorized as less.

1. Introduction

Maintenance of damaged roads at this time is not a problem because with the development of construction technology there is a heavy equipment called the Cold Milling Machine. This Cold Milling Machine is used for stripping the old asphalt surface to be repaired into small pieces [1]. Factors affecting machine productivity are the ability of the mechanical operator to run the Cold Milling Machine. The ability of the operator in operating the machine is needed so that the tool can operate optimally [2]. Operator expertise and skills can be obtained from education and skills. The ability of the operator can be obtained by looking at the operator efficiency factor. Because the efficiency factor is one of the factors that can affect the machine in addition to machine capacity and cycle time [3]. The conditions that exist in toll road maintenance projects, the operators who operate cold milling equipment consist of unskilled operators who have little work time and skilled operators with years of service, these unskilled operators replace the previous operators duc to resignation, so that only know how to operate which is obtained from the instructions of skilled operators so that his knowledge is limited. Therefore, the importance of determining the work productivity factor of cold milling machine operators on road maintenance projects is to determine the factors that can increase operator efficiency in operating cold milling machine tools.

1.1 Literature review

1.1.1 Productivity. Productivity is the amount of output (output) of a particular volume of work produced by time unity [4]. Productivity is defined as the ratio between output and input or the ratio between production output with the total resources used, resources including labor, materials, methods, and tools [5].

1.1.2 Operator Heavy equipment operator is a job that requires special skills to run a variety of tools such as Cold Milling Machines, mastery of these tools is highly needed by highly skilled professionals in their fields, mastery of heavy equipment, knowledge, and operator attitudes about equipment will affect other aspects [6].

1.1.3 The Important Role of a Heavy Equipment Operator Productivity is defined as the ratio between output and input or the ratio between production output and the total resources used [7]. Resources include labor, materials, money, methods and tools. Productivity is the ability of the tool in units of time (m^3 / hr) that depends on the capacity, tool cycle time, and tool efficiency [8]. Operators who operate heavy equipment must be capable or competent (preferably certified) through rigorous selection. We recommend that each device and operator remain, do not change operators too often without sufficient reason [9].

1.1.4 Efficiency Factor Factors that affect work productivity, it is very difficult to determine the amount of work efficiency at the operator, but with the training gained and the experiences gained by the operator. In order to obtain a value that is close to reality on the ground, a correction factor must be included that is suitable for working conditions.

Table 1. Operator Efficiency

Skills	**Efficiency**
Very good	0,90 – 1,00
Well	0,83
Less	0,5 – 0,60

1..1.5 Cold Milling Machine Cold Milling Machine is a heavy equipment used in road maintenance projects with asphalt concrete or rigid pavement construction which functions to damage asphalt pavement layers (surface course and bunder course) with a certain depth and width so that flat surfaces are obtained (level) [10]. The specifications of a Cold Milling Machine are determined by the width, depth of cutting capability, and cutting speed. Cold Milling Machine per hour production capacity is:

$$Q = v \times b \times Fa \times t \times 60 \ m^3 \quad (1)$$

Explanation:
b = excavation / demolition width capacity (m)
t = thickness of excavation / demolition (m)
v = disassembly rate (m / min)
Fa = work efficiency factor
60 = hour to minute conversion

Cold Milling Machine is one type of working tool used to peel the asphalt pavement layer. This tool is very helpful for time and cost efficiency. Cold milling movement in the direction of traffic and scrapping results are directly loaded into the dump truck, at this stage of the work the data is collected once every 1 hour by measuring the results of cold milling excavation.

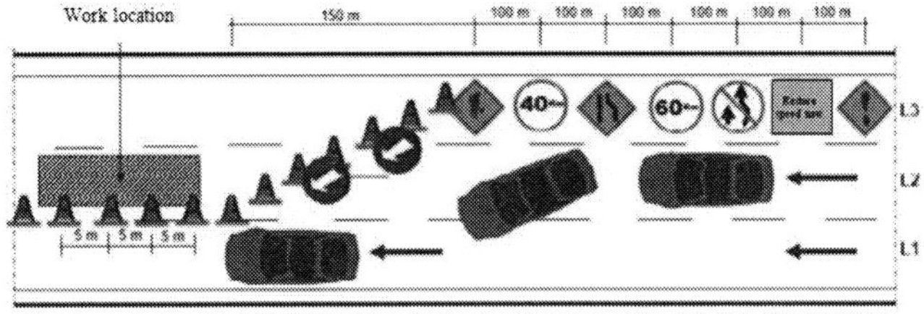

Figure 1. Traffic Handling

Figure 2. Job description for cold milling

2. Methodology

This research uses a comparative quantitative approach by conducting literature studies, surveys and observations in the field to obtain productivity data. Each operator, observations were made on 2 lanes namely the Surabaya-Gempol toll road, from the samples obtained were categorized as correlated samples so that the test used a t-test and to determine the value of the operator efficiency factor using a comparison of theoretical productivity with actual productivity. There are two data needed, namely primary and secondary data. Primary data consists of excavated volume data, observation data and direct interviews with operators. Whereas secondary data consists of shop drawing, device specifications, and operator profiles.

3. Result and Discussion

Road Conditions

$$K \quad (\%) = \frac{P_k \; x \; L_k}{P_r \; x \; L_r} \; x \; 100 = 15{,}83 \; \%$$

Pr : Length of section (m)
Lr : Section width (m)
Pk : Damage length (m)
Lk : Damage width (m)

Cold Milling Machine Operator

Table 2. Operator Profiles

	Operator A	Operator B
Age	37	21
Gender	Male	Male
Education	Vocational High School	Vocational High School
Years of service	12	0

Operator Productivity A and Operator B

Table 3. Operator Productivity A and Operator B

Numb	Operator A	Operator B
1	40,71	14,8
2	56,06	12,85
3	55,97	27,83
4	53,73	47,55
5	69,12	29,61
6	51,64	56,73

Sample Calculation, Standard Deviation and Variants

Table 4. Sample Calculation Results, Standard Deviation and Variants

	Operator A	Operator B
Sample	6	6
Average	54,54	31,56
Standard Deviation	9,14	17,54
Variant	83,45	307,58

The hypothesis to be tested is based on the same n. But the second variant of the homogeneity of the sample or not, then the variance homogeneity needs to be tested first with the F test.

$$F = \frac{\textit{The biggest variant}}{\textit{The smallest variant}} = 3{,}686 \quad (2)$$

Because n1 = n2, the t-test can use the separated variant or polled variant formula.

$$t = \frac{\overline{X}_1 - \overline{X}_2}{\sqrt{\frac{(n_1-1)s_a^2 + (n_2-1)s_t^2}{n_1+n_2-2}\left(\frac{1}{n_1}+\frac{1}{n_2}\right)}} = 2{,}846$$

Theoretical Productivity

$$\frac{\text{tonase}}{\text{specific gravity}} = \frac{150}{2{,}3} = 65{,}22\ m^3/day \quad (3)$$

Based on field conditions that work at night and cold milling equipment that only operates for 4 hours, the planned production value of 65.22 m^3 / day is divided by operating the equipment for 4 hours. with the result productivity of 16.304 m^3 / hour.

Operator Efficiency Value

Q = v x b x Fa x t x 60 m^3 = 0,697

In determining the efficiency value of the operators, the actual productivity of each operator is divided by terrorist productivity as shown in table 5.

Table 5. Operator Efficiency Values A and B

Day	Operator A	Operator B
1	0,624	0,227
2	0,860	0,197
3	0,858	0,427
4	0,824	0,729
5	1,060	0,454
6	0,792	0,855
Average	**0,836**	**0,481**

4. Conclusions

The conclusions obtained from this study are:

1. There are significant differences in productivity of cold milling operators between operators who have skills based on years of service.
2. The productivity of cold milling equipment operated by the operator is used as a benchmark in finding the value of the operator efficiency factor using a method by comparing the theoretical productivity with the actual productivity of the two cold milling machine operators. operator efficiency factor for cold milling machine is very good = 0.9-1, good = 0.8 and less = 0.5-0.6 results that have been obtained from the analysis conducted, the operators who have skills based on work period have a value of 0.84 in. in the very good category while operators who do not have skills based on years of service score 0.48 in the poor category.
3. The performance of the two operators in this study after the analysis was not included in the category very well. Therefore, operators should be disciplined, improve work ethic and the company provides motivational support that will help in increasing operator productivity.

5. References

[1] Kementerian Pekerjaan Umum, 2013. *Katalog Alat Berat Konstruksi*. ISBN 978-602-70342-04. Jakarta.
[2] Muis, A. R, 2007. *Penentuan Faktor Efisiensi Kerja Operator Alat Berat Wheel Loader*. Dimensi., Vol 1: 98-101.
[3] Rostiyanti, F. S, 2008. *Alat Berat Untuk Proyek Konstruksi (Edisi Kedua)*, Jakarta: Rineka Cipta.
[4] Sokop, Ronald Martin. 2018. *Analisa Perhitungan Produktivitas Alat Berat Gali – Muat (Excavator) dan Alat Angkut (Dump Truck) Pada Pekerjaan Pematangan Lahan Perumahan* Residence Jordan Sea. Jurnal Tekno, Volume 16, No 70, ISSN: 0215-9617.
[5] Ervianto, W. I, 2005. *Manajemen Proyek Konstruksi*. Yogyakarta: Andi Offset.
[6] Badan Pembinaan Konstruksi dan Sumber Daya Manusia. 2007. *Mengoperasikan Cold Milling Machine Sesuai Dengan Prosedur*. Jakarta: Departemen Pekerjaan Umum.
[7] Wulfram I, Ervianto, 2008. *Pengukuran Produktivitas Kelompok Pekerja Bangunan Dalam Proyek Konstruksi (Studi Kasus Gedung Bertingkat di Surakarta)*. Jurnal Teknik Sipil, Volume 9: No 1.
[8] Rostiyanti, F. S, 2008. *Alat Berat Untuk Proyek Konstruksi (Edisi Kedua)*, Jakarta: Rineka Cipta.
[9] Asiyanto, Susy, MBA, Ir. 2007. *Manajemen Alat Berat untuk Konstruksi*. PT Pradnya Paramita. Jakarta.
[10] Badan Pembinaan Konstruksi dan Sumber Daya Manusia. 2007. *Pelatihan Operator Cold Milling Machine (Cold milling machine operator)*. Jakarta: Departemen Pekerjaan Umum.

Technology acceptance model for online cinema ticketing among moviegoers in java island Indonesia: an empirical study on tix id application

Y Palumian[1], S C K Jayanti[1], R Indriyani[1], Z Tarigan[1]

[1]Petra Christian University, Surabaya, East Java, Indonesia

ypalumian@petra.ac.id

Abstract. Based on the technology acceptance model, factors that form the intention to use of technology product of performed by individual users are perceived ease of use and perceived usefulness. The purpose of this study is to examine the effect of these two variables along with the attitude toward using as a part of the theory of planned behavior as the intervening variable. This research is conducted under quantitative with partial least square (PLS) method. The results show that perceived ease of use has no effect on other variables while perceived usefulness has a significant effect 0.191 on attitude toward using and effect 0.490 intention to use. In addition, there is a significant effect influence perceived usefulness as 0.581to attitude toward using and effect as 0.338 intention to use.

1. Introduction

Internet technology has become aa primary need for the wider society in Indonesia. Based on the report written by Kemp[1] regarding the penetration of internet users in Indonesia in 2020, currently it has reached 175.4 million people who use internet and will surely continue to grow. This indicates that 65.05% of the total population in Indonesia has been touched by this advanced technology. Growing align with this advancement, film industry is one of the areas affected by the widespread influence of the internet, particularly in terms of ticket purchases. According to the data held by Cinema XXI Indonesia, online ticket sales transactions compared to offline ticket sales in the second quarter of 2019 successfully reached more than 40%[2] By this increasing number of movie viewers in theaters, many cinema e-ticketing applications continue to appear. The most downloaded and popular e-ticketing cinema application in Indonesia according to the Entertainment Top Charts on the Apple Store is TIX ID[3]. TIX ID is the most recent e-ticketing cinema application among other cinema e-ticketing applications and has received the attention of the Indonesian movie watchers.

Online cinema ticketing can be beneficial and promote ease of use when it is used with technology acceptance attitude. This technology acceptance attitude is usually known as the Technology Acceptance Model (TAM). TAM is a widely recognized and widely used theoretical model that predicts whether a technology is accepted by users. There are two primary variables of TAM, namely perceived ease of use and perceived usefulness[4]. This study aims to examine the effect of the two main factors of TAM on intention to use with factors from Theory of Planned Behavior, namely attitude toward behavior, more precisely attitude toward using as an intervening variable. Theory of Planned Behavior (TPB) was developed by Ajzen in 1988. This theory has become one of the most popular and most influential conceptual frameworks for the study of human action[5].

One of the closest approaches to technology acceptance model can be seen through intention to use. In their research, Suki and Suki[6] stated that the definition of intention to use is a measure of the emergence of an individual's intention to use a technology product. Chi[4] also explains that intention to use can be affected by attitude toward using. According to Drew and Alharbi[7] the intention to use variable is also directly influenced by the variable perceived ease of use and perceived usefulness.

Attitude toward using is an attitude that refers to the evaluation or judgment of the user about the desire to use a specific information system application[8]. In addition, attitude toward using can be affected by perceived ease of use and perceived usefulness. Perceived ease of use is a user's perception that describes the level of effort that is felt when using a technology, while the definition of perceived usefulness is the level of one's belief that using a technology will improve job performance[4].

1.1 Research Questions and Research Objective
The formulation of the problem proposed based on the background of this research can be describe as follows:

RQ1. Does perceived ease of use affect attitude toward using among TIX ID users?
RQ2. Does perceived usefulness affect attitude toward using among TIX ID users?
RQ3. Does attitude toward using affect the intention to use of TIX ID users?
RQ4. Does perceived ese of use affect the intention to use of TIX ID users?
RQ5. Does perceived usefulness affect the intention to use of TIX ID users?

The purpose of this study is to analyze the effect of perceived ease of use, perceived usefulness, and attitude toward using on intentional behavior of using TIX ID application among users in Java Island in Indonesia.

1.2 Literature Review
1.2.1 Technology Acceptance Model (TAM) In their research, Kanchanatanee, Suwanno, and Jarernvongrayab[9] explains that the Theory of Reasoned Action (TRA) developed by Davis in 1989 was applied to the field of information technology and developed it into a Technology Acceptance Model (TAM). Chi[4] explains that TAM is a theoretical model that has been widely recognized and utilized and predicts whether an information technology is accepted by its users. TAM refers to the behavior of information technology adoption through two primary factors, namely perceived ease of use and perceived usefulness.

1.2.2 Perceived Ease of Use Perceived Ease of Use can be described as the level of effort performed by users when using or accessing information technology. The easier information technology, the more consumers accept that technology[9]. According to Hubert, Blut, Brock, Backhaus, and Eberhardt [10], perceived ease of use refers to a person's level of confidence that using or accessing information systems will be easy as without effort. From the explanation above, it can be concluded that perceived ease of use is the level of one's trust when using or accessing technology or information systems feel easy as without effort to use it. Hubert et al. [10]state that there are four items to measure perceived ease of use: an application must be easy to learn, an application must be easy to use on the smartphone, an application must support users to master it easily, and an application must be easy to use shop mobile.

1.2.3 Perceived Usefulness According to Chi[4]perceived usefulness cab be defined as the level of a person's belief that using technology will improve his job performance. In the context of online shopping, this definition refers to consumers' belief that using the internet as a media can increase their productivity or performance so that it will improve their shopping experience[11]. From these definitions, it can be concluded that perceived usefulness is the perception or level of confidence of users and potential users that using internet technology will increase their productivity and performance.

Chi[4] explains that there are four items to measure perceived usefulness of an application. Firstly, increasing performance which refers to the extent to which a technology can increase the shopping ability of its users. Secondly, promoting productivity which means usefulness in making purchases and users can carry out their daily activities productively. Thirdly, giving more convenient. A technology is perceived more convenient when it has become a better option for purchasing compared to other options such as buying directly in the cinema. Finally, effective. the refers to which an application can help users to achieve goals precisely through a series of alternatives or choice of means.

1.2.4 Attitude Toward Using Attitude refers to a person's process of determining the value of an object that leads to developing trust in that object[12]. Attitude toward using is an attitude that is related to the user's assessment of the desire to use a specific information application technology[13]. In another description, Theo and Noyes[14] describe attitude toward using is a person's positive or negative feelings when the individual uses an object, such as technology. According to Lee et al. [13] attitude toward using can be measured by four indicators, namely: evaluation of an object which refers to evaluation that can occur when an individual has assessed the object; feelings or emotions that arise when using a certain object; attraction, feeling attracted to use, reuse, or continuously use an object; goodness and positive values, a feeling in the form of good things and have a positive impact that is felt when or after using a certain object.

1.2.5 Intention to Use Lee[15] explains intention as a level of strength in measuring an individual's willingness to exert effort when carrying out certain behaviors. Yoganathan and Kajanan[16] explain that intention to use has a definition as a measure of the strength of a person's intention to actually use a technology. According to [17], if users are provided with a specific technology that meets their expectations, they will give an assessment and feel positive emotions about the technology. As a result, satisfied users are more likely to intend to reuse the technology. According to Lee[15] intention to use has two indicators. Firstly, willingness to use which means how much is an individual willing to use a technology in the future. Secondly, plan to use. This defines the measurement of the interest of an individual in using a technology in the future.

1.2.6 Hypothesis Development

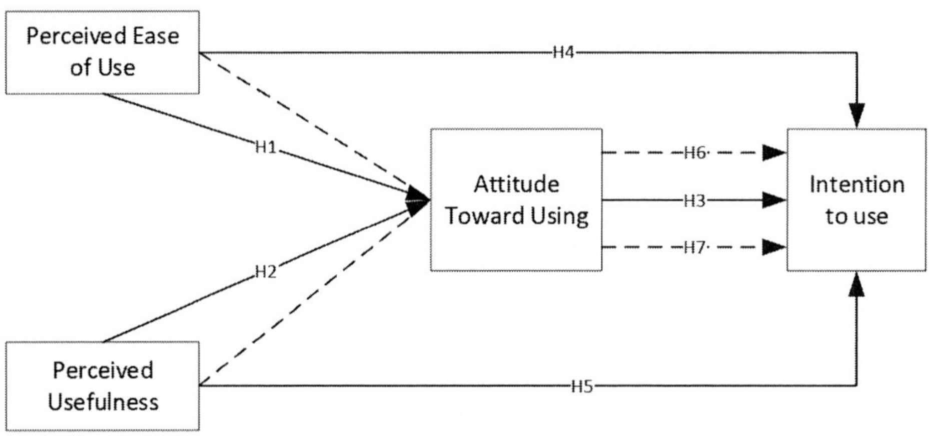

Figure 1 Conceptual Framework

H_1: Perceived ease of use significantly gives impact to attitude toward using.
H_2: Perceived usefulness significantly gives impact to attitude toward using.
H_3: Attitude toward using significantly gives impact intention to use.
H_4: Perceived ease of use significantly gives impact to intention to use.
H_5: Perceived usefulness significantly gives impact to intention to use.
H_6: Perceived ease of use has a significant influence on intention to use through attitude toward using.
H_7: Perceived usefulness has a significant influence on intention to use through attitude toward using.

2. Research Method

This study was conducted under the quantitative approach using five-score Likert scale questionnaire. Answer for score 1 represents strongly disagree and 5 (the highest) represents strongly agree. Then, the answer collected was included to the data pre-processing step. In addition to setting all the variables as latent variables, we placed attitude toward using as the intervening variable while perceived ease of use and perceived usefulness as the independent variable. Finally, we set intention to use as dependent variable. Furthermore, for the multivariate data analysis, partial least square (PLS) were chosen to test the hypothesis. The path diagram accordingly drawn to the hypothesis development can be described as follows:

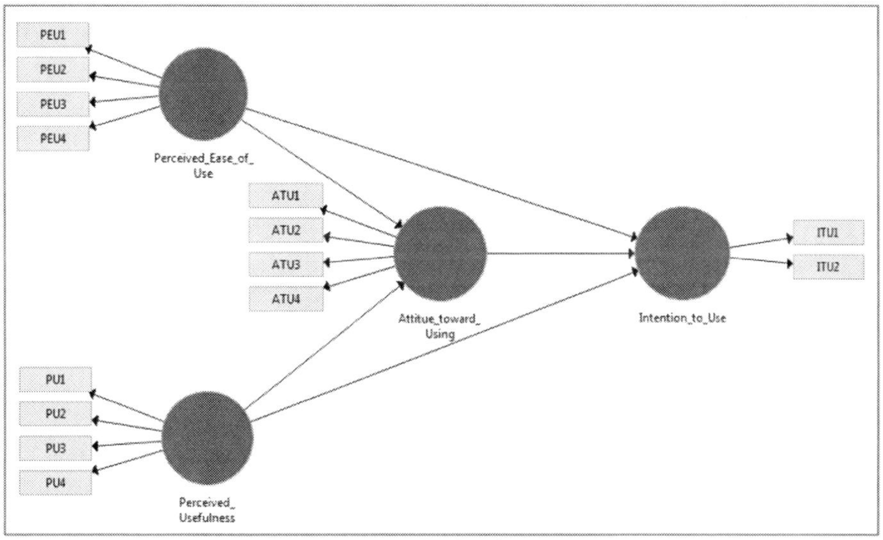

Figure 2 Path Diagram

This study focuses on TIX ID application users who live on the Indonesian island of Java, which is the most populous island in Indonesia. The number of respondents who have filled in the questionnaire online through the online form application is 123 respondents with valid answers consisting of 54 men and 69 women. Based on the data collection, there are 101 respondents aged 17-24 years while the rest are aged 25-35 years, totaling 22 people. Seeing from occupation perspective, student is the dominating occupation out of all the respondents which is 87 respondents. The rest, 36 respondents, were divided into several types of work, namely, self-employed, private employees, and civil servants.

3. Result

The first test conducted prior to the inferential testing validity and reliability test. These tests are to determine the validity and reliability of measuring instruments (indicators) in research. To see the results of data validity, it can be seen through the loading value. Indicators that are considered valid are those that have a loading value of > 0.50[18]. Meanwhile, reliability is considered consistent if the value obtained > 0.70 [19].

Using Smart PLS 3.0 and according to table 1, it can be concluded that the AVE value of all variables has met the minimum limit, namely 0.5 with the respective values of 0.725, 0.584, 0.914, 0.742 for PEU, PU, ITU, and ATU. Outer loading of all indicators has exceeded the minimum limit of 0.5 and is declared valid and can be used to measure the latent variables. In addition, Cronbach's Alpha value where the reliability value of 0.6 is considered, 0.7 is accepted and 0.8 is good [19]. The Cronbach's Alpha value of all variables in this study exceeded 0.7 which means that it is declared reliable.

Table 1. Outer Model Evaluation

Variable and AVE Score	Cronbach's Alpha	Indicator	Description	Factor Loading	Remark
Perceived Ease of Use (PEU)	0.875				Reliable
(AVE = 0.725)		PEU01	Learning TIX ID application is easy	0.759	Valid
		PEU02	TIX ID application gives me an easy way to book seats in the cinema with movie I want	0.906	Valid
		PEU03	TIX ID application is easy and quick to use	0.85	Valid
		PEU04	I feel so easy to book the ticket mobile	0.884	Valid
Perceived Usefulness (PU)	0.762				Reliable
(AVE = 0.584)		PU01	Using TIX ID application, I feel my ticket purchase performance improved	0.717	Valid
		PU02	Buying movie ticket from TIX ID does disturb my productivity	0.702	Valid
		PU03	It is easier to book the movie tickt online than to buy in on the spot.	0.809	Valid
		PU04	TIX ID application gives effectiness when buying movie ticket.	0.821	Valid
Intention to Use (ITU)	0.906				Reliable
(AVE = 0.914)		ITU01	I will always use TIX ID for buying movie ticket in the future	0.956	Valid
		ITU02	I plan to continue using TIX ID when I want to buy movie ticket	0.956	Valid
Attitude toward Using (ATU)	0.884				Reliable
(AVE = 0.742)		ATU01	I think that using TIX ID is useful	0.854	Valid
		ATU02	I feel happy because I have downloaded and used TIX ID	0.868	Valid
		ATU03	I feel attracted by TIX ID	0.884	Valid
		ATU04	I feel the positive effect after buying ticket	0.84	Valid

The next step is to interpret the coefficient determination or R^2 value. Running the bootstrapping results, the R2 values of intention to use and attitude toward using are 0.587 and 0.528. This means that intention to use is influenced by attitude toward using, perceived ease of use, and perceived usefulness of 58.7%, the remaining 41.3% is explained by other variables outside the proposed model. In addition, this also indicates that the attitude toward using is influenced by 52.8% of perceived ease of use and perceived usefulness, while the remaining 47.2% is explained by other variables unmentioned in this study.

According to table 2, with a path coefficient of 0.581, perceived usefulness has a positive and significant effect on attitude toward using since the p-value is 0.000 and the T-statistic is 5.481, where the p-value is < 0.05 and the T-statistic is > 1.96, then H_2 is supported. Furthermore, with a path coefficient of 0.490 and 0.338, attitude toward using and perceived usefulness respectively have a positive and significant effect on intention to use indicated by the p-value is <0.05 and the T-statistic is> 1.96. Therefore, it can be concluded that H_3 and H_5 are supported.

Table 2. Inner Model Evaluation - Direct Effect

Hypothesis	**Direct Effect**	***Path Coefficient***	***T Statistics***	***P Values***	**Remarks**
H_1	PEU→ATU	0.191	1.679	0.094*	Supported*
H_2	PU→ATU	0.581	5.481	0.000	Supported
H_3	ATU→ITU	0.490	5.748	0.000	Supported
H_4	PEU→ITU	0.004	0.043	0.965	Not supported
H_5	PU→ITU	0.338	3.402	0.001	Supported

*: Significat at $\alpha = 10\%$

On the other hand, at a significance level of 10%, perceived ease of use has a direct positive and significant effect on attitude toward using because the p-value is 0.094 and the T-statistic is 1.679, where the p-value is < 0.1 and the T-statistic is > 1.96. Then, H_1 is accepted when using $\alpha = 10\%$. Finally, based on the calculation results, perceived ease of use has no direct effect on intention to use because the p-value is 0.965 and the T-statistic is 0.043, where the p-value is > 0.05 and the T-statistic is < 1.96, then H4 is not supported.

Table 3. Inner Model Evaluation - Indirect Effect

Hypothesis	**Indirect Effect**	***Path Coefficient***	***T Statistics***	***P Values***	**Remarks**
H_6	PEU→ATU→ITU	0.094	1.584	0.114	Not Supported
H_7	PU→ATU→ITU	0.285	4.103	0,000	Supported

Based on table 3, with a path coefficient of 0.285, perceived usefulness has a positive and significant effect on intention to use through attitude toward using because the p-value is 0.000 and the T-statistic is 4.103 where the p-value is <0.05 and the T-statistic is> 1.96, then H_7 accepted. However, perceived ease of use has no effect on intention to use through attitude toward because the p-value is 0.114 and the T-statistic is 1.584, where the p-value is> 0.05 and the T-statistic is <1.96, then H_6 is supported.

4. Discussion

In this study, the TAM concept used as a variable is consumer perceptions that are directly related to the use of the TIX ID application. Perceived ease of use and perceived usefulness are variables that have been widely used in previous research. In that study, it was proven that ease of use did not affect the intention to use of TIX ID users. In other words, it is not the convenience that builds technology acceptance among moviegoers to use application-based ticketing. This means that users do not see the ease of TIX ID as something that needs attention, but rather the use of the application which has a stronger appeal which encourages users to be able and continue to use it. This finding contrasts with what was stated by[10] confirming that ease of use has a significant effect on the intention to use on smartphone-based applications. The use of the application that is considered more important by users is the features of TIX ID that can provide benefits for prospective moviegoers, including ordering cinema tickets, describing cinema seats that are still empty, and seeing movie showtimes that can be defined as all in the menu. a glance[19].

5. Conclusion

What can be concluded in this research is that the perceived ease of use of the TIX ID application cannot increase or support the formation of an attitude toward using or intention to use, either directly or indirectly. Conversely, what can promote technology acceptance is perceived usefulness which can build an attitude toward using the TIX ID application which will later form an intention to use among moviegoers. Further research that can be suggested is to involve perceived sacrifice or perceived risk from the use of mobile-based ticketing technology as well as expanding the research sample to include other ticketing applications.

6. References

[1] Digital 2020: Indonesia — DataReportal – Global Digital Insights, https://datareportal.com/reports/digital-2020-indonesia (accessed 27 October 2020).

[2] Pembelian Tiket Bioskop Online Terus Meningkat Tiap Tahun, https://mediaindonesia.com/read/detail/271701-pembelian-tiket-bioskop-online-terus-meningkat-tiap-tahun (accessed 27 October 2020).

[3] App Store Top Charts Ranking for iPhone Apps | AppFollow, https://appfollow.io/rankings (accessed 27 October 2020).

[4] Understanding Chinese consumer adoption of apparel mobile commerce: An extended TAM approach - ScienceDirect, https://www.sciencedirect.com/science/article/abs/pii/S0969698918303187 (accessed 27 October 2020).

[5] Ajzen I. Perceived Behavioral Control, Self-Efficacy, Locus of Control, and the Theory of Planned Behavior1. *Journal of Applied Social Psychology* 2002; 32: 665–683.

[6] Mohd Suki N. Exploring the relationship between perceived usefulness, perceived ease of use, perceived enjoyment, attitude and subscribers' intention towards using 3G mobile services. *Journal of Information Technology Management* 2011; 22: 1–7.

[7] Alharbi S, Drew S. Using the Technology Acceptance Model in Understanding Academics' Behavioural Intention to Use Learning Management Systems. *International Journal of Advanced Computer Science and Applications* 2014; 5: 13.

[8] Lee C-Y, Tsao C-H, Chang W-C. The relationship between attitude toward using and customer satisfaction with mobile application services: An empirical study from the life insurance industry. *Journal of Enterprise Information Management* 2015; 28: 680–697.

[9] Kanchanatanee K, Suwanno N, Jarernvongrayab A. Effects of Attitude toward Using, Perceived Usefulness, Perceived Ease of Use and Perceived Compatibility on Intention to Use E-Marketing. *Journal of Management Research* 2014; 6: 1.

[10] Hubert M, Blut M, Brock C, et al. Acceptance of Smartphone-Based Mobile Shopping: Mobile Benefits, Customer Characteristics, Perceived Risks, and the Impact of Application Context. *Psychology and Marketing* 2017; 34: 175–194.

[11] Cho Y, Sagynov E. Exploring Factors That Affect Usefulness, Ease Of Use, Trust, And Purchase Intention In The Online Environment. *International Journal of Management & Information Systems (IJMIS)* 2015; 19: 21.

[12] Ajzen I. Martin Fishbein's Legacy: The Reasoned Action Approach. *The ANNALS of the American Academy of Political and Social Science*. Epub ahead of print 10 February 2012. DOI: 10.1177/0002716211423363.

[13] Lee C-Y, Tsao C-H, Chang W-C. The relationship between attitude toward using and customer satisfaction with mobile application services: An empirical study from the life insurance industry. *Journal of Enterprise Information Management* 2015; 28: 680–697.
[14] Teo T, Noyes J. An assessment of the influence of perceived enjoyment and attitude on the intention to use technology among pre-service teachers: A structural equation modeling approach. *Computers & Education* 2011; 57: 1645–1653.
[15] Seifermann R. Factors influencing the adoption of internet banking: An integration of TAM and TPB with perceived risk and perceived benefit, https://www.academia.edu/39104426/Factors_influencing_the_adoption_of_internet_banking_An_integration_of_TAM_and_TPB_with_perceived_risk_and_perceived_benefit (accessed 27 October 2020).
[16] Yoganathan D, Kajanan S. *What Drives Fitness Apps Usage? An Empirical Evaluation*. 2014.
[17] Park E. The role of satisfaction on customer reuse to airline services: An application of Big Data approaches. *Journal of Retailing and Consumer Services* 2019; 47: 370–374.
[18] Hair JF, Black WC, Babin BJ, et al. *Multivariate Data Analysis: Pearson New International Edition*. 7th Edition. Pearson, 2013.
[19] Research Methods For Business: A Skill Building Approach, 7th Edition | Wiley. *Wiley.com*, https://www.wiley.com/en-id/Research+Methods+For+Business%3A+A+Skill+Building+Approach%2C+7th+Edition-p-9781119266846 (accessed 27 October 2020).

IOP Conf. Series: Materials Science and Engineering 1010 (2021) 012038 doi:10.1088/1757-899X/1010/1/012038

Risk management in shipbuilding using bayesian network with noisy-or

R Asdi[1,*] and M Basuki[2]

[1] Department of Industrial Engineering, Adhi Tama Institute of Technology, Jl. Arif Rahman Hakim 100 Surabaya 60117, Indonesia
[2] Department of Naval Architecture and Shipbuilding Engineering, Adhi Tama Institute of Technology, Jl. Arif Rahman Hakim 100 Surabaya 60117, Indonesia

*asdi.ridzky@gmail.com

Abstract. This paper presents a method for risk management applicable in shipbuilding projects. We focused this study specifically to assess delay in project completion by analysing the production processes. The production processes were grouped into three major parts, namely the work of design, material procurement, and production processes. Risk data were collected from risk documentation of fast patrol boat constructions in PT. PAL Indonesia (the company). Bayesian network was used to assess the probability of project completion delay. Noisy-OR canonical model was used to define parameters needed for the Bayesian network model. Based on the current condition, the Bayesian network model predicted that there is a probability of delay of 0.67761739 that the project completion will be delayed. The Bayesian network model was also used to assess mitigation actions to reduce the probability of delay. Three mitigation action sets were provided based on the complexity of mitigation action: (1) less complex mitigation action, (2) more complex mitigation action, and (3) most complex mitigation action. The Bayesian network model predicting that the most complex mitigation action would provide the best result in mitigating the project completion delay.

1. Introduction

As a maritime nation, shipbuilding industry plays a key role in Indonesia's industrialization. Shipbuilding industries act as a backbone for the national economy due to its function as the main supporting industries. In general, Indonesian shipbuilding industries are still far behind when compared to foreign shipbuilding industries such as Japan, China, and Korea. Indonesian shipbuilding industry only gained 0.35% of the global market share [1]. This number is minuscule when compared to 35.67% that of Korean shipbuilder. The shipbuilding industries in Indonesia can be characterized as follows: capital-intensive, labour-intensive, slow yield, low added value, complex value chain, non-competitive, high-risk, low demand, high-tech, requires highly skilled personnel for ship design and fabrication, high import level with a low level of local material content, low experience levels, and long-term delivery [2].

The nature of shipbuilding industries allows for inherent production risk. Despite the high-risk, implementation of risk management in Indonesian shipbuilding industries is still lacking. Risk management is needed in capital-intensive industries to improve competitiveness and prevent financial loss. A well-managed risk can ultimately contribute to competitive advantage for shipbuilding industries. One major problem that currently faced by Indonesian shipbuilding is the delay in ship delivery. It has been shown in the work of Basuki et al. [3] that in general, any delay in production processes can result in financial loss.

Diagrams such as cause and effect diagrams can be used as methodologies for predicting probabilities production delay, thus can be used as methodologies for risk assessment. A Bayesian network is a graphical model that presents probabilistic relationships among a set of variables by determining the causal relationships among them. In determining a Bayesian network, it is necessary to compile conditional probabilities as a quantitative means to represent causal relationships among nodes. Compiling conditional probabilities can be difficult especially when the network consists of many nodes. That is why it has not been widely used in risk management procedures. This study presents a means for assessing risk using Bayesian network for predicting the delay in ship production.

1.1. Literature review

1.1.1. Risk management. In managing risk there are two essential phases, namely: risk identification to identify all potential risks, and risk assessment to assess each risk so that managers can establish priority in mitigation actions. In the risk identification phase, there are several methodologies used: brainstorming, document review, Delphi technique, and historical data. Crispim et al. [4] utilized questionnaires with Delphi technique to obtain unbiased risk data. Mousavi et al. [5] utilized a brainstorming session with project stakeholder and risk management expert and document review in identifying risks in a large engineering project. In several cases, risks can readily be identified according to technical criteria or specifications as in the case in Felderer and Ramler [6]. Felderer and Ramler [6] utilized a methodology called risk-based testing to assess risk in software development. Risk-based testing is an integrated methodology to identify and assess risks. In assessing risk, Felderer and Ramler based their calculation on the combination of business criteria and technical criteria. In general, for risk identification, the most used source of information has been the opinion of experts or project managers, collected through interviews, questionnaires, brainstorming, or Delphi. Only a few literatures utilized historical data as risk identification.

In the risk assessment phase, traditional methods (namely, risk matrix, weighted risk factor) are still being used widely due to their simplicity, despite they have often been criticized. Thomas et al. [7] argued that there is no empirical evidence showing that the risk matrix improves risk management or the outcomes of decision making. Zhang and Chu [8] argued that in Failure Mode and Effect Analysis method, several risk items may have the same risk priority number, despite having different combinations of occurrence, severity, and detection values. This may hide the risk implications of different risk items. Other than traditional methods, Fuzzy logic, and combined with other methods has been the most used risk assessment technique. This can be seen in the work of Abdul-Rahman et al. [9], Aliahmadi et al. [10], Bi and Wang [11], Jozi et al. [12], and Li and Zou [13]. Bayesian network as a means of assessing risk has not been widely used. Only several literatures utilized Bayesian network. This can be seen in Crispim et al. [4], Basuki et al. [2], and Lee et al. [14].

1.1.2. Bayesian network. Bayesian network is an acyclic directed graph with nodes that represent random variables and edges that represent causal relationships among nodes. Initially, Bayesian network was developed as an inference engine in the field of artificial intelligence. As the method grows in popularity, many researchers developed Bayesian network in other applications such as diagnosis, forecasting, automated vision, sensor vision, sensor fusion, and manufacturing control [15]. Other applications including disease diagnosis [16], transportation [17], and ecosystem and environmental management [18]. The first application on risk assessment can be found in the work by Fan and Yu [19]. There are many advantages of Bayesian network such as suitability for small and incomplete data sets, structural learning possibility, combination of different sources of knowledge, explicit treatment of uncertainty and support for decision analysis, and fast responses [18].

The major difficulty in model building using Bayesian network is the need for quantifying all conditional probabilities needed [20]. The complete conditional probability tables for a binary variable with n binary causes in a network requires 2^n independent parameters [21]. Hence the number of parameters needed to specify a Bayesian network grows exponentially with the number of causes, which can prohibit its use in modelling. The Noisy-OR model [21], provides a logarithmic reduction in the number of parameters required to specify a Bayesian network. The Noisy-OR model can be thought of as a probabilistic extension of OR gate. Consider a variable Y which has several causes $X = \{X_1, X_2, \cdots X_n\}$; by using two assumptions: (1) each of the cause X_i has a probability of affecting Y despite the absence of all other causes, (2) each cause X_i is sufficiently independent of other causes, then Noisy-OR model can be described using the following equation:

$$p_i = P(y|x_i) = P(y|\overline{x_1}, \overline{x_2}, \cdots x_i, \cdots \overline{x_{n-1}}, \overline{x_n}) \quad (1)$$

The probability p_i is called the link probability and it represents 'the probability of y given that X_i is present and other causes are not.' The link probability p_i can be thought of as the causal strength of cause X_i in affecting Y. Once all necessary link probabilities are found, then the rest of conditional probabilities of y given a subset X_p (x_i present) can be approximate using the following equation [21]:

$$P(y|X_p) = 1 - \prod_{x_i \in X_p} (1 - p_i) \quad (2)$$

Consider a node that has 10 causes, by using Noisy-OR model one would only need 10 parameters as opposed to $2^{10} = 1024$ if not using Noisy-OR model. A study by Anand and Downs [16] has shown that the Noisy-OR formulation does not sacrifice prediction accuracy when compared to network with traditional conditional probability tables parametrization.

2. Research methods

Figure 1 shows the proposed method for managing risk. The risk management method used in this study can be divided into three phases: (1) risk identification phase, (2) risk assessment phase, and (3) risk analysis and mitigation phase.

In the risk identification phase, the risks which affect the project performance were identified. The focus was to identify all risks that might affect the delay in project completion. The sources for the identification phase are from documentation regarding previous projects of the same vessel type.

Once all production processes are listed, the next step is to assess each process probability of delay and consequence. Risk items' probabilities and consequences are needed to construct all parameters needed for defining Bayesian network. This study focuses on assessing delays in production processes; therefore, consequence is measured in days and it represents 'how many days the project completion will be delayed if the underlying process facing a delay'. Once the probabilities and consequences are assessed, the consequences of processes must be converted into link probabilities. Table 1 shows how to convert consequence into link probability. Once all the parameters have been established, then the Bayesian network can be constructed. Bayesian network can be used to predict the probability of risk events, i.e., delay in shipbuilding completion.

Production processes that affect the project completion delay can be classified according to their sources of delay. These sources can be internal (the company), owner, supplier, etc. In general, the more parties involved, the more complex the mitigation action will be. Therefore, it is important to classify mitigation actions based on their complexity. Once mitigation action has been established, Bayesian network that has been constructed in the risk assessment phase can be used to assess the effectiveness in reducing the probability of delay.

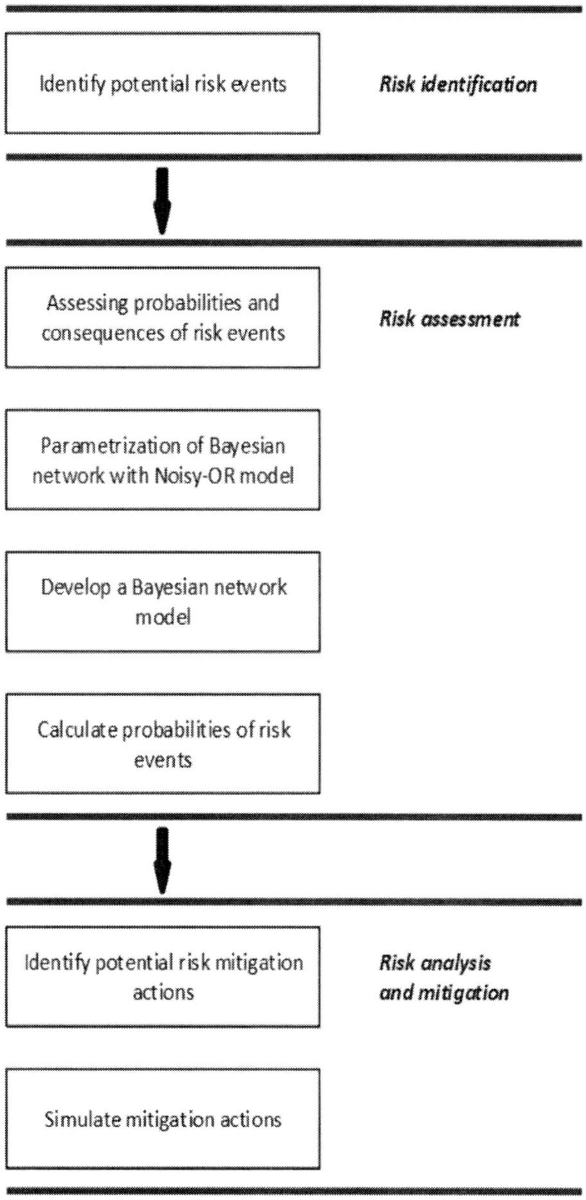

Figure 1. Research method.

Table 1. Converting consequence into link probability.

Consequence Criteria	Measurement of Criteria	Link Probability
Insignificant	Delay in project completion of < 10 days	0.2
Minor	Delay in project completion of 10 - 20 days	0.4
Moderate	Delay in project completion of 20 - 50 days	0.6
Major	Delay in project completion of 50 - 100 days	0.8
Catastrophic	Delay in project completion of > 100 days	1

3. Analysis and results

3.1. Risk identification phase

Risks are identified according to each process in ship production. Processes in ship production can be classified into three groups, namely: (1) design, (2) material procurement, and (3) production process. Each group has its processes, in the design group, there are five processes, namely: (1) basic and performance, (2) key plan, (3) yard plan, (4) production drawing, and (5) test procedure and finished plan. The list of all processes can be seen in table 2.

Table 2. Identification of production processes with their probabilities of delay, consequences, and link probabilities.

No.	Process	Probability of Delay	Consequence (days)	Consequence Criteria	Link Probability
1	**Design**				
	Basic and performance	0.099	9.30	Insignificant	0.2
	Key plan	0.120	11.16	Minor	0.4
	Yard plan	0.129	12.09	Minor	0.4
	Production drawing	0.129	12.09	Minor	0.4
	Test procedure and finished plan	0.081	7.44	Insignificant	0.2
2	**Material**				
	Electric outfitting	0.729	67.89	Major	0.8
	Machinery outfitting	0.828	77.19	Major	0.8
	Hull outfitting	0.729	67.89	Major	0.8
	Painting and corrosion control	0.702	65.10	Major	0.8
	Hull construction	0.717	66.96	Major	0.8
3	**Production**				
	Work preparation and general	0.462	42.78	Moderate	0.6
	Hull construction	0.492	42.57	Moderate	0.6
	Painting and corrosion protection	0.462	42.78	Moderate	0.6
	Piping system	0.480	44.64	Moderate	0.6
	Machinery system	0.474	43.71	Moderate	0.6
	Steelwork outfitting	0.468	43.71	Moderate	0.6
	Interior and furniture	0.468	43.71	Moderate	0.6
	Electric system	0.474	43.71	Moderate	0.6
	Electronic system	0.462	42.78	Moderate	0.6

3.2. Risk assessment phase

Table 2 also shows the result of process identification along with the probability of delay, consequence, and the resulting link probability of each process. Once all parameters needed are obtained, then Bayesian network can be constructed, and the project completion delay can be simulated. Figure 2 shows the resulted Bayesian network. With the current condition, there is a probability of 0.67761739 that the project completion will be delayed. The group with the highest probability of delay is material procurement with 0.98903659 of delay probability, followed by production with 0.94980732, and design with 0.17426336. All the processes that fall under the material procurement group are facing a high probability of delay of more than 0.7.

IOP Conf. Series: Materials Science and Engineering 1010 (2021) 012038 doi:10.1088/1757-899X/1010/1/012038

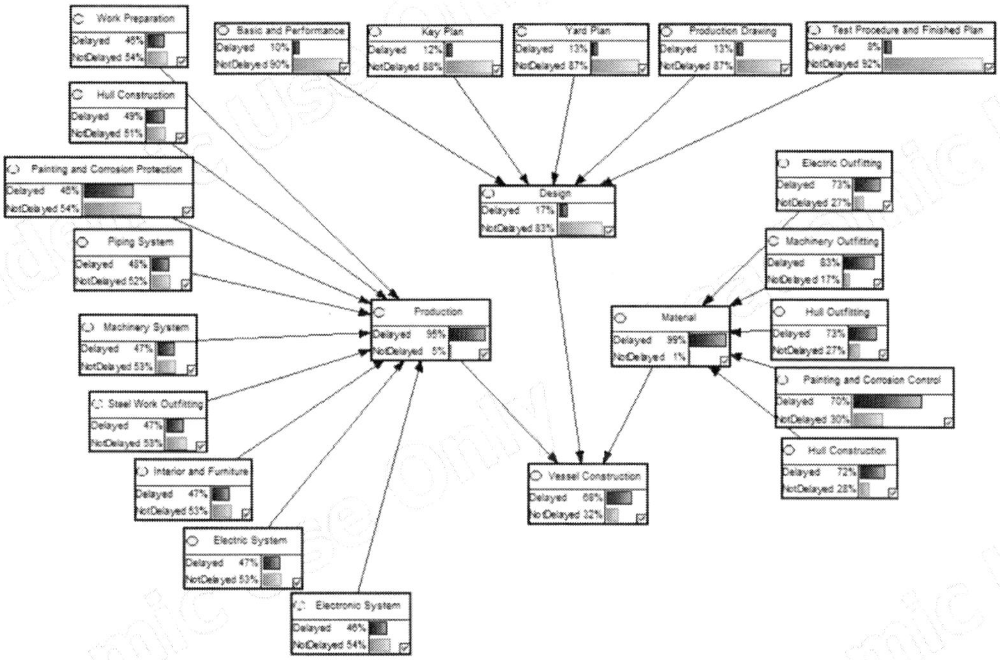

Figure 2. Risk network model.

Table 3. Sensitivity analysis of each process. Item number 1 – 19 is sensitivity for each of the process. Item number 20 – 22 is sensitivity for set of processes.

No.	Processes	Reduced by	Reduction in Project Completion
1	Basic and Performance	10%	-0.00170218
2	Key Plan	12%	-0.00424872
3	Yard Plan	13%	-0.00458471
4	Production Drawing	13%	-0.00458471
5	Test Procedure and Finished Plan	8%	-0.00138759
6	Electric Outfitting	30%	-0.00370418
7	Machinery Outfitting	30%	-0.00457317
8	Hull Outfitting	30%	-0.00370418
9	Painting and Corrosion Control	30%	-0.00352168
10	Hull Construction	30%	-0.00362079
11	Work Preparation	30%	-0.00021153
12	Hull Construction	30%	-0.00021693
13	Painting and Corrosion Protection	30%	-0.00021153
14	Piping System	30%	-0.00021474
15	Machinery System	30%	-0.00021366
16	Steel Work Outfitting	30%	-0.00021259
17	Interior and Furniture	30%	-0.00021259
18	Electric System	30%	-0.00021366
19	Electronic System	30%	-0.00021153
20	**Less Complex Mitigation Action**	30%	**-0.01778355**
21	**More Complex Mitigation Action**	30%	**-0.05966343**
22	**Most Complex Mitigation Action**	30%	**-0.06621022**

3.3. Risk analysis and mitigation phase

In general, risk mitigation is done by reducing the probability of delay of certain process(s). To know which process(s) should be mitigated, we must be aware of each process sensitivity. This is done by reducing the probability of a certain process by 30% and then record the reduction in the probability of delay of the project completion. Table 3 shows the result of sensitivity analysis. For any process under the design group, which already has a very low probability of delay, the sensitivity analysis was done by reducing the probability of delay until it becomes 0.

However, as can be seen in table 3, mitigating just one process would not give a significant result. A more generalized reduction can be obtained by mitigating a set of actions. Mitigation action can be categorized into three levels according to their complexity: (1) less complex mitigation action, (2) more complex mitigation action, (3) most complex mitigation action. The complexity of mitigation action is based on the party involved in the process. Table 4 shows each level of mitigation action and the processes involved. In the less complex mitigation action for example, only involves internal and external party that has close relations. The most complex mitigation action has the most parties involved, and thus this level of mitigation action is the most difficult approach.

Table 4. Mitigation action category.

Mitigation Level	Parties Involved	Set of Processes
1. Less complex mitigation action	Internal Owner	Design - basic and performance Design - key plan Design - yard plan Design - production drawing Design - test procedure and finished plan
2. More complex mitigation action	Internal Supplier	Material - electric outfitting Material - machinery outfitting Material - hull outfitting Material - painting and corrosion control Material - hull construction
3. Most Complex mitigation action	Internal Owner Supplier Subcontractor Customs	Material - electric outfitting Material - machinery outfitting Material - hull outfitting Material - painting and corrosion control Material - hull construction Production - work preparation and general Production - hull construction Production - painting and corrosion protection Production - piping system Production - machinery system Production - steelwork outfitting Production - interior and furniture Production - electric system Production - electronic system

4. Conclusion

This study has provided a method for risk management applicable to shipbuilding projects. The resulting framework is simple and understandable. Generally, the method shown in this study is applicable to any shipbuilding project. However, there might be some adjustments needed for different kinds of shipbuilding projects. The risk identification and probability assessment are based on past data, therefore little to no bias is introduced. The resulting Bayesian network has provided visual interpretation of causes and effects, which greatly help in analysis. The Noisy-OR model helps in greatly reducing the parameters needed to define Bayesian network. Moreover, this paper has shown that the resulting Bayesian network can be used for sensitivity analyses, thus providing means for planning and simulating mitigation action.

The developed method has several restrictions. Historical data from past projects may not be readily available. The network provided did not consider dependency between processes. In future development, dependency between processes can be considered as this possibly provide better accuracy. Based on the results, the material procurement processes have the highest probability of delay. In this study, we only consider immediate supplier to the project. In future research it is interesting to analyse risks that may arise from the whole shipbuilding supply chain.

5. References

[1] Bagenda W, Sule E T, Febrian E dan Sari D 2018 The influence of unique capability, functional strategy on company performance of shipbuilding industry in Indonesia *Int. J. Bus. Glob.* **20** 273–91

[2] Basuki M, Manfaat D, Nugroho S dan Dinariyana A A B 2014 Probabilistic risk assessment of the shipyard industry using the Bayesian method *Int. J. Technol.* **5** 88–97

[3] Basuki M, Manfaat D, Nugroho S dan Dinariyana A 2012 Improvement of the process of new business of ship building industry *J. Econ. Business, Account. | Ventur.* **15** 187–204

[4] Crispim J, Fernandes J dan Rego N 2020 Customized risk assessment in military shipbuilding *Reliab. Eng. Syst. Saf.* **197** 106809

[5] Mousavi S M, Tavakkoli-Moghaddam R, Hashemi H dan Mojtahedi S M H 2011 A novel approach based on non-parametric resampling with interval analysis for large engineering project risks *Saf. Sci.* **49** 1340–8

[6] Felderer M dan Ramler R 2014 Integrating risk-based testing in industrial test processes *Softw. Qual. J.* **22** 543–75

[7] Thomas P, Bratvold R dan Bickel J 2013 The risk of using risk matrices *SPE Econ. Manag.* **6**

[8] Zhang Z dan Chu X 2011 Risk prioritization in failure mode and effects analysis under uncertainty *Expert Syst. Appl.* **38** 206–14

[9] Abdul-Rahman H, Wang C dan Lee Y L 2013 Design and pilot run of Fuzzy Synthetic Model (FSM) for risk evaluation in civil engineering *J. Civ. Eng. Manag.* **19** 217–38

[10] Aliahmadi A, Sadjadi S J dan Jafari-Eskandari M 2011 Design a new intelligence expert decision making using game theory and fuzzy AHP to risk management in design, construction, and operation of tunnel projects (case studies: Resalat tunnel) *Int. J. Adv. Manuf. Technol.* **53** 789–98

[11] Bi X dan Wang G L 2011 Application of fuzzy comprehensive evaluation method based on entropy to evaluation risk of an EPC project *2011 IEEE 18th International Conference on Industrial Engineering and Engineering Management, IE and EM 2011* (Institute of Electrical and Electronics Engineers (IEEE)) pp 55–8

[12] Jozi S A, Shoshtary M T dan Zadeh A R K 2015 Environmental risk assessment of dams in construction phase using a Multi-Criteria Decision-Making (MCDM) method *Hum. Ecol. Risk Assess.* **21** 1–16

[13] Li J dan Zou P X W 2011 Fuzzy AHP-based risk assessment methodology for PPP projects *J. Constr. Eng. Manag.* **137** 1205–9

[14] Lee E, Park Y dan Shin J G 2009 Large engineering project risk management using a Bayesian belief network *Expert Syst. Appl.* **36** 5880–7

[15] Heckerman D, Mamdani A dan Wellman M P 1995 Real-world applications of Bayesian networks *Commun. ACM* **38** 24–6

[16] Anand V dan Downs S M 2008 Probabilistic asthma case finding: a Noisy-OR reformulation *AMIA ... Annu. Symp. proceedings. AMIA Symp.* **2008** 6–10

[17] Ülengin F, Önsel Ş, Ilker Topçu Y, Aktaş E dan Kabak Ö 2007 An integrated transportation

decision support system for transportation policy decisions: The case of Turkey *Transp. Res. Part A Policy Pract.* **41** 80–97
[18] Uusitalo L 2007 Advantages and challenges of Bayesian networks in environmental modelling *Ecol. Modell.* **203** 312–8
[19] Fan C F dan Yu Y C 2004 BBN-based software project risk management *J. Syst. Softw.* **73** 193–203
[20] Oniśko A, Druzdzel M J dan Wasyluk H 2001 Learning Bayesian network parameters from small data sets: Application of Noisy-OR gates *Int. J. Approx. Reason.* **27** 165–82
[21] Pearl J 1988 *Probabilistic Reasoning in Intelligent Systems* (Elsevier)

IOP Conf. Series: Materials Science and Engineering 1010 (2021) 012039 doi:10.1088/1757-899X/1010/1/012039

Design concept of reinforced concrete beams with large web openings

D Pertiwi[1], I Komara[1], R Fristian[1]

[1] Department of Civil Engineering, Faculty of Civil, environmental, and planning, Institut Adhi Tama Surabaya, East Java, Indonesia.

* indrakomara@itats.ac.id

Abstract. This paper presents numerical investigations in studying large web openings. Based on the previous guidelines, the web openings in reinforced concrete beams are being investigated. The simple verification design procedure is followed by ACI-318. In general, web openings have to consider a certain position so that each chord sides imply a sufficient area to expand the ultimate compression in flexure and give the appropriate reinforcement. In this case, two openings are placed nearly to one-half from support on each side. The analysis follows the recommended procedure for elastic bending moments and shear forces while the deflections are evaluated both in the short and long-term periods. It is found that the beam with a large web opening meets the required standard in all aspects, but it is indicative of having a large moment.

1. Introduction

In most of the tall buildings, utilities of pipes and ducts are accommodated in the space within the floor ceiling and beams. Implementing these utilities through transverse openings in the floor beams saves a significant amount of headroom and result in a more compact and economical design [1]. The total height also can be reduced due to the presence of the transfer openings that allow the passage of service utilities [2], [3]. However, the design provision with a hole in flexural members changes their simple mode behavior to a complex and indeterminate mode. In general, the evaluation of members with a different type of openings becomes underhanded and requires much effort and time which may not be justified in many projects. In this particular case, research work will give a reasonable procedure for the analysis and design of flexural members with different web openings.

The previous studies have been proposed for the strength design of beams with large web openings under various conditions. Mansur, et al [1], [4] investigated the large rectangular of web openings reinforced concrete (RC) under pure tension within a different of opening sizes and their locations. This study was also carried out by Tan et al [5] which analyzed the behavior of large openings based on different reinforcement criteria and design assumptions. The assumption of a typical free body diagram is presented in Figure 1. In addition, Rodríguez, et al [6] and Alsaeq [7] evaluated the effect of opening shape on the RC beams by using numerical analysis. The previous research works concluded that the opening locations have more effect on the structural strength that the opening shape. Notwithstanding, these methods require the magnitudes of forces and moments at the center of the opening to be predetermined and can, therefore, be readily applied to statically determinate beams only. In continuous beams that generally occur in practice, reduction in stiffness due to the provision of openings through webs causes a redistribution of internal forces and moments, the amount of which needs to be evaluated before a design can proceed. To support the strength and deformation of the openings some modifications into RC beams are also necessary such as adding fiber reinforce concrete [8], [9] and engineered cementitious composite [10]–[13]. Another simulation is also conducted to verify experimental works [14]

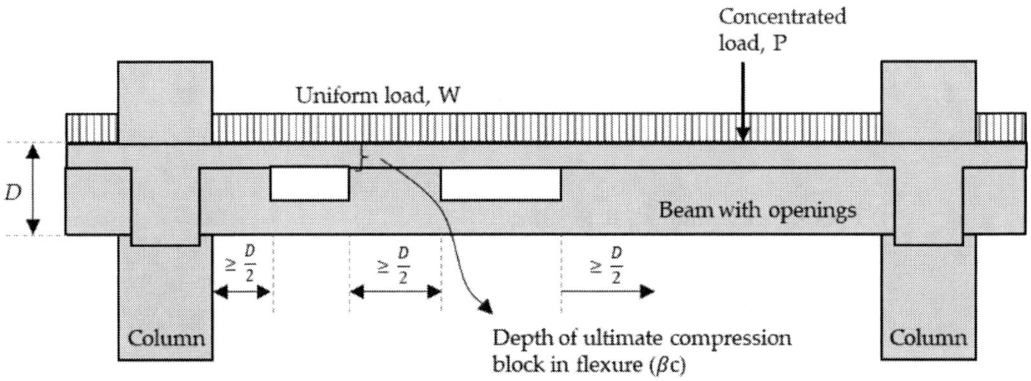

Figure 1. Typical condition of the beam with openings considering guidelines of Tan and Mansur [5]

1.1. Research Significance

This paper provides a case study within the implementation procedure to complete analysis and design the RC beam with the large web openings. General information for the placement of openings is given and supported by the numerical analysis. The procedure check for the ultimate strength, flexure, and shear. It also ensures that the capacity of the section fills the short and long-term deformation.

2. Design Method

The guidelines procedure is taken from a brief review from several works of literature [8], [9], [15]–[17] which considers its behavior and strength from this case to facilitate the selection of the size and location of the openings. In this special condition, the flexural member is used as a rectangular beam. The openings are placed at mid-depth of the section and exercise to in precise dimension to provide sufficient concrete cover as the support of reinforcement for the chord members below and above the opening. This is concerned to have an acceptable concrete area in terms of developing the ultimate compression block in flexure and have adequate depth to give effective shear reinforcement.

Another procedure that should carry is openings location. The openings should omit the position which is closer than one-half the beam depth *D* to the supports to avoid the critical region for shear failure. Openings depth should also limit to fifty percent of overall beam depth. The factor that limits the length of the openings are the stability of the chord members, the compression chord, and the deflection based on the serviceability required criteria. In this case, the opening location is remained in the same position (single opening) with a variety of opening sizes.

Furthermore, the design procedure herein is used ACI-318 [18] which involves structural analysis, strength design, and serviceability design. All procedures are to determine the distribution of shear forces and moments due to ultimate loads, wherein the critical sections are designed for ultimate strength in bending and in shear. When it comes to serviceability, the performance level is used as a design parameter using direct displacement-based design to ensure the structure performs its intended functions satisfactory under the given loads.

2.1. Design of stiffness of beam segments

The member containing an opening is considered as a no prismatic beam. The beam with no openings is also considered into account as a parameter control. In this stiffness calculations, gross section of the concrete is evaluated based on ACI-318 [18]. The modulus of elasticity of the concrete is according to Eq. (1a) where the shear modulus can be taken as Eq. (1b).

$$E_c = 4730\sqrt{f'_c} \quad (1a)$$

$$G = \frac{E_c}{2(1+v)} \quad (1b)$$

where f'_c is the cylinder compressive strength in MPa and v is the Poisson's ratio which is taken 0.2 follows the recommendation by BS-8110 [19]. In the design parameter, the equivalent flexural stiffness of an opening segment is considered as $(EI)_{eq.}$ the moment of inertia of concrete section minus the openings should be applied.

The equivalent of shear stiffness for an opening segment and the effective length of its openings can be obtained as Eq. (2a) and (2b).

$$(GA)_{eq} = \frac{12E_c\left(I_{gt}+I_{gb}\right)}{{l_e}^2} \quad (2a) \qquad l_e = \frac{l_o}{1-\left(\frac{d_o}{D}\right)^{1.5}} \quad (2b)$$

Where I_{gt}, I_{gb} l_o, d_o and D are the gross moment of inertia for top and bottom chord members, opening length, opening depth, and beam depth, in order. The shearing deformation of the solid segments is generally ignored.

2.2. Design of bending moment and shear force envelopes

The ultimate design bending moment M_m and shear force V_m at the middle of the opening segment from bending moment and shear force envelopes, and calculate axial forces N_t and N_b (positive for compression) acting, respectively, in the top and bottom chords which can be obtained from Eq. (3a) and (3b), where z is the distance between the plastic centroids of the top and bottom chords.

$$N_t = \frac{M_m}{{l_e}^2} \quad (3a) \qquad N_b = -N_t \quad (3b)$$

Further, to distribute the applied shear between the top and the bottom chords are calculated in Eq. (4a-b), where $I_{gt,}$ and I_{gt} are the gross moment of inertia of the top and bottom chords, respectively. For the evaluation of moments at the ends of chord members, the analysis can refer to Eq. (5a-d). Where ω the uniformly distributed load acting directly on the top chord and M is a moment. Subscripts 1, 2, 3 and 4 designate the opening corners, as illustrated in Figure 2.

$$V_t = V_m\left(\frac{I_{gt}}{I_{gt}+I_{gb}}\right) \quad (4a) \qquad V_t = V_m\left(\frac{I_{gb}}{I_{gt}+I_{gb}}\right) \quad (4b)$$

$$M_1 = -\frac{\omega {l_o}^2}{8} - \frac{V_t l_o}{2} \quad (5a) \qquad M_2 = -\frac{\omega {l_o}^2}{8} + \frac{V_t l_o}{2} \quad (5b)$$

$$M_3 = -\frac{V_b l_o}{2} \quad (5c) \qquad M_4 = \frac{V_b l_o}{2} \quad (5d)$$

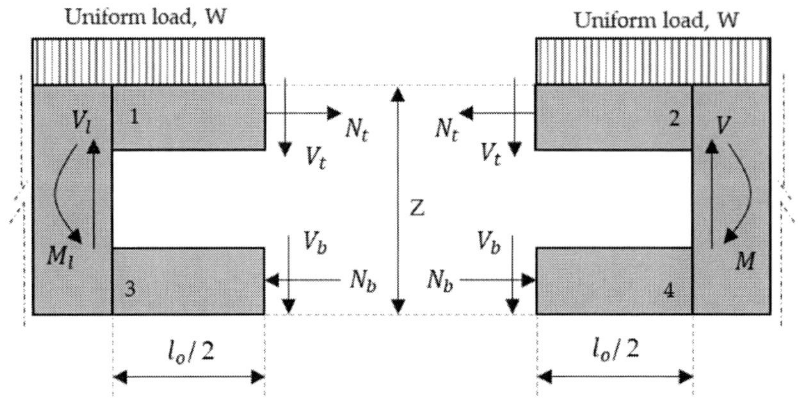

Figure 2. Detail of opening section [5]

2.3. Design of longitudinal and shear reinforcement

The longitudinal reinforcement in the bottom and the top of the section adjacent to the opening should be continued throughout the opening segments. Additional reinforcement required to resist the combined moment and axial force in each chord member is design considering idealized column interaction diagrams using the strain compatibility method. All combinations of bending moment and axial load should be evaluated within the appropriate interaction diagram to ensure sufficient reinforcement. As well as the capacity of flexure on the top chord. On the other hand, an updated reinforcement is necessary.

The shear forces transmitted to the bottom and the top chords are given by Eq. (4a-b). The condition to design is based on the knowing forces that are typically as same as to RC beam and slabs. Nonetheless, according to ACI-318 [18], the effects of axial forces in the chord members must be evaluated in design.

2.4. Design of deflections

The indirect way of satisfying the serviceability following the required deflection is by limiting the effective span depth ratio but it is not accurate for a beam with openings. Hence, the count of the actual service load deflection is necessary. In this case, the beam at the ultimate load may be used for the analysis since the detail of reinforcements is obvious. Then, the identical shear stiffness of the openings following the Eq. (2a-b) can be used.

3. Result and Discussion

One type, RC beam containing a web opening in the interior span is illustrated in Figure (3a). This represents a typical feature of a residential building (apartment) in which utility ducts run along the corridor between the rows of rooms. The full RC without opening will also be evaluated as the parameter control. The beam carries a dead load $G_{k,,}$ not including a self-weight of 10.66 kN/m of the and a uniformly imposed load Q_k of 3.75 kN/m. The material properties are f'_c = 30 MPa. Longitudinal and transverse steel, f_y, 400 MPa, and 240 MPa, respectively. The parameter location is cited in Surabaya with typically hard soil sites with the spectrum design $S_s = 0.705$ and $S_1 = 0.305$. Following the proposed method using DDBD for the beam with particular openings is given as follows.

3.1. Structural analysis

From previous Eq. (1a-b), E_c=2600 MPa and G = 10800 MPa. Then the following section information to support the evaluation are obtained respectively, $I_{gs} = 9.81\times10^9$ mm^4, $I_{gt} = 4.22\times10^8$ mm^4, $I_{gb} = 3.91\times10^8$ mm^4 and the effective length of the opening l_o = 1114.5 mm. The cross-section area of solid segment $A_{so} = 3.6\times10^5$ mm^2 and $E_c I_{gs} = 2.67\times10^5$ mm^2 and $GA_{so} = 3.89\times10^9$ N whereas for the opening one $(EI)_{eq} = 2.54\times10^{14}$ Nmm^2 and $(GA)_{eq} = 2.03\times10^8$ N. All the calculations are referred to Eq. (2a-b).

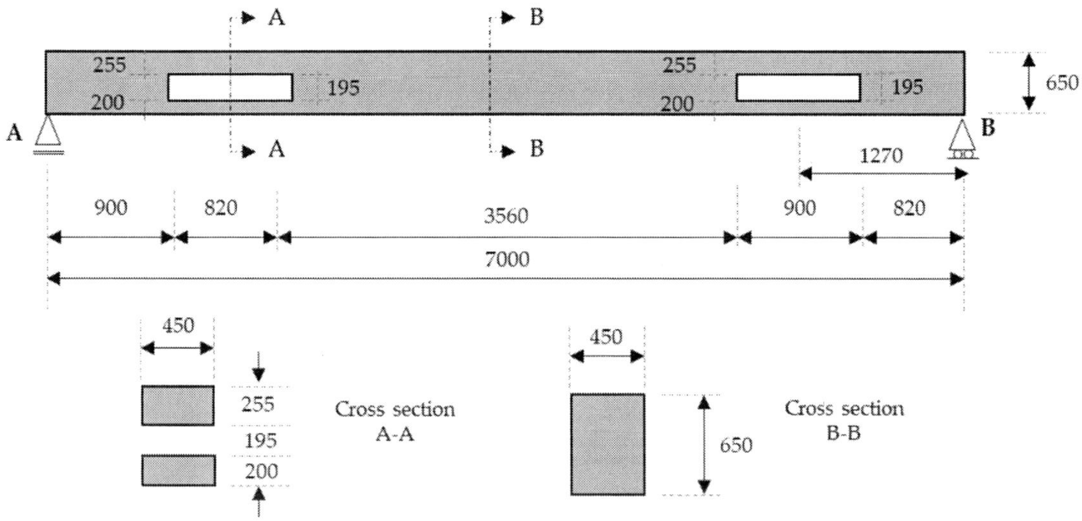

Figure 3. Detail of RC beam with large web opening

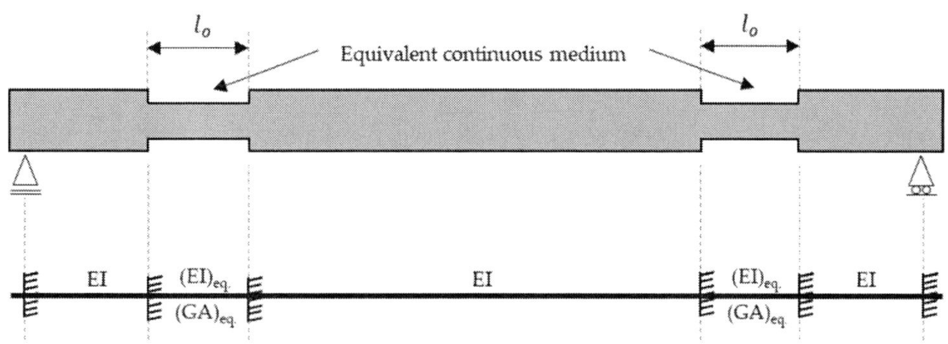

Figure 4. Equivalent beam with equivalent restrained structure scheme [unit: mm]

Table 1. Summary of design for critical sections – Beam with and without WO

Type of beam	Unit weight (kg)	Unit loss (%)	Moment (kNm)	Shear (kNm)	Axial (kNm)
Beam without WO	4914	-	- 294.76	-151.19	0
Beam with WO	297	6.04	- 243.31	151.12	27.04

The method of analysis to evaluate the load combinations is presented in Figure 4 and the segmented RC beams which are handed in the analysis are illustrated in Figure 5. Where the result is shown in Figure 6 as redistribution moment at the ultimate condition.

3.2. Strength design analysis

The result of the solid section was implemented in the usual manner. Both of RC with and without opening are summarized in Table 1. Furthermore, for the opening beam, the axial loads and shear forces referred to Figure 2 where bending moment M_m and shear force V_m at the center of the opening using Eq. (3a-b) and (4a-b). The result is shown in Table 2, considering z = 400 mm while the secondary moments using Eq. (5a-d).

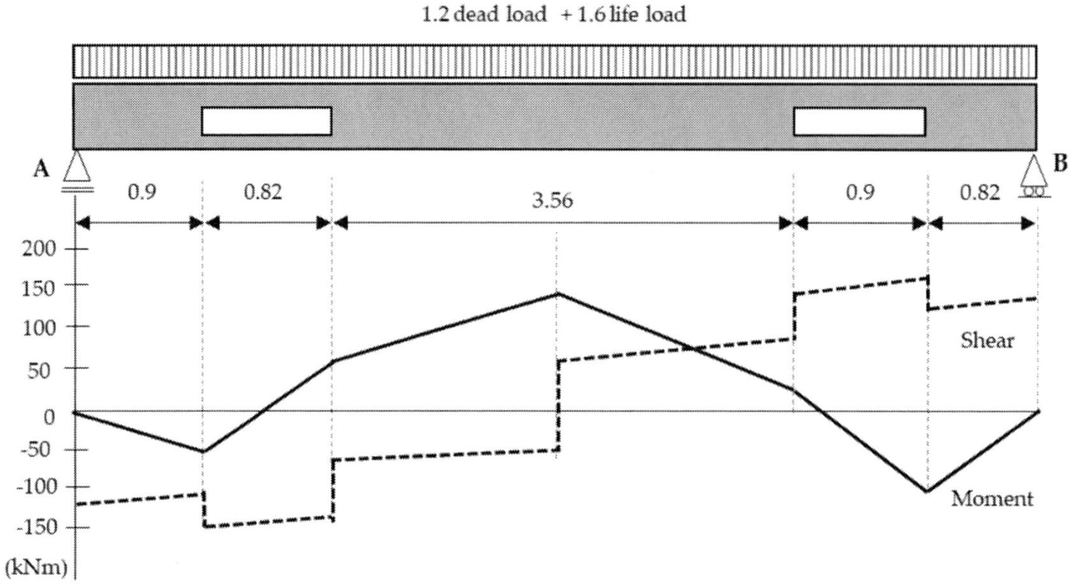

Figure 5. Redistribution of internal forces, moments, and shear in the beam with large web openings [unit: m]

From Table 1, the longitudinal reinforcement is informed. It is known that the solid beam adjacent to the opening is provided with 4D22 (four high-strength deformed bars of 22-mm diameter) and 2D22 (two high-strength deformed bars of 22-mm diameter). Hence, for the opening arrangement of reinforcement, the condition is illustrated in Figure 3 within 3D22 and 2D22 for the bottom and top chord.

The maximum shear of 95 kN at the opening center occurs and refer to Figure 4. Therefore, from Table 2, shear at top and bottom chord, respectively, V_t = 49.8 kN and V_b = 45.8 kN. The shear strength is evaluated by ACI-381 within the space limit is $d/2$ = 192/2 96 mm., therefore, provide stirrups of d10 (mild steel plain bars of 10-mm diameter) at 90-mm spacing.

$$\phi\, V_c = 0.85 \times 0.17 \left[1 + 0.073 \frac{(58)}{300 \times (250)}\right]$$
$$\sqrt{30} \times 300 \times 192 = 45.6\, kN \; < 45.9\, kN$$

3.3. Deflection

In the first parameter as a short-term deflection cracked moment is assumed from the inertia of the beam. The beam is evaluated using a service load of (G_k + G_k) = 25.2 kN/m. The maximum deflection is informed typically for bot side span AB and CB with a value of nearly 10 mm due to the combination of dead and life load. According to ACI-381, the allowable short-term deflection is likely to be designed as $L/360$ = 7000/360 = 19.4 mm > 10 mm; thus, the result is satisfactory.

In the following parameter, in long-term deflection (5 years period), the assumption is considered that 20 percent of the live load is sustained due to unpredicted conditions. Additionally, shrinkage and creep may also be determined following the indication of immediate deflection by a factor $\lambda = 2/(1{+}50\rho')$ where ρ' is the compression steel ratio at a maximum location when a deflection occurs. Seeing that ρ' = 945/(300×542) = 0.0058, then λ = 1.7. On the other hand, the sustained load is given ($1.0G_k + 0.2Q_k$) = 16.3 kN/m while the result of the calculation for the maximum deflection is 8.15 mm; hence, the long-term deflection is counted = 10 + 1.55×8.15 = 22.63. The result has to be compared to the allowable long-term deflection according to ACI-318 of $L/240$ = 7000/240 = 29.2 mm. It is concluded that the result is satisfactory.

4. Conclusions

This study is indicated the calculation for large web openings in the RC beam. The calculation is made based on the previous investigation which is followed by the instruction guidelines. The investigated is made by a solid member and the opening member using the direct stiffness method. The predicted results found that deflection and support reactions have a good agreement with the required code ACI-318. The method also gives a reasonable data to be implemented the incorporating openings.

5. References

[1] Mansur M A, 2006 Design of Reinforced Concrete Beams with Web Openings in *6th Asia-Pacific Structural Engineering and Construction Conference* **6** p. 104–120.

[2] Jassim N Q and Jarallah H K, 2018 Performance Enhancement of R.C. Beams with Large Web Openings by Using Reactive Powder Composite: An Experimental Study *Al-Nahrain J. Eng. Sci.* **21**, 3 p. 405–416.

[3] Ehmann S and Schnellenbach-Held M, 2001 Behavior and design of reinforced concrete beams with large openings in *Fracture Mechanics of Concrete Structures, de Borst et al (eds)* p. 903–909.

[4] Mansur M A Huang L M Tan K H and Lee S L, 1992 Deflections of reinforced concrete beams with web openings *ACI Struct. J.* **89**, 4 p. 391–397.

[5] Tan K H and Mansur M A, 1996 Design procedure for reinforced concrete beams with large web openings *ACI Struct. J.* **93**, 4 p. 404–411.

[6] Ahmed A Fayyadh M M Naganathan S and Nasharuddin K, 2012 Reinforced concrete beams with web openings: A state of the art review *Mater. Des.* **40** p. 90–102.

[7] Alsaeq H M, 2013 Effects of opening shape and location on the structural strength of RC deep beams with openings *Int. J. Civ. Environ. Eng.* **7**, 6 p. 494–499.

[8] Osman B H Wu E Ji B and S Abdelgader A M, 2016 A state of the art review on reinforced concrete beams with openings retrofitted with FRP *Int. J. Adv. Struct. Eng.* **8**, 3 p. 253–267.

[9] Hassan N Z Sherif A G and Zamarawy A H, 2017 Finite element analysis of reinforced concrete beams with opening strengthened using FRP *Ain Shams Eng. J.* **8**, 4 p. 531–537.

[10] Oktaviani W N Tambusay A Komara I Sutrisno W Faimun F and Suprobo P, 2020 Flexural Behaviour of a Reinforced Concrete Beam Blended with Fly ash as Supplementary Material *IOP Conf. Ser. Earth Environ. Sci.* **506** p. 012042.

[11] Komara I Tambusay A Sutrisno W and Suprobo P, 2019 Engineered Cementitious Composite as an innovative durable material: A review *ARPN J. Eng. Appl. Sci.* **14**, 4 p. 822–833.

[12] Mooy M Tambusay A Komara I Sutrisno W Faimun and Suprobo P, 2020 Evaluation of Shear-Critical Reinforced Concrete Beam Blended with Fly Ash *IOP Conf. Ser. Earth Environ. Sci.* **506** p. 012041.

[13] Bastian M A Tambusay A Komara I Sutrisno W Irawan D and Suprobo P, 2020 Enhancing the Ductility of a Reinforced Concrete Beam using Engineered Cementitious Composite *IOP Conf. Ser. Earth Environ. Sci.* **506** p. 012044.

[14] Musmar M, 2020 Nonlinear Finite Element Flexural Analysis of RC Beams *Int. J. Appl. Eng. Res.* **13**, 4 p. 2014–2020.

[15] Redwood R G, 2011 *Design of Composite Beams With Web Openings.* April .

[16] Torunbalci N, 2002 Behaviour and design of large rectangular openings in reinforced concrete beams *Archit. Sci. Rev.* **45**, 2 p. 91–96.

[17] Ramadan O M Metwally K G and Shaban W M, 2015 Proposed recommendations for the design of reinforced concrete beams with openings in *Advances in Structural Engineering and Mechanics (ASEM15)* p. 1–19.

[18] American Concrete Institute (ACI 318-99), 1999 *Building Code Requirements for Structural Concrete* **2007** .

[19] BRITISH STANDARD, 1985 *Structural use of concrete Part 2: Code of practice for special circumstances* **2** .

IOP Conf. Series: Materials Science and Engineering 1010 (2021) 012040 doi:10.1088/1757-899X/1010/1/012040

High porous carbon nanofiber derived from lignosulfonate material

M F Rois[1], W Widiyastuti[1] and H Setyawan[1]

[1]Department of Chemical Engineering, Institut Teknologi Sepuluh Nopember, Surabaya, East Java, Indonesia

widi@chem-eng.its.ac.id

Abstract. In this study, electrospinning technology, iodine treatment, thermal stabilizing, and carbonization processing were applied to produce lignosulfonate-based carbon nanofibers. The porous structure of the produced lignosulfonate-based carbon nanofibers primarily contained mesopores and a relatively small amount of micropores. Moreover, the increasing amount of sodium lignosulfonate (SLS) as lignin source was also studied. The diameter of fibre was impacted by the additional of SLS that may cause by the alkali content of SLS itself. In electrospun nanofiber phase, the presence of SLS reduce the specific surface area, but contrary increase the specific surface area after the nanofibers were carbonized into carbon nanofiber. Lignosulfonate-based activated carbon fibres can be used as a highly efficient adsorption and filtration material, and further development of its applications would be valuable.

1. Introduction

As a mesoporous and microporous structured material, activated carbon has many promiscuous usage, due to its absorptivity, specific surface area, and hydrophilicity, activated carbon is commonly used as an adsorbent material in food and environment application [1, 2]. Furthermore, it has porous structure that able to interact with the adsorbed subjects chemically or physically to eliminate odorous, coloured or harmful substances. It makes activated carbon is currently used in food, medical, solvent recovery, water treatment, waste gas, fuel cell industries and or even energy storage [3].

Based on the shape and usage, activated carbon can be divided into powdered activated carbon, granular activated carbon, and activated carbon fibre [4]. Activated carbon fibre or usually known as carbon nanofiber (CNF) is commonly prepared by activating the nanofiber that synthesized using common polymer material such as PAN, PVA, cellulose acetate, etc. CNF has many favourable characters such as large surface area, uniform pore size, high adsorptive ability, regenerable, and satisfactory mechanical properties [5, 6]. Industrial applications of general-purpose and high-performance CNF have been widely explored. Researchers have attempted to develop bio-based CNF precursors that are low-cost and environmentally friendly. Among such precursors, lignin is introduced as inexpensive and has a favourable structure as a precursor for the production of CNF [7].

Lignin is one of the natural biopolymers that exists in the cell wall of plants. It is a natural, aromatic, high carbon content, heterogeneous biomacromolecule that has a total worldwide production of approximately 70 million tons/year [8]. However, conventionally, in the pulp and paper industry, chemicals are applied to eliminate lignin during pulping, and technical lignin are produced as byproducts. These are generally trashed or used as a fuel source, but in a very small quantity is used in commercial applications. For these reasons, lignin is the renewable raw materials that are largely used for the preparation of nanofiber.

The use of technical lignin as precursors to produce CNF has been extensively studied [4, 9]. Particularly, lignosulfonate-based CNF were fabricated through a series of electrospinning and carbonization processes. However, those previous study used acid-based lignosulfonate to synthesize the nanofiber. It makes after carbonization, an activation process needs to be implemented to generate submicron or nanoscale lignin-based Activated Carbon Nanofiber (ACNF) with small diameters and large surface areas. Whereas some previous study provided data that by using alkali lignin, the product of carbon can have high surface area with less separated activation process.

In work, the sodium lignosulfonate were used as lignin source for the electrospinning process to produce nanofiber. The effect of lignosulfonate mass ratio solution was studied to the specific surface area.

2. Experimental

2.1. Materials
The precursor of the Carbon nanofiber was a mixture of Sodium Lignosulfonate (Gac-NaLS grade 3, Green Agrochem) and poly-vinyl alcohol (PVA, Mr = 60,000, Sigma Aldrich). And for stabilisation, technical grade of iodine crystal purchased from UD SIP Indonesia was used. All of the materials were used as received without purification.

2.2. Solution Preparation
The PVA was stirred in distilled water for an hour at a temperature of 85°C. After all of the PVA was diluted, the sodium lignosulfonate (SLS) was then added to the solution to obtain black viscous solution with ratio of PVA and the SLS in the water was 15:0,1.5, and 3% wt/vol. After thawed at room temperature for several hours, the solution was then filled into 50ml scaled modified syringe before ready to be installed in electrospinning device.

2.3. Electrospinning
Electrospinning was performed in the horizontal direction (Figure 1) with electrospinning equipment was based on our previous research [6, 10]. The black solution was carried out in a Terumo plastic 50 ml syringe with needle an inner diameter 0.8 mm (21 gauge). The steady flow rate was maintained by the Cole Parmer Instrument Co. syringe pump at 1 ml per hour. A high voltage power supply was connected the needle and the collector. The applied voltage at 15 kV was applied to the polymer solution and using 10 kV was applied onto a rotary collector device. Previously, the rotary collector device was wrapped by a 15 cm x 30 cm of aluminium foil around rotating cylinder. The distance from needle to collector was 10 cm. The rotary collector was rotated with a speed of 100 rpm in a room temperature for 7 h, respectively, and the measured relative humidity percentage was between 55 and 65%.

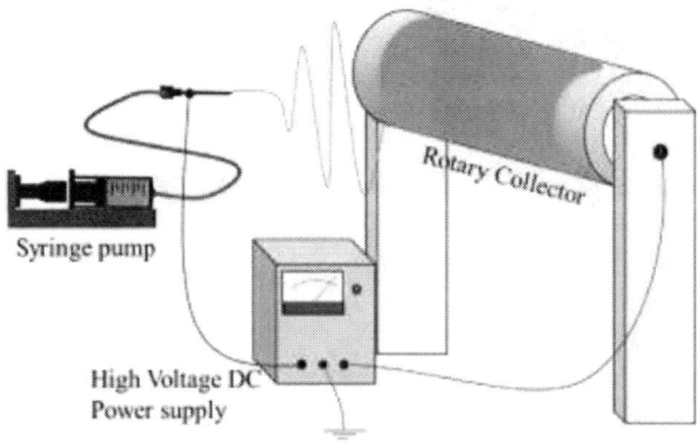

Figure 1. The illustration of electrospinning device

2.4. Iodine Treatment, Thermal Stabilization and Carbonization

After peeled off from the aluminium foil on rotary collector, the electrospun obtained was then placed with iodine in the sealed bottle and heated at 80 °C for 24 h before the carbonization process. After that, the iodine-treated electrospun lignosulfonate nanofiber mat was thermally stabilized by gradually heated in the air at temperatures of 100, 150, and 200°C for an hour at each temperature. It was then followed by carbonization under nitrogen flow with a flow rate of 25 mL/ minute at 400 °C for an hour, and then 700°C for an hour following the highest temperature of activated carbon from lignin in previous study [11]. The generated activated carbon nanofibers were washed with the mineralized water until the washing water was neutralized.

2.5. Characterization

To examine the morphology of the nanofiber and carbon nanofiber, scanning electron microscope (SEM, HITACHI FLEXSEM-100) equipped with Dispersive Energy X-Ray Spectroscopy (EDS) was used. The specific surface area was obtained from the Brunauer-Emmett-Teller (BET) method by measurement of nitrogen adsorption-desorption isotherms (Quantachrome instrument, NOVA1200e).

3. Result and Discussion

Before applied as precursor in synthesizing nanofiber, we need firstly know the possibility of SLS itself to be treated as activated carbon. Therefore, the SLS was treated following our previous research [11]. And the result is described as below (Figure 2)

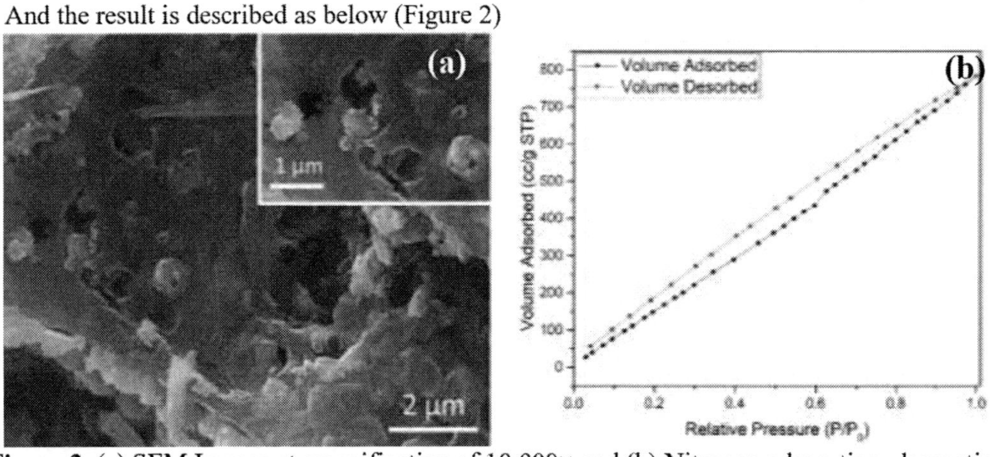

Figure 2. (a) SEM Images at magnification of 10,000× and (b) Nitrogen adsorption–desorption isotherms of carbon derived from Sodium Lignosulfonate (SLS) calcined at 700°C

As can be seen above (Figure 2(a)), the carbon morphology obtained from SLS show rough surface that comprises numerous hole with average diameter 400 nm that represent active side, describing that the carbon has been activated. This is further provided by nitrogen adsorption-desorption data (Figure 2(b)) that showing BET graph type II that indicates that the pore is focused on the first layer. The specific surface area obtained was 1168.241 m^2/g with pore volume of 1.211 cc/g and averaged pore diameter of 4.15 nm. As supporting data, Table 1 showing that high amount of alkali that may increase the active side of carbon derived from SLS.

Table 1. Content of activated carbon from SLS obtained from EDS

Element	Wt%
C-O	21.47
S	5.95
Alkali	35.10
Cl	21.08
Impurities	16.4

The homogenous of carbon content in solution is one of important factor in electrospinning polymer/inorganic composite. In other study that involve particulate material as composited additive to polymer, sonication is an effective way for mixing polymer/inorganic composite [6]. Since SLS is water-soluble material, sonication is no need to make the mixture become heterogenous. The electrospun mats obtain brown smooth fibre while only PVA fibre mats had white colour. The brown colour of 1.5 wt.% and 3 wt.% of SLS was darker by the increasing of SLS content.

Figure 3 shows the results of SEM observations on 15 % w/v PVA samples added with various concentration of SLS. The addition of 1.5 wt.% has been able to change the morphology of PVA fibres to be thinner from average 650 nm to 280 nm. The distribution of diameter fiber is becomes wider as well. Increasing of SLS to 3 wt.% produces fibre that is still relatively heterogeneous in diameter but numbers of diameter distribution is more concentrate to 300 nm.

Figure 3. SEM images of electrospun PVA/SLS fibres with various SLS contents of: (a) 0 wt. %, (b) 1.5 wt.%, and (c) 3 wt.%

Figure 4 is the typical nitrogen physisorption isotherm plot for the nanofiber developed in this study. It is all most similar to type IV(a) [12], although the knee part is not as bent as a typical IV(a) isotherm. This indicates that the material mainly consists of mesopores and a relatively small amount of micropores. Based on the hysteresis loop shape, the lignin-based fibre can be categorized as the H2-b type.

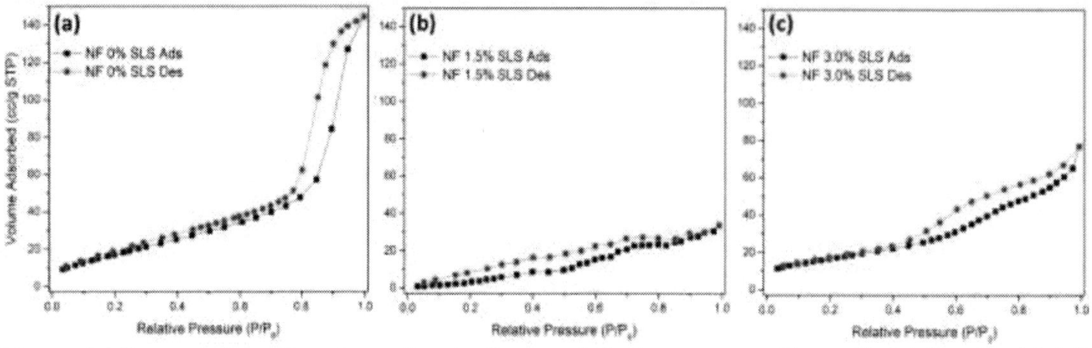

Figure 4. Nitrogen adsorption–desorption isotherms of electrospun nanofiber with (a) 0 wt.%, (b) 1.5 wt.%, and (c) 3 wt.% of sodium lignosulfonate content.

However as shown from the volume adsorbed (Figure 4(b) and (c)), followed by the specific surface area provided by nitrogen adsorption-desorption isotherm (Table 2), the increasing of SLS content decrease the pore of fibre itself. Furthermore, compared to PVA nanofiber, the pore diameter and pore volume of electrospun nanofiber that contain SLS describe significant reduction. It may caused by the characteristic of alkali content in SLS that become a natural surfactant so it reduce the surface tension that later become the pore potential. That is why the micropore side of the nanofiber is slightly reduced as figured in the graph above.

Table 2. Nitrogen adsorption-desorption of nanofiber at various SLS concentration

Sample	Specific surface area (m^2/g)	Pore diameter (nm)	Pore volume (cm^3/g)
Nanofiber 0.0 wt.% SLS	72.414	12.3492	0.2236
Nanofiber 1.5 wt.% SLS	70.218	2.6889	0.0519
Nanofiber 3.0 wt.% SLS	60.698	7.8262	0.1188

After passing iodine treatment, thermal stabilizing, and carbonisation process at 700°C, it can be shown in Figure 5, The adsorption– desorption isotherm curve of the carbon nanofiber (CNF) sample indicates the Type IV-a isotherm for CNF with SLS content of 1.5 wt.% and Type I isotherm for CNF with SLS content of 3 wt.% according to the IUPAC classification that the curve is concave to the relative pressure axis. The appearance of a nearly horizontal plateau indicates a small external surface area compared to the total surface area.

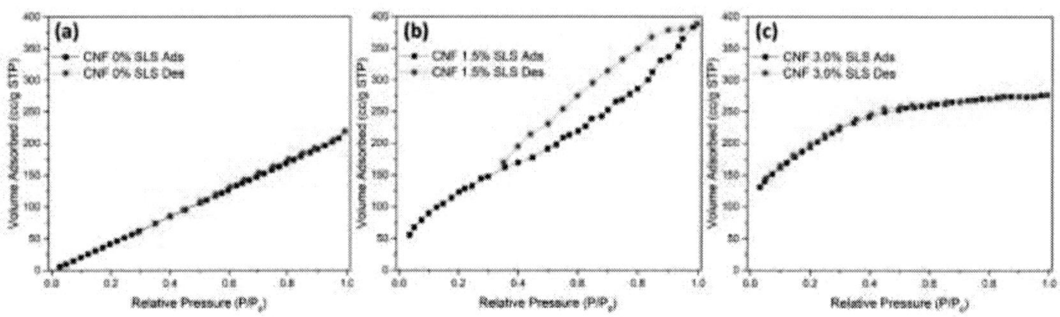

Figure 5. Nitrogen adsorption–desorption isotherms of Carbon nanofiber with (a) 0 wt.%, (b) 1.5 wt.%, and (c) 3 wt.% of sodium lignosulfonate content.

Table 3. Nitrogen adsorption-desorption of carbon nanofiber at various SLS concentration

Sample	Specific surface area (m^2/g)	Pore diameter (nm)	Pore volume (cm^3/g)
CNF 0.0 wt.% SLS	292.986	3.5358	0.3384
CNF 1.5 wt.% SLS	496.547	4.8445	0.6014
CNF 3.0 wt.% SLS	703.615	2.4373	0.4287

Interestingly, contrary to the previous phenomenon, after transformed into carbon nanofiber, the specific surface area of CNF got dramatically increase by the additional and increasing content of SLS. And as described in Table 3, the pore size of CNF supported by 3 wt.% of SLS perform slightly micropore rather than others. It may cause by the conversion of SLS into activated carbon support the increasing of micropore side of nanofiber.

4. Conclusions

In summary, a study about carbon nanofiber was prepared via electrospinning using Sodium Lignosulfonate as the source of lignin and PVA as the source of polymer. The SLS as affordable lignin- based material was proved as promising material to be used as lignin source, since the high of alkali content can provide a high porous of carbon nanofiber. The increasing content of SLS will reduce the surface area of nanofiber and the average fibre diameter. But contrary increase the porous after it was converted into carbon nanofiber. Furthermore, this research is potentially continue to some of application such as a pigment adsorbent or even more the carbon source of energy storage or battery application.

5. References

[1] Zhu K, Shang YY, Sun PZ, et al. Oil spill cleanup from sea water by carbon nanotube sponges. *Frontiers of Materials Science* 2013; 1: 170–176.

[2] Titirici MM. *Hydrothermal Carbons: Synthesis, Characterization, and Applications*. Elsevier Ltd, 2012. Epub ahead of print 2012. DOI: 10.1016/B978-0-08-097744-7.00012-0.

[3] Hu S, Hsieh Y Lo. Lignin derived activated carbon particulates as an electric supercapacitor: Carbonization and activation on porous structures and microstructures. *RSC Advances* 2017; 7: 30459–30468.

[4] Chang FC, Yen SH, Wang SH. Developing lignosulfonate-based activated carbon fibers. *Materials*; 11. Epub ahead of print 2018. DOI: 10.3390/ma11101877.

[5] Marsh H, Rodríguez-Reinoso F. *Activated Carbon*. 1st ed. Netherland: Elsevier, 2006. Epub ahead of print 2006. DOI: 10.1016/b978-008044463-5/50018-2.

[6] Widiyastuti W, Fahrudin Rois M, Suari NMIP, et al. Activated carbon nanofibers derived from coconut shell charcoal for dye removal application. *Advanced Powder Technology*. Epub ahead of print 2020. DOI: 10.1016/j.apt.2020.06.012.

[7] Du B, Chen C, Sun Y, et al. Lignin bio-oil-based electrospun nanofibers with high substitution ratio property for potential carbon nanofibers applications. *Polymer Testing* 2020; 89: 106591.

[8] Liu H, Dai Z, Cao Q, et al. Lignin/Polyacrylonitrile Carbon Fibers: The Effect of Fractionation and Purification on Properties of Derived Carbon Fibers. *ACS Sustainable Chemistry & Engineering* 2018; 6: 8554–8562.

[9] Yen S-H, Chang F-C. Effects of Fiber Processing Conditions on the Yield, Carbon Content, and Diameter of Lignosulfonate-based Carbon Fibers. *BioResources* 2016; 11: 10158–10172.

[10] Yuniar RA, Widiyastuti W, Setyawan H, et al. Formation of Carbon Fibres from Polymer Poly(vinyl alcohol)/Acetylene Black using Electrospinning Method. *IOP Conference Series: Materials Science and Engineering*; 543. Epub ahead of print 2019. DOI: 10.1088/1757-899X/543/1/012030.

[11] Widiyastuti W, Rois MF, Setyawan H, et al. Carbonization of Lignin Extracted from Liquid Waste of Coconut Coir Delignification. *Indonesian Journal of Chemistry* 2020; 20: 842.
[12] Thommes M, Kaneko K, Neimark A V., et al. Physisorption of gases, with special reference to the evaluation of surface area and pore size distribution (IUPAC Technical Report). *Pure and Applied Chemistry* 2015; 87: 1051–1069.

Acknowledgement
The authors are grateful for the financial support provided by the "Deputi Bidang Penguatan Riset dan Pengembangan Kementerian Riset dan Teknologi/Badan Riset dan Inovasi Nasional" through the research grant under Contract No. No: 3/E1/KP.PTNBH/2020

Determining the weight of the hammer based on expert experience for estimating load-carrying capacity

M K Wardani[1], M F N Aulady[1], W Frido[1], and S Hendri[1]

[1]Institut Teknologi Adhi Tama Surabaya, East Java, Indonesia

milakusuma@itats.ac.id

Abstract. Inexperience geotechnical engineer has difficulty in specifying the best type of weight of the hammer. Here, this study aims to help inexperience geotechnical engineers to decide the optimum weight of the hammer on the pile driving machine. The Bayes method is applied to decide between drop hammer and diesel hammer pile driving machines. Furthermore, we interviewed the expert by several questions about the criteria of the best pile driving machine using a Likert scale. The result suggests that the best type of pile driving is Drop Hammer. The weight of the hammer required to carry out the piling process is 0.9 ton. Recommendations for allowable bearing use the Meyerhof 1976 method. This condition meets the workload that more 141.14 tons at a depth of 26 m when using a square pile dimension of 35 x 35 and 40 x 40 cm. The dimensions are 45 x 45 cm, the recommended depth of piles is 26 m. the allowable bearing carrying when compared with the dynamic capacity using the Hilley, ENR, and WIKA methods meet the requirement. This is given the result of the allowable load-carrying capacity of 170 tons that less than 280 tons until 390 tons at dynamic capacity.

1. Introduction

Foundation is one of the crucial parts of building work. The foundation work is tied to the soil condition at the construction location. A construction location with a deep hard soil layer is potentially needed attention by planning engineers. The planning engineers might use a pile foundation for foundation work in this situation [1]. Hence, to start pile foundation work in construction work, we must pay attention to several factors that affect to pile foundation. The factor such as installation method, selection of pile driver, and loading conditions are some of the essential parts that possibly affect the project successfully and efficiently [2].

The project successfully and efficiently is also determining by many factors. One of them is the experience of the engineer levels. For planning, engineers with less experience determining the pile driver might solve the complex problem because it will affect the time and cost of the project. Piles that are driven into place are generally considered to be low displacement or high displacement, depending on the amount of soil that must be pushed out of the way as the pile is driven [3]. Thus, the selection of a pile driver should be considered at the same time when planning the project.

This paper aims to help less experience planning engineers chose the optimum pile driver based on expert experience using the Bayes method. Bayes method is one of the methods that can help to make decisions based on complex criteria [4]. In 2016, Waluyo et al. [5] study about using the Bayes method for selecting long of pile foundation based on the risk analysis. The result indicates that the Bayes method successfully helps the engineer for the complexity of selecting long of the pile. In 2017, Yulianto [6] studied about selecting a pile foundation based on the noise of the pile driver and vibration of the pile driver. The result is related to the previous study that the Bayes method is could help to solve the foundation problem. However, this study focusses on selecting pile driving. Either, we are continuing to calculate the pile foundation using the pile driver expert judgment. For calculating the pile foundation, we used the static method and dynamic method. In completion, we confirm our calculation result with expert experience.

2. Method

This study starts with a questionnaire about pile driver selection. The questionnaire was given to the expert with various positions related to foundation work. The questionnaire involves the Site Manager, Site Engineer, Quality control engineer, Quality surveyor engineer, and Supervisor engineer in two cities, Surabaya and Gresik, East Java province, Indonesia. The expert was selected with the experience for foundation work for more than five years. Four factors that we determined based on our literature study.

The factor is Environmental impact, cost impact, productivity impact, and location accessibility. However, the expert respondent fills the questionnaire using a Likert scale from 1 to 5; the number indicates that the low impact of the pile driver to the factor that was determined into high impact. Afterward, we calculated the total value using the Bayesian formula, as shown in equation 1.

$$\text{The total value I} = \sum_{j=1}^{m} \text{ij } (Criteria[\text{ij}]) \quad (1)$$

Where:

Total Value i	= *the* final total value of the i-th alternative
ij value	= the first alternative value for j criteria
Criteria [ij]	= level of importance quality value for j-criteria
i	= 1,2, 3,n ; n = number of alternative
j	= 1,2,3...m; m = number of criteria

Next, we calculated the pile foundation. We collected data of standard penetration test (SPT) also the specification of the pile driver based on expert judgment. SPT was performed during the advancement of soil boring to obtain a disturbing drive sample (split barrel type) of the soil being penetrated and an approximate measure of its dynamic resistance. NSPT test obtained results relative density (Dr (%)) and the ground shear angle (ϕ) based on the value of the number of blow (N). If energy efficiency (Ef) is measured, then the energy-corrected SPT N-value adjusted to 60% efficiency (N60). N-values measured in the field should be corrected to N60 for all soils [7]. Theoretically, it is mandatory to measure Ef to get the proper correction to N60. In the absence of data, AASHTO (2004 with 2006 Interims) [8] recommends Ef = 60 for rope and cathead systems, i.e., donut and safety hammers and Ef = 80 for automatic hammer systems. Others, The NSPT value, must be corrected for the groundwater level and the overburden stress.

The prediction of pile load-carrying capacity can be achieved using different methods such as pile load test, dynamic analysis, static analysis based on soil properties from laboratory tests, and static analysis utilizing the results of in situ tests such as cone penetration test or standard penetration test. According to Meyerhof, when a pile is driven into loose sand, its density is increased, and the horizontal extent of the compacted zone has a width of 6-8 times pile diameter [9, 10] The principle of the load-carrying capacity of the pile foundation, including the Static method (Using the principles of classical soil mechanics), Dynamic and loading test (full-scale load test). The ultimate load-carrying capacity of a pile is given by equation 2 [11]:

$$Qu = Qp + Qs \quad (2)$$

where:
Qp = capacity of the pile point
Qs = load-carrying frictional resistance (skin friction) derived from the soil–pile interface

Meyerhof N-SPT is a more actual method to use in foundation planning because the carrying capacity value is closest to the static loading test result. Later, we calculated the end bearing pile used the following equation:

$$Qu = 40\ Nb\ Ap + 0{,}2\ N\ Ap \tag{3}$$

Further, Single pile load-carrying capacity for cohesive soil and laboratory data found can used formula based on *[10] :*

$$Qb = Ab \times fb \tag{4}$$
$$fb = cb\ Nc + Pb \tag{5}$$

where,
Qb = point bearing single pile (kN)
fb = end bearing resistance of single pile (kN/m^2)
Ab = End Pile Area (m2)
Cb = Cu value at the end of pile
Nc = load-carrying capacity factor (Nc = 9) [14]
Pb = Po, overbudden pressure

On the other hand, for the static method, we used the Meyerhof formula [9]. This Load-carrying capacity Method observed the effect of piling the foundation in different soil layers [9]. It can be described using this illustration in Figure 1 and Figure 2.

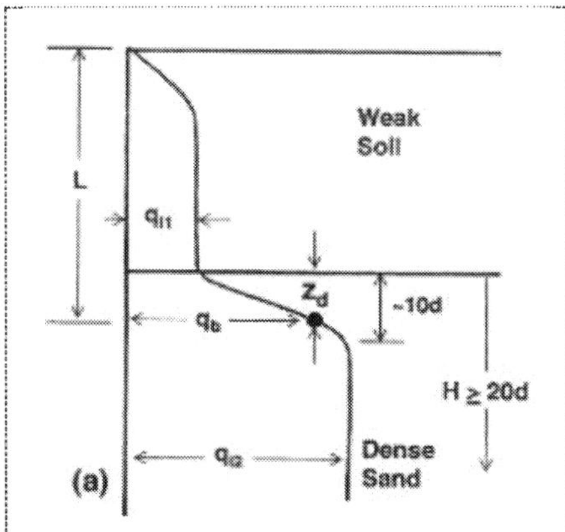

Figure 1. Resistance Pile tip and depth of sand between soft soils on the Thick sand layer.

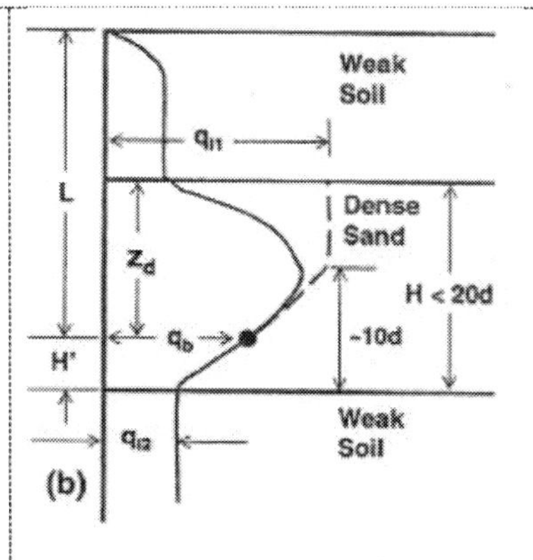

Figure 2. Resistance Pile tip and depth of sand between soft soils thin sand layer

End bearing resistance with Single pile ultimate load-carrying capacity pays attention to the depth factor. The point bearing pile capacity of a pile is given by equation 6 and 7:

$$Qb = Ab\,(38N) * Lb/d) \leq 380\,N\,(Ab) \tag{6}$$
$$fb = 0.40N60'\left(\frac{L}{d}\right)\sigma r \leq 4N60'\sigma r\,(\frac{kN}{m^2}) \tag{7}$$

where,
fb = end bearing single pile (kN/m^2)
N_{60} = The corrected N-SPT affects field procedures only and overbudden pressure
σr = stress references = 100 kN/m^2
L = depth (m)
D = pile diameter (m)

The average N value is calculated using 4D below the pile tip and 4D above the pile tip. Lb/D is the depth ratio; it can less then L/d if the soil condition has a variation Layer. Then, Ab is the area of the end pile. If we have a maximum value from the equation $L/d \geq 7.5$, frictional pile capacity used assumed massive pile displacement [12]. The frictional pile capacity of a pile for cohesive soil is given by equation 8:

$$fs = \frac{1}{50}\,\sigma r\,N60\,(\frac{kN}{m^2}) \tag{8}$$

where,
fs = frictional single pile (kN/m^2)
N_{60} = The corrected N-SPT affects field procedures only
σr = stress references = 100 kN/m²

The factor of safety generally used ranges from 2.5 to 4, depending on the uncertainties surrounding the calculation of the ultimate load *[13]*. The piles were installed by driving (hammering and vibration) and boring techniques. The results of influence of installation methods, show load-carrying capacity increments of 10% in bored piles, 20-22% in hammered driven piles, and 20-30% in vibrated driven piles *[14]*. Diesel hammer has a more exact shape compared to other hammers. The advantages of Diesel hammer are easy to move, provide big hit carrying capacity, good tool ability, and low fuel cost. The disadvantage is that the work is slow; there are a sound and splash from the lubricating oil, the big hammer tools. The drop hammer system requires a certain height because this system uses a massive hammer. The drop hammer method has the advantages of simple tools; falling height can be checked easily, with little difficulty and low operating costs. In this method, because there is a heavy hammer, it can cause the pile head to be easily damaged, the length of the driving is limited, often becomes a slow eccentricity. The selection of the weight of the pile stacker can be calculated using equation 9:

$$B = (0{,}5\,x\,P) + 600\,kg \tag{9}$$

Where B is the weight of hammer (kg), and P is the weight of pile (kg). Pile driving equations, such as the Engineering News Formula, were developed that related the pile capacity to the energy of the pile driving hammer and the average net penetration of the pile per blow of the pile hammer [13].

Several equations have been developed to calculate the ultimate capacity of a pile during driving. These dynamic equations are widely used in the field to determine whether a pile has reached a satisfactory bearing value at the predetermined depth. One of the earliest such equations commonly referred to as the Engineering News (EN) Record formula. This method uses derived from the work-energy theory. The pile penetration, S, is usually based on the average value obtained from the last few driving blows. The factor of safety FS = 6 was recommended for estimating the allowable pile capacity.

$$Qu = \frac{E\,WRh}{S+C}\;\frac{Wr + n^2\;Wp}{Wr + Wp} \tag{10}$$

where,
E = efficiency of hammer
C = 2.54 mm if the units of S and h are in mm
Wp = weight of the pile
n = coefficient of restitution between the ram and the pile cap

3. Result and Discussion

Table 1. Results Analysis based on Hammer Selection for Pile construction using Bayes Method

Alternative	Criteria				The Value Alternative Score
	Environmental Impact	Implementation Cost	Pile Productivity	Location Topography	
Drop Hammer	4	4	4	3	3,75
Diesel Hammer	4	3	2	3	3,05
Score Criteria	0,3	0,2	0,25	0,25	

The result of the Bayes method is illustrated in Table 1. The Table shows that on the environmental impact criteria, the two piles have a high score. This is because of both harm erection and the noise pollution caused by the pile driver. In the cost criteria analysis, there is a slight difference where the drop hammer has higher points than the diesel hammer. This condition is influenced by the fuel requirements of the drop hammer a little bit more than the diesel hammer. The advantage of diesel hammers is because diesel hammer easier to move places. It is a better productivity criterion than drop hammers. As for topography and location, the two alternative piling tools are easy to access to the project. Then from that value is continued by giving the Quality weight value. The most significant weight value is given to the environmental impact criteria. The second is given on the productivity criteria and the topography of the location. The quality weight value is given to the implementation cost. Table 2 shows the composition and calculating the decision matrix based on equation 1. From the matrix that has been compiled, it is found that the total weight is 3.75. This value of a drop hammer is selected in the use of the piling process.

Table 2. Calculation of the Decision Matrix Using the Bayes Method

Alternative	Score Criteria	Quality Score	Alternative Score
1	2	3	4=2*3
Drop Hammer	4	0.3	1.2
	4	0.2	0.8
	4	0.25	1
	3	0.25	0.75
		Total value	3.75
Diesel Hammer	4	0.3	1.2
	3	0.2	0.6
	2	0.25	0.5
	3	0.25	0.75
		Total value	3.05

From the results of the Bayes method analysis, then calculate the impact weight of the pile tool. In this case, the pile weight used according to the service piling company is 486 kg / m. From these results, the weight of the hammer required for the piling process is calculated and based on the pile driver specification, namely selected with the Kubota V1505 Turbocharged engine type. Then the weight of the hammer required to carry out the piling process is 0.9 ton. The calculation of the allowable load-carrying capacity o for a single pile based on the consultant's recommendation at the first point soil investigation is 28 m deep with a Q_{allow} value = 120 tons. Whereas at the second point, a soil investigation depth of 24-26 m is recommended. So, we analyse two Static load-carrying capacity.

Table 3. Comparison Allowable load-carrying capacity, Working load and Dynamic Capacity

Points	Depth (m)	Pile dimension cm	Meyerhof 1956	Meyerhof 1976	Working Load ton	Dynamic capacity (ton)		
						Hilley	ENR	WIKA
	24		70.45	113.80				
DB1	26	35 x 35	74.25	111.09		332,45	293,63	391,12
	24		132.72	169.34				
DB2	26	35x35	85.56	129.03				
	24		87.25	116.32				
DB1	26	40x40	90.45	141.18	141.14	326,74	288,58	384,40
	24		165.12	174.20				
DB2	26	40x40	101.76	162.53				
	24		106.29	119.18				
DB1	26	45 x 45	108.81	175.29		320,67	283,23	377,26
	24		201.84	179.71				
DB2	26	45 x 45	120.12	200.50				

Table 3 shows a comparison between the two static methods at the two points of soil investigation. Besides, the calculation results are also compared to workload and the calculation of dynamic capacity. At two points of investigation, the recommended depth of piles is the same at a depth of 24-26 m. It is known that in the Meyerhof 1956 method, the load-carrying capacity value is smaller than the working load. Except at the DB2 point of 24 m depth with dimensions of 40x40 and 45x45 cm, the value of the load-carrying capacity meets the working load. Whereas in Meyerhof 1976 method, it is obtained that the allowable load-carrying capacity value meets the two points of investigation. The dimensions of 40x40 and 45x45 with a depth of 24 m at point DB1 and point DB2. The allowable load-carrying capacity is almost the same as the load obtained at point DB1 with dimensions 35x35 cm, depth of 26 m, namely 141.18 tons. The coefficient value influences this at the pile tip and the NSPT calculation from the pile tip. When using the Meyerhof 1956 method, the most massive load-carrying capacity value is at the DB2 point with dimensions of 45 x 45 cm and a depth of 24 m. However, at a depth of 26 m, the load-carrying capacity decreased in all pile dimensions—the conditions of the subgrade influence this from clay silt to sand. Overall, the value of the carrying capacity of sand in the Meyerhof method is smaller than the type of clay soil. The calculation method in clay is influenced by the Cu value, while the Nq value influences the sand due to the shear angle of the soil. Overall, comparisons between static methods versus dynamic methods are satisfactory. The allowable load-carrying capacity less or equal with dynamic load-carrying capacity. Thus, the drop hammer selection from the recommended Bayes method can be used for this case,

4. Conclusions

Drop hammer was selected from the results of the Bayes method analysis using a questionnaire on the expert. The quality weight value is 3.75 higher than the diesel hammer. Drop hammer selected, namely Kubota V1505 Turbocharged engine type. Then the weight of the hammer required to carry out the piling process is 0.9 ton. The static allowable load-carrying capacity, which is close to the working load value, is Meyerhof 1976 method. At the DB2 point, the allowable load-carrying capacity value only differs from 3% - 6% if the pile dimensions increase. At point DB1, the load-carrying capacity is still under the working load. The value of the Meyerhof 1976 static carrying capacity fulfills the requirements in planning because it is smaller than the dynamic carrying capacity. This study, though, needs improvement. In this study, we did not calculate the time and cost to confirm the efficiency. However, it just confirms that we help inexperience engineering with expert decisions. In the future, this study continuing try to make another questionnaire to determine how to select the piling method for another case pile foundation implementation

5. References

[1] Gunaratne M. *The foundation engineering handbook.* CRC Press, 2013.
[2] Poulos HG, Davis EH. *Pile foundation analysis and design.* 1980.
[3] Day RW. *Foundation Engineering Handbook: Design and Construction with the 2009 International Building Code.* McGraw-Hill, https://books.google.co.jp/books?id=ui9ZswEACAAJ (2010).
[4] Marlina A. Metode Bayes Untuk Menentukan Kelayakan Calon Tenaga Kerja Luar Negeri [Bayes Method To Determine Eligibility of Prospective Overseas Workers]. *Moneter: Jurnal Keuangan dan Perbankan*; 1.
[5] Waluyo B, Sulistiyono H, Murtiadi S. Metode Bayes Untuk Pemilihan Panjang Tiang Pancang Beton Berdasarkan Analisis Risiko Pada Jembatan Banyumulek Lombok Barat [Bayes Method for Selecting Concrete Pile Length Based on Risk Analysis at Banyumulek Bridge, West Lombok]. *Spektrum Sipil* 2017; 3: 156–166.
[6] Yulianto J. Pemilihan alat pancang menggunakan expert choice [The choice of stake using expert choice]. *Jurnal Riset Rekayasa Sipil* 2017; 1: 50–58.
[7] Samtani NC, Nowatzki EA. *Soil and Foundations Reference Manual - Volume 1.* FHWA-NHI-06-088, Washington DC: National Higway Institute.
[8] American Association of State Highway and Transportation Officials. *AASHTO LRFD Bridge Design Specifications, Customary U.S. Units: 2006 Interim Revisions*. American Association of State Highway and Transportation Officials, https://books.google.co.jp/books?id=b2omAAAACAAJ (2006).
[9] Mayerhof G. Bearing capacity and settlemtn of pile foundations. *Journal of Geotechnical and Geoenvironmental Engineering*; 102.
[10] Meyerhof G. Compaction of sands and bearing capacity of piles. *Transactions of the American Society of Civil Engineers* 1959; 126: 1292–1322.
[11] Nadella K, Wardani MK, DS AI, et al. Perbandingan Daya Dukung Tiang Pancang dengan Metode Statis dan Dinamis pada Proyek SBE Plant PT. Ecooils Jaya Indonesia. 2019, pp. 545–550.
[12] Hardiyatmo HC. Fondasi II. *Edisi Kelima Yogyakarta: Universitas Gajah Mada.*
[13] Das BM. *Principles of foundation engineering*. Cengage learning, 2015.
[14] Adejumo T. Research Article Effects of Shape and Technology of Installation on the Bearing Capacity of Pile Foundations in Layered Soil.

Effect of hole diameter and basin size on the vortex gravity system

F A Herbhakti[1] and R Hantoro[1,2]

[1]Engineering Physics, Sepuluh Nopember Institut of Technology ITS
[2]Ocean Engineering, Sepuluh Nopember Institut of Technology ITS

kobarbekti@gmail.com

Abstract. The vortex gravity system consists of a tube basin, a canal (worksection) and a water tank. Water is pumped into the upper tanker and flowed into the canal by opening the valve which then hits the tube basin and creates a whirlpool. The water that hits the tube and enters the outlet hole goes down to the lower tank and is circulated back to the upper tank. The outlet hole in the tube has 4 sizes and the tube size has 3 sizes. This study is based on experimental tests to determine the flow performance of the outlet diameter of the tube against the formation of vortices in the vortex tube. From several existing studies, the type of outlet hole in this tube has not been thoroughly studied. From the analysis results, it can be seen the difference in the results of each outlet size in the tube basin. With this question, this study aims to determine the strength of the vortex of the parameters that must be looked for, namely Water Flow Discharge (m3 / s), velocity (m / s), and Total Head Water (m). The vortex gravity system on a laboratory scale, based on previous studies or studies varying the diameter of the vortex tube and tube outlet hole, this study examined the effect of differences on the diameter of the vortex tube and the outlet hole of the tube basin. From the calculation of the vortex strength at the outlet hole of 25 cm, the results of the vortex strength are 7.53 m2 / s; 22.60 m2 / s; 45.21 m2 / s; 52.75 m2 / s. After that, proceed with the calculation of the outlet hole of 30 cm, the results of the vortex strength are 12.24 m2 / s; 28.26 m2 / s; 45.21 m2 / s; 56.52 m2 / s. Furthermore, the calculation at the 35 cm outlet hole obtained the vortex strength of 14.13 m2 / s; 30.14 m2 / s; 42.39 m2 / s; 60.28 m2 / s. the occurrence of fluctuation in the vortex strength results due to the energy available in the extracted water flow after hitting the outlet hole in the vortex tube. The graph produced in this calculation shows that the highest vortex strength occurs in a tube with a diameter of 35 cm.

1. Introduction

The Water Vortex Gravity Power Plant (GWVPP) is a generating system that generates electricity from a vortex water turbine with a low height and low flow rate. The vortex gravity system consists of a tube basin, a canal (work section) and a water tank. Water is pumped into the upper tank and flowed into the canal by opening the valve which then hits the tube basin and creates a whirlpool. The water that hits the tube and enters the outlet hole goes down to the lower tank and is circulated back to the upper tank [1]. Despite the fact that the outlet holes in the tubes are of different sizes, the perfect vortex formation can occur in certain tube sizes. From several studies that there have been various sizes of the tube and the outlet hole of this tube has not been studied in depth. To increase the velocity and the formation of the eddy strength, a study of the flow of the open channel is needed to determine the sizing characteristics of the tube and the tube outlet.

IOP Conf. Series: Materials Science and Engineering 1010 (2021) 012042 doi:10.1088/1757-899X/1010/1/012042

Given this background, this study aims to design tools and conduct experiments on vortex gravity models and determine the effect of the diameter ratio of the vortex tube and the outlet of the vortex tube to the height and velocity of the vortex flow. Furthermore, checking the strength of the vortex generated around the outlet hole of the tube requires a simulation experiment using the Inventor software as a visual representation of the laboratory-scale vortex gravity system model.

Hydroelectric power is one way to take advantage of renewable energy sources, but its utilization is still on a small scale and uses simple technology as a generating system that can meet the required electrical energy. This type of hydroelectric power is often called microhydro or often called Pico hydro depending on the electricity output generated by the generator. Microhydro which usually uses waterfalls with high water falls. Meanwhile, the height and fall are not maximally utilized. This is a reference for utilizing river flow by converting it into a vortex flow.

A researcher from Germany Viktor Schauberger developed vortex flow technology (vortex) to be applied to water turbine modelling. Vortex flow, also known as pulsating or vortex flow, can occur in a fluid flowing in a channel that undergoes sudden changes. The phenomenon of vortex flow is often found in airplane wing modelling, vortex flow tends to be considered a loss in a fluid flow. Starting from the conditions above, it is necessary to conduct research to see the effect of tube dimensions and discharge on vortex formation.

Sarkardeh conducted experimental modelling to investigate the effects of head walls and trash shelves on vortices [2]. Acoustic Doppler Velocimetry is used to determine velocity and circulation profiles. The authors determine the relationship between strength and vortex type. Recently, Yang conducted dimensional analysis and experimental modelling to determine the effect of the intake-entrance profile on critical submergence [3]. The authors tested eight entrance profiles and found that the square edge profile produced the highest local head losses. Li and Wang, and all successfully used particle imaging techniques to investigate air core vortices in hydraulic intakes [4] [5]. Carriveau conducted extensive studies on the evolutionary mechanisms of free-surface vortices as well as axial stretching [6].

Further researchers who became the research literature "Experimental and Numerical Analysis of Free-Surface Turbulent Vortex Flows with Strong Circulation" vortex system determines the vortex strength which is influenced by differences in several variations of outlet holes in the tube basin. This data will be used as a reference to determine vortex strength [7].

The research of Fajar Sumantri and Muhamad Fitri aims to design a vortex test device that determines the working system of the tool so that it can be used to observe free vortices and forced vortices in water. This vortex tester is designed with a water outflow velocity of 2.425 m / s and a flow rate of 73.8 liters / minute [8].

Subsequent research Javed Ahmad Chattha The Water Vortex Turbine generator determined that the performance of the propeller rotor that is positioned is higher than the rotor located below it, thus showing that undistorted surface eddies have more energy available for power generation. To build multi-staging in GWVT, this study suggests using the same propeller and optimal setting distance [9].

1.1. Overall design
In this study, a vortex system is used without using a turbine. Prior to data collection, a set-up was carried out by connecting each component so that it became a vortex system.

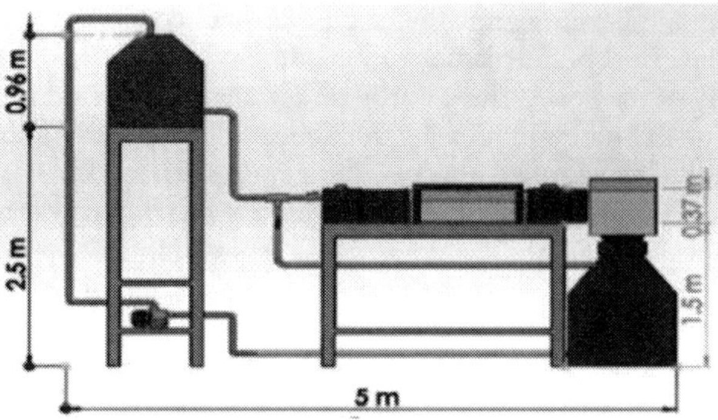

Figure 1. Vortex System Set-Up

Carried out in this study is to take data from tools that have been assembled on several variations of the diameter of the vortex tube with the appropriate measuring instrument. The parameters required include discharge, velocity, head worksection (channel) on the quality of vortex formation and the strength of the vortex produced. Based on the experimental results of this study, it can be used as a reference in further studies in laboratory scale vortex systems.

2. Method

Experimental tests of the vortex system using 4 experimental variations in the vortex tube outlet holes were carried out next to the Energy Engineering Laboratory, Mechanical Engineering Department, Faculty of Engineering Physics, Faculty of Industrial Technology and Systems Engineering. Measurements carried out in this study include measuring the velocity of the vortex in the vortex tube using a Flow Meter from JDC Flowatch.

Before testing the vortex system and collecting data, checking is first carried out on several installations and equipment, which includes:

a. Check the water discharge in the lower tank and upper tank
b. Check of the connecting pipe between the feed pump and the gutter.
c. Vortex tube examination according to the size of the tube diameter and the outlet of the vortex tube.
d. Feed pump check.

After the inspection procedure of some of the above installations and equipment has been carried out and the inspection is confirmed to be in standby condition, the testing procedure can be started. The experimental test procedure for the vortex system with tubes with a diameter of 25, 30, 35 cm

2.1 Retrieval of Data

This research begins with experimental tests carried out on a canal (work section) and an acrylic vortex tube that can be changed by variations in the diameter of the tube and the outlet diameter of the vortex tube.

Table 1. Variation Table on acrylic vortex tubes

Variation of tube diameter (cm)	Variation of the outlet diameter of the vortex tube (cm)
25	3
30	6
35	9
	12

In this experimental test stage, to obtain various flow rates, the velocity measurement on the vortex tube was carried out several variations in table 1 using a Flow Meter from JDC Flowatch as shown in the figure.

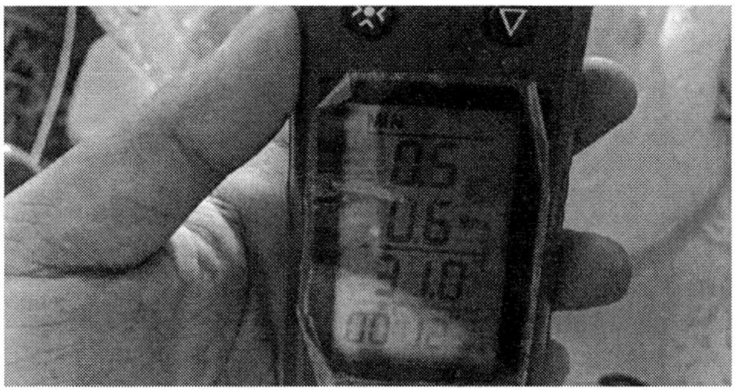

Figure 2. JDC Flowatch for Flow Meter with 12 samples

Flow velocity sampling was carried out in 12 variations. The flow velocity obtained was 4 variations of the outlet diameter of the vortex tube in the tube diameter of 35 cm and the other two were the diameter of 30 cm and 25 cm as shown in Table 2. After determining the variation that will be used for testing, the next step is to measure the worksection area (channel) and the vortex tube. In this research, using worksection (canal) and vortex tube made of acrylic with predetermined dimensions to perform experimental tests. The form of worksection (canal) and vortex tube made of acrylic used in this study is shown in Figure 3.

Figure 3. Acrylic Vortex Tube and Canal System.

3. Results and discussion

3.1 Flow discharge test on channel (worksection)

Flow discharge over triangular sharp lintels designed to meet the requirements can be calculated based on the following equation.

$$Q = v\,h\,b \quad (1)$$

Information:

Q = discharge (m^3/s)
v = flow velocity (m/s)
h = flow height (cm)
b = channel width (worksection)

Table 2. Canal discharge on tubes with diameter 35 cm, 30 cm, 25 cm.

Hole Size	Velocity of the tube 25	Velocity of the tube 30	Velocity of the tube 35	Flow of the tube 25	Flow of the tube 30	Flow of the tube 35	b	Discharge of tube 25	Discharge of tube 30	Discharge of tube 35
3 cm	0.8	1.3	1.5	6	6	6	20	96	156	180
6 cm	1.2	1.5	1.6	5.5	5.5	5	20	132	165	160
9 cm	1.6	1.6	1.5	5	5	4.5	20	160	160	135
12 cm	1.4	1.5	1.6	4.5	4.5	4	20	126	135	128

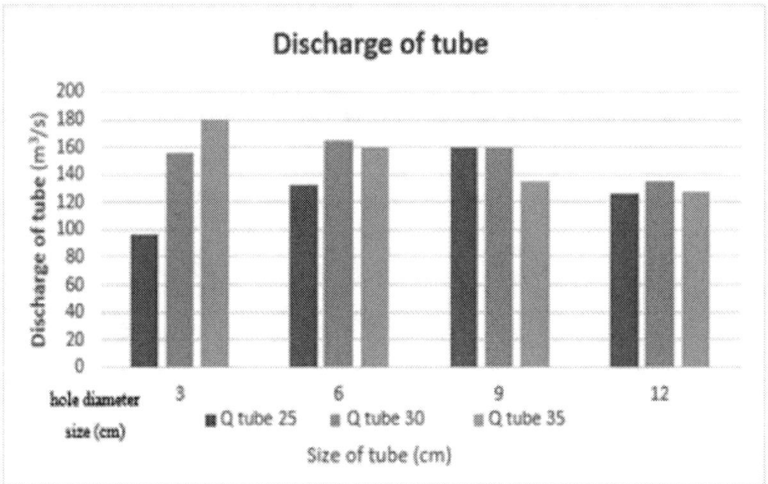

Figure 4. Graph of Discharge on each tube diameter

3.2 Velocity Test

The measurement in this study aims to determine the velocity generated in several variations in each size of the vortex tube. This measurement is carried out 4 data retrieval which will later be averaged so that it can be used as a comparison chart. In one data collection on one size of the vortex tube, 4 variations were made of the outlet diameter of the vortex tube so that 12 measurements were obtained. On variations in the outlet diameter of the vortex tube.

In order to analyze the directional velocity of the flow, a measuring line is set in position on each part of the water surface. To find a suitable outlet diameter, the tangential velocity and radial velocity are calculated. The distribution of the velocity vectors on the measuring line is shown in Figure 5 which is a top view of the vortex housing. The distribution of velocity is shown on line 1, line 2, line 3, and line 4. Computerized fluid dynamics simulations run with no slip on the wall and with exit pressure at the exhaust hole.

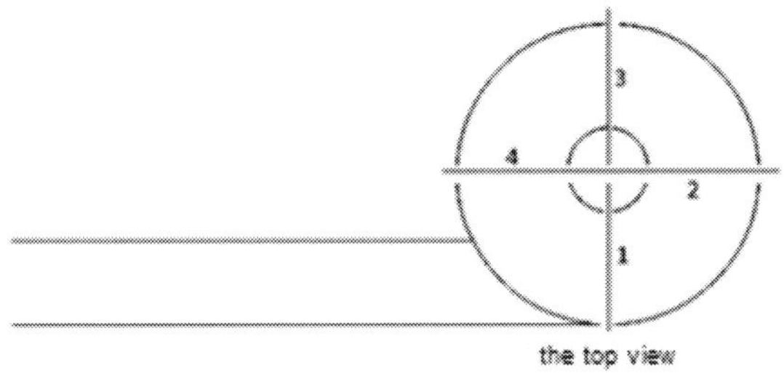

Figure 5. The top view of the tube basin where the position of the measuring line on each part is shown at point 1, point 2, point 3, and point 4

Table 3. Velocity measurement on the outlet diameter of the vortex tube

Hole Size	Tube Diameter 25 cm				Tube Diameter 30 cm				Tube Diameter 35 cm			
	v1	v2	v3	v4	v1	v2	v3	v4	v1	v2	v3	v4
3 cm	0.7	0.6	0.5	0.4	0.4	0.6	0.8	0.7	0.6	0.7	0.9	1
6 cm	0.8	0.6	0.5	0.7	1.7	1.1	1.9	0.8	0.8	0.6	0.7	0.7
9 cm	0.8	1.1	0.8	0.9	1.1	1	1.2	1.1	0.5	0.6	0.7	0.8
12 cm	0.9	0.8	0.9	0.8	1.4	1.2	1.3	1.5	0.7	0.6	0.4	0.7

Based on Figure 5 above, it can be concluded that the greatest fluid velocity is at the outlet hole 3 cm, then 6 cm, 9 cm and the next 12 cm. Comparison of the average speed at each exhaust can be seen in Table 4 and Figure 6 Graph below.

Table 4. Average measurements of the outlet diameter of the vortex tube.

Hole Size (cm)	Tube Diameter 25 cm	Tube Diameter 30 cm	Tube Diameter 35 cm
3 cm	0.55	0.625	0.8
6 cm	0.65	1.375	0.7
9 cm	0.9	1.1	0.65
12 cm	0.85	1.35	0.6

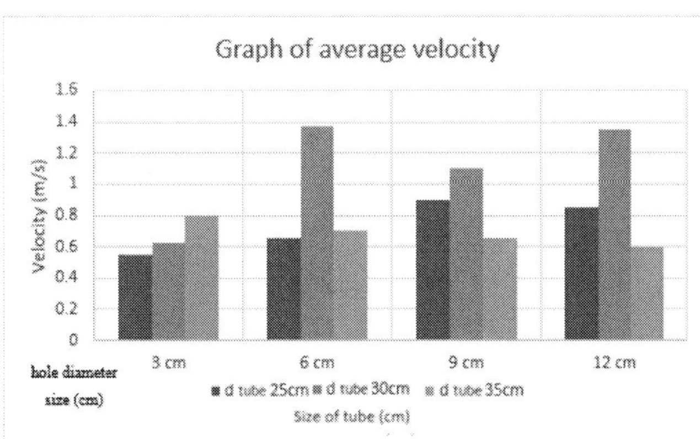

Figure 6. Graph of average Velocity on the vortex system testing

After doing 3 data retrieval it can be concluded with the average results shown in Figure 6 The rotational force (torque) generated by the shaft acting on the water flow which hits the outlet diameter of the vortex tube causes a whirlpool in the tube.

For the four vortex tube outlet diameter conditions, the inlet flow rate is lower than the outflow rate when a new vortex is formed. Therefore, the height of the vortex will decrease and remain constant when the outlet flow rate drops to the value of the inlet flow rate. The inflow rate is greater than the outflow rate when a new vortex is formed. Therefore, the vortex height increases, the vortex strength increases, and the outlet flow rate remains almost constant. This explains why the vortex height can be very different as the inflow rate increases.

Figure 6 shows that velocity has increased in tubes with a diameter of 35 cm and fluctuates in tubes with a diameter of 30 cm and 25 cm. The velocity graph in this test shows that the highest velocity occurs in a tube with a diameter of 30 cm. The calculation result of velocity is used to determine the vortex strength calculation.

3.3 Vortex strength analysis

Vortex Strength is the meter squared per second of water that forms a vortex in the basin hitting the outlet hole.

$$rv = \frac{2\pi rinQ}{bhin} \tag{2}$$

$$rv = 2\ phi\ r_{in}\ Q/b\ h_{in}$$
$$= 2phi\ r_{in}\ A\ V/\ bh$$
$$= 2phi\ r_{in}\ bh\ V/\ bh$$
$$rv = 2phi\ r_{in}\ V$$

Where:

a	= p1. p2 (m2)
Q	= discharge = a. v
v	= velocity (m / s)
b	= channel width (m)
Deff	= outlet diameter (m)
hin	= water level in the canal (m)
r_{in}	= radius inlet (m) = $\frac{D+b}{2}$
h	= water level in the canal (m)

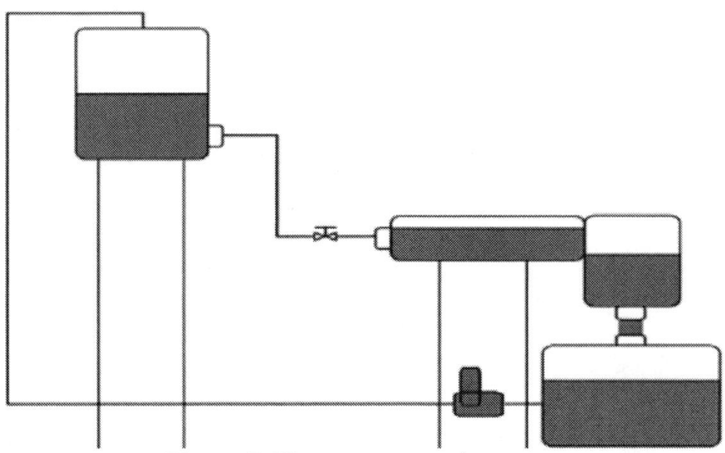

Figure 7. Vortex system schematic.

$$rv = \frac{2\pi rinQ}{bhin}$$
$$\frac{2\pi rinAV}{bhin}$$
$$\frac{2\pi rinbhV}{bh}$$
$$rv = 2phi\ r_{in}\ V$$

By using the formula above, the results of the vortex strength chart below are obtained.

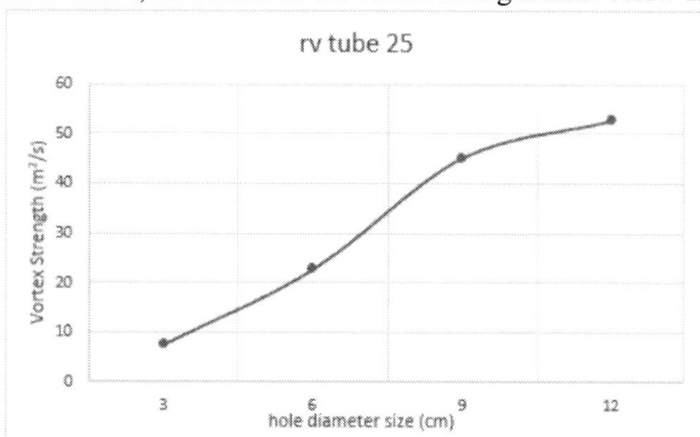

Figure 8. Vortex Strength chart tube diameter 25cm.

In Figure 8 above we can see the results of vortex strength graphs in each variation of basin tube sizes and outlet holes. For vortex strength it can be explained that there is a decrease in the large outlet hole to the small hole. The test was continued by changing the tube outlet holes from 6, 9, 12 to a drastic drop in the 3 cm outlet hole. This test can determine the results of the vortex strength at each variation in the outlet diameter of the vortex tube. From the calculations for each outlet hole 25 cm, the vortex strength results are 7.53 m^2 / s; 22.60 m^2 / s; 45.21 m^2 / s; 52.75 m^2 / s.

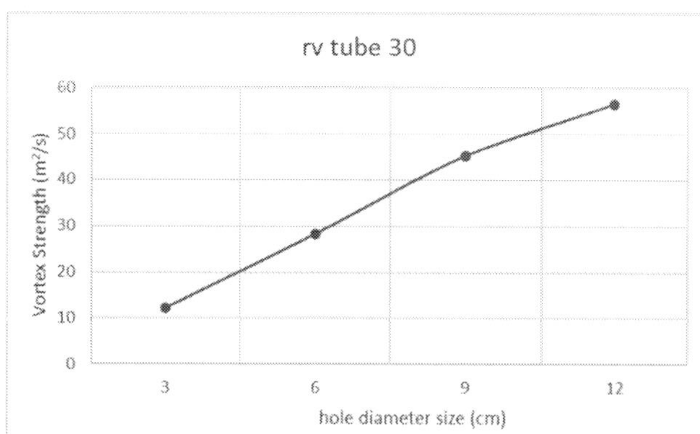

Figure 9. Vortex Strength graph of 30cm tube diameter.

In Figure 9 above, you can see the results of the vortex strength graph in each variation in the size of the tube basin and outlet hole. For the strength of the vortex, it can be explained that there is a decrease in the large outlet hole to the small hole. The test was continued by changing the tube outlet holes from 6, 9, 12 to a drastic drop in the 3 cm outlet hole. This test can determine the results of the vortex strength at each variation in the outlet diameter of the vortex tube. From the calculations for each outlet hole 30 cm, the vortex strength results are 12.24 m2 / s; 28.26 m2 / s; 45.21 m2 / s; 56.52 m2 / s.

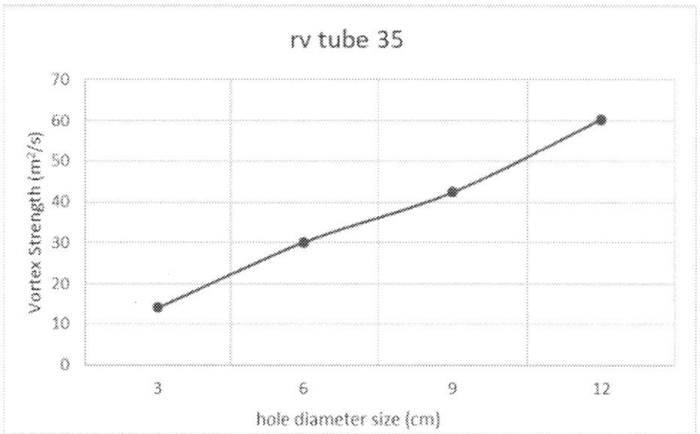

Figure 10. Vortex Strength chart tube diameter 25cm.

In Figure 10 above, you can see the results of the vortex strength graph in each of the variations in the size of the tube basin and outlet hole. For vortex strength it can be explained that there is a decrease in the large outlet hole to the small hole. The test was continued by changing the tube outlet holes from 6, 9, 12 to a drastic drop in the 3 cm outlet hole. This test can determine the results of the vortex strength at each variation in the outlet diameter of the vortex tube. From the calculations for each outlet hole 35 cm, the vortex strength results are 14.13 m2 / s; 30.14 m2 / s; 42.39 m2 / s; 60.28 m2 / s.

Table 5. Canal vortex strength table on tubes 35cm, 30cm, 25cm in diameter.

Hole Size (cm)	Q tube 25	Q tube 30	Q tube 35	Rv tube 25	Rv tube 30	Rv tube 35
3 cm	96	156	180	7.536	12.246	14.13
6 cm	132	165	160	22.608	28.26	30.144
9 cm	160	160	135	45.216	45.216	42.39
12 cm	126	135	128	52.752	56.52	60.288

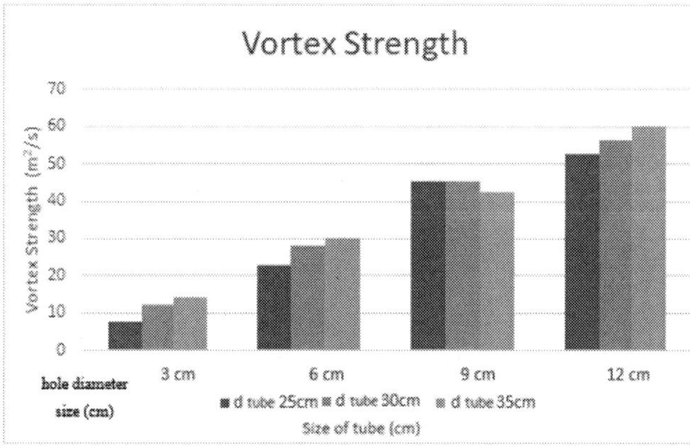

Figure 11. Graph of Vortex Strength on each tube diameter.

In table 5 above, we can see the results of the calculation of the vortex strength in each tube basin and outlet hole size variation. For the strength of the vortex, it can be explained that there is a decrease in the large outlet hole to the small hole. The test was continued by changing the tube outlet holes from 12, 9, 6 to a drastic drop in the 3 cm outlet hole. This test can determine the results of the vortex strength at each variation in the outlet diameter of the vortex tube.

The inlet width b becomes so small that the viscous effect limits the inlet velocity or the velocity increases towards a supercritical value resulting in energy dissipation via hydraulic jumps downstream of the inlet. The result of this variation changing the hole in the vortex tube outlet will give a significant reduction in the formation of a vortex tube in the vortex system in one circulation. The inlet radius r

becomes so large that the local tangential velocity in the far field is dissipated and fails to affect the near field.

Where Q = the discharge entering the vortex system, R_{in} is the radius of the outlet, where b is the width of the channel, H_{in} is the water level in the channel, Q is a multiplied by velocity and R_{in} is half of the sum of d and b. Experimentally, testing the vortex system using flowing water to form a vortex flow. Water will flow through the canal and fill the tank and exit through the exhaust hole until it forms a vortex flow. Based on equation (2), it can be seen that the results of the vortex strength are in table 5 where if the size of the outlet hole in the tube basin is large, the resulting vortex strength is also large. Conversely, if the size of the outlet hole in the tube basin is small, the resulting eddy strength is also small.

4. Conclusions

From the experimental tests on circular vortex basin tubes with dimensions of height 50 cm and tank diameters 35, 30, 25cm and variations in the diameter of the exhaust hole 12, 9, 6 cm and 3 cm. Based on the results of research and analysis that has been done, it can be concluded that:

a) A laboratory scale miniature vortex canal system has been made and it has been tested using various predetermined variations.
b) From the calculation of the vortex strength at the outlet hole of 25 cm, the results of the vortex strength are 7.53 m2 / s; 22.60 m2 / s; 45.21 m2 / s; 52.75 m2 / s. After that, proceed with the calculation at the 30cm hole outlet, the results of the vortex strength are 12.24 m2 / s; 28.26 m2 / s; 45.21m2 / s; 56.52 m2 / s. Furthermore, the calculation of the 35cm hole outlet results in the vortex strength of 14.13 m2 / s; 30.14 m2 / s; 42.39 m2 / s; 60.28 m2 / s.
c) From the results obtained vortex strength where if the size of the outlet hole in a large basin tube then the strength of the whirlpool produced is also large. Conversely, if the size of the outlet hole in the tube basin is small, the resulting eddy strength is also small.
d) From the results of testing variations in the outlet diameter of the vortex tube of 3 cm resulted in a longer and perfectly formed whirlpool.

5. Reference

[1] Achmad. K H, 2017 Pengaruh Sudut Inlet Notch Pada Turbin Reaksi Aliran Vortex Terhadap Daya Dan Efisiensi *J. Tek. Mesin*.
[2] Sarkardeh H Reza Zarrati A Roshan R and Graduate Student P, 2010 Effect of intake head wall and trash rack on vortices *J. Hydraul. Res.* **48**, 1 p. 108–112.
[3] Yang J Liu T Bottacin-Busolin A and Lin C, 2014 Effects of intake-entrance profiles on free-surface vortices *J. Hydraul. Res.* **52**, 4 p. 523–531.
[4] Li T Grace J and Bi X, 2010 Study of wall boundary condition in numerical simulations of bubbling fluidized beds *Powder Technol.* **203**, 3 p. 447–457.
[5] Wang Y K Jiang C B and Liang D F, 2010 Investigation of air-core vortex at hydraulic intakes *J. Hydrodyn.* **22**, 5 SUPPL. 1 p. 696–701.
[6] Cen H Carriveau R … D T-V O A and 2015 undefined, EFFECT OF MASS RATIO ON THE HYDRODYNAMIC RESPONSE OF A FLEXIBLE CYLINDER *pdfs.semanticscholar.org*.
[7] Mulligan S, 2015 Experimental and Numerical Analysis of Three-Dimensional Free-Surface Turbulent Vortex Flows with Strong Circulation p. 1–299.
[8] Sumantri F and Fitri M, 2017, PERANCANGAN ALAT UJI VORTEX BEBAS DAN VORTEX PAKSA.
[9] Ullah R Cheema T A Saleem A S Ahmad S M Chattha J A and Park C W, Oct. 2019 Performance analysis of multi-stage gravitational water vortex turbine *Energy Convers. Manag.* **198** p. 111788.

The development of software for calculating green open space adequacy to absorb CO_2 in Bangkalan city

Rosa Canina Pissera [1], Yulfiah[1]

[1] Environmental Engineering Master Program, Institut Teknologi Adhi Tama Surabaya
Jl. Arief Rachman Hakim 100 Surabaya
E-mail: rosa.sugiharto@gmail.com

Abstract. Bangkalan Regency, located at the west of Madura Island covering the area 1,260.14 Km^2 , has developed tremendously in some recent years, particularly in the post-development of Suramadu Bridge. To anticipate detrimental effects on the environment due to Bangkalan Regency improvement, the researchers developed software for calculating total CO_2 emissions of vehicles that could be absorbed by Green Open Space (GOS) in this city. The software was developed through Hypertext Preprocessor (PHP) based on the results of empirical study upon emission load and total trees calculation required for absorbing emission residue of vehicles. The software yielded from this research was successfully developed as indicated by the appropriateness of results between manual calculation and software one.

1. Introduction

The rapid development in Bangkalan Regency, not only gives positive effects to many sectors, but also contributes negative ones to the environment. If the Local Government is considered unable to prepare strategies to anticipate the existing changes, particularly the environmental management at the city area, the detrimental effects will vividly exist. One of development effects is the increasing activities of transport, industry, service, and others which eventually raise the emission residue of those activities in the air and cause air pollution.

Air pollution signified by high content of CO_2 will enable unhealthy environment and decrease people's health. Therefore, reducing CO_2 by developing Green Open Space (GOS) belongs to one of solutions to decrease total emission of carbon dioxide.

This research aimed at designing software for calculating: (a) total CO_2 emissions of vehicle fumes that could be absorbed by Green Open Space in the urban area, and (b) the coverage adequacy of urban Green Open Space.

1.1 Leterature Review

1.1.1 Green Open Space. Green Open Space (GOS) refers to a stretched area or path and/or a clumped area for open use, a place where trees grow naturally or consciously planted [1]. Zhang, Biao et al. [2] revealed that GOS was proven to have ability in lowering air temperature and mitigating hot regions due to urbanization process, particularly in Beijing.

The Regulation of Minister of Public Works Number 05/PRT/M/2008 concerning The Principles of Green Open Space Provision and Utilization in the city areas mentioned several functions of GOS as follows.

1. Ecological Function, GOS belongs to a part of air circulation system (lung of city); as the regulator of micro climate so that the circulation system of water and air can last naturally; as shady maker; and as the oxygen producer and rain water absorber.
2. Socio-cultural function, GOS depicts local cultural expression, interpersonal communication media, recreation site, and media as well as object of education, research, and training for learning nature.
3. Economic function, GOS refers to the sources of products that can be sold such as flower, fruit, vegetable, and it can be a part of businesses in agriculture, plantation, forestry, and so on.
4. Aesthetic function, GOS improves convenience, beautifies city environment, both micro-scale in terms of house yard and housing environment, but also macro-scale such as whole city landscape; stimulates the creativity and productivity of city people; as the factor shaping architectural beauty; and creates harmonious and balance between developed area and no-developed one.

1.1.2 Calculation of Green Open Space. Factor of Emission (FE) is defined as value/score representing the unit/quantity of pollutants emitted to the atmosphere by a certain activity. FE value is stated in pollutant mass per weight unit, volume, and distance or duration of a certain activity emitted the pollutant. In several conditions, FE represents facilities having air pollution controller that has been mounted. However, the yielded value must be considered as a value reflecting a condition when the pollutant controller is attached and functional. To get estimation that can be applied in general and representing emission in the long term, a regular monitoring is indispensable.

This approach provides better estimation value compared to temporary emission measurement. The approach of material balance will be used when the material percentage released to the atmosphere is great enough (for instance sulphur content of fuel). However, data will get high if the material is consumed, reacts chemically within the process, or if the emission released in the atmosphere covers only a small part of total materials being used. Factor of emission can be obtained through modelling, for example empirical model based on the existing data format.

Based on the data of SAMSAT (*Sistem Administrasi Manunggal Satu Atap* or One-stop Administration Services Office) at Bangkalan Regency in 2015, total vehicles in this city around 125,482 units consisting of 115,813 motorcycles, 123, 491 gasoline cars, and 1,991 solar cars.

The load of emission is calculated based on the following equation [3].

$$Emisi = (n \times L \times f \times \rho) \div FE \quad (1)$$

Notes:

Emission : CO_2 emission load (tons/year)
n : total vehicles (vehicles/hour)
L : road length (km)
f : actor of emission
FE : fuel economy (km/L)
ρ : gasoline density 0.63 kg/L and solar 0.7 kg/L

Meanwhile, the absorption capacity of CO_2 employs the following equation:

The absorption capacity of trees = the absorption capacity of CO_2 x total trees (2)

The projection of vehicles would be based on the calculation of correlative coefficient. According to Ribka [4], the equation became:

$$Pn = Po\ (1 + r)^{dn} \tag{3}$$

Notes:
Pn : total vehicles at the end of period
Po : total vehicles of initial projection
r : average annual population growth
dn : projection period

The calculation of emission residue would be based on the equation:

$$CO_2 \text{ emission residue} = A - B \tag{4}$$

Notes:
A : Total emissions of CO_2 (g/hour)
B : Total absorption capacity of CO_2 by GOS of road(g/hour)

1.1.3 Review of Related Studies. Ribka Regina Roshintha and Sarwoko Mangkoedihardjo [4] conducted a study concerning The Analysis of Green Open Space Adequacy as Carbon dioxide (CO_2) Gas Emission Absorber at Campus Area of ITS Sukolilo Surabaya. This research calculated total existing vehicles and vegetation at campus area of ITS Sukolilo Surabaya as the bases for determining GOS adequacy to absorb CO_2 emission. The results of analysis indicated that five zones had satisfied GOS adequacy in 2021 and three zones had not yet.

Sukentyas Estuti Siwi [5] revealed that the reduction of GOS area ensued in Depok City for 11 years (in 2000 - 2011) based on the data of Landsat 7 ETM+ and SPOT 4. The data were analysed by referring to radio metrical and geometrical corrections upon by applying Normalized Difference Vegetation Index (NDVI) algorithm, specifically to separate vegetation objects from non-vegetation ones.

Moreover, Rulli Pratiwi Setiawan [6] has proven that to review the comparison of carbon dioxide emissions resulted from the settlement uses at urban and urban fringe areas in Gerbangkertosusila. In the case, urban areas comprised the cities of Surabaya, Mojokerto, and Sidoarjo, whereas urban fringe areas involved the regencies of Gresik, Bangkalan, Mojokerto, and Lamongan. The results of research depicted that the use of settlement area in urban fringe area gave high contribution toward carbon dioxide emission though the housing areas and total households in urban fringe area were extremely fewer than the ones in urban area.

The research of Nurhayati Abdul Malek, et al. [7] was successful in validating the pattern scales of GOS uses in Malaysia including the patterns of passive and active activities. The results indicated that unsuitability between every pattern and theory. The patterns were used consistently for measuring the aspect of people's participation along the green path.

The paper of Collins Adjei Mensah, et al. [8] justifies that green space bestows social, economic, and environmental benefits which then ultimately improves welfare in terms of physical, psychological, social, and individual material so that the living standard can automatically be increased. Unfortunately, the efforts to actualize all of them are often focused on the social-economical phases by restricted consideration on the green space aspect. This research aimed at bridging the research gap in tracking the relationship between green space and life quality. It also was dedicated to investigate how that relationship could inform the development of policy for assisting the government to obtain positive outcomes.

2. Method

This research was carried out in Bangkalan City exactly in Burneh and Bangkalan Districts. The former district consisted of one urban village and two villages, while the later one had seven urban villages and one village. The data were meant to calculate total motorcycles, gasoline and solar cars, trucks, and buses which were collected for one week at 07.00-08.00; 12.00-13.00; and 16.00-17.00.

To calculate GOS adequacy, the researchers inputted the area data for determining the zone of software being developed. The process of inputting the data of plants and their absorption capacity toward CO_2 was addressed to calculate the plant capacity for absorbing emission. Data of vehicles, total vehicles, road length, emission factor, fuel economy of vehicles, and fuel density were inputted to gain emission load. The inputs of total vehicles at the end of period, total vehicles at the initial projection, average population growth, and projection duration were necessary to count the projection of vehicles. Meanwhile, the inputs of total CO_2 emission and total CO_2 absorption were fruitful to evaluate GOS.

3. Findings and Discussions

The results of field measurement demonstrated total vehicles per week in Burneh District in Table 1 and the ones in Bangkalan District in Table 2.

Table 1. Total Vehicles per Week in Burneh District

No	Location	Motorcycle	Car (Gasoline)	Car (Solar)	Truck	Bus
1	Langkap	184,544	6,688	8,696	2,108	-
2	Burneh	327,656	40,352	39,504	4,732	2,388
3	Tunjung	310,752	37,248	38,096	4,444	2,388

Table 2. Total Vehicles per Week in Bangkalan District

No	Location	Motor cycle	Car (Gasoline)	Car (Solar)	Truck	Bus
1	Martajasah	2,808	880	0	0	0
2	Mlajah	641,160	91,224	80,560	5,340	0
3	Kemayoran	616,400	76,568	68,664	5,340	0
4	Pengeranan	162,976	16,848	15,072	476	0
5	Pejagan	417,464	45,600	49,904	1,576	0
6	Kraton	269,720	36,360	19,352	420	209
7	Demangan	282,288	45,120	24,904	688	241
8	Bancaran	208,832	16,368	12,696	0	0

The total vehicles in Tables 1 and 2 were dominated by motorcycles and some regions had no bus. The biggest number of vehicles passed through Burneh Village, whereas Mlajah and Kemayoran Urban Villages were also the areas mostly crossed by many motorcycles.

Table 3. Total Emissions in Burneh and Bangkalan Districts

No	Location	Motorcycle	Car (Gasoline)	Car (Solar)	Truck	Bus	Total Emissions (kg/week)
1	Langkap	13,204	1,136	1,700	1,064	0	17,104
2	Burneh	23,444	6,857	7,721	2,388	1,515	41,925
3	Tunjung	22,234	6,329	7,446	2,243	1,515	39,767
4	Martajasah	201	150	0	0	0	350
5	Mlajah	45,875	15,501	15,746	2,695	0	79,817
6	Kemayoran	44,103	13,011	13,421	2,695	0	73,230
7	Pengeranan	11,661	2,863	2,946	240	0	17,710
8	Pejagan	29,870	7,749	9,754	795	0	48,167
9	Kraton	19,298	6,178	3,782	212	1,333	29,604
10	Demangan	20,198	7,667	4,868	347	153	33,232
11	Bancaran	14,942	2,781	2,482	0	0	20,205

The biggest total emissions occurred in Mlajah and Kemayoran Urban Villages as presented in Table 3.

Table 4. The Absorption Capacity of Trees (kg/week) per Urban Village

No	Location	Total Trees	Total absorption capacity of trees per week (kg)
1	Langkap Village	255	4,299.1
2	Burneh Village	58	624.0
3	Tunjung Urban Village	81	679.1
4	Martajasah Village	63	948.1
5	Mlajah Urban Village	109	10,228.8
6	Kemayoran Urban Village	61	7,995.5
7	Pengeranan Urban Village	661	14,510.4
8	Pejagan Urban Village	211	8,446.1
9	Kraton Urban Village	86	1,208.3
10	Demangan Urban Village	225	1,431.4
11	Bancaran Urban Village	118	1,147.2

Table 4 demonstrates that the tree absorption against the biggest emission happened in Mlajah and Pengeranan Urban Villages. Meanwhile, the biggest emission residue in both Urban Villages can be seen in Table 5. The data indicated that the highest absorption value did not correlate with total trees as can be seen in Martajasah Village due to limited shade trees having high absorption capacity for emission. On the contrary, trees with low absorption capacity of emission occupied great numbers. In the other words, there were many shade trees in Martajasah Village but they had low capacity to absorb emission. The ability of a plant to absorb carbon is determined by three factors i.e. leaf size, total leaves, and tree canopy. In addition, emission residue in Martajasah Village was the fewest because the activities of vehicles in this area were extremely low. As a result, all emissions of vehicles could be absorbed perfectly by GOS.

Table 5. The Calculation of Emission Residue (kg/week)

No	Location	Total Emission (kg/week)	Total Tree Absorption per Week (kg)	Emission Residue (kg/week)
1	Langkap Village	17,104.0	4,299.1	12,805.0
2	Burneh Village	41,924.6	624.0	41,300.6
3	Tunjung Urban Village	39,767.1	679.1	39,088.0
4	Martajasah Village	350.4	948.1	-597.7
5	Mlajah Urban Village	79,816.9	10,228.8	69,588.2
6	Kemayoran Urban Village	73,229.8	7,995.5	65,234.3
7	Pengeranan Urban Village	17,710.0	1,208.3	16,501.7
8	Pejagan Urban Village	48,167.5	1,431.4	46,736.1
9	Kraton Urban Village	29,603.9	14,510.4	15,093.5
10	Demangan Urban Village	33,232.2	8,446.1	24,786.1
11	Bancaran Urban Village	20,204.8	1,147.8	19,057.6

Based on the data of all tables, the researchers developed software by inputting the data of research location, data collection time, total vehicles, total vehicle emission, and total emission absorption of trees. Data input aimed at calculating GOS adequacy so that engineering could minimize the value of emission residue. Figure 1 shows the display of software that was successfully developed.

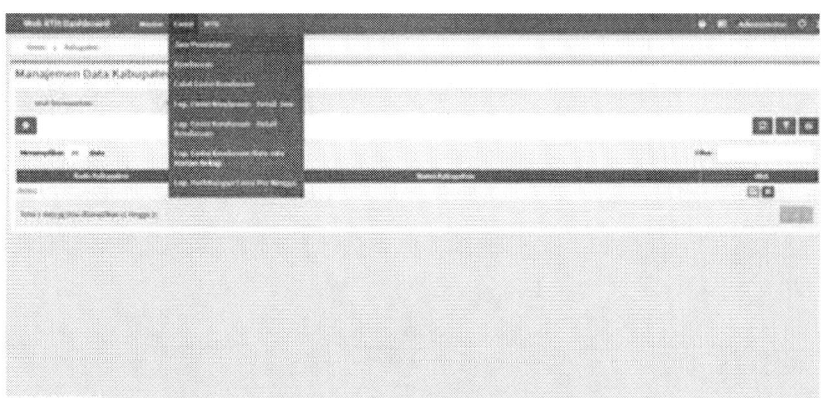

Figure 1. The Application Display

4. Conclusions

1. Out of eleven locations of measurement, one location obtained the lowest emission residue and all emissions could be absorbed by GOS namely Martajasah Village. Besides, two GOSs gained the biggest shade trees i.e. Keraton and Demangan Urban Villages.

2. The results of GOS adequacy calculated manually were suitable with the ones calculated through the developed software. Accordingly, the usability of software contributed significantly as it could ease the process for calculating the coverage adequacy of urban GOS, thereby saving time.

5. References

[1] Undang-Undang Republik Indonesia Number 26 Year 2007 tentang Penataan Ruang.

[2] Zhang, Biao *et al.*, 2014, *The Cooling Effect of Urban Green Spaces as A Contribution To Energy-Saving and Emission-Reduction: A Case Study in Beijing, China.*

[3] Merry J. Pasaribu, Bieby F. Tangahu, 2015, *Kajian Kecukupan Ruang Terbuka Hijau Publik untuk Menyerap CO_2 Udara Ambien dari Transportasi Darat di Jalan Perak Barat dan Jalan Perak Timur, Surabaya,* Jurnal Teknik ITS Vol. 5, No. 2

[4] Ribka R. R., Sarwoko M., 2016, *Analisis Kecukupan Ruang Terbuka Hijau Sebagai Penyerap Emisi Gas Karbon Dioksida (CO_2) Pada Kawasan Kampus ITS Sukolilo*, Surabaya, Jurnal Teknik ITS Vol. 5 No. 2.

[5] Sukentyas Estuti Siwi, 2012, *Kemampuan Ruang Hijau dalam Menyerap Gas Karbon Dioksida (CO2) di Kota Depok*, Fakultas Matematika dan Ilmu Pengetahuan Alam, Departemen Geografi Program Pasca Sarjana Ilmu Geografi Universitas Indonesia.

[6] Rulli P. S., 2014, *Perbandingan Emisi Karbon Dioksida Dari Penggunaan Lahan Permukiman di Kawasan Urban dan Peri-Urban di Wilayah GERBANGKERTOSUSILA*, Jurnal Tata Kota dan Daerah. Vol. 6 No. 2

[7] Nurhayati Abdul Malek, Manohar Mariapan, Nik Ismail Azlan Ab Rahman, 2015, *Community Participation in Quality Assessment for Green Open Spaces in Malaysia*, Science Direct, Procedia – Social and Behavioral Sciences, 168 (2015), 219 – 228

[8] Collins Adjei Mensah, Lauren Andres, Upuli Perera, Ayanda Roji, 2016, *Enhancing Quality of Life Through The Lens of Green Spaces: A Systematic Review Approach*, International Journal of Wellbeing, 6 (1), 142-163

Acknowledgments
We thank to the Environmental Engineering Master Program of ITATS and the Government of Bangkalan Regency for their research assistance.

The application of the Six Sigma method in reducing the defects of welding on the steel material

Agin Viakri Dagmar[1], Zeplin Jiwa Husada Tarigan[2*]

[1]Postgraduate Student, Master of Industrial Engineering Department, Institut Teknologi Adhi Tama Surabaya, Indonesia
Arif Rachman Hakim, Klampis Ngasem, Surabaya 60117, Indonesia
[2]Lecturers, Faculty of Business and Economics, Petra Christian University,
Jl. Siwalankerto 121–131, Surabaya 60236, Indonesia.

zeplin@petra.ac.id

Abstract. Quality control is an essential factor in producing products that are following what consumers want. The high quality of the product significantly affects the efficiency of costs incurred during the production process. This study aims to determine the types of defects and their causing factors as well as actions taken to reduce the number of defects to increase the efficiency of production costs. The measurement results obtained the number of productions of 12499 pcs with the number of defects of 2712 pcs of 21.6% with a sigma level of 3.1100 DPMO. Identification of defects in welding is porosity defects of 724 pcs, undercut defect of 613 pcs, slag defect of 435 pcs, crack defect of 491 pcs, and spatter defect of 449 pcs. The actions taken are periodic maintenance and repair of welding machines and connectors, supervising the condition and cleanliness of the material, providing periodic directions to welding operators by explaining reasonable welding procedures and arranging a welding field development program for welding operators through a job training program. Improvements have been implemented and found a reduction of defects up to 15% and a reduction of 6.6%.

1. Introduction

Product quality is a must for industries in producing products following specifications achieved as outputs produced by manufacturing companies. The products produced must meet the requirements stipulated and fall within the quality ranges set by external customers and government regulations (Patel and Desai, 2018) [1]. The quality that is maintained needs to be controlled in such a way as to produce quality products. Quality control is crucial for the company and needs to be realized so that the company can overcome quickly and agile any deviations in the production process [2]. The ability to manufacture companies to detect quality deviations in a short time will reduce defects in the final product so that they can reduce waste and costs incurred by the company [1]. The process systems approach that businesses use to be able to find and eliminate the causes of defects and errors, to reduce product manufacturing cycles and reduce production costs, improve product performance, and provide quality products to customers and better use of company assets is an implementation of six Sigma [3]. Product quality control is a technique and activity, or planned action carried out to achieve, maintain, and improve the quality of a product and service so that it is under predetermined standards and can meet consumer satisfaction [4]. The company uses a statistical approach to control the process quality, product quality, and service quality in a systematic and organized manner as a form of performance improvement [5].

Quality control is to use the Six Sigma method. Six Sigma can be used as a measure of the performance of industrial systems that allows companies to improve performance and be able to reduce complaints from customers because the products produced by the company have met the required quality to reduce defect rates [6]. Six Sigma can also be viewed as a customer-focused industrial process control concerning the company's production process capabilities. Six Sigma that has been used so far is using the DMAIC (Define-Measure-Analyze-Improve-Control) method [7], which in the implementation practice in companies there will be different definitions and approaches according to conditions depending on employee understanding and knowledge transfer from consultants. The implementation of six Sigma in companies will be able to reduce product variations or the results of the production process by determining the process failure rate of 3.4 products per million or two parts per billion of products produced, and analysis is carried out with normal distribution [8]. The six sigma implementation program on project innovation can provide improvements in company performance through increased customer satisfaction for current customers, and incremental innovation [3].

Freitas et al. [9] stated that the company's ability to implement lean six Sigma (LSS) would be able to provide sustainability in the project when the implementation can have an impact on the company's financial profit. Lean Six Sigma for the company is an important thing when the company can increase the competitive advantage of its products and increase the market for the company's products, which are the goals of the company [9,10]. Companies engaged in steel construction is trusted as subcontractors for the larger-scale companies, which is assigned by the Government of Indonesia to build the infrastructure. The suppliers to the main contractor have high standard requirements or specifications, especially for the work specification on bridge construction that requires the welding process with a very high specification. Any defective product that exceeds the tolerance limit will incur higher costs. The total cost will increase, and the selling price will be higher as well. Products will be unable to compete with competitors with lower selling prices and better quality for the same type of product. The application of the Six Sigma method is expected to reduce the level of defective products to the lowest point or even reduce them until the production process runs towards perfection (zero defect) so that the company will be able to increase its profits [3]. Besides, the company can maintain its survival and even improve its market position in the face of hypercompetitive competition. The defective condition of the company during the period of July 2019 to June 2020 is indicated in Table 1

Table 1. Percentage of Defective Products

No.	Month	Product (Pcs)	Rejected Product	Percentage Rejected
1	July 2019	644	154	24%
2	August 2019	1022	194	19%
3	September 2019	974	182	19%
4	October 2019	734	152	21%
5	November 2019	849	156	18%
6	December 2019	1119	215	19 %
7	January 2020	1109	231	21%
8	February 2020	1235	276	22%
9	March 2020	1331	311	23%
10	April 2020	1254	298	24%
11	May 2020	1109	264	24%
12	June 2020	1119	279	25%
	Total	12,499	2,712	22%
	Mean	1,042	226	22%

Initial observations made by researchers at steel construction companies by paying attention to the results of welding products for a year obtained 2712 pieces of defective products and an average of 226 pcs per month, and the percentage of defects obtained at 22%. Based on these data, it is necessary to implement six Sigma in companies to reduce product defects. Based on this result, this research was conducted to measure and make quality improvements in order to reduce defects during the welding process by using DMAIC (define-measure-analyze-improve-control) at the Six Sigma approach.

1.1 Literature Review

The quality of the products that the company produces should focus on customer requirements [2]. Companies should produce quality products by applying the concept of zero defect, or zero error rate as the goal of quality. This concept leads to the lowest possible product error rate, even until there is no error. The quality is the overall characteristics of a product or service that supports the ability to satisfy needs [11]. Quality control can be defined as activities carried out to monitor activities and ensure actual performance [10]. Control and supervision are activities carried out to ensure that production and operation activities carried out per as planned, and if deviations occur, these deviations can be corrected to achieve the expected target. Quality control is an effort to maintain the quality/quality of the goods produced to meet predetermined product specifications [1]. Quality control is a technique and activity, or planned action carried out to achieve, maintain, and improve the quality of a product and service to achieve predetermined standards and can meet consumer satisfaction. The production process needs to be designed to facilitate quality control [12]. Quality control cannot be separated from production control because quality control is part of production control. Production control, both in quality and quantity, is a significant activity in a company.

1.1.1 Six Sigma. Six Sigma has a strategic goal to reduce the occurrence of defects in a production process with the ultimate goal of reducing process variations through special improvement programs and structured methods [13]. The defect is a deviation from predetermined specifications. The sigma quality level is usually also used to describe the output of a process [4]. Six sigma results are associated with a lower level of tolerance given to a product or service through the improvement of the process capability. Six Sigma is a tool or systematic method for process improvement and new product development based on statistical and scientific methods to reduce the number of defects defined by consumers [8]. Six Sigma levels are often associated with process capabilities defined in terms of defects per million opportunities (DPMO). The process of improvement in Six Sigma is known as DMAIC (Define, Measure, Analyze, Improve, Control). DMAIC is a process for continuous improvement towards Six Sigma targets. DMAIC is carried out systematically, based on science and facts. The successful implementation of programs on improving Six Sigma quality is shown through the improvement of process capabilities in producing products towards zero failure rates [7]. The concept of calculating process capability is fundamental to understand and implement the Six Sigma program.

Define is a step determined by the company to determine problem selection and analyze the benefits obtained. Define is the first step in the Six Sigma quality improvement program [6]. At this stage we need to identify several things related to the six sigma project selection criteria, the roles and responsibilities of people who will be involved in the Six Sigma project, training needs for people involved in the six sigma project, the processes involved determine the quality of the product and the specific requirements of the customer. This stage also determines the calculation of the initial sigma level for a particular data period [5]. The measuring stage is carried out with three main issues, firstly, establishing a CTQ (Critical to Quality) which is directly related to the specific needs of the customer which is derived directly from the output and service requirements, secondly investigates and observes to obtain data, the third stage is to set goals setting on output and outcome levels. The Analyze stage is used by analyzing the stability and process capability, determining the performance targets of the key quality characteristics (CTQ) that will be improved in the Six Sigma project, and identifying the root causes of defects or failures [6].

The fifth stage is to improve by making action plans (action plans) regarding the allocation of resources as well as priorities and alternatives to implement the plan. The supervision and efforts through data collection and analysis when implementing a plan must also be planned at this stage, to decide what should be achieved (related to the targets set). The action plan describes the plan in terms of what, where, when, how the action will be implemented. The action plan also indicated how much it would cost and the associated benefits. The last stage is controlling by adjusting the process management and control system so that there is continuous improvement. Designing a process system that may be carried out and maintained so that the system becomes standardized and documented [12].

2. Research Method

This method is a stage of mapping the process by searching, obtaining, collecting, or recording data, both in the form of primary data and secondary data to analyze factors related to the condition of the company [11]. The quality control is performed by checking periodically at each stage of the construction project with the DMAIC stage (Define, Measure, Analyze, Improve, Control) approach [6,7]. This process determines the quality of steel welding under the criteria set by the consumers [7]. The companies are companies that provide physical infrastructure. This study has a flow chart shown in Figure 1.

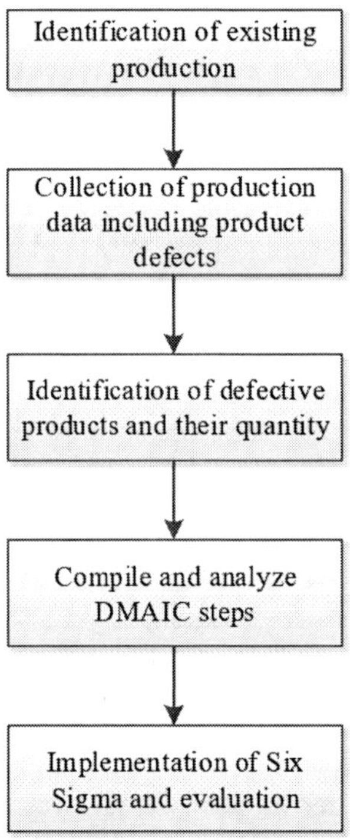

Figure 1. Flowchart of Six Sigma Implementation

3. Six Sigma Project and Discussion

The research analysis uses the six-sigma method, which consists of five stages, namely define, measure, analyze, improve, and control in a welding company

3.1. Define Stage

The stage is the stage of defining quality problems in the welding result. Six Sigma focuses on defects and variations by starting with the identification of the Critical to Quality (CTQ) elements of the product. At this stage, what causes the product to experience a defect is defined as the cause as well. Critical to quality are the attributes of welding products that are considered essential to customers. The results of observation and analysis of secondary data collection in the welding process identified the five causes of the highest defect products, namely porosity, undercut, slag, crack, and spatter defects. In this stage, the goals and objectives of improving six sigma quality are based on observations by reducing or reducing the defected product from 21.6% to 15%.

3.2. Measuring stage

The measuring is done by creating a check sheet. Check sheets are used for the data collection process and data analysis. Also, it is useful for knowing the problem area based on the frequency of the type or cause and making decisions to make improvements or not. Table 2 shows the cause of the defect in the welding process from June 2019 to June 2020

Table 2. Data of defect type

No	Month	Products (Pcs)	Type of Defect				
			Porosity	Undercut	*Slag*	*Crack*	*Spatter*
1	July 2019	644	51	43	24	31	22
2	August 2019	1022	52	36	29	35	42
3	September 2019	974	43	40	36	32	31
4	October 2019	734	47	26	23	34	22
5	November 2019	849	42	28	26	29	31
6	December 2019	1119	46	37	36	39	39
7	January 2020	1109	61	36	41	52	36
8	February 2020	1235	74	68	45	43	46
9	March 2020	1331	86	79	47	47	52
10	April 2020	1254	81	76	43	49	49
11	May 2020	1109	65	72	38	52	37
12	June 2020	1119	68	74	42	56	39
	Total	12,499	716	615	430	499	446
	Mean	1,042	59.67	51.25	35.83	41.58	37.17

Table 2 demonstrated the biggest flaw in porosity with the number of 716 pcs and followed by the undercut of 615 pcs, slag of 430 pcs 499 pcs, and the crack of the last spatter 44 pcs. The defects were found in the first stage of the process. All pieces passed to the next process have met the criteria. Further processing is carried out by setting the welding result in the sigma value, as shown in Table 3.

Table 3. The setting of six sigma level

Month	Production (Pcs)	Number of Defects (Pcs)	DPMO Value	Sigma Level
July 2019	644	154	59,782.61	3.0566
August 2019	1022	194	47,455.97	3.1700
September 2019	974	182	46,714.58	3.1776
October 2019	734	152	51,771.12	3.1279
November 2019	849	156	45,936.40	3.1856
December 2019	1119	215	48,033.96	3.1642
January 2020	1109	231	52,073.94	3.1251
February 2020	1235	276	55,870.45	3.0904
March 2020	1331	311	58,414.73	3.0682

Month	Production (Pcs)	Number of Defects (Pcs)	DPMO Value	Sigma Level
April 2020	1254	298	59,409.89	3.0597
May 2020	1109	264	59,513.07	3.0589
June 2020	1119	279	62,332.44	3.0355
Average	1,041.58	226	53,942.43	3.1100

From the results of the calculations in Table 3, the production results in welding have a sigma level of 3.1100. This result is, of course, a considerable loss if not handled, because the more welding products are not standard, of course, result in swelling of production costs.

3.3. Analyze stage

Stages The next stage is to determine the stability and capability of the process. Determine the performance targets of the key quality characteristics (CTQ) that will be improved in the Six Sigma project. Identify the root causes of defects or failures. After knowing the types of defects that occur, corrective steps are needed to prevent similar damage. The important thing that must be done and traced is to find the cause of the damage. As a tool to find the cause of the porosity failure, a cause and effect diagram is used and the cause of the defect is found due to the machine is a less standard inverter setting, current connector, defects caused by the material are corrosion, lack of material cleanliness, the welding arc is too long, the welding current is too low, and the travel speed is too high. The cause of the method is the relatively global standard operating procedure, the electrodes used are still damp or exposed to water, and the implementation of the method does not work well. The causes of porosity defect are shown in the fish bone diagram (Figure 2).

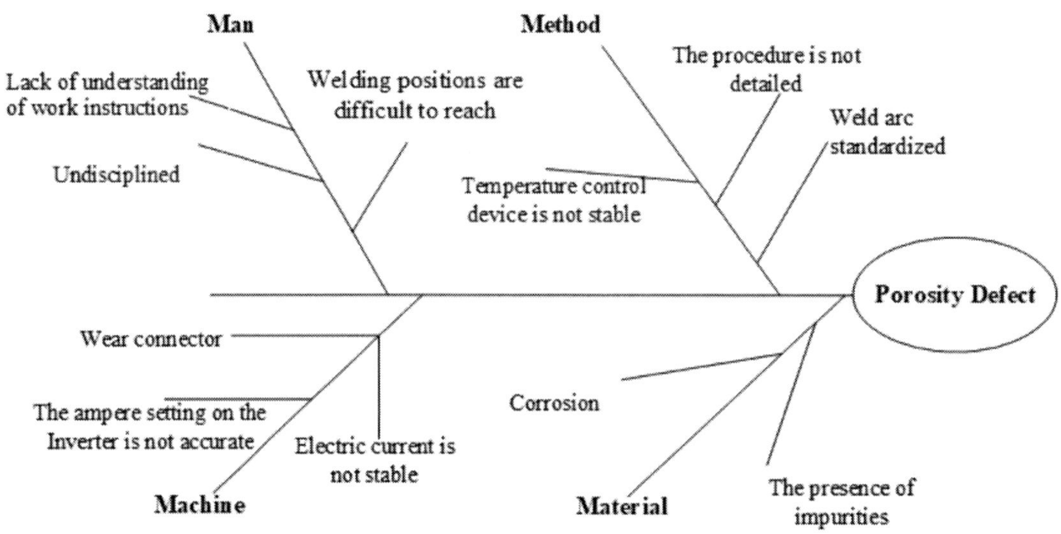

Figure 2. Fish Bone Diagram for Porosity Defect

Undercut defects are caused by several factors, namely the material is too high the electrode, the arc length as the material is too high, the welding current applied by operator is too high, the travel speed/welding speed is too high. The defects caused by the method are the uneven hand swing, the swing time on the side is too fast, and the standard hand restraint needs to be made so that the hand does not shake.

There are two kinds of Crack welding results, Hot Crack defect (hot crack) is a crack in the welding where the crack occurs after the welding process is complete or when the weld metal solidification process. Cold Cracking (cold cracking) is a crack that occurs in the weld area after some time (it takes time, can be 1 minute, 1 hour, or one day); the welding process is complete. Usually, to check for cracks, a test is carried out, namely the Penetrant Test or Magnetic Test. Crack defects of the material are caused by dirty material and the wrong type of electrode, while for machines, the current is too high, and the travel speed is too high. While the spatter defect is caused by the material, the amperage of the instrument is too high, and the travel speed is too high. Meanwhile, what is caused by the method is that the electrodes used are still damp or exposed to water, and the hand swing is too fast. The occurrence of Welding Defect Slag Inclusion is a defect that occurs in the area in the weld. This defect is in the form of slag (melting flux) that is in the weld, which often occurs in the stop and run area (beginning and stopping of the welding process). To see this defect, we must perform radiographic or bending tests.

3.4. Stages Improve

Action plan to implement Six Sigma quality improvement. After knowing the cause of product defects from welding results, a recommendation or recommendation for general corrective action is prepared in an effort to reduce the level of product damage. The porosity failure was carried out by improving the standard operating procedure, replacing the current connector component and setting the inverter according to the standard, and cleaning the material before the welding process. The action plan for undercut defects is carried out by training employees to understand better, establishing work instructions in carrying out work, establishing appropriate standard operating procedures, adjusting welding currents, taking into account standardized amperes, and wrapped in electrodes or WPS (Welding Procedure Specification), setting settings. Re-welding current and welding travel and setting the arc length to be shortened or as high as 1.5 times the electrode diameter.

The action plan for crack defects is to use electrodes that are compatible with WPS or Low Hydrogen which have high tensile properties, conduct heat treatment, set the recommended current to match the standard, travel speed of welding is not too fast concerning WPS, determine the carbon equivalent material above 0.40 must be preheated. Meanwhile, the action plan caused by a spatter defect is that the current is lowered according to the standards set in the work instructions, the electrode is oven according to the handbook (especially low hydrogen welding wire), and the arc length is set at 1.5 times the electrode diameter. Finally, the action on slag defects is that the Ampere quantity is measured in advance and adjusted to the procedure, ensuring that the material is spotless of slag before being carried out before re-welding, and the arc as material needs to be adjusted.

3.5. Control

This stage is the final stage of the Six Sigma project, which emphasizes documenting and disseminating the actions that have been taken. The documentation includes maintaining and repairing welding machines and connectors regularly, monitoring the condition and cleanliness of materials, providing periodic guidance to welding operators with explaining adequate welding procedures, and develop a welding field development program for welding operators through a job training program. With the implementation of the use of the Six Sigma method, the calculation result of the sigma level is 3.1100, where the initial condition of the defect level of the production is 21.6%. Quality control dramatically affects the productivity level of the production process. This result is evidenced by a decrease in the defect rate of the welding products to 15%.

4. Conclusions

The number of welding process results include 5 types of defects, namely 724 pcs porosity defects, 613 undercut defects, 435 pcs slag defects, 491 pcs crack defects, 449 pcs spatter defects. The findings have a sigma level of 3.1100 DPMO. The factors that cause welding defects are due to imperfect standard operating procedures, the welding method used does not follow welding standards, the material is not well controlled, especially the electrodes used when welding, and the lack of maintenance of the welding machine. The stages of repairs carried out are socializing standard operating procedures, maintaining the cleanliness of raw materials before repeated welding and welding, regular checking of welding equipment and raw materials, and maintaining the workshop in conditions that meet the requirements of the welding process.

5. References

[1] Patel, M. and Desai, D.A. (2018). Critical review and analysis of measuring the success of Six Sigma implementation in manufacturing sector. *International Journal of Quality & Reliability Management*, 35(8), 1519-1545, DOI 10.1108/IJQRM-04-2017-0081.

[2] Tarigan, Z.J.H., Suprapto, W., and Basana, S.R. (2019). The effect of procedure change, TQM and ERP implementation to company performance on manufacturing industries. *IOP Conference Series: Materials Science and Engineering*, 473, doi:10.1088/1757-899X/473/1/012052 012052

[3] Parast, MM (2011). The effect of six sigma projects on innovation and firm performance. *International Journal of Project Management*, 29, 45–55, DOI: 10.1016 / j.ijproman.2010.01.006

[4] Kokkranikal, J., Antony, J., Kosgi, H., and Losekoot, E. (2013). Barriers and challenges in the application of Six Sigma in the hospitality industry: Some observations and findings.International *Journal of Productivity and Performance Management*, 62(3), 317-322, DOI 10.1108/17410401311309203

[5] Costa, JP, Lopes, IS & Brito, JP (2019). Six Sigma application for quality improvement of the pin insertion process. *Procedia Manufacturing*, 38, 1592–1599, 10.1016 / j.promfg.2020.01.126

[6] Mast, JD & Lokkerbol, J. (2012). An analysis of the Six Sigma DMAIC method from the perspective of problem-solving. *International Journal Production Economics*, 139, 604-614. DOI: 10.1016/ j.ijproman.2010.01.006.

[7] Sin. B.A., Zailani, S., Iranmanesh, M., and Ramayah, T. (2015). Structural equation modelling on knowledge creation in Six Sigma DMAIC project and its impact on organizational performance. *International Journal of Production Economics*, 168, 105-117, https://doi.org/10.1016/j.ijpe.2015.06.007.

[8] Aldowaisan, T., Neurelfatah, M., & Hassan, J. (2015). Six sigma performance for non-normal processes. *European Journal of Operational Research*, 247, 968-977. http://dx.doi.org/10.1016/j.ejor.2015.06.036.

[9] Freitas, JGD, Costa, HG, and Ferraz, FT (2017). Impacts of lean six Sigma over organizational sustainability: A survey study. *Journal of Cleaner Production*, 156, 262-275, http://dx.doi.org/10.1016/j.jclepro.2017.04.054

[10] Anthony, J., Snee, R., and Hoerl, R. (2017). Lean Six Sigma: Yesterday, today and tomorrow. *International Journal of Quality & Reliability Management*, 34(7), 1073-1093, DOI 10.1108/IJQRM-03-2016-0035.

[11] Yadav, G. and Desai, T.N. (2016). Lean Six Sigma: a categorized review of the literature. *International Journal of Lean Six Sigma*, 7(1), 2-24. https://doi.org/10.1108/IJLSS-05-2015-0015

[12] Tarigan, Z.J.H., Suprapto, W., and Basana, S.R. (2018). Enterprise resources planning project manager competency on improving organizational performance through process design and quality performance. *ICEBT 2018: Proceedings of the 2nd International Conference on E-Education, E-Business and E-Technology*, 153–157, https://doi.org/10.1145/3241748.3241777

[13] Gutierrez-Gutierrez, L., Barrales-Molina, V., Fernandez-Giordano, M., and Lopez-Morales, B. (2020). *International Journal of Lean Six Sigma*, 11 (1), 35-56, DOI 10.1108 / IJLSS-10-2018-0115.

IOP Conf. Series: Materials Science and Engineering 1010 (2021) 012045 doi:10.1088/1757-899X/1010/1/012045

Household electricity network monitoring based on IoT with of automatic power factors improvement using neural network method

D. C. Nugroho[1] , Y. Mayaratri[1] , M. Syai'in[1] , M. K. Hasin[1] , N. H. Rohiem[2], N. P. U. Putra[2], A Soeprijanto[3]

[1]Politeknik Perkapalan Negeri Surabaya (Shipbuilding Institute of Polytechnic Surabaya), Indonesia
[2] Institut Teknologi Adhi Tama Surabaya (ITATS), Indonesia
[3]Institut Teknologi Sepuluh Nopember (ITS), Surabaya, Indonesia

Email: danangchandra@student.ppns.ac.id

Abstract. The development of installed capacity in power plants in 2018 was 41.696 MW or increased from the previous year in 2017 amounted to 39.652 MW (PLN Statistics, 2018). This has an impact on the reduced availability of fuel due to overexploitation. The highest energy sold per customer group in 2018 was the household customer group that was 41.7% higher than the industrial sector customer group by 32.8% (PLN Statistics, 2018). At present the use of electronic equipment for household needs is increasingly diverse. Many equipment that is often used in daily life is electronic equipment that is inductive load. Inductive load causes the value of the power factor to fall so that the power usage (Watt) becomes less than optimal. To overcome the problem caused by the large number of inductive loads a reactive power compensator is needed which is to use a capacitor. In this final project, a system designed to be able to measure and correct power factors automatically uses the Neural Network method and can monitor power online based on IoT. The results of testing the power factor improvement system were 97.8% successful in the trained electric load and 94.8% in the untrained electrical load.

1. Introduction

The increasing demand for electrical energy has led to an increase in the power generation industry in Indonesia. The development of installed capacity in power plants in 2018 amounted to 41,696 MW or an increase from the previous year, in 2017 which was 39,652 MW [1]. The highest energy sold per customer group in 2018 was the household customer group, which was 41.7% higher than the industrial sector customer group at 32.8% [1]. Many of the equipment that is often used in everyday life is electronic equipment that is inductive load. The inductive load is known as the lagging load (current lagging the voltage). To overcome the problems caused by the large number of inductive loads, a reactive power compensator is needed to compensating a lagging current by a leading current, through connecting capacitance to the supply [2]. The reactive power compensator must be able to improve the power factor automatically under varying loads. By controlling the combination of capacitor circuit so that the capacitance value varies and requires a smaller number of capacitors. The Neural Network based power factor improvement is an accurate methods of power factor correction meanto induce the necessary capacitance in the system [3]. In 2018 Y. Yuniarto and his team have performed power factor correction for household electrical installation. The results of this study indicate that the electric current flowing before the capacitor is installed is greater than the electric current when the capacitor is installed. As for the power factor, there was an increase in the value of the power factor, which was originally from 0.34 to 0.36 to 0.90 to 0.95. [4].

Information the use of electrical energy can be accessed via a web page using the MQTT protocol. In the use of the MQTT protocol there is a term Quality of Service (QoS). MQTT protocol become very useful because work asynchronously in real time and ensure their functioning in hostile communication environments [5]. MQTT is a protocol for popular Iot systems and can be applicable in some fields such as environmental monitoring [6] [7], industrial sector [8], energy management [9], and other sectors like transportation [10]. With the addition of IoT facilities, it will make it easier to control the use of electrical energy, which can increase modern public awareness of the importance of making efforts to save electrical energy.

2. Methodology

The purpose of this project is to improve power factor automatically. This project also and can monitor household electricity power online based on IoT. Here is the explanation about Hardware, System and the overall steps of this research can be explained as follows.

2.1. Hardware Design

This project uses several component and hardware. The hardware component and can be explain as follows:

1. Power Input
   This device is used to connect main supply of this system hardware.
2. Current Tape Sensor
   This device is used to sense electrical current. And the signal will be processed by pzem-004t power meter module.
3. Liquid Crystal Display
   It used to display information and data on the system.
4. Controller Board.
   Used as the main controller that processes data input and output system.
5. Load Power Socket
   It used as the electrical power source to be connected to the load.
6. Capacitor
   Used as reactive power compensator used to improve the power factor.

All of this component that have been explain in above shown on figure 1.

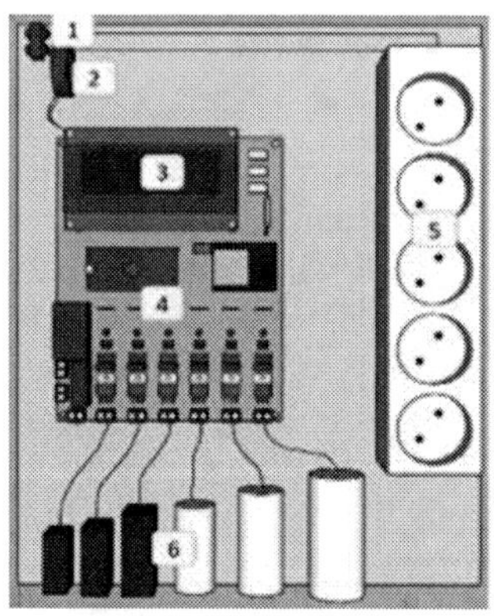

Figure 1. Hardware Design System

Figure 2 will show the diagram block of the project system.

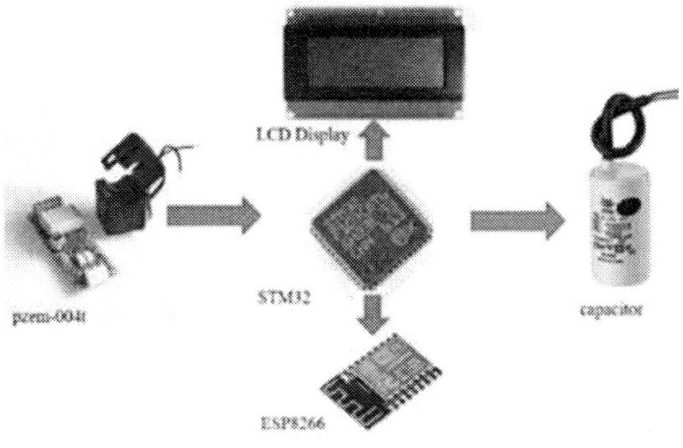

Figure 2. Diagram Block of the Hardware System

2.2. IoT System Design

The device to be made functions as a publisher on the MQTT protocol, periodically the device will send data on a certain topic to the broker. While the device used for monitoring functions as a subscriber, the way it works is by first sending a request for certain topic and then receiving data according to the requested topic. The data that has been received is then processed and displayed so that users can find out information on the use of electric power in the installation. The diagram as shown on figure 3.

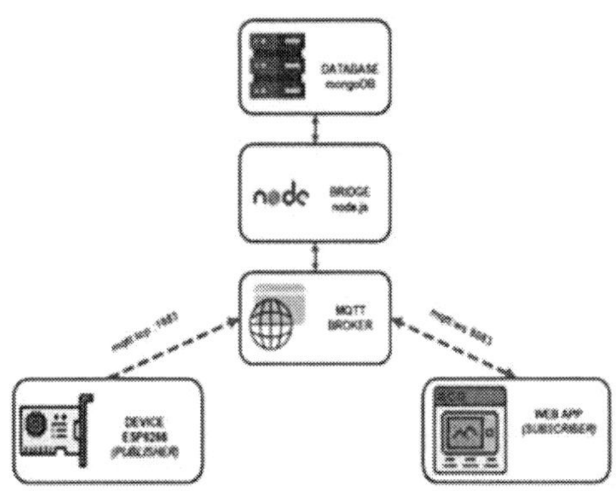

Figure 3. IoT System Architecture

2.3. Neural Network Algorithm

Neural network method will function to determine the value of capacitor compensation needed to improve the power factor value. In this step, the design of the neural network algorithm is divided into several stages, namely, the formation of sample data, the training process, weight extraction, and the formation of the neural network structure. Several steps in the design of a neural network algorithm are described as follows:

1) Step 1 Create Data Sample

Data sample formation is done by taking data on the 5 predefined electrical equipment. The electrical equipment is switched on alternately to form different load variations. Then performed data collection, namely power and power factor as input to the neural network algorithm. While the output is the calculated capacitor value according to their needs. The result of the formation of sample data for the training process.

2) Step 2 Training Process

The training data process uses the NN (Feed Forward) toolbox in Matlab by using 2 hidden layers with 1 hidden layer having 12 neurons and 1 hidden layer having 1 neuron. The training process is carried out until a good performance is obtained. Then the weight extraction process is carried out to take the weight values after the training process is successful. The training data process can be seen in figure 4 below.

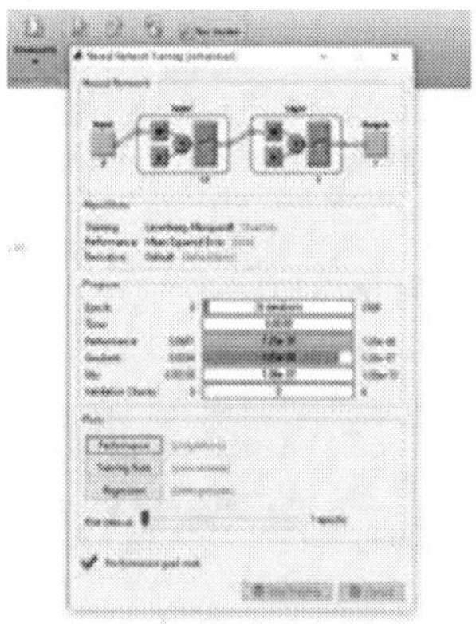

Figure 4. Training Process on MATLAB

3) Neural Network Structure

Making the network structure is done to facilitate the creation of a neural network algorithm on the microcontroller. To create a network structure, it is necessary to consider the number of neurons in each layer and the bias. The structure of the neural network is shown in figure 5 below.

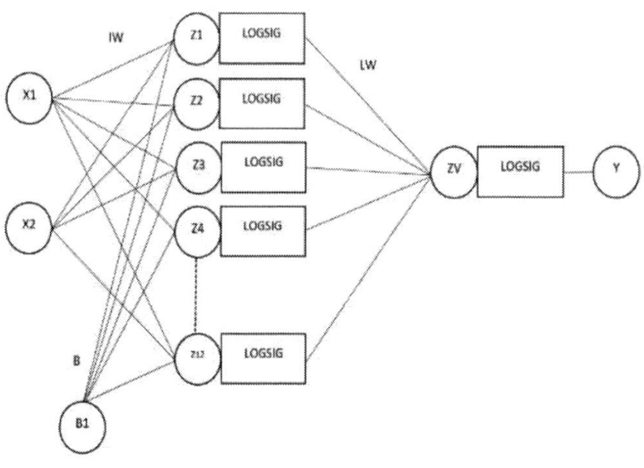

Figure 5. Neural Network Structure

Neural network structure with input X1 and X2, while Y is the output. This network has 2 hidden layers with 12 neurons in the first hidden layer and 1 neuron in the second hidden layer.

IOP Conf. Series: Materials Science and Engineering 1010 (2021) 012045 doi:10.1088/1757-899X/1010/1/012045

3. Test Result

In the IoT based monitoring test, users can view data information about electrical parameters, namely voltage, current, power, power factor, energy and frequency on the web page. Users can also view additional information in the available additional information (InfoBar) as shown in figure 6 below.

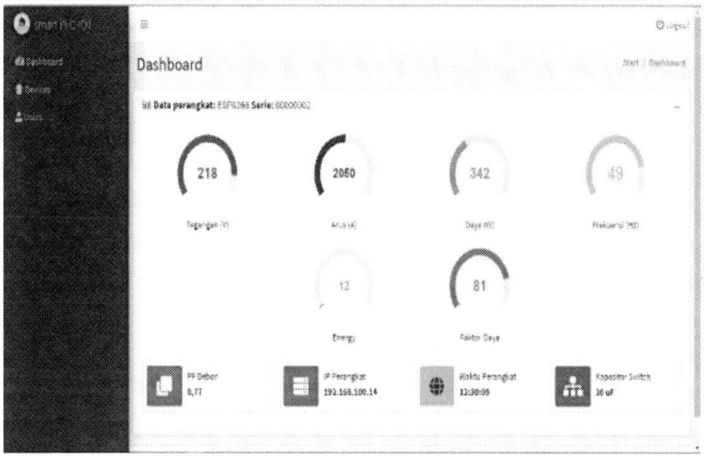

Figure 6. IoT Based Monitoring System

At the power factor improvement testing stage, it aims to determine whether the circuit can work as expected or not. Tests are carried out on various types of loads for which the power factor value will be corrected and to observe whether there is an increase in the power factor value before and after connecting the capacitor. The load test is the load that has been used to create training data for the neural network algorithm and the second stage is to test the load that is not used to create training data. The load used and the test result can be seen in table 1 and table 2.

Table 1. Load Test

Load Code	Load Name	Type
KK	Small Fan	Cosmos7 LDA
KB	Large Fan	Cosmos 46W
LK	Fluorescent	Sonora 10W
LB	Fluorescent	Sonora 15W
MX	Mixxer	Phillips 170W

Table 2. Load Test Result

No	Load Code	C	PF before	PF after	Error (%)
1	MX	1	0,88	0,98	2
2	TB	6	0,35	0,98	2
3	TK	3	0,37	0,97	3
4	KB	1	0,90	0,98	2
5	KK	1	0,65	0,98	2
6	MX+TB	5	0,67	0,97	3
7	MX+TK	3	0,77	0,99	1

No	Load Code	C	PF before	PF after	Error (%)
8	MX+KB	2	0,93	0,98	2
9	MX+KK	2	0,87	0,97	3
10	TB+TK	7	0,36	0,99	1
11	TB+KB	5	0,59	0,97	3
12	TB+KK	5	0,42	0,97	3
13	TK+KB	4	0,69	0,97	3
14	TK+KK	4	0,46	0,97	3
15	KB+KK	2	0,83	0,98	2
16	MX+TB+TK	7	0,61	0,99	1
17	MX+TB+KB	6	0,76	0,97	3
18	MX+TB+KK	5	0,67	0,99	1
19	MX+TK+KB	4	0,83	0,97	3
20	MX+TK+KK	4	0,74	0,98	2
21	MX+KB+KK	2	0,91	0,98	2
22	TB+TK+KB	8	0,53	0,99	1
23	TB+TK+KK	8	0,61	0,98	2
24	TK+KB+KK	5	0,68	0,98	2
25	TB+KB+KK	6	0,61	0,97	3
26	MX+TB+TK+KB	8	0,70	0,97	3
27	MX+TB+TK+KK	8	0,61	0,98	2
28	MX+TK+KB+KK	5	0,82	0,97	3
29	TB+TK+KB+KK	6	0,76	0,99	1
30	MX+TB+KB+KK	8	0,56	0,99	1
31	MX+TB+TK+KB+K KK	8	0,71	0,97	3
			Average of Error		2,2

From the results of the first stage testing on trained loads (according to sample data) it can be concluded that the tool made is able to improve power with an average error of 2.2%. In this second testing stage, aims to determine the level of success of the neural network method for improving the power factor at unknown loads. The load test in this stage can be seen in table 3 and the results are shown in table 4.

Table 3. Load Test

Load Code	Load Name	Type
LH	Fence Lamp	INS 45W

Table 4. Test Result

No	Load Code	C	PF before	PF after	Error (%)
1	LH	2	0,71	0,97	3
2	LH	2	0,71	0,97	3
3	LH	1	0,71	0,86	14
4	LH	2	0,71	0,97	3
5	LH	1	0,71	0,86	14
6	LH	2	0,71	0,97	3

No	Load Code	C	PF before	PF after	Error (%)
7	LH	2	0,71	0,97	3
8	LH	2	0,71	0,97	3
9	LH	2	0,71	0,97	3
10	LH	2	0,71	0,97	3
			Average of Error		5,2

From the results of the second stage of testing, it can be concluded that for untrained loads or not training data, it can be concluded that the tool is made to improve power with an average error of 5.2%.

4. Conclusions

The Neural Network method is able to improve the power factor with a success of 97.8% for trained electrical loads and 94.8% for untrained electrical loads. The use of the MQTT protocol can optimize data communication on an IoT-based monitoring system so that it can monitor directly on the website.

5. Reference

[1] PT. PLN (Persero) 2019 *Statistik PLN 2018* Sekretariat Perusahaan PT PLN (Persero) Jakarta
[2] Y. Kabir, Y. M. Mohsin and M. M. Khan 2017 Automated power factor correction and energy monitoring system *2017 Second International Conference on Electrical, Computer and Communication Technologies (ICECCT)* Coimbatore pp. 1-5
[3] V. R. Ananthapadmanabhan, I. Tenison and T. N. Shanavas 2018 Automated Power Factor Improvement Based on Artificial Neural Networks *2018 International Conference on Intelligent Autonomous Systems (ICoIAS)* Singapore pp. 170-174
[4] Yuniarto, Eko Ariyanto 2018 *Automatic Power Factor Corrector in Household Electrical Installations GEMA TEKNOLOGI* vol 19 no 4
[5] F. Panchi, K. Hernández and D. Chávez 2018 MQTT Protocol of IoT for Real Time Bilateral Teleoperation Applied to Car-Like Mobile Robot *2018 IEEE Third Ecuador Technical Chapters Meeting (ETCM)* Cuenca pp. 1-6
[6] S. Chanthakit and C. Rattanapoka 2018 MQTT Based Air Quality Monitoring System using Node MCU and Node-RED *2018 Seventh ICT International Student Project Conference (ICT-ISPC)* Nakhonpathom pp. 1-5
[7] R. K. Kodali and V. S. K. Gorantla 2017 Weather tracking system using MQTT and SQLite *2017 3rd International Conference on Applied and Theoretical Computing and Communication Technology (iCATccT)* Tumkur pp. 205-208
[8] R. A. Atmoko and D. Yang 2018 Online Monitoring & Controlling Industrial Arm Robot Using MQTT Protocol *2018 IEEE International Conference on Robotics, Biomimetics, and Intelligent Computational Systems (Robionetics)* Bandung, Indonesia pp. 12-16
[9] P. Thomas and J. P. Kumar 2019 Cloud Based Dynamic Energy Management in Power System Using Message Queuing Telemetry Transport (MQTT) Protocol *2019 3rd International conference on Electronics, Communication and Aerospace Technology (ICECA)* Coimbatore, India pp. 1301-1305
[10] M. Swarnamugi and R. Chinnaiyan 2019 Smart and Reliable Transportation System based on Message Queuing Telemetry Transport Protocol *2019 International Conference on Intelligent Computing and Control Systems (ICCS)* Madurai, India pp. 918-922

AUTHOR INDEX

Institute of Physics
Dirac House, Temple Back
Bristol BS1 6BE UK

ISSN: 1757-8981
ISBN 978-1-7138-3319-2

CURRAN ASSOCIATES INC.
proceedings.com